Sustainable Development Goals: Their Impacts on Forests and People

Forests provide vital ecosystem services crucial to human well-being and sustainable development, and have an important role to play in achieving the 17 Sustainable Development Goals (SDGs) of the United Nations 2030 Agenda. Little attention, however, has yet focused on how efforts to achieve the SDGs will impact forests and forest-related livelihoods, and how these impacts may, in turn, enhance or undermine the contributions of forests to climate and development. This book discusses the conditions that influence how SDGs are implemented and prioritised, and provides a systematic, multidisciplinary global assessment of interlinkages among the SDGs and their targets, increasing understanding of potential synergies and unavoidable trade-offs between goals. Ideal for academic researchers, students and decision-makers interested in sustainable development in the context of forests, this book will provide invaluable knowledge for efforts undertaken to reach the SDGs. This title is available as Open Access via Cambridge Core.

Pia Katila is Senior Research Scientist at the Natural Resources Institute Finland (Luke). She is the coordinator and editor-in-chief of the International Union of Forest Research Organizations' Special Project 'World Forests, Society and Environment' (IUFRO WFSE), a large international research network.

Carol J. Pierce Colfer is Senior Associate at the Center for International Forestry Research (CIFOR), Bogor, Indonesia, and a Visiting Scholar, Southeast Asia Program, at Cornell University, Ithaca, NY, US.

Wil de Jong is Professor at the Center for Southeast Asian and Integrated Area Studies, Kyoto University, Japan and appointed top level foreign expert at Renmin University of China.

Glenn Galloway is Director of the Master of Sustainable Development Practice Program, at the Center for Latin American Studies/Center for African Studies, University of Florida, USA. He is also Chair to the IUFRO WFSE Special Project's Steering Committee.

Pablo Pacheco is Global Forest Lead Scientist at the World Wildlife Fund (WWF), Washington, DC, and Senior Associate at the Center for International Forestry Research (CIFOR), Indonesia.

Georg Winkel is Head of Bonn Office and the Resilience Research Programme at the European Forest Institute (EFI), Bonn, Germany, and Associate Professor at both the University of Freiburg and the University of Bonn, Germany.

Sustainable Development Goals: Their Impacts on Forests and People

Edited by

PIA KATILA
Natural Resources Institute Finland

CAROL J. PIERCE COLFER
Center for International Forestry Research and Cornell University

WIL DE JONG
Kyoto University

GLENN GALLOWAY
University of Florida

PABLO PACHECO
World Wildlife Fund and Center for International Forestry Research

GEORG WINKEL
European Forest Institute

CAMBRIDGE
UNIVERSITY PRESS

CAMBRIDGE
UNIVERSITY PRESS

University Printing House, Cambridge CB2 8BS, United Kingdom

One Liberty Plaza, 20th Floor, New York, NY 10006, USA

477 Williamstown Road, Port Melbourne, VIC 3207, Australia

314–321, 3rd Floor, Plot 3, Splendor Forum, Jasola District Centre, New Delhi – 110025, India

79 Anson Road, #06–04/06, Singapore 079906

Cambridge University Press is part of the University of Cambridge.

It furthers the University's mission by disseminating knowledge in the pursuit of education, learning, and research at the highest international levels of excellence.

www.cambridge.org
Information on this title: www.cambridge.org/9781108486996
DOI: 10.1017/9781108765015

When citing this work, please include a reference to the DOI 10.1017/9781108765015

First published 2020

A catalogue record for this publication is available from the British Library.

ISBN 978-1-108-48699-6 Hardback

The development of this publication was made possible by the financial support of the Ministry for Foreign Affairs of Finland and Natural Resources Institute Finland, and the in-kind contributions from many universities and research and development organizations. The views expressed in this book do not necessarily present the views of these organizations.

Contents

Contributors

Emmanuel Acheampong
Department of Silviculture and Forest Management, Kwame Nkrumah University of Science and Technology, Kumasi, Ghana

Shruti Agarwal
Centre for Science and Environment, New Delhi, India

Jaime Amezaga
School of Engineering, Newcastle University, UK

Alexander S. Antonarakis
Department of Geography, University of Sussex, UK

Seema Arora-Jonsson
Department of Urban and Rural Development, Swedish University of Agricultural Sciences, Uppsala, Sweden

Bas Arts
Department of Environmental Sciences, Wageningen University, The Netherlands

Rebecca Asare
Nature Conservation Research Centre, Accra, Ghana

Stibniati Atmadja
Center for International Forestry Research, Addis Abeba, Ethiopia

Toe Toe Aung
Forest Department, Ministry of Natural Resources and Environmental Conservation, Naypyidaw, Myanmar

Nur Bahar
Faculty of Forestry, University of British Columbia, Vancouver, BC, Canada

Robert Bailis
Stockholm Environment Institute, Somerville, MA, USA

John Barry
School of History, Anthropology, Philosophy and Politics, University of Belfast, UK

James Bathurst
School of Engineering, Newcastle University, UK

Shonil Bhagwat
Faculty of Arts and Social Sciences, The Open University, Milton Keynes, UK

Laxmi Dutt Bhatta
International Centre for Integrated Mountain Development, Kathmandu, Nepal

Agni K. Boedhihartono
Faculty of Forestry, University of British Columbia, Vancouver, BC, Canada,

Jana Brauer
Collaborating Centre on Sustainable Consumption and Production, Wuppertal, Germany

Jessica Budds
School of International Development, University of East Anglia, Norwich, UK

Justin Bull
Sauder School of Business, University of British Columbia, Vancouver, BC, Canada

Colin D. Butler
National Centre for Epidemiology and Population Health, Australian National University, Australia; Health Research Institute, University of Canberra, Australia; College of Arts, Humanities and Social Sciences, Flinders University, Australia

Zhaohua Cindy Cheng
Faculty of Forestry, Center for Interactive Research on Sustainability, University of British Columbia, Vancouver, BC, Canada

Guéladio Cissé
Department of Epidemiology and Public Health, Swiss Tropical and Public Health Institute, Basel, Switzerland

Carol J. Pierce Colfer
Center for International Forestry Research, Indonesia; Cornell University, Ithaca, NY, USA

Abu Conteh
Department of Biological Sciences, University of Alberta, Edmonton, Canada

Charlotte Cross
Faculty of Arts and Social Sciences, The Open University, Milton Keynes, UK

Dean Current
Center for Integrated Natural Resource and Agricultural Management, University of Minnesota, St Paul, MN, USA

Ronnie de Camino Velozo
Fundación Costa Rica para la Innovación, San José, Costa Rica; Forest and Climate Change Fund, Luxemburg, Luxemburg

Wil de Jong
Center for Southeast Asian and Integrated Area Studies, Kyoto University, Kyoto, Japan

Ahmad Dermawan
Center for International Forestry Research, Bogor, Indonesia

Tahia Devisscher
Faculty of Forestry, Center for Interactive Research on Sustainability, University of British Columbia, Vancouver, BC, Canada

David P. Edwards
Department of Animal and Plant Sciences, University of Sheffield, Sheffield, UK

Marlène Elias
Bioversity International, Rome, Italy

Daniel A. Friess
Department of Geography, National University of Singapore, Singapore

Mónica Gabay
Escuela de Política y Gobierno, Universidad Nacional de San Martín, Buenos Aires, Argentina

Glenn Galloway
University of Florida Center for African Studies, University of Florida, Gainesville, FL, USA

Maya Gislason
Faculty of Health Sciences, Simon Fraser University, Burnaby, BC, Canada

Marta Gruca
Food and Agriculture Organization of the United Nations, Rome, Italy

Kerryn Higgs
Club of Rome; School of Geography and Environmental Studies, University of Tasmania, Australia

Mark Hirons
Environmental Change Institute, School of Geography and the Environment, University of Oxford, Oxford, UK

Pierre Horwitz
School of Science, Edith Cowan University, Perth, Australia

David Humphreys
Faculty of Arts and Social Sciences, The Open University, Milton Keynes, UK

Mark Huxham
School of Applied Sciences, Edinburgh Napier University, Edinburgh, UK

Amy Ickowitz
Center for International Forestry Research, Bogor, Indonesia

Andrés Iroumé
Facultad de Ciencias Forestales y Recursos Naturales, Universidad Austral de Chile, Valdivia, Chile

Pamela Jagger
School for Environment and Sustainability, University of Michigan, Ann Arbor, MI, USA

Julia Jones
College of Earth, Ocean, Atmospheric Science, Oregon State University, Corvallis, OR, USA

Peter Kanowski
Fenner School of Environment & Society, The Australian National University, Canberra, Australia

Pia Katila
Natural Resources Institute Finland, Helsinki, Finland

Sean P. Kearney
Faculty of Forestry, University of British Columbia, Vancouver, BC, Canada

Rodney J. Keenan
School of Ecosystem and Forest Sciences, University of Melbourne, Victoria, Australia

Stephanie Keene
Rights and Resources Initiative, Washington DC, USA

Kaysara Khatun
Environmental Change Institute, School of Geography and the Environment, University of Oxford, Oxford, UK

Noah Kittner
Environmental Sciences and Engineering, Gillings School of Global Public Health, University of North Carolina at Chapel Hill, NC, USA

Daniela Kleinschmit
Chair of Forest and Environmental Policy, University of Freiburg, Freiburg, Germany

Ryo Kohsaka
Graduate School of Environmental Studies, Nagoya University, Japan

Cecil Konijnendijk
Faculty of Forestry, University of British Columbia, Vancouver, BC, Canada

Rajan Kotru
International Centre for Integrated Mountain Development, Kathmandu, Nepal

Robert Kozak
Faculty of Forestry, University of British Columbia, Vancouver, BC, Canada

Priya Kurian
School of Social Sciences, University of Waikato, Hamilton, New Zealand

James Langston
Centre for Tropical Environmental and Sustainability Science, James Cook University, Cairns, Australia

Anne M. Larson
Center for International Forestry Research, Lima, Peru

Kathleen Lawlor
Economics Department, University of North Carolina Asheville, Asheville, NC, USA

Jennifer Lenhart
Worldwide Fund for Nature Sweden/Chile, Las Condes, Región Metropolitana, Chile

Liwei Lin
North Carolina State University, Raleigh, NC, USA

Bas Louman
Tropenbos International, Wageningen, The Netherlands

Catherine Lovelock
School of Biological Sciences, University of Queensland, Brisbane, Australia

Shuaib Lwasa
Department of Geography, Makerere University, Kampala, Uganda

Kenneth G. MacDicken
Monroe, WA, USA

Ryan McCord
Carolina Population Center, University of North Carolina at Chapel Hill, NC, USA

Constance L. McDermott
Environmental Change Institute, School of Geography and the Environment, University of Oxford, Oxford, UK

Rosemary A. McFarlane
Faculty of Health, University of Canberra, Canberra, Australia

Kathleen McGinley
International Institute of Tropical Forestry, Rio Piedras, PR, USA

Mary Menton
Sussex Sustainability Research Programme, University of Sussex, Brighton

Gavyn Mewett
University of British Columbia, Vancouver, BC, Canada; Albert Ludwigs Universität Freiburg, Freiburg, Germany

Iliana Monterroso
Center for International Forestry Research, Bogor, Indonesia

Jean Pierre Morales
Tropical Agricultural Research and Higher Education Center, Turrialba, Costa Rica

Giulia Muir
Food and Agriculture Organization of the United Nations, Rome, Italy

Nibedita Mukherjee
College of Life and Environmental Sciences, University of Exeter, Penryn, Cornwall, UK

Rodd Myers
Dala Institute, Jakarta, Indonesia

Lauren Nerfa
Faculty of Forestry, University of British Columbia, Vancouver, BC, Canada

Lorien Nesbitt
Faculty of Forestry, University of British Columbia, Vancouver, BC, Canada

Helen Newing
University of Oxford, Oxford, UK

Giang Huu Nguyen
Faculty of Economic and Rural Development, Thai Nguyen University of Agriculture and Forestry, Thai Nguyen City, Vietnam; School of Science, Edith Cowan University, Joondalup, Australia

Isilda Nhantumbo
Independent consultant, Mozambique

Giulia Rota Nodari
Bioversity International, Rome, Italy

Fiona Nunan
School of Government and Society, University of Birmingham, Birmingham, UK

Alida O'Connor
Faculty of Forestry, University of British Columbia, Vancouver, BC, Canada

Jane O'Sullivan
School of Agriculture and Food Sciences, University of Queensland, Brisbane, Australia

Pablo Pacheco
World Wildlife Fund, Washington DC, USA

Mahesh Poudyal
Forest Action, Kathmandu, Nepal

Rebecca A. Riggs
College of Science and Engineering, James Cook University, Cairns, Australia

Jack N. Saddler
Faculty of Forestry, University of British Columbia, Vancouver, BC, Canada

Subhashis Sahu
Department of Physiology, University of Kalyani, West Bengal, India

Fabio Salbitano
Department of Agricultural, Food, Environmental, and Forestry Sciences and Technologies, Università di Firenze, Florence, Italy

Sigit Sasmito
Research Institute for the Environment and Livelihoods, Charles Darwin University, Darwin, Australia

Poshendra Satyal
School of International Development, University of East Anglia, Norwich, UK

Jeffrey Sayer
Faculty of Forestry, University of British Columbia, Vancouver, BC, Canada

Patrick Schröder
The Royal Institute of International Affairs, Chatham House, London, UK

Abidah Setyowati
School of Regulation and Global Governance, Australian National University, Canberra, Australia

Douglas Sheil
Faculty of Environmental Sciences and Natural Resource Management, Norwegian University of Life Sciences, Ås, Norway

Bimbika Sijapati Basnett
Center for International Forestry Research, Bogor, Indonesia

Erin Sills
Department of Forestry and Environmental Resources, North Carolina State University, Raleigh, NC, USA

Benjamin Singer
United Nations Forum on Forests Secretariat, New York, NY, USA

Almeida A. Sitoe
Faculty of Agronomy and Forestry, Eduardo Mondlane University, Maputo, Mozambique

Roy Smith
School of Arts and Humanities, Nottingham Trent University, Nottingham, UK

Karnjana Songwathana
School of Economics, Bangkok University, Pathumthani, Thailand

Elaine Springgay
Food and Agriculture Organization of the United Nations, Rome, Italy

Dietmar Stoian
World Agroforestry, Bonn, Germany / Bioversity International, Montpellier, France

Terry C. H. Sunderland
Faculty of Forestry, University of British Columbia, Vancouver, BC, Canada; Center for International Forestry Research, Bogor, Indonesia

Joleen Timko
Faculty of Forestry, University of British Columbia, Vancouver, BC, Canada

Maria Fernanda Tomaselli
Faculty of Forestry, University of British Columbia, Vancouver, BC, Canada

Yuta Uchiyama
Graduate School of Environmental Studies, Nagoya University, Nagoya, Japan

Matilda van den Bosch
Faculty of Forestry, University of British Columbia, Vancouver, BC, Canada

Susan Van Dyk
Faculty of Forestry, University of British Columbia, Vancouver, BC, Canada

Guangyu Wang
Faculty of Forestry, University of British Columbia, Vancouver, BC, Canada

Camilla Wildmark
Department of Forest Economics, Swedish University of Agricultural Sciences, Umeå, Sweden

Georg Winkel
European Forest Institute, Bonn, Germany

Stephen Wyatt
Faculté de Foresterie, Université de Moncton, Edmundton, Canada

Dollie Yao
Fenner School of Environment & Society, The Australian National University, Canberra, Australia

Xinxin Zhu
Faculty of Forestry, University of British Columbia, Vancouver, BC, Canada

Preface

The world's leaders agreed on the Sustainable Development Agenda, or Agenda 2030, in September 2015, and it officially came into force on 1 January 2016. The agenda is embodied in 17 Sustainable Development Goals (SDGs) and 169 associated targets to be achieved by 2030. Agenda 2030 applies to all countries and is now the major framework for guiding development policies and efforts across local to global scales. It calls for transformative changes to increase human well-being and prosperity while addressing environmental protection and climate change.

Human survival and well-being ultimately rest on the natural resources of the planet. Forests cover about a third of the world's land area and provide a wide range of ecosystem services that are crucial for human well-being and sustainable development worldwide. How forests and trees are included in Agenda 2030 and how the efforts undertaken by different sectors to advance towards the 17 SDGs will impact forests, forest ecosystem services, forest-related livelihoods and human well-being are thus important questions. Little attention, however, has yet focused on these issues, or on how the potential impacts, in turn, will support or undermine the contributions of forests to climate and sustainable development. Understanding the potential impacts of the SDGs on forests and forest-related livelihoods and development as well as the related trade-offs and synergies is crucial for efforts undertaken to reach these goals. It is especially important for reducing potential negative impacts and to leverage opportunities to create synergies that will ultimately determine whether comprehensive progress towards the SDGs will be accomplished.

Realisation of the lack of discussion on the potential and likely impacts of the SDGs on forests and forest-related livelihoods and the related synergies and trade-offs motivated the International Union of Forest Research Organizations (IUFRO) Special Project World Forests, Society and Environment (WFSE) to develop this book. It provides a systematic scientific assessment of potential and anticipated impacts of efforts to achieve the SDGs on forests, related socio-economic systems and forest-related development. It discusses the conditions that influence how SDGs are implemented and prioritised, and how these conditions and SDG implementation influence these impacts. Furthermore, it considers the important interconnections and linkages among the SDGs and the potential or anticipated trade-offs and synergies among the SDGs from the perspective of forests and related socio-economic systems, shedding light on how SDG implementation may transform existing forest-related development scenarios and affect the roles of forests in sustainable development in the future.

WFSE is a wide, open, independent network of experts and scientists coordinated by the Natural Resources Institute Finland (Luke). WFSE supports sustainable natural resource management, sustainable development and livelihoods, and related policy processes. It focuses on topics in the forest, society and environment interface that are recognised by the scientific community as important and having significant policy implications, but which appear not to be receiving adequate attention from the policy community. It addresses these topics in a holistic, interdisciplinary and collaborative manner, producing science-based, future-oriented, policy-relevant information.

The development of this book started by identifying and inviting a core group of lead authors to address the above-mentioned questions from the point of view of each of the SDGs. The lead authors were further tasked to form an international team of authors to collaborate in writing the SDG chapters.

In the course of the development of this book we organised two large workshops that brought together the lead authors of the SDG chapters and the editors of the book. The first workshop was organised in collaboration with the Food and Agriculture Organization of the United Nations (FAO) in Rome, Italy, in March 2018. The event brought together the lead authors and editors of the volume and colleagues from the FAO to discuss forest and SDGs interactions, especially the potential and likely impacts of the SDGs and their implementation on forests and related socio-economic systems.

The second workshop for developing this publication was organised in conjunction with the European Forest Institute's (EFI) Annual Conference and Scientific Seminar in Sardinia, Italy, in September 2018. This workshop concentrated on the main findings from the SDG chapters and the contextual conditions that influence how the SDGs are taken up and prioritised.

Furthermore, the editors of this book convened at the University of Florida, Gainesville, USA, in December 2018 to develop and discuss the findings and conclusions of the book.

Acknowledgements

The development of this book was a collaborative effort involving 6 editors and 114 authors from universities and research and development organisations from different parts of the world (see 'Contributors' list for authors and their affiliations). Many of the authors contributed to this book on their own time, in addition to their primary duties and responsibilities. We sincerely thank all of them for the quality of their contributions, their commitment and the outstanding efforts that made this publication possible. Furthermore, we are grateful for the support of the authors' host organisations for the in-kind contribution they provided by supporting the authors' work.

The SDG chapters (Chapters 1–17) were reviewed by renowned scientists and experts. We are very grateful for their generous contributions to the development of this book. Their comments and guidance were crucial for improving the quality of this publication.

Reviewers

Simone Borelli
Joanna Bourke-Martignoni
Duncan Brack
Timothy Cadman
Jeffrey Chow
Marius Claassen
David Ellison
Shelley Feldman
Arturo Gianvenuti
Lukas Giessen
Sarah Giroux
Sarah Hearn
Juha Hiedanpää
Syed Ainul Hussain
Pierre Ibisch
Soo-Yeon Laura Jin
Valerie Kapos
Marko Katila
Steven Lawry
Qiang Ma
Mazlin Bin Mokhtar
Irmeli Mustalahti

Till Neeff
Symphorien Ongolo
Shonali Pachauri
Maureen G. Reed
Mika Rekola
Sandra Rodríguez Piñeros
Chris Seijger
Mila Sell
Gill Shepherd
Markku Simula
Peter Speldewinde
Andreas Waaben Thulstrup
Antonio Tomao
Anne Toppinen
Ingrid Visseren-Hamakers
Sven Wunder

We gratefully acknowledge the financial support from the Ministry for Foreign Affairs of Finland that made the development and publishing of this book possible. We are grateful to the Natural Resources Institute Finland for providing the coordination of the WFSE project, and to the IUFRO Secretariat for continuous support in administrative issues.

We are also grateful to the FAO, the EFI and the University of Florida for the in-kind and practical support that made the organisation of the WFSE workshops possible. Furthermore, we are grateful to Amelia Pope and Ree Sheck for language editing and guidance in the technical editing of this book.

The editors,
Pia Katila, Carol J. Pierce Colfer, Wil de Jong,
Glenn Galloway, Pablo Pacheco, Georg Winkel

Executive Summary

In 2015, 193 countries adopted Agenda 2030 for Sustainable Development and its 17 Sustainable Development Goals (SDGs). The SDGs build on the Millennium Development Goals (MDGs), but there are significant differences between them and the processes leading up to their adoption. The process leading up to the adoption of the SDGs involved considerably broader participation. The SDGs expanded the focus by integrating a wider development policy agenda addressing many aspects of economic, social and environmental sustainability. In addition, while the MDGs were mainly relevant for developing countries, the SDGs apply to all countries.

The 17 SDGs and 169 related targets form an overarching development framework meant to guide government and non-state actor efforts at different scales, from global to local, until 2030. The SDGs and their targets form a complex, integrated system with clear sectoral emphases, but also strong interlinkages among goals and targets. The agenda does not explicitly address these interlinkages, or the synergies and trade-offs among targets.

Forests provide ecosystem services that are crucial for human well-being and, as such, are critical for reaching the SDGs. Yet, forests are only explicitly mentioned in two SDGs. SDG 15 (Life on Land) focuses on the protection, restoration and sustainable use of terrestrial ecosystems and halting the loss of biodiversity. The other, SDG 6 (Clean Water and Sanitation), calls for the protection and restoration of forests in one of its targets: Target 6.6 aims at protecting and restoring water-related ecosystems, including forests. Due to the interrelated nature of the SDGs and targets, the implementation of the SDG agenda will inevitably influence forests and forest-related livelihoods and the possibilities to achieve the forest-specific targets. Understanding the potential impacts of SDGs on forests, forest-related livelihoods and forest-based options to generate progress towards achieving the SDGs, as well as related trade-offs and synergies, is crucial for efforts undertaken to reach these goals. It is especially important for reducing potential negative impacts and to leverage opportunities to create synergies, which will ultimately determine whether comprehensive progress towards the SDGs is accomplished.

No Poverty – SDG 1

SDG 1 seeks to 'end poverty in all its forms everywhere', specifically by ensuring that the poor are covered by social protection systems; by securing their rights to economic resources, access to basic services and property ownership; and by building their resilience to economic, social and environmental shocks. The relationship between poverty reduction and forests varies across regions, decades, forest transition stage and degree of market access. The empirical literature shows that more secure property rights – especially for community land – and social protection in the form of cash transfers can support forest conservation, given the right contexts and conditionalities. As demonstrated by programmes that reforest hillsides and re-establish mangroves to prevent natural disasters, policies designed to reduce vulnerability can promote ecosystem-based adaptation, including expansion of forest cover. This is consistent with the evidence that forests are both a mainstay of rural livelihoods and a buffer and source of natural insurance. However, if poverty alleviation and national development strategies continue to be based on infrastructure and agricultural development, they are likely to remain in conflict with the conservation and sustainable management of forests.

Zero Hunger – SDG 2

Pressure to increase food production augments with population growth. Agriculture dominates landscapes around the world, and more food is being produced than ever before. Yet a large part of the population is undernourished. Concomitantly, much of the agricultural expansion related to achieving global food security is at the expense of forest ecosystems, which are critical for biodiversity and the provision of ecosystem services. SDG 2 seeks to 'end hunger, achieve food security and nutrition and promote sustainable agriculture'. A 'business-as-usual' approach to food production will continue to cause mass deforestation. This would be detrimental for biodiversity, impacting forest-dwelling communities who depend on forests for the direct provision of food. With the loss of forests comes the loss of far-reaching ecosystem services, vital for many facets of food production relied on by the wider population. SDG 2 and five of its targets (2.1–2.5) are closely related to forests. Targets 2.1 and 2.2 strive to end malnutrition and make nutritious food accessible to all. Investing in small-scale farmers and encouraging operations that grow a diversity of crops (Targets 2.3 and 2.5) are necessary for making Targets 2.1 and 2.2 a reality. Target 2.4 calls for sustainable and resilient agricultural practices. These five targets underscore the reciprocity between forests and SDG 2. Forest biodiversity is integral for nutrition and the ability to grow and harvest diverse crops. In turn, investing in small-scale farming systems and sustainable farming techniques can help conserve forests and enhance the integration of trees into landscapes. If we are to achieve SDG 2 sustainably, we need a reimagined food system that does not polarise agricultural production and the conservation of forest resources. This calls for land management that promotes the maintenance of biodiversity and integrated land-use planning. This is especially evident when examining the relationship between SDG 2 and the other SDGs, the majority of which are concomitantly contingent on each other.

Health and Well-Being – SDG 3

The achievement of SDG 3 depends on many other SDGs, yet there are also potential conflicts and trade-offs. Forests are of crucial importance to global health and well-being. In contrast, short-term economic and human health gains from further forest conversion (e.g. deforestation for food production) will create direct and indirect health risks for humans, as well as for other biota. Controlling indiscriminate burning and clearing of forests can reduce significant harm to health and well-being via improved quality of water, soil and air (a transnational issue), by reducing exposure to some infectious diseases, through the preservation of traditional (and future) medicines and by supporting other forest resources and services, including climate regulation. Many infectious diseases are associated with forest disturbance and intrusions, and some may be prevented or modified through forest management. Universal access to sexual and reproductive healthcare services, including for family planning, is a critical SDG 3 target to decrease demographic pressures on forests at local, regional and global scales and to enhance human well-being. Greater exposure to green space, including urban forests, has been linked to many benefits for mental, social and physical health for the increasingly urban global population. More broadly, forests play important roles in enriching cultural and religious well-being.

Quality Education – SDG 4

Education has been characterised as 'the golden thread' that runs through all 17 SDGs. SDG 4 broadens the depth and breadth of education to people of all ages and expands its scope to a lifelong process spanning formal, non-formal and informal settings. It emphasises quality of educational access, particularly for girls and women and marginalised groups. Literature exploring pro-environment behaviour informs our consideration of how progress towards SDG 4 might impact forests, forest ecosystem services and forest-related livelihoods. The concept of pro-forest behaviour describes those elements of pro-environmental behaviour related to forests; encouraging and enabling pro-forest behaviour is the basis of building a positive relationship between SDG 4 and forests. Inclusive education that builds and reinforces positive attitudes towards forests, as well as relevant knowledge and competencies, and that helps individuals and communities feel or stay connected to forests, will foster and sustain pro-forest behaviours. This requires that education systems respect, nurture and enable Indigenous and traditional knowledge; promote forest-related environment and sustainability education; strengthen forest-related professional, technical and vocational education and capacity development; and capitalise on the power of both established and new media that will continue to evolve and emerge over time.

Gender Equality – SDG 5

Taking SDG 5 seriously in relation to forests brings to the forefront what is usually taken for granted in forest debates: people and their relationships, to one another and to forests. These relationships determine forest outcomes. Forest governance and everyday management are upheld by a superstructure of gendered forest relations (invisible to mainstream forestry) that often disadvantages women as a social group. Systemic and contextual factors such as health, gender-based violence and unpaid care work are crucial to the welfare of forest-dependent peoples and forests. So far, little progress has been made in implementing SDG 5 targets within forestry. Political will is needed to transform unequal relationships and to support demands for forest justice. There is a need to challenge privilege based on sex, class, ethnicity or caste and to destabilise inequitable micro- and macroeconomic structures such as commodification and to support democratic forest governance to work towards greater sustainability. It is also important to keep in mind that well-intentioned efforts, such as gender programmes, can have adverse effects if not cognisant of contextual power relations. The welfare and dignity that achieving SDG 5 would bring to forest peoples and livelihoods is essential to ensuring better managed and sustainable forests; however, the gender-neutral framing of some SDG goals undermines efforts towards achieving the outcomes called for in SDG 5.

Clean Water and Sanitation – SDG 6

Predicting the impact of SDG 6 on forests and people requires a balanced understanding of the relationship between forests and water. Notable aspects are that forest cover reduces annual runoff but correlates well with water quality, and that evapotranspiration from forests is important for downwind precipitation. Within this context a target-by-target review of SDG 6, informed by South American examples, suggests that SDG 6 is unlikely to exert a major influence on forest cover. Targets 6.1 and 6.2, concerning water and sanitation provision (likely to be the major focus of SDG 6), will have relatively little impact on forests except through a demand for hydrological ecosystem services and the use of wastewater in forestry. Within the four water resources targets (6.3–6.6) significant impacts may be limited to water efficiency considerations (Target 6.4) restricting plantations in water-stressed areas and Integrated Water Resources Management (Target 6.5) driving a more integrated view of catchments and their management. SDG 6 impacts will depend on the context of water–forest relationships (illustrated using the Hindu Kush Himalayas as an example), the extent to which SDG 6 is implemented and its alignment with forest policies. This alignment must be guided by a shared understanding of the complex relationships between water and forests and their impacts on both forest-dependent peoples and the communities downstream, and possibly downwind.

Affordable and Clean Energy – SDG 7

SDG 7 aims to ensure access to affordable, reliable, sustainable and modern energy for all. Forests contribute to SDG 7 through four pathways: sustainable use of traditional woodfuels, processed woodfuels, liquid biofuels and biopower. We hypothesise that the role of traditional woodfuels (e.g. firewood and charcoal) in household energy portfolios will decline in most low- and middle-income countries, but will not be completely replaced with modern fuels. In the transition to affordable clean fuels, processed woodfuels (e.g. pellets), liquid biofuels produced from forest feedstock and biopower will play an increasing role in energy service provision. How forest-based transitions to clean energy will fare relative to other renewable energy technologies including solar, wind and micro-hydro will depend on how renewable energy policy evolves, and on relative costs and storage capacity. Reaching SDG 7 through the promotion of large-scale hydro and agricultural commodity derived biofuels can threaten forests and forest-based livelihoods. In general, promoting transitions to sustainable forest-based clean energy supports the realisation of other SDGs, highlighting the potential for forests to play a significant role in discourse and action on the SDGs.

Decent Work and Economic Growth – SDG 8

Diverse combinations of predominant development paradigms (modernisation, economic growth, basic needs, sustainable development) that shape the agendas of governments, private sector, civil society and investors lead to differentiated prioritisation of SDG 8 targets, with mixed impacts on forests and forest-dependent livelihoods. At the country level, significant trade-offs are expected where growth policies and strategies focus on sectors competing with forestry for space and resources, such as agriculture, energy and mining.

Combined, such policies and strategies lead to global trade-offs by exacerbating climate change. In these cases, decoupling economic growth from environmental degradation will be a major challenge. Synergies between SDG 8 and forests exist where sustainable development is explicitly sought in the forest sector, focusing on tree plantations, timber and non-timber forest products from natural forests, eco-tourism and environmental services. Enhanced enabling environments help minimise trade-offs and maximise synergies by reconciling government policies and private sustainability standards, formalising community stewardship of tropical forests, addressing informality in forest-product value-chains and providing incentives for youth to become involved in forest-based economic activities.

Industry, Innovation and Infrastructure – SDG 9

SDG 9 and its 8 targets and 12 indicators will have multiple impacts on forests, forest-based livelihoods and forest-based economies. Major trade-offs are anticipated between SDG 9 and SDG 15 (Life on Land), especially if economic expansion and increasing planetary impacts remain coupled. More specifically, the implementation of Target 9.1 and its corresponding indicators (road, infrastructure and transportation expansion) may lead to irreversible and widespread forest degradation and deforestation. As such, the short- and long-term environmental and social costs of this goal need to be better assessed, especially in light of the fact that other SDG 9 targets, e.g. small-scale industry expansion (Target 9.3) and access to information and communications technology (Target 9.C), may have diverse consequences for forests and livelihoods, depending on how they are applied. We call for reforms of SDG 9 to promote and support alternative socio-economic models that are not based on indefinite economic growth nor reliant on the ongoing expansion of infrastructure, but, rather, necessitate forests and terrestrial ecosystem services to be essential building blocks of a green and sustainable economy.

Reduced Inequalities – SDG 10

SDG 10 calls for reducing inequalities within and among countries. Considerable synergies and complementarities can be found between the SDG 10 targets and the goals of environmental justice, which comprise three interrelated dimensions: representational, recognition and distributive justice. However, the disjuncture between SDG 10 and environmental goals within the SDGs may undermine efforts to promote environmental justice. Trade is not included in SDG 10; this is an important gap as markets for forest products can drive forest resource extraction, exacerbating inequalities among actors within global production networks. If SDG 10 addresses structural inequalities, it is also likely to support distributive, representational and recognition justice for forest-dependent populations. However, the myopic translation of its aspirational targets into easily measurable indicators may dampen the potential effects of addressing SDG 10 in advancing environmental justice. Addressing migration-related targets and indicators is likely to elevate the importance of these issues in forestry policy and research, while also prompting a rethinking of some of the underlying assumptions informing existing research in forestry. Managing migration requires incorporating a better understanding of the net effects of migration on environmental justice and the multiple drivers that contribute to positive outcomes for forest-dependent populations.

Sustainable Cities and Communities – SDG 11

Cities have become critical drivers of global socio-economic, behavioural and environmental changes far beyond urbanised borders. Their transformative force has been recognised with the endorsement of SDG 11 to 'make cities and human settlements inclusive, safe, resilient and sustainable'. The capacity to address global urban challenges through the implementation of SDG 11 depends on how cities prioritise resources and urban planning strategies over the next decade. This prioritisation is context specific and depends on socio-economic development trajectories, spatio-temporal urbanisation patterns and strategic urban visions. The implementation of SDG 11 will have effects on forests and forest livelihoods near and far from urban centres. The strategic inclusion of urban and peri-urban forests in city agendas and planning may help manage potentially adverse effects, emphasising the role forests play in delivering ecosystem services to urban and rural people, and fostering productive rural–urban relationships. If SDG 11 implementation aims at fostering people–nature connections in cities, it can help to avoid the negative consequences the 'urbanisation of minds and attitudes' may have on forests and forest-based livelihoods. Currently, many cities prioritise SDG 11 targets focused on basic services such as housing, transport, waste management and sanitation. Less attention is given to SDG 11 targets encouraging inclusive access to urban forests, protecting cultural and natural heritage or improving urban–rural linkages. SDG 11 shows synergies with all the other SDGs, creating opportunities for human health and well-being, green justice, resilience and adaptive capacity in and around cities. These synergies, delivered through sound urban forestry approaches for example, could benefit not only urban dwellers, but also forest communities by reducing pressure on forest resources. The potential role of urban forests in achieving SDG 11 implementation may be enhanced through the New Urban Agenda and global networks that help create multi-scale bridges for collective stewardship involving a large range of government and other actors. The benefits that greener and more resilient cities may have on forests and forest-dependent livelihoods will largely depend on integrated governance and territorial planning.

Responsible Consumption and Production – SDG 12

The focus of SDG 12 – sustainable consumption and production (SCP) – has been part of the international policy discourse for more than four decades, but the uptake of SCP has not been smooth and has tended to be biased towards relatively weak measures. The inclusion of SCP in the SDG framework gives hope that it will receive stronger attention in international efforts for sustainable development. Although SDG 12 targets and indicators make no direct reference to forests or forest communities, achieving the targets will result in positive contributions towards forest conservation and will support forest-dependent livelihoods. SDG 12 targets can contribute to reducing trade-offs among other SDGs: in particular, Target 12.3 – aimed at reducing food waste and food losses – can limit trade-offs between SDG 2 (Zero hunger) and SDG 15 (Life on Land). SDG 12 can contribute to creating enabling conditions for advancing a more responsible and sustainable supply of timber and other forest commodities, also linked to more responsible demand. SDG 12 has its limitations, including the lack of absolute limits to consumption of forest products or products that place pressures on forests leading to deforestation and forest degradation. The main players for achieving SDG 12 targets with positive outcomes for forests will comprise national governments, large companies and consumers involved in global value chains. A thorough integrative SCP approach that addresses systemic issues is required to achieve sustainable forest management and land use associated with responsible consumption.

Climate Action – SDG 13

Climate change causes changes in forests, their ecological functions and ecosystem services. Many of these changes will negatively impact people, plants, animals and microorganisms that depend on forests. SDG 13 aims to reduce greenhouse gas emissions that cause climate change and to drive adaptation actions. Current commitments are insufficient to reach the Paris Agreement goals of restricting global warming to less than 2°C and increasing the resilience of vulnerable communities. Better forest and land management can contribute up to 20 per cent of the Paris goals while increasing community and ecosystem resilience, and can therefore help progress towards reaching the

Paris Agreement goals. Strong synergies between SDG 13 and forests can drive investment in sustainable forest management, forest restoration and forest conservation. However, achieving these synergies is challenged by unsustainable forest exploitation and pressures to develop land for agriculture, urban areas and infrastructure. Maximising potential synergies between forests and SDG 13 requires long-term finance and local collaboration; currently, only 3 per cent of climate finance is dedicated to forest actions, and much less is used for local implementation. Improved forest management and conservation can be achieved through a more efficient use of finances, increased investment from public and private sectors and stronger commitment to local actions.

Life below Water – SDG 14

The targets of SDG 14 address the world's oceans, covering more than 70 per cent of the planet; they also address the coastal zones, where a range of coastal forests are located. In this chapter we investigate the potential negative consequences of SDG 14 on forest resources, using the example of coastal mangrove forests. SDG 14 is likely to have negative impacts on forest resources because it focuses primarily on fisheries, potentially excluding other coastal natural resources. Many SDG 14 targets are more appropriate for oceanic areas rather than the complex governance arrangements found in the coastal zone. This means that coastal forests such as mangroves may be neglected, inadvertently impacted or fall through the 'policy gap' between terrestrial and marine legislation or between different levels of governance. This has impacts on the human populations that rely on the ecosystem services provided by mangrove forests, and has implications for environmental justice. To minimise the impacts of SDG 14 on mangrove forests and associated coastal communities, we recommend that SDG 14 indicators should be broadened to encompass other coastal and oceanic natural resources, that decentralisation of coastal zone governance should continue to be encouraged and that management regimes should include coastal communities and enshrine principles of environmental justice.

Life on Land – SDG 15

SDG 15 requires the maintenance of life on land and endorses priorities already established through international conventions and agreements. The scale and complexity of tropical forest loss and biodiversity decline versus the limited resources for conservation and forestry poses many challenges. The main innovation of SDG 15 is that decision-makers will see this goal as one to integrate with other SDGs. The risk, however, is that short-term priorities and a 'business-as-usual' approach will undermine this. There will be trade-offs between SDG 15 and other SDGs resulting from competition for land, but there are also opportunities for synergies and opportunities that require recognition. Greater cross-sectoral integration, not just sectoral policy reform, is essential to advancing SDG 15. We encourage conservation and development professionals to engage with those responsible for all the Agenda 2030 targets to ensure that SDG 15 is a priority in all SDG-related processes.

Peace, Justice and Strong Institutions – SDG 16

SDG 16 addresses three broad thematic areas: (1) peace and the reduction of armed conflict; (2) the rule of law, accountability, transparency and access to justice; and (3) inclusiveness and participation. Research on peace and armed conflict reveals highly variable effects on forests and people. Conflict may prevent the conversion of forests to agriculture, or drive illicit crop production; it may foster migration into or out of forested areas. Peace may be accompanied by state-supported mining and expansion of commercial agriculture, and/or may improve enforcement of environmental laws. In general, laws in many countries favour the political elite, large-scale industry actors and international trade, and thus the focus of SDG 16 on the 'rule of law' risks reinforcing existing inequalities. The goals of SDG 16 may best be served by legal reforms that strengthen local rights to land and resources, and by greater involvement of non-state actors and institutions at multiple scales – from traditional governance systems to global-scale initiatives. While there has been much recent progress in promoting participatory forest management, this is often tightly controlled by the state, contributing to local administrative burdens without redistributing power and benefits. In sum, the impacts of SDG 16 on forests and people depend on how its interpretation and implementation shape power and resource distribution.

Partnerships for the Goals – SDG 17

Successful attainment of SDG 17 is essential for implementing the other 16 SDGs, all of which depend upon secure means of implementation and durable partnerships. Funding for forests from official development assistance and other sources has trended upwards since 2000, providing reason for cautious optimism. However, REDD+ finance is declining. Private sector investment remains important. The idea of impact investment, which aims to solve pressing environmental and social problems while providing a return for investors, could make a significant contribution to the SDGs. However, not all sustainable development finance promotes forest conservation. Increasing funding for agricultural production often incentivises the conversion of forests to agricultural land while resulting in deforestation. The policy of zero net deforestation is leading to the creation of partnerships to promote deforestation-free commodity supply chains for four forest-risk commodities (palm oil, soy, beef and timber). Some innovative partnerships have been created to promote sustainable development involving intergovernmental organisations, the private sector, research institutes, non-governmental organisations and grassroots organisations. However, such partnerships exist within a neoliberal global economic order in which there are net financial flows from the Global South to the Global North that negate financial flows for sustainable development.

Synergies, Trade-offs and Contextual Conditions

Findings across the SDGs indicate that the ones that can be linked to deforestation or forest degradation are primarily SDG 9 (Industry, Infrastructure and Innovation), SDGs 1 (No Poverty), SDG 2 (Zero Hunger), and SDG 7 (Affordable and Clean Energy). The largest threat to forests linked to SDG 9 is undoubtedly infrastructure expansion, which often encourages mega agro-industrial projects. The impact of SDGs 1 and 2 on forests is projected to occur through an expansion of agricultural production, which leads to deforestation and forest degradation. Land-use change driven by agricultural expansion is mostly linked to SDG 2. While SDG 1 is not directly focused on food production, it is likely to lead to similar impacts on forests because a high proportion of the global poor live in rural areas, and supporting their emergence from poverty can most readily be achieved by boosting income from agriculture or other natural resource-based activities. SDG 7 implementation may have negative and positive impacts on forest cover and forest condition. Its successful implementation will reduce consumption of woodfuels, offset by increases in the use of hydrocarbon-based fuels or other cleaner energy sources. Future energy trends include turning to improved woodfuels, such as wood pellets, and the use of liquid biomass fuels, such as palm oil-based biodiesel, whose production may happen at the expense of forests.

The undesirable impacts described above resulting from trade-offs implicit in the pursuit of different SDGs are counteracted by SDG 13 (Climate Action) and SDG 15 (Life on Land). The implementation of these two SDGs is primarily expected to have positive impacts on forests, while the impact on forest peoples is less clear.

The impacts of the SDGs on forests and people, as well as the positive and negative interactions among SDGs and how those will affect forests and people, are influenced by contextual conditions. These include a country's national development status and trajectory and the overall condition of forests. Many of these conditions are actually targeted by the SDG agenda. Like the SDGs, these conditions do not operate in isolation, but, rather, interact in complex ways. This results in a web of interactions of SDGs and contextual conditions leading to observed and projected impacts on forests and people.

In the analysis of the findings across the SDG chapters, two broad groups of SDGs emerge. One includes SDGs that primarily focus on institutional, governance and social conditions (1 No Poverty; 3 Good Health and Well-being; 4 Quality Education; 5 Gender Equality; 10 Reduced Inequalities; 12 Responsible Consumption and Production; 13 Climate Action; 16 Peace, Justice and Strong Institutions). These contribute to an enabling environment for inclusive forest management and conservation with associated livelihood benefits. A second group of SDGs concern land use directly and thus are expected to impact forests directly (2 Zero Hunger; 6 Clean Water and Sanitation; 7 Affordable and Clean Energy; 8 Decent Work and Economic Growth; 9 Industry, Innovation and Infrastructure; 14 Life below Water; 15 Life on Earth). Progress in the first group of SDGs results in synergistic interactions and positive outcomes for forests and peoples. Among the second group of SDGs, the potential for trade-offs is high, with important repercussions for forests and people. Understanding the potential for these trade-offs is essential in order to avoid implementation pathways that favour a small subset of these SDGs at the expense of the others.

Key Lessons

The key lessons that emerge from this volume, drawing on a reflection process among the editors and on deliberations among lead authors during a collaborative workshop, can be articulated as follows: (1) forests are often a foundation for sustainable development, and thus need to be fully considered in decision-making processes related to the SDGs; (2) the implementation of the SDGs will impact forests and people dependent on them in many ways, with the exact impact being contingent on the specific ecological, socio-economic and political context; (3) within the SDGs, partially conflicting visions for forests and people are implicit that correspond to distinct values and interests, making it necessary to consider trade-offs and set priorities when implementing them; (4) there are fundamental values and principles that should guide sustainable development related to forests and peoples regardless of context, such as respect for basic human rights, the importance of intergenerational equity, the recognition of temporal dimensions of forest ecosystem conservation, the need to detect and address trade-offs, and applying acknowledged tenets of good governance; (5) implementation of the SDGs and associated goals and targets will require continuous learning and adaptation of implementation strategies and approaches, but possibly also targets, taking into consideration observed outcomes. Creative and forward-looking human engagement at the forest-people interface is urgently needed to ensure that sustainable development benefits both forests and peoples.

Abbreviations

APEC	Asia Pacific Economic Cooperation
CATIE	Tropical Agricultural Research and Higher Education Centre
CBD	Convention on Biological Diversity
CDM	Clean Development Mechanism
CIAT	International Centre for Tropical Agriculture
CIFOR	Center for International Forestry Research
CITES	Convention on International Trade in Endangered Species
CPF	Collaborative Partnership on Forests
DFID	Department for International Development
EFI	European Forest Institute
EIA	Environmental Impact Assessment
EKC	Environmental Kuznets Curve
EU	European Union
FAO	Food and Agriculture Organization of the United Nations
FLEGT	Forest Law Enforcement, Governance and Trade
FLR	Forest Landscape Restoration
FPIC	Free, Prior and Informed Consent
FSC	Forest Stewardship Council
GCF	Green Climate Fund
GDP	Gross Domestic Product
GEF	Global Environment Facility
GHG	Greenhouse Gas
HLPF	High-Level Political Forum
ICT	Information and Communication Technology
IFPRI	International Food Policy Research Institute
IIED	International Institute for Environment and Development
ILO	International Labour Organization
IMF	International Monetary Fund
IPBES	Intergovernmental Science-Policy Platform on Biodiversity and Ecosystem Services
IPCC	Intergovernmental Panel on Climate Change
ITTO	International Tropical Timber Organization
IUCN	International Union for Conservation of Nature
IUFRO	International Union of Forest Research Organizations
LAC	Latin America and Caribbean
LDC	Least Developed Countries
LMIC	Low- and Middle-Income Country
LULUCF	Land Use, Land-Use Change and Forestry

MDG	Millennium Development Goal
MRV	Monitoring, Reporting and Verification
NDC	Nationally Determined Commitments
NGO	Non-Governmental Organization
NTFP	Non-Timber Forest Product
NYDF	New York Declaration on Forests
ODA	Official Development Assistance
ODI	Overseas Development Institute
OECD	Organisation for Economic Co-operation and Development
PEFC	Program for the Endorsement of Forest Certification
PES	Payment for Ecosystem Services
R&D	Research and Development
RECOFTC	The Centre for People and Forests
REDD+	Reducing Emissions from Deforestation and Forest Degradation and the role of conservation, sustainable management of forests and enhancement of forest carbon stocks in developing countries
RRI	Rights and Resources Initiative
RSPO	Roundtable on Sustainable Palm Oil
SCP	Sustainable Consumption and Production
SDG	Sustainable Development Goal
SFM	Sustainable Forest Management
SIDS	Small Island Developing States
SME	Small- and Medium-Sized Enterprises
SMFE	Small and Medium Forest Enterprises
SPP	Sustainable Public Procurement
UN	United Nations
UNCCD	United Nations Convention to Combat Desertification
UNCLOS	United Nations Convention on the Law of the Sea
UNCTAD	United Nations Conference on Trade and Development
UN DESA	United Nations Department of Economic and Social Affairs
UNDP	United Nations Development Programme
UNEP	United Nations Environment Programme
UNECE	United Nations Economic Commission for Europe
UNFCCC	United Nations Framework Convention on Climate Change
UNFF	United Nations Forum on Forests
UNISDR	Currently known as UNDRR, United Nations Office for Disaster Relief Reduction
UNSD	United Nations Statistics Division
UNU-WIDER	United Nations University World Institute for Development Economics Research

USD	US Dollar
VNR	Voluntary National Review
VPA	Voluntary Partnership Agreement
WBCSD	World Business Council for Sustainable Development
WHO	World Health Organization
WRI	World Resources Institute
WTO	World Trade Organization
WWF	World Wide Fund for Nature
ZND	Zero Net Deforestation

Introduction

Pia Katila, Carol J. Pierce Colfer, Wil de Jong, Glenn Galloway, Pablo Pacheco and Georg Winkel

The Aim of This Book

In 2015, 193 countries adopted Agenda 2030 for Sustainable Development and its 17 Sustainable Development Goals (SDGs). Each goal is composed of a number of targets to be achieved by 2030. The goals and the 169 targets cover a wide range of social, economic and environmental issues addressing crucial global challenges, including ending hunger and poverty, protecting life below water and on land, advancing sustainable production and consumption, and guaranteeing well-being to all with reduced inequalities. Agenda 2030 forms an overarching framework that is expected to guide government and non-state actor efforts at different scales, from global to local, until 2030. The global indicator framework to follow and periodically review the progress towards the SDG targets was adapted in 2017 (UN 2017). The main responsibility to reach the SDG targets rests with national governments, but the agenda calls for cooperation and global partnerships that bring together governments, civil society, the private sector, the United Nations (UN) system and other social actors.

The SDGs (Table I.1) and their targets form a complex, integrated system with clear sectoral emphases but also strong interlinkages among goals and targets. The agenda does not explicitly address these interlinkages, nor the synergies and trade-offs among targets. In many instances, efforts to advance one target can directly or indirectly contribute to the advancement of other targets. In other instances, progressive measures for one target can hinder the achievement of others.

Natural resources are the fundamental basis for life and human well-being. Many of the efforts to achieve the SDGs and specific targets will have direct or indirect impacts on natural resources, the services they provide and the ways they are used, along with the distribution of their benefits. Forests cover about one-third of the world's land area and are crucial for fundamental ecological processes and human well-being – from climate regulation and

Table I.1 Sustainable Development Goals
Goal 1. End poverty in all its forms everywhere
Goal 2. End hunger, achieve food security and improved nutrition and promote sustainable agriculture
Goal 3. Ensure healthy lives and promote well-being for all at all ages
Goal 4. Ensure inclusive and equitable quality education and promote lifelong learning opportunities for all
Goal 5. Achieve gender equality and empower all women and girls
Goal 6. Ensure availability and sustainable management of water and sanitation for all
Goal 7. Ensure access to affordable, reliable, sustainable and modern energy for all
Goal 8. Promote sustained, inclusive and sustainable economic growth, full and productive employment and decent work for all
Goal 9. Build resilient infrastructure, promote inclusive and sustainable industrialization and foster innovation
Goal 10. Reduce inequality within and among countries
Goal 11. Make cities and human settlements inclusive, safe, resilient and sustainable
Goal 12. Ensure sustainable consumption and production patterns
Goal 13. Take urgent action to combat climate change and its impacts*
Goal 14. Conserve and sustainably use the oceans, seas and marine resources for sustainable development
Goal 15. Protect, restore and promote sustainable use of terrestrial ecosystems, sustainably manage forests, combat desertification, and halt and reverse land degradation and halt biodiversity loss
Goal 16. Promote peaceful and inclusive societies for sustainable development, provide access to justice for all and build effective, accountable and inclusive institutions at all levels
Goal 17. Strengthen the means of implementation and revitalize the Global Partnership for Sustainable Development

* Acknowledging that the United Nations Framework Convention on Climate Change is the primary international, intergovernmental forum for negotiating the global response to climate change.

Source: UN 2015

pollination services to provision of timber and non-timber forest products. They also provide habitat for a vast array of plants and animals. The future of the world's forests is thus critical for sustainable development at all scales, from global to local.

Yet, forests are explicitly mentioned in only two SDGs. SDG 15 (Life on Land) focuses on the protection, restoration and sustainable use of terrestrial ecosystems and halting the loss of biodiversity. SDG 6 (Clean Water and Sanitation) calls for the protection and restoration of forests in one of its targets: Target 6.6 aims at protecting and restoring water-related ecosystems, including forests. Due to the interrelated nature of the SDGs and targets, the implementation of the SDG agenda will inevitably affect forests and forest resources, thus impacting the possibilities to achieve these forest-specific targets. This will further affect the capacity and potential of forests to provide important ecosystem services (e.g. climate regulation, soil protection and formation, biodiversity protection, water regulation and supply, and an environment for recreation) and consequently to contribute to achieving the SDGs and supporting human well-being and sustainable development in the future.

This book presents a comprehensive interdisciplinary assessment of potential and anticipated impacts of efforts towards attaining the different SDGs on forests and forest-related livelihoods. It identifies possible synergies and trade-offs associated with efforts to achieve the SDGs and the goods and services provided by forests. The assessment places special attention on interactions among the goals and their impacts on forests, forest use and forest-related livelihoods and economies, as well as how the goals and their interactions affect policies and governance relevant to forests. We expect that this book will thus contribute to the formulation of more integrated and coherent policies for reaching the SDGs and targets – policies that would leverage beneficial synergies and minimise the inherent trade-offs among the targets. By analysing the interactions among the SDGs through a forest lens, this book provides an analysis of the SDG framework from the point of view of this crucial natural resource base on which human well-being depends.

The book addresses the following questions:

1. What are the possible and anticipated impacts of efforts to achieve the 17 SDGs and related targets on forests and forest-related livelihoods? What are the contextual conditions that determine how SDGs are implemented and prioritised, and how do these conditions and related SDG implementation pathways influence impacts on forests and related livelihoods?

2. What are the important interconnections and interlinkages among the SDGs and related trade-offs and opportunities for beneficial synergies vis-à-vis

forests, forest ecosystem services and forest-related livelihoods in different contexts? How may the implementation of the SDGs transform existing rural well-being scenarios and forest-dependent economies?

Here, we introduce the SDGs and the processes that led to their adoption. We also summarise the importance of forests for human well-being and the crucial role and contributions of forests towards reaching the SDGs.

The Road to SDGs

The challenge of maintaining environmental sustainability in the context of economic growth and material well-being entered global discussions in the UN Conference on the Human Environment in Stockholm in 1972. Several years later, the report of the Brundtland Commission, *Our Common Future*, placed the concept of sustainable development into the global environmental and development agenda. It defined *sustainable development* as 'development that meets the needs of the present without compromising the ability of future generations to meet their own needs' (UN 1987).

Since that time, sustainable development has been the overarching theme and guiding principle of global development. It was the main concept for the UN Conference on Environment and Development held in Rio de Janeiro in 1992: the summit adopted Agenda 21 – a programme of action for sustainable development worldwide. Building on the declaration from the Stockholm conference, it presented the first international plan of action for global sustainable development into the twenty-first century (UN 1994).

The 2002 World Summit on Sustainable Development in Johannesburg, South Africa, reaffirmed this commitment and included a reference to the three pillars of sustainable development by assuming 'a collective responsibility to advance and strengthen the interdependent and mutually reinforcing pillars of sustainable development – economic development, social development and environmental protection – at the local, national, regional and global levels' (UN 2002: 1).

In 2000, the UN Millennium Summit adopted the Millennium Declaration with eight time-bound targets, the Millennium Development Goals (MDGs), with a deadline of 2015. It committed all countries to reduce extreme poverty and set a road map towards the implementation of the MDGs, which focused on eradicating extreme poverty and hunger; achieving universal primary education; promoting gender equality and empowering women; reducing child mortality; improving maternal health; combatting HIV/AIDS, malaria and other diseases; ensuring environmental sustainability; and developing a global partnership for development (UN 2000).

The MDGs became widely accepted goals, but were mostly relevant for developing countries. While remarkable progress has been made towards these goals, especially in eradicating poverty and improving access to primary education, progress has been uneven within and across countries (UN 2012). The MDG framework's format – based on a limited number of concrete end goals and numerical targets – has been acknowledged as one of its main strengths (UN 2012). However, this approach has also been criticised for addressing complex development challenges with only eight concrete goals and reducing the development agenda to meeting basic material needs. The resulting narrow focus on selected indicators has accordingly been criticised for ignoring issues that are not captured with specific, quantifiable indicators, thus potentially leading to shifting priorities when implementing the MDGs (Fukuda-Parr 2017).

While the SDGs build on the MDGs, there are significant differences between them and the processes leading to their adoption. The process leading to the acceptance of the MDGs was criticised for being top-down, led by technocrats with limited consultations with other sources of knowledge and expertise. The process leading to the adoption of the SDGs was based on considerably broader participation. The SDGs were negotiated mainly through the Open Working Group of the UN General Assembly, whose work paralleled the work of the High-Level Panel on the Post-2015 Development Agenda set up by the UN Secretary General. The process included an extensive global consultation and negotiation involving the interests of specific social groups such as women, Indigenous peoples, farmers and non-governmental organisations (Dodds et al. 2017, Fukuda-Parr 2016, Stevens and Kanie 2016).

The SDGs broadened the focus from that of the MDGs towards a wider development policy agenda addressing many aspects of economic, social and environmental sustainability. While the MDGs were mainly relevant for developing countries, the SDGs apply to all countries. They also address issues that were not included in the MDGs and ones that the MDGs were criticised for only partially including, such as inequality, gender, political and human rights, economic development and climate change.

Agenda 2030 is a universal plan of action that aims at guiding development efforts and national development policies until 2030. It recognises that the SDGs are 'integrated and indivisible and balance the three dimensions of sustainable development: the economic, social and environmental' (UN 2015: 1). Universal development and ending poverty in the spirit of 'no one will be left behind' are the central principles of Agenda 2030. 'Eradicating poverty in all its forms and dimensions, including extreme poverty' is considered the 'greatest global challenge and an indispensable requirement for sustainable development' (UN 2015: 1).

SDGs as a Complex, Integrated System

While the SDGs are stated to form a complex, integrated system of goals and interrelated targets that cut across traditional administrative sectors, many of the goals have an overall sectoral focus. Furthermore, the references to sectors other than the ones specifically addressed in a specific goal are not systematically considered (Boas et al. 2016). There is great variation in the nature and scope of the SDGs, and the related targets have different functions. Some targets are ends in themselves, while others are means towards reaching other targets by supporting the development of an enabling environment or providing resources that support the achievement of other targets (Elder et al. 2016, Holden et al. 2017). The relationship among the more intermediate targets can be synergistic, and a lack of progress in one may often hinder progress within another. In other instances, however, efforts towards achieving a specific target can undermine progress towards another one (Elder et al. 2016). While synergies among the targets can increase the effectiveness of implementation, incompatible targets lead to trade-offs. In addition, the interactions among the SDGs are non-linear. Deficient performance in one crucial goal or target can potentially undermine progress in the overall agenda. Similarly, the implementation of one target at the expense of a non-compatible one can reinforce the trade-offs.

The interlinkages and interactions among the SDGs and related targets are receiving increasing attention. Le Blanc (2015) conducted a network analysis of the links among SDGs and targets (except those related to implementation) based on their wording. The results show that the SDGs are unequally connected. Some goals are connected through multiple targets, while others have weak connections to other goals. Sustainable consumption and production (SDG 12), reducing inequality (SDG 10), eliminating poverty (SDG 1) and promoting economic growth and employment (SDG 8) are directly or indirectly linked to at least ten other goals. Life on land (SDG 15) is linked to six other goals. The results of the network analysis were compared to previous studies that had used a nexus approach to analyse the interconnections among climate, land, energy and water targets. This showed that most of the relevant interactions identified in nexus studies are not explicitly captured in the wording of the SDGs and targets.

SDG interactions have also been analysed through classifying and clustering the goals and their interactions. Waage et al. (2015) have proposed a framework whereby the SDGs are grouped into three domains and represented by three nested circles: well-being (comprising SDGs 1, 3, 4, 5, 10, 16) in the inner circle, infrastructure (comprising SDGs 2, 6, 7, 8, 9, 11, 12) in the middle circle and environment (comprising SDGs 13, 14, 15) in the outer circle. The potential interactions among SDGs are closely related to their position in the framework. The authors argue that there is potential for synergies among the goals in the inner level as they focus on different aspects of human well-being. Similarly, the outer-level environmental goals are interrelated and potentially synergistic. The infrastructure goals in the middle contribute to the achievement of the well-being goals, but compete for limited environmental resources such as land for agriculture, forestry or energy. The trade-offs between the inner- and outer-level goals need to be addressed within the infrastructure goals, which requires effective governance of these goals (Waage et al. 2015).

Different approaches have been proposed to address the interlinkages among the goals and targets. In this connection the nexus approach has gained renewed attention. For instance, the UN Prototype Global Sustainable Development Report (UN 2014: 21) advocates for the climate–land–energy–water–development nexus as a 'pragmatic approach to integrated assessment for selected clusters of strongly interlinked issues'. In general, the nexus approach aims at increasing policy coherence across sectors. The water–energy–food nexus has received special attention as a research agenda and development paradigm (Biggs et al. 2015, Bizikova et al. 2013, FAO 2014, Hoff 2011, Leck et al. 2015, Weitz et al. 2016). The downside of the nexus approach is that it specifically focuses on the interactions among a chosen, limited number of SDGs and targets, and thus fails to acknowledge possible important linkages to other SDGs that could be crucial for the specific SDGs under study.

As an attempt to develop a more comprehensive approach, Nilsson et al. (2016) present a framework consisting of a typology of interactions, where the interactions and relationships among SDG targets are assigned scores ranging from +3 (indivisible), +2 (reinforcing), +1 (enabling) and 0 (neutral) to –1 (constraining), –2 (counteracting) and –3 (cancelling). This framework is suggested as an approach to systematically assess the target-level interactions. It recognises that the scoring is context specific and is affected by timescale, governance arrangements, technology and geography.

As these analyses demonstrate, the SDGs and related targets form a complex, interconnected set of different kinds of goals. However, these interconnections are neither systematically recognised nor addressed in the SDG agenda: 'A tendency to ignore interlinkages among sectors and across national borders has meant that success in one area or location has all too often come at the expense of increasing problems elsewhere' (UN 2014: 21). Understanding the interactions among the targets is thus fundamental for making comprehensive progress towards the targets and ensuring that progress towards a specific target is not impeding the achievement of other targets (Griggs et al. 2017).

Forests and SDGs

Human welfare is ultimately based on global natural resources and biodiversity; the sustainable use of these resources within environmental limits is the foundation for sustainable development (Holden et al. 2017). Forests cover about one-third of global land area, totalling nearly 4000 million ha (FAO 2016a). However, ecological, social and economic conditions vary greatly in different parts of the world, among countries and even within countries, leading to great variations in forest–human interactions, the importance and potential contributions of forests for achieving the SDGs, and the potential and likely impacts of policies and measures for progressing towards the SDGs.

Brief Overview of the World's Forests

The global forest area is made up of 3695 million ha of natural forest and 291 million ha of planted forest.[1] Natural forest area continues to decline and planted forest area to increase. From 2010 to 2015, the natural forest area worldwide decreased by 6.5 million ha per year, while the reported planted area increased by 3.3 million ha per year. Most planted forests are located in Asia, with 129 million ha, followed by Europe, with 83 million ha (FAO 2016a).

Most of the world's forests are located in tropical countries (44 per cent), followed by temperate (26 per cent), boreal (22 per cent) and subtropical countries (8 per cent). Forest loss has also been concentrated in tropical countries, particularly in South America and Africa: from 2010 to 2015, tropical forest area declined by 5.5 million ha per year. In general, forest area has declined in Central America, South America, South and Southeast Asia and in Africa, while increasing in Europe, North America, the Caribbean, East Asia and Western-Central Asia (Keenan et al. 2015).

The pressures on forests are related to population growth through the increasing demand for forest-based products and services and through increasing competition for the use of land currently forested. The world's population has increased concurrently with global forest loss, leading to declining per capita forest area. While on average it declined from 0.8 ha to 0.6 ha per capita from 1990 to 2015, during this period the forest area per capita in the tropics nearly halved and declined by more than 35 per cent in the subtropics (FAO 2016a, Keenan et al. 2015).

[1] FAO (2012) definition of forest: Land spanning more than 0.5 hectares with trees higher than 5 metres and a canopy cover of more than 10 per cent, or trees able to reach these thresholds in situ.

Forest ownership and management rights are an essential part of forest governance; they are important in shaping the ways forests are used and managed and for the distribution of related costs and benefits. Most of the world's forests (76 per cent) are publicly owned, and about 20 per cent are privately owned; for the rest, information is not available (FAO 2016a). Public ownership is highest in Western and Central Africa (99 per cent), Western and Central Asia (98 per cent) and South and Southeast Asia (90 per cent).

The share of privately owned forests is highest in East Asia and Oceania (42 per cent) and North America (33 per cent) (FAO 2016a). The category of privately owned forests includes forests legally held by individuals, communities and firms. Individual, smallholder and family forest ownership is dominant in many European countries; it is increasing in countries with formerly centrally planned economies due to privatisation and restitution policies (Forest Europe 2015). Forest tenure reforms in China and Vietnam have led to forestland being allocated, leased or contracted to individuals and families, increasing the forest area under some degree of private management (Hou et al. 2017, Yasmi et al. 2017). In the United States, 58 per cent of forests are under private ownership, including private corporate ownership, which covers 19 per cent of forests.

According to the Rights and Resources Initiative data from 58 countries, covering 92 per cent of global forest area, private individuals and firms own 11 per cent and Indigenous and local communities 12 per cent of the total forest area in the analysed countries (RRI 2018). The areas owned by Indigenous and local communities totalled 447 million ha in 2017. Furthermore, Indigenous and local communities hold legally designated rights to 80.5 million ha of the publicly owned forests, meaning that 'national law recognises Indigenous Peoples' and local communities' rights to access and withdrawal, as well as to participate in the management of forests or to exclude outsiders' (RRI 2018: 8). Taken together, most forest area either legally owned by or designated for Indigenous and local communities is located in Latin America, followed by Asia and then Africa (RRI 2018).

Governments maintain legal and administrative authority over publicly owned forest areas; a large part of this is used and managed by Indigenous and local communities on the basis of customary, community-based tenure systems without formal government recognition. This, combined with the fact that large areas of government-administered forests have been granted to companies and investors under concession and license agreements without acknowledging the existing customary rights, has led to conflicts and disputes over forestland (RRI 2018). This situation has had serious negative livelihood implications for peoples residing in and around forests.

Forests' Contributions to SDGs

Forests provide ecosystem services that are crucial for human welfare. The contributions of forests to achieving the SDGs were explored before the SDGs were formally adopted and have since been further analysed and discussed (Brack 2014, FAO 2016b, 2018; Grazer and Keeton 2017, Sunderland et al. 2013, Vira et al. 2015). Forests can be directly or indirectly linked to each of the SDGs. Forests provide plant- and animal-based products that are important as foods and medicines, contributing directly to ending hunger (SDG 2) and ensuring health and well-being (SDG 3). Forestry employment can contribute to providing decent work (SDG 8) and forest-based incomes can contribute to ending poverty (SDG 1), and these incomes can be used to buy food, which further contributes to food security (SDG 2). Forests also provide clean water and influence hydrological cycles and downstream water supply, contributing to water and sanitation (SDG 6). Forest biomass can contribute to reducing global dependence on fossil fuels for energy (SDG 7), and forests can contribute to responsible consumption and production (SDG 12) by providing renewable materials to substitute non-renewable ones. Forests can also support industrial development and innovation (SDG 9). Some forest communities are among the most equitable globally, relating to gender (SDG 5) and equality (SDG 10), while participatory forest management approaches contribute to inclusive societies and building inclusive institutions (SDG 16). Forests are crucial for carbon storage and for regulating climate (SDG 13). Forests also provide supporting services, such as nutrient cycling and crop pollination, that are fundamental for sustainable agricultural production. Furthermore, mangroves (SDG 14) provide coastal protection benefits, strengthening coastal community resilience to climate-related hazards. Forest-related cultural ecosystem services include recreational, spiritual, religious and other non-material benefits. These benefits are important for rural and urban populations and contribute to learning and physical and mental well-being (SDG 4 and SDG 3) and more resilient and sustainable cities (SDG 11). Furthermore, most of the world's terrestrial biodiversity is found in forests (SDG 15).

Forest ecosystem services contribute to human welfare at different scales. At a global scale, all people benefit from the climate change mitigation and crop pollination services of forests. A large share of the global population also benefits from forest-based products such as wooden furniture or timber for housing. In addition, it is estimated that 350 million rural inhabitants are highly dependent on forests for food security, livelihoods and energy, while an estimated 60 million Indigenous peoples are totally dependent on forests for their subsistence (World Bank 2008).

Impacts of SDGs on Forests and Livelihoods

As already noted, despite the profound importance of forests for human well-being and their important role in reaching the SDGs, forests are only mentioned in two: in single targets under SDGs 6 (Clean Water and Sanitation) and 15 (Life on Land). Due to the interrelated nature of the SDGs and targets previously discussed, the implementation of the SDG agenda will inevitably influence the possibilities to achieve these forest-specific targets. In addition, SDG implementation will likely affect the capacity and potential of forests to provide the important ecosystem services described and consequently sustain forests' contributions to achieving the SDGs.

The implementation of the SDGs is principally in the hands of national governments: 'Targets are defined as aspirational and global, with each Government setting its own national targets guided by the global level of ambition but taking into account national circumstances. Each Government will also decide how these aspirational and global targets should be incorporated into national planning processes, policies and strategies' (UN 2015: 13). From the viewpoint of international law, the SDGs are considered as norms at the 'softest end of the soft law–hard law continuum' (Persson et al. 2016: 60). SDG 17 focuses specifically on the means of implementation, concentrating on finance, technology, capacity-building, trade and systemic issues, including policy and institutional coherence, multistakeholder partnerships and data, monitoring and accountability. In addition, specific targets under each SDG relate to the implementation of the given SDG. Yet many targets are rather vague, aspirational-outcome targets that do not specify the ways or conduct by which they should be achieved: different pathways can be taken (Persson et al. 2016). Moreover, the implementation targets do not address the interdependencies and interlinkages among the goals and targets (Stafford-Smith et al. 2017). Beyond the key role national governments play in Agenda 2030, it is important to recognise the emerging importance of regional and local governments, communities and private-sector entities in SDG implementation.

The implementation of the SDGs varies according to economic prosperity, political stability, social cohesion and national circumstances, such as geographic features, natural resource base, level of technological development, and policy, institutional and social frameworks. National circumstances and development priorities will influence how the SDGs are prioritised and how they are included in national policies and strategies. In this connection, the extent and condition of forest resources and the role of forests in industrial development and for livelihoods, as well as social and cultural perspectives and voices pertaining to forest use and conservation, are important factors in shaping the policies and strategies that relate to forests and the relative importance given to forests in relation to other land uses.

Understanding the potential impacts of SDGs on forests, forest-related livelihoods and forest-based options to generate progress towards achieving the SDGs, as well as the related trade-offs and synergies, is crucial to the efforts undertaken to reach these goals. It is especially important for reducing potential negative impacts and to leverage opportunities to create synergies that will ultimately determine whether comprehensive progress towards the SDGs will be accomplished.

Little attention, however, has yet focused on possible and likely impacts that efforts undertaken by different sectors to advance towards the 17 SDGs will have on forests, forest ecosystem services, forest-related livelihoods and human well-being, or on how these impacts, in turn, will contribute to or undermine the contributions of forests in achieving the SDGs. Efforts to achieve the SDGs may result in an emphasis on forest management and production or on forest protection, or a combination of both. Where the emphasis lies will affect the ecosystem services provided by forests and the associated benefits accruing to different segments of society. In this light, efforts to achieve some of the SDGs will likely drive additional land-use change and deforestation. Developments in sectors such as agriculture, transport, mining and energy may have crucial implications for forest-related development. Policies and actions to improve governance and build effective institutions for natural resource management may affect the conditions shaping forest resource use, benefit-sharing and sustainable production, while also affecting gender concerns and wider issues of social equity and equality.

Impacts of SDG implementation can vary according to geographical scale. National policies favouring more gender-equitable land tenure, for instance, can wind up being ignored at the local level. Progress on a particular SDG in one location can lead to harmful impacts in other locations. For example, a study on seven developing countries that have experienced forest transition – a shift from net deforestation to net increase in forest cover – found that in most cases reforestation was accompanied with deforestation in other countries through trade in timber and agricultural products (Meyfroidt et al. 2010).

The impacts can also vary according to the time horizon. Intensifying the use of forest products (e.g. collection of non-timber forest products) can support livelihoods and increase incomes in the short term, but in the long term may lead to depletion of the resource base and reduced availability of these products, resulting in declining incomes.

Assessing the impacts of the SDGs and related policies on forests and people is not a trivial undertaking. It requires a thorough look into the SDGs and the inherent trade-offs and synergies among them, the contextual factors that shape how the SDGs are prioritised and implemented, and the consequent impacts on forests and people.

Chapters 1 to 17 focus on the potential impacts of the implementation of the SDGs on forests and forest-related livelihoods and economies, taking into account the considerations discussed. Each chapter is dedicated to the analysis of one of the 17 SDGs. The analyses focus especially on people who live in or near forests and who depend on forests for their material, social, cultural and emotional well-being. In the following chapters these impacts are generally referred to as impacts on 'forests and people'. Based on existing evidence, these chapters identify the most important contextual conditions that guide or determine how a specific SDG is prioritised or pursued and discuss the possible impacts of its implementation on forests and the goods and services forests provide in different contexts. They also consider the linkages among the SDGs, identifying important opportunities for synergies and drawing attention to possible or unavoidable trade-offs.

Chapter 18 synthesises the findings from Chapters 1 to 17. Chapter 19 concludes with broad conclusions on a few key lessons learnt, with a view to providing guidance for the future co-evolution of people and forests in a changing world.

References

Biggs, E. M., Bruce, E., Boruff, B. et al. 2015. Sustainable development and the water–energy–food nexus: A perspective on livelihoods. *Environmental Science & Policy* 54:389–97.

Bizikova, L., Roy, D., Swanson, D., Venema, H. D. and McCandless, M. 2013. *The Water–Energy–Food Security Nexus: Towards a practical planning and decision-support framework for landscape investment and risk management*. Winnipeg: International Institute for Sustainable Development Report.

Boas, I., Biermann, F. and Kanie, N. 2016. Cross-sectoral strategies in global sustainability governance: towards a nexus approach. *International Environmental Agreements* 16:449–464. doi:10.1007/s10784-016-9321-1

Brack, D. 2014. *Sustainable Development Goals and Forests. A summary of UN Open Working Group debates and country reflections*. UK: IIED.

Dodds, F., Donoghue, D. and Roesch, J. L. 2017. *Negotiating the Sustainable Development Goals. A transformational agenda for an insecure world*. London: Routledge.

Elder, M., Bengtsson, M. and Akenji, L. 2016. An optimistic analysis of the means of implementation for Sustainable Development Goals: Thinking about goals as means. *Sustainability* 8:962. doi:10.3390/su8090962.

FAO 2012. *FRA 2015 terms and definitions*. Forest Resources Assessment Working Paper 180. Rome: FAO.

FAO 2014. *The water–energy–food nexus: A new approach in support of food security and sustainable agriculture*. Rome: FAO.

FAO 2016a. *Global forest resources assessment 2015. How are the world's forests changing?* 2nd ed. Rome: FAO.

FAO 2016b. *Forest-related indicators: Monitoring and reporting progress related to the achievement of the Sustainable Development Goals*. Committee on Forestry, 23rd session, Rome, 18–22 July. COFO/2016/5.2.

FAO 2018. *State of the world's forests 2018: Forest pathways to sustainable development*. Rome: FAO.

Forest Europe 2015. *State of Europe's forests 2015*. Madrid: Ministerial Conference on the Protection of Forests in Europe.

Fukuda-Parr, S. 2016. From the Millennium Development Goals to the Sustainable Development Goals: Shifts in purpose, concept, and politics of global goal setting for development. *Gender & Development* 24(1):43–52. doi:10.1080/13552074.2016.1145895.

Fukuda-Parr, S. 2017. *Millennium Development Goals: Ideas, interests and influence*. New York: Routledge.

Gratzer, G. and Keeton, W. S. 2017. Mountain forests and sustainable development. The potential for achieving the United Nations' 2030 Agenda. *Mountain Research and Development* 37(3):246–53.

Griggs, D. J., Nilsson, M., Stevance, A. and McCollum D. (eds.) 2017. *A Guide to SDG interactions: From science to implementation*. Paris: International Council for Science (ICSU).

Hoff, H. 2011. *Understanding the nexus*. Background Paper for the Bonn 2011 Conference: The Water, Energy and Food Security Nexus. Stockholm: Stockholm Environment Institute.

Holden, E., Linnerud, K. and Banister, D. 2017. The imperatives of sustainable development. *Sustainable Development* 25:213–26.

Hou, J., Yin, R. and Wu, W. 2017. Intensifying forest management in China: What does it mean, why, and how? *Forest Policy and Economics* 98:82–9.

Keenan, R. J., Reams, G. A., Achard, F. et al. 2015. Dynamics of global forest area: Results from the FAO Global Forest Resources Assessment 2015. *Forest Ecology and Management* 352:9–20.

Le Blanc, D. 2015. Towards integration at last? The Sustainable Development Goals as a network of targets. *Sustainable Development* 23(3):176–87.

Leck, H., Conway, D., Bradshaw, M. and Rees, J. 2015. Tracing the water–energy–food nexus: Description, theory and practice. *Geography Compass* 9(8):445–60.

Meyfroidt, P., Rudel, T. K. and Lambin, E. F. 2010. Forest transitions, trade, and the global displacement of land use. *Proceedings of the National Academy of Sciences* 107(49):20917–22.

Nilsson, M., Griggs, D., Visbeck, M., Ringler, C. and McCollum, D. 2016. Introduction. A framework for understanding Sustainable Development Goal interactions. In Griggs, D. J., Nilsson, M., Stevance, A. and McCollum D. (eds.), *A guide to SDG interactions: From science to implementation*. Paris: International Council for Science, pp. 18–30.

Persson, A., Weitz, N. and Nilsson, M. 2016. Follow-up and review of the Sustainable Development Goals: Alignment vs. internalisation. *RECIEL* 25(1):59–68.

RRI (Rights and Resources Initiative) 2018. *At a crossroads: Consequential trends in recognition of community-based forest tenure from 2002–2017*. Washington, DC: RRI.

Siry, J. P., Cubbage, F. W., Potter, K. M. and McGinley, K. 2018. Current perspectives on sustainable forest management: North America. *Current Forestry Reports* 4(3):138–49.

Stafford-Smith, M., Griggs, M., Gaffney, O. et al. 2017. Integration: The key to implementing the Sustainable Development Goals. *Sustainability Science* 12:911–19. doi:10.1007/s11625-016-0383-3.

Stevens, C. and Kanie, N. 2016. The transformative potential of the Sustainable Development Goals (SDGs). *International Environmental Agreements: Politics, Law and Economics* 16(3):393–6.

Sunderland, T., Powell, B., Ickowitz, A. et al. 2013. *Food security and nutrition. The role of forests*. Discussion Paper. Bogor, Indonesia: CIFOR.

UN 1987. *Report of the World Commission on Environment and Development: Our common future*. New York: United Nations.

UN 1994. *Agenda 21. Programme of Action for Sustainable Development*. Rio Declaration on Environment and Development. Statement of Forest principles. The final text of agreements negotiated by Governments of the United Nations Conference on Environment and Development (UNCED), 3–14 June 1992, Rio de Janeiro, Brazil.

UN 2000. *Resolution adopted by the General Assembly*. A/RES/55/2. 55/2 United Nations Millennium Declaration. 18 September 2000.

UN 2002. *Report of the World Summit on Sustainable Development*. Johannesburg, South Africa, 26 August–4 September 2002 A/CONF.199/20.

UN 2012. *Realizing the future we want for all*. Report to the Secretary-General. UN, New York 2012. Available at: www.un.org/en/development/desa/policy/untaskteam_undf/untt_report.pdf (Accessed 20 July 2018).

UN 2014. *Prototype global sustainable development report*. New York: United Nations Department of Economic and Social Affairs, Division for Sustainable Development, July 2014. Available at: http://sustainabledevelopment.un.org/globalsdreport/ (Accessed 20 July 2018).

UN 2015. *Resolution adopted by the General Assembly on 25 September 2015*. Transforming our world: the 2030 Agenda for Sustainable Development. UN A/RES/70/1. Available at: www.un.org/ga/search/view_doc.asp?symbol=A/RES/70/1&Lang=E (Accessed 20 July 2018).

UN 2017. *Resolution adopted by the General Assembly on 6 July 2012. 71/313*. Work of the Statistical Commission pertaining to the 2030 Agenda for Sustainable Development. A/RES/71/313. Available at: https://undocs.org/A/RES/71/313 (Accessed 22 February 2019).

Vira, B., Wildburger, C. and Mansourian, S. (eds.) 2015. *Forests, trees and landscapes for food security and nutrition: A global assessment report*. IUFRO World Series no. 33. Vienna: International Union of Forest Research Organizations.

Waage, J., Yap, C., Bell, S. et al. 2015. Governing Sustainable Development Goals: Interactions, infrastructures, and institutions. In Waage, J. and Yap, C. (eds.) *Thinking beyond sectors for sustainable development*. London: Ubiquity Press, pp. 79–88. doi:http://dx.doi.org/10.5334/bao.i.

Weitz, N., Strambo, C., Kemp-Benedict, E. and Nilsson, M. 2016. Closing the governance gaps in the water–energy–food nexus: Insights from integrative governance. *Global Environmental Change* 45:165–73.

World Bank 2008. *Forests sourcebook: Practical guidance for sustaining forests in development cooperation*. Washington, DC: World Bank. Available at: http://documents.worldbank.org/curated/en/356731468155739082/Forests-sourcebook-practical-guidance-for-sustaining-forests-in-development-cooperation (Accessed 14 February 2019).

Yasmi, Y., Ram Dahal, G. and De Bruyn, T. 2017. *Forest tenure in Cambodia, Nepal and Viet Nam*. Regional Office for Asia and the Pacific, Food and Agriculture Organization of the United Nations, Bangkok: FAO.

Chapter 1 SDG 1: No Poverty – Impacts of Social Protection, Tenure Security and Building Resilience on Forests

Kathleen Lawlor*, Erin Sills*, Stibniati Atmadja, Liwei Lin and Karnjana Songwathana

Key Points

- The relationship between poverty reduction and forests varies across regions, decades, stage in the forest transition and degree of market access.
- Achieving the specific targets of SDG 1, such as social protection and secure land tenure, can have positive effects on forests, especially if benefits are conditional on forest conservation.
- The overall effect on forests of pursuing SDG 1 depends on which poverty reduction policies are pursued; for instance, allocating forest land to poor farmers has very different implications for land use than targeting payments for ecosystem services to poor farmers.
- Exposure and vulnerability to environmental shocks in coastal areas and near steep slopes can be mitigated by forests – if the poor retain access to forest products and ecosystem services.

1.1 Introduction

SDG 1 seeks to 'end poverty in all its forms everywhere'. Poverty is increasingly recognised as a multidimensional concept. For example, the UN Multidimensional Poverty Index (used in the UN Human Development Reports) considers multiple deprivations that people might experience in the domains of health, education and standard of living (UNDP 2018). The World Bank's (2001) 'attacking poverty' framework is another widely used multidimensional approach to poverty (see Lawlor et al. 2013 for an application to forests). Building on the work of Sen (1999), this framework focuses on how opportunities, security and empowerment interact to promote human well-being. While quantifying all of these dimensions remains a challenge, the World Bank (2018) reports multidimensional indices of poverty that encompass multiple SDGs through measures of educational achievement (SDG 4),

* Lead authors.

access to drinking water and sanitation (SDG 6) and access to electricity (SDG 7), as well as considering whether income is sufficient to satisfy basic needs (SDG 1). Thus, the SDGs taken together embody the multidimensional approach to human development and poverty reduction.

Target 1.1 of SDG 1 focuses squarely on consumption poverty by calling for eradication of extreme poverty as defined by a monetary threshold (Table 1.1). However, SDG 1 also engages with other dimensions of poverty, recognising 'poverty in all its dimensions according to national definitions' (Target 1.2). Targets 1.3–1.5 specify how to reduce poverty: by ensuring that the poor are covered by social protection systems; by securing the rights of the poor to economic resources, access to basic services and property ownership; and by building their resilience to economic, social and environmental shocks.

In this chapter, we focus on the three targets under SDG 1 that specify strategies for reducing poverty, allowing us to draw on the existing evidence base about how those strategies affect forests. Specifically, we examine the potential consequences for forests of (1) implementing social protection systems that cover the poor and vulnerable (Target 1.3), (2) increasing the land tenure

Table 1.1 SDG 1 Targets

1.1	Eradicate extreme poverty for all people everywhere, currently measured as people living on less than $1.25 a day
1.2	Reduce at least by half the proportion of men, women and children of all ages living in poverty in all its dimensions according to national definitions
1.3	Implement nationally appropriate social protection systems and measures for all, including floors, and achieve substantial coverage of the poor and the vulnerable
1.4	Ensure that all men and women, in particular the poor and the vulnerable, have equal rights to economic resources, access to basic services, ownership and control over land and other forms of property, inheritance, natural resources, appropriate new technology and financial services, including microfinance
1.5	Build the resilience of the poor and those in vulnerable situations and reduce their exposure and vulnerability to climate-related extreme events and other economic, social and environmental shocks and disasters
1.A	Ensure significant mobilization of resources from a variety of sources, including through enhanced development cooperation
1.B	Create sound policy frameworks at the national, regional and international levels, based on pro-poor and gender-sensitive development strategies

Source: Adapted from https://sustainabledevelopment.un.org/sdg1

security of the poor (Target 1.4) and (3) reducing the vulnerability of the poor and building their resilience to shocks (Target 1.5). Recognising that more than 75 per cent of the global poor live in rural areas (World Bank 2016) and that poverty and forest cover are coincident in many parts of the world (Sunderlin et al. 2008), our analysis of these three targets focuses on the rural poor and their relationships with forests.[1] These relationships are moderated by market access – e.g. through its effect on land rents and therefore incentives for deforestation (Angelsen 2010, Pfaff et al. 2007) – as well as through its effects on employment opportunities, credit availability and insurance against environmental shocks.

We set the stage for our analysis of these relationships by examining the correlation between forests and extreme poverty (defined by a global income/consumption standard) at the cross-country level (as relevant to Target 1.1) and the role of forests in national poverty reduction strategies (as relevant to Target 1.2). We conclude the chapter by relating our analysis to the means of implementation for SDG 1, suggesting that the implications for forests depend on whether national policymakers recognise the role of forests in rural livelihoods.

1.2 Relationship between Forests and Poverty

The relationship between poverty and forests is the subject of a large body of literature. Household-level studies have demonstrated how forests support rural livelihoods – as a source of subsistence, a safety net and a potential pathway out of poverty (Cheng et al. 2017) – through ecosystem products (Angelsen et al. 2014, Shackleton et al. 2011) and services (Daw et al. 2011). In the other direction, poverty or income level is often included as a potential driver of deforestation in models at both the micro- and macro-scale (Atmadja and Sills 2015, Busch and Ferretti-Gallon 2017). Atmadja and Sills (2015) conclude that evidence on the relationship is mixed, with studies of Latin America more likely to find an environmental Kuznet's curve (Choumert et al. 2013) or win–lose relationship (i.e. correlated poverty reduction and deforestation). The environmental Kuznet's curve is one possible explanation for the 'forest transition', or the widely observed empirical regularity that forest cover declines until a turning point or transition, after which gains in forest cover due to natural regeneration and plantations overtake losses due to deforestation (Angelsen and Rudel 2013). The mechanisms underlying this common path vary across regions and reflect the mutual effects forests and economic development have on each other (Rudel et al. 2005). The effect on

[1] Due to greater purchasing power, a poor urban family may have a greater impact on forests than a rural family living in extreme poverty. However, given that the goal is to eradicate poverty, we choose to focus on the more numerous rural poor.

forests of external aid to reduce poverty is likely to vary depending on the stage of the forest transition, possibly accelerating forest loss in early stages and encouraging the transition to forest recovery in later stages.

To provide empirical context we plot national poverty rates against forest cover across three decades and four regions. While acknowledging that trends and patterns in poverty vary depending on the dimensions considered and the thresholds applied (World Bank 2018), we consider the percentage of the population living in extreme poverty, as measured by the USD 1.90 per day threshold established by the World Bank in 2015 and consistent with SDG Target 1.1.[2] As shown in Figure 1.1, in East Asia and Latin America an

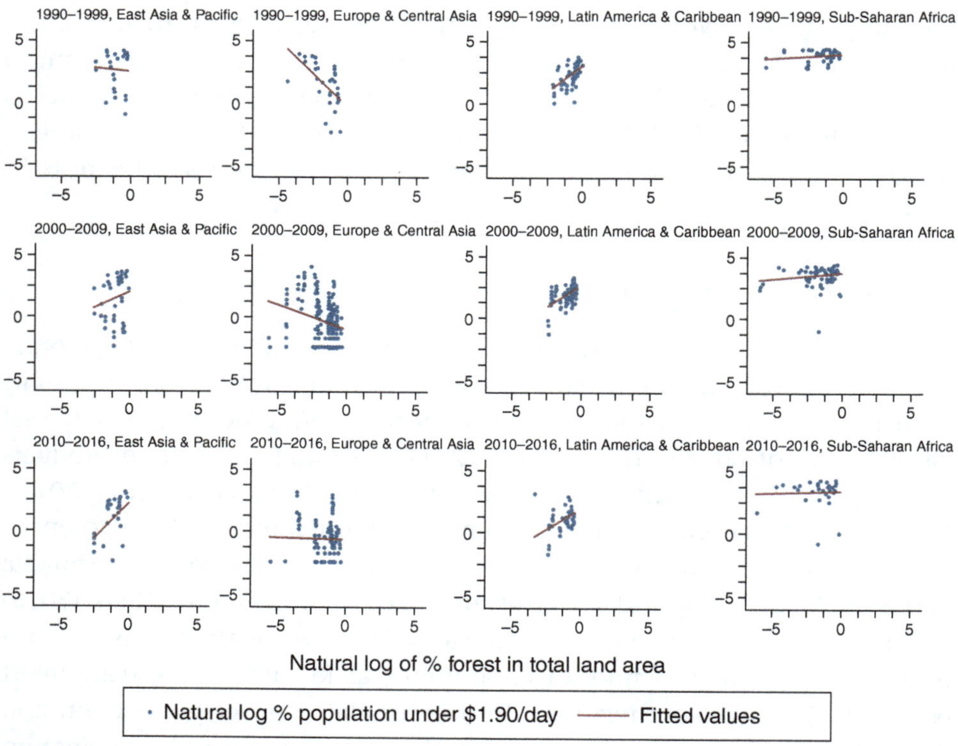

Figure 1.1 Relationship between forest cover and poverty, by decades and regions. Data source: World Bank.[3]

[2] See Ferreira et al. (2015) for an explanation of why, despite using different numbers, the SDG USD 1.25 per day and World Bank USD 1.90 per day poverty thresholds are consistent.

[3] Total land size by country: https://data.worldbank.org/indicator/AG.LND.TOTL.K2 Total population by country and year: https://data.worldbank.org/indicator/SP.POP.TOTL Poverty headcount earning less than USD 1.90/day (percentage of total population) by country and year (not all years are available for each country): https://data.worldbank.org/indicator/SI.POV.DDAY Forest area (sq km): https://data.worldbank.org/indicator/AG.LND.FRST.K2

inverse relationship between poverty and forests emerges over time: by the most recent decade, lower poverty rates are clearly associated with higher forest cover. In Europe and Central Asia, higher poverty rates are associated with a higher percentage of forest cover in earlier decades, with no apparent relationship in the most recent decade. In Africa, there is no relationship evident in any decade. While these plots only show correlations, they suggest that the effects on forests of pursuing SDG 1 are likely to vary across regions and time, and that there may not be any systematic relationship in the region with the highest levels of extreme poverty (sub-Saharan Africa). It could be that there are nonlinearities (e.g. kinks, reversed relationships) across the income continuum that are poorly characterised by this simple analysis. For example, those living far above the USD 1.90 per day threshold may have a very different relationship with forests than those living far below it.

To provide policy context we consider the role of forests in Poverty Reduction Strategy Papers (PRSPs). The International Monetary Fund (IMF) considers these policy documents to preserve national ownership of poverty reduction strategies and provide flexibility reflecting the particular circumstances of countries supported through their Poverty Reduction and Growth Trust (IMF 2018). PRSPs reveal whether national governments consider forests and poverty reduction to be 'friends' (i.e. synergistic relationship), 'foes' (i.e. competitive relationship) or completely independent of one another (i.e. no relationship). Pursuit of SDG 1 is likely to lead to better outcomes for forests in countries where conservation and sustainable management of forests are considered synergistic with poverty reduction. Of the 12 low-income or low-middle-income countries with the highest forest cover per capita, 9 have PRSPs. Table 1.2 summarises these, noting both specific references to forests and the overall stance towards forests, revealing policy priorities and political realities.

The PRSPs reflect different national positions on the role of forest conservation and sustainable management in poverty reduction, which we categorise as: (1) no role – forest protection is a responsibility unrelated to poverty reduction; (2) implicit – environmental protection (including forest protection) is a cross-cutting theme but few explicit actions related to forests are included; (3) supporting – forest protection and sustainable management is expected to contribute to poverty reduction, e.g. through ecosystem services; and (4) major – better governance of forests could be a key source of economic growth and thus critical for poverty reduction. Many PRSPs suggest that forests play an implicit or supporting role in poverty reduction. In contrast, the PRSPs consistently identify transportation infrastructure and agricultural development as important means of poverty reduction, both of which are tied to deforestation. This reflects conflicting policy priorities in some countries, while in other countries (especially those with relatively abundant forests and

Table 1.2 Forests in the PRSPs in countries with high forest cover

Country, Year of publication[a]	Forest (ha) per capita	Reference to forests	Likely impact on forest	Role of forest	Poverty indicator used
Republic of Bolivia (2001)	5.3	Increase rural employment through better roads, irrigation and electrification infrastructures, and access to land for agriculture and settlement; increase export competitiveness of agricultural products; increase non-agricultural income, such as rural tourism; actions to reduce levels of risk and vulnerability to water scarcity and natural disasters include reforestation and environmental conservation through integrated, sustainable natural resource management.	Negative	Implicit – Not a major economic sector; mainly to ensure rural resilience	Income to purchase a basket of goods and services
Rep. of Congo (2012)	5.0	Improve forestry and wood industry through specialised schools and industrialisation, improved forest management, and improved utilisation of non-wood forest products; forest conservation through participatory approaches and strategies such as REDD+, PES and forest certification.	Positive	Major – Forestry a source of national wealth (5.6% of GDP, 10% of foreign trade, 16 000 jobs)	Multidimensional (employment, access to services, quality of governance, income)

Mongolia (2003)	4.6	Forests need to be protected and expanded, mainly to provide ecosystem services to support other sectors; notably, livestock forests are acknowledged for their potential to generate jobs for the poor, e.g. through tree planting and forest utilisation.	No impact	Implicit – Mainly to support livestock, avoid desertification and provide informal jobs	Income
Royal Government of Bhutan (2004)	3.6	The 2020 target includes maintaining 60% of Bhutan's land area under forest coverage in perpetuity (from 72% forest cover in 2004), increase access to roads, increase income and employment; 'preserving and promoting cultural heritage and environment conservation' is one of the plan's 5 main objectives.	Negative	Supporting – Strong baseline environmental policies and environmental conservation is 1 of 5 pillars of long-term economic development	Household expenditure
Republic of Zambia (2006)	3.3	The overall strategy of broad-based wealth and job creation through economic infrastructure and human development focuses on rural development and agriculture (irrigation, food security, roads, livestock, microfinance); forests are a part of the natural resources sectoral plan as the main provider of household energy, with untapped potential for generating income from wood industries and tourism.	Negative	Implicit – Forests contribute 3.7% of GDP via charcoal and firewood production	Income level; forest loss/ degradation an indicator and result of poverty

Table 1.2 (cont.)

Country, Year of publication[a]	Forest (ha) per capita	Reference to forests	Likely impact on forest	Role of forest	Poverty indicator used
Lao PDR (2006)	2.8	Effective poverty reduction is achieved through increased agricultural productivity and better access to markets via better roads.; sustainable and participatory forest management mentioned as a subcomponent in 1 of 5 strategies for reducing rural poverty; poverty reduction can reduce environmental degradation, and economic growth can encourage environmental conservation if accompanied by education and training and development of scientific and technological capacities.	Negative	No role – Co-location: most poverty found in remote highlands, where forests are located	Includes lack of agricultural land
Dem. Rep. of Congo (2006)	2.2	As 1 of 6 sectors with growth potential, forestry is targeted with actions to improve forest management and institutions; 1 pillar of poverty reduction is improved governance, notably in forestry and mining sectors.	Positive	Major – Better governance needed to tap into this income	Peace, access to public services and productive capital, governance and meeting basic needs

Rep. of Guinea-Bissau (2011)	1.3	Environmental management and protection is a subcomponent of promoting inclusive, sustainable economic development, which is the last of 6 core areas in poverty reduction; the focus is on building capacity to address natural disasters including forest degradation; strategy for targeting the very poor focuses on revitalising agriculture, notably cashew and rice production.	Negative	Supporting – Small part of 1 of 4 core areas for poverty reduction related to sustainable economic development	Monetary and non-monetary (housing, sanitation, safe drinking water, consumer durables)
United Rep. of Tanzania (2010)	1.0	Alleviate income poverty by focusing on identified growth areas in agriculture, tourism, manufacturing and mining, and cross-sectoral drivers (e.g. roads, energy, water); forestry and forest products are one of 7 agricultural sub-sectors targeted for growth by 2015, as part of reducing income poverty; forest is a factor of production that needs to be used more efficiently for productivity gains and value addition.	Negative	Supporting – via tourism, rural job creation	Income, well-being, and good governance

[a] Not included due to lack of PRSP: Solomon Islands, Vanuatu and Zimbabwe

relatively low income) it reflects the view that forest resources are a means of economic development (Maini 2003). In sum, while the PRSPs suggest a wide range of approaches to forest policy, their almost universal call to alleviate rural poverty through agricultural and infrastructural development is likely to result in forest loss.

1.3 Social Protection

Target 1.3 calls for implementing nationally appropriate social protection systems and measures for all, including floors, and achieving substantial coverage of the poor and the vulnerable. Tirivayi et al. (2017) point out that forest-dependent peoples are typically poor and vulnerable, and therefore in particular need of social protection. Social protection systems, including programmes such as unemployment insurance and pensions for the elderly, are designed to help people cope with shocks and meet their basic needs. Over the past 20 years, cash transfer programmes have become a prominent component of social protection systems throughout the developing world (Handa et al. 2017, Hulme et al. 2012).

In addition to protecting recipients' human capital, cash transfer programmes can also affect households' economic production. The intent of most cash transfer programmes is to break the intergenerational transmission of poverty to children and build their human capital by increasing their consumption of food, health services and education (Baird et al. 2014, Bastagli et al. 2016, Manley et al. 2013). But cash transfers can also affect the economic production of households, especially family farmers who are otherwise cash constrained. These changes in households' production could impact forest resources both positively and negatively. For example, infusions of cash could increase pressure on forests if they enable households to expand their agricultural operations. Or, transfers could decrease pressure on local ecosystems if they enable migration to cities or the establishment of non-farm businesses.

There are multiple other pathways through which cash transfers could affect forests. A regular cash flow could make households less vulnerable to income shocks and thus less likely to rely on forests as 'natural insurance' (Pattanayak and Sills 2001) through harvesting and selling forest products. Cash transfers can enable increased consumption, with significant effects on deforestation locally or through markets for products that drive deforestation, such as beef, milk, soy and palm oil. Tracking the associated supply responses across space and time is challenging, making it difficult to quantify the full causal impacts of social protection systems on forests.

Many studies examine the effects of cash transfers on agricultural productivity. They typically find that in addition to building children's human capital, cash transfers help households increase their agricultural output (Tirivayi et al. 2016). This empirical regularity is likely associated with the other consistent finding that transfers increase purchases of agricultural inputs, such as chemical fertiliser. This could mean transfers are promoting agricultural intensification (increases in agricultural output without corresponding increases in hectares farmed), although increases in the area farmed (i.e. extensive agriculture) are also possible. Among these studies, there are a few that directly consider impacts on land use, including forests.

Our search of the literature uncovered ten studies of how cash transfers to the rural poor affect natural resources, including one study of remittances rather than government transfers (López-Feldman and Chávez 2017). Table 1.3 summarises the nine studies that estimate impacts on land use (the tenth study – Gilliland et al. 2018 – focuses on fisheries). All of the study sites are in the early stage of the forest transition, i.e. forest loss is ongoing. The recipients of the cash transfers generally have limited market access. Two of the studies (Alix-Garcia et al. 2013, Ferraro and Simorangkir 2018) combine household survey data with geospatial data to identify impacts on forests; the remaining seven report impacts on land used for farming. Two of the studies (Lawlor 2015, López-Feldman and Chávez 2017,) also examine impacts on forest product harvesting. Finally, two of the studies (Alix-Garcia et al. 2013, Lawlor 2015) explore how variations in market access affect transfers' impacts on natural resources. Overall, this literature finds that both conditional and unconditional cash transfer programmes have significant impacts on consumption and production in the short run (e.g. after only two years of payments).

Specifically, there is evidence that cash transfers might be encouraging land intensification in Lesotho (Daidone et al. 2014) and among farmers with larger landholdings in Mexico (Gertler et al. 2012, Todd et al. 2010). Transfers are promoting agricultural expansion among smallholders and those receiving agricultural subsidies in Mexico (Todd et al. 2010), Malawi (Asfaw et al. 2016b) and Zambia (Lawlor 2015). Cash is enabling the previously landless to farm in Mexico (Gertler et al. 2012, Todd et al. 2010) and increasing the number of farmers in Ethiopia (Asfaw et al. 2016a) and in Zambia among households living more than 10 km from markets (Lawlor 2015). In Ethiopia, transfers are reducing the likelihood of leaving land fallow (Asfaw et al. 2016a). The only evidence that cash transfers can decrease the likelihood of participating in agriculture comes from the López-Feldman and Chávez (2017) study of remittances in Mexico. However, their sample

Table 1.3 Studies evaluating the effect of cash transfers on land use

Authors (year)	Country, programme	Type of income support*	Study design	Impacts of cash transfers on …	
				Land use	Non-farm business and wage labour
Todd et al. (2010)	Mexico, *Oportunidades*	CCT for extremely poor households with children	Treatment-comparison with randomised data and matching weights	– Increases likelihood of farming land among previously landless – Increases per capita area farmed by smallholders; no change for large landholders – Increases in area farmed, largest for those receiving agricultural subsidies	Not reported
Gertler et al. (2012)	Mexico, *Oportunidades*	CCT for extremely poor households with children	Randomised treatment – control	– Increases likelihood of farming land among previously landless – No impact on land area used among landed farmers	– Increases likelihood of owning non-farm business
Alix-Garcia et al. (2013)	Mexico, *Oportunidades*	CCT for extremely poor households with children	Instrumental variable	– Increases deforestation due to increased consumption of beef and milk – Impacts appear larger in isolated communities	Not reported
Daidone et al. (2014)	Lesotho, Child Grant Program	UCT for poor households with children	Randomised treatment – control	– No impact on the probability of growing crops or area farmed	– Reduces non-farm businesses for labour-constrained households – Reduces participation in wage labour

Asfaw et al. (2016a)	Ethiopia, Tigray Social Cash Transfer Pilot Programme	UCT for extremely poor, labour-constrained households	Matched treatment – comparison	– Increases probability of growing crops – Decreases likelihood of leaving land fallow	– Reduces non-farm businesses for female-headed households – Reduces participation in wage labour
Asfaw et al. (2016b)	Malawi, Social Cash Transfer Program	UCT for extremely poor, labour-constrained households	Randomised treatment – control	– Increases area farmed – Increases adoption of sustainable farming practices	– Reduces non-farm businesses for labour-constrained and female-headed households – Reduces charcoal/ firewood businesses and increases petty trade enterprises – Increases number of days adult males spend earning wage income
López-Feldman and Chávez (2017)	Mexico	Remittances	Instrumental variable	– Decreases likelihood of participating in agriculture or natural resource extraction as well as reliance on environmental income	– Increases likelihood of earning wage income

Table 1.3 (cont.)

Authors (year)	Country, programme	Type of income support*	Study design	Impacts of cash transfers on …	
				Land use	Non-farm business and wage labour
Ferraro and Simorangkir (2018)	Indonesia, Program Keluarga Harapan	CCT for extremely poor households with children	Matched treatment-comparison	– Reduces village forest loss	Not reported
Lawlor (2015)	Zambia, Child Grant Programme	UCT for households with a child under age 5	Randomised treatment – control	– No impacts on fuelwood or bushmeat Close to markets (< 10 km) – increases use of charcoal – no impact on decision to farm – increases area farmed Far from markets (> 10 km) – no impact on charcoal use – increases likelihood of farming – increases area farmed	– Increases likelihood of owning non-farm business

*CCT: Conditional Cash Transfer

UCT: Unconditional Cash Transfer

excludes the country's smallest villages, with perhaps the most limited market access. Across the studies there is no evidence that cash transfers promote afforestation/reforestation.

In addition to agricultural impacts, two studies estimate impacts on households' natural resource use. In Mexico, remittances decreased harvesting of natural resources (firewood, timber and wild fruits, plants and meat) as well as reliance on environmental income (López-Feldman and Chávez 2017), supporting the hypothesis that transfers can replace natural insurance. However, in Zambia, cash transfers had no impact on consumption of bushmeat or fuelwood, and for households living within 10 km of markets, transfers increased the likelihood of using charcoal by 10 percentage points (Lawlor 2015). This is especially notable because charcoal is the principal driver of forest loss in Zambia (Day et al. 2014) and many other African countries, highlighting the importance of coupling poverty reduction programmes with clean energy initiatives (see Chapter 7).

Some of the studies reviewed also estimate impacts on livelihood strategies beyond farming and natural resource use. For example, in Mexico transfers increased the number of households owning small businesses by 67 per cent (Gertler et al. 2012), and remittances increased the likelihood of earning wage income by 14 percentage points (López-Feldman and Chávez 2017). In Zambia, transfers promoted diversification into non-farm businesses, with much larger impacts on those living close to markets (increased likelihood by 23 percentage points) than those living far from markets (increased likelihood by 11 percentage points) (Lawlor 2015). The magnitude of these impacts on non-farm businesses is quite large, especially considering these impacts are estimated after only two years of cash transfers. Taken together, these results provide evidence that cash transfers can help households diversify livelihood strategies, and may decrease pressure on forests over the long run by decreasing reliance on agriculture for income.

The two studies that harness geospatial data to identify the impacts of cash transfers on forests are most relevant to the question of how reducing poverty will impact forests because they capture not only how beneficiary households respond, but also spillovers to other households. Cash transfers could have significant multiplier effects, raising non-beneficiaries' consumption and production while avoiding inflation (Handa et al. 2017, 2018). This could have implications for land use, for instance, if beneficiaries' increased demand for food is met by increased food production by their neighbours or neighbouring communities. However, the two studies provide contradictory results. Alix-Garcia et al. (2013) find that transfers increased deforestation in Mexico, whereas Ferraro and Simorangkir (2018) find that cash transfers decreased deforestation in Indonesia. Both studies examine impacts after five years of payments. In Mexico, cash transfers increased deforestation rates by 15–33 per cent. The authors investigate how impacts vary according to market access and observe the largest impacts in isolated communities, which they conclude is because better market access diffuses the supply response across other communities. In Indonesia, on the other hand, transfers reduced village forest loss by 20 per cent (3.63 ha). The authors find some heterogeneity of impacts across forest governance institutions, with the largest reductions in forest loss in community forests, followed by concessions and protected areas. This raises the question of whether the cash transfers influence collective action, as well as household livelihood strategies. Further research is needed to understand the causal mechanisms.

Taken together, what can we say about the potential impact on forests of expanding the coverage of social protection systems? Clearly, the impacts of cash transfers vary by region due to differences in access to markets for land, labour, inputs and outputs, as well as differences in forest clearance costs and land tenure. The Indonesia study is the only one that finds clear positive impacts on forests. The remaining studies suggest that in the short run, rural households invest part of the transfers in their farms and that this results in the expansion of farmed area. Furthermore, households increase their food consumption, which elicits a supply response that increases pressure on forests. In the long run, some households living close to markets may be able to shift out of agriculture to non-farm businesses or wage labour, decreasing pressure on forests (cf. Sierra and Russman 2006). This could be encouraged by making cash transfers conditional on forest conservation, as in payments for ecosystem services (PES) (Alix-Garcia and Wolff 2014, Rodríguez et al. 2011). PES are often presented as a means to reduce both rural poverty and ecosystem degradation, although the targeting rules that maximise poverty reduction are likely to differ from the rules that maximise conservation benefits (James and Sills 2018). Both experience and field experiments have demonstrated the importance of local institutions in moderating the effects of PES, including effects on poverty (Sills and Jones 2018). The moderating effect of community tenure in the Indonesia case (Ferraro and Simorangkir 2018) suggests this may also be true of cash transfers.

1.4 Land Tenure

Target 1.4 calls for ensuring equal rights to economic resources, as well as access to basic services, ownership and control over land and other forms of property, inheritance, natural resources, appropriate new technology and financial services. Access depends fundamentally on transportation infrastructure (SDG 9) and basic services including housing (SDG 11), water (SDG

6) and energy (SDG 7), all of which can generate demand for ecosystem products and/or services, with implications for forests. In this section we focus on ownership and control over land, in particular on the security of private and communal land tenure.

Land tenure encompasses the institutions and policies that determine how land and its resources are accessed, who can hold and use these resources, for how long and under what conditions (Robinson et al. 2014). Tenure regimes can be characterised both by *who* holds rights (an individual, a community, a private entity, the state or, in the case of open access, no one) and by the *security* of those rights (defined by degree of protection against eviction and ability to exclude others). Elbow (2014) notes that tenure security can be achieved through public recognition of customary or Indigenous rights, certificates that secure the rights to use or manage resources, or titling of community-managed land or individual property rights. This is reflected in Indicator 1.4.2, which measures both the fraction of the population with legally recognised documentation of land tenure and the fraction of the population who perceive their rights to land as secure. Land tenure security is the perception that rights will be upheld by society (Sjaastad and Bromley 2000) or the certainty 'that a person's rights to land will be recognised by others and protected in cases of specific challenges' (Land Portal 2018). Securing land tenure has long been recognised as a tool for reducing poverty and enhancing economic development since it can both encourage investment and facilitate access to credit.

The security of land tenure affects forests through several channels. First, the need to secure tenure by demonstrating investment in the land may encourage deforestation in some contexts (Alston et al. 2000) and tree planting in others (Barbier and Tesfaw 2013). In many contexts, insecure tenure creates disincentives to invest in reforestation or forest management since land users have no assurance of reaping future benefits (Chazdon et al. 2016). Additionally, when forest users perceive their rights to the resource as time-limited or insecure, they are incentivised to harvest as much of the resource as fast as possible. Numerous studies have found that insecure land tenure promotes faster timber harvesting (Dorner and Thiesenhusen 1992, Puppim de Oliveira 2008, White and Martin 2002). Where the agents of deforestation are external to the customary occupants of the land, tenure insecurity for those occupants means that they do not have clear rights or incentives to defend forests from the external agents. Finally, access to technical assistance and direct conditional incentives to conserve forest (such as REDD+) may require secure tenure (Larson et al. 2013). On the other hand, increasing land tenure security can increase deforestation if it encourages investment in profitable agricultural activities that replace forest, such as plantation crops, or sale of land to agents that plan to clear the forest (Liscow 2013). In sum, the literature reports cases of tenure security both reducing and increasing deforestation, with differences related to livelihood strategies, socio-cultural institutions, tenure type, level of inequality (in communal tenure systems) and numerous other factors.

In a recent systematic review, Robinson et al. (2014) identify 36 publications that report 118 relationships between specific forms and security of tenure and (sub)tropical deforestation. All major regions of the tropics are represented in their sample. They categorise studies based on whether tenure security has a causal relationship with positive forest outcomes (defined as slowing deforestation or maintaining/regenerating forests) or negative forest outcomes (defined as accelerating deforestation). They find that communal (but not private or customary/traditional) tenure increases the likelihood of positive forest outcomes. Tenure security is consistently associated with positive forest outcomes across all types of tenure. This contrasts with the findings of a meta-analysis of spatially explicit econometric studies of deforestation by Busch and Ferretti-Gallon (2017). Based on 27 studies that estimate the effect of tenure security (defined as land ownership, legal title or duration of occupancy), they conclude that there is no systematic relationship between tenure security and deforestation.

To update the findings of Robinson et al. (2014) and Busch and Ferretti-Gallon (2017), we searched for recent studies (published in 2014 or later) of how tenure security affects forests, focusing on private and communal tenure. Like Robinson et al. (2014), we only include studies that give some indication of the degree to which tenure rights are secure. Following Robinson et al. (2014), we define forest outcomes as either positive or negative and only include studies that give some indication of the degree to which tenure rights are secure. The results of seven recent studies are summarised in Table 1.4. All of these studies estimate the effects of circumstances or interventions that increase tenure security.

L'Roe et al. (2016) find that formalising individual land claims in the eastern Brazilian Amazon by mapping and recording them in a state-run registry decreases deforestation on medium-sized properties (100–300 ha). Registration of land claims, however, has no impact on deforestation of larger properties. Holland et al. (2017) find the titling of private lands around a reserve only reduces deforestation when accompanied by 'forest friendly' restrictions.

In Uganda, Call et al. (2017) find that households are more likely to engage in tree-planting if they have secure tenure, are educated and live in isolated communities. In China, Lin et al. (2018) find that these types of investments are more likely when households can obtain logging permits but are not affected by tenure security.

Table 1.4 Effects of securing tenure on forests across tenure type: 2014–2017 studies

	Common-Property Regime			Individual Property		
	Positive Outcome for Forests	Negative Outcome for Forests	No Impact	Positive Outcome for Forests	Negative Outcome for Forests	No Impact
Brazil			BenYishay et al. (2017)	L'Roe et al. (2016)		
China						Lin et al. (2018)
Ecuador			Buntaine et al. (2015)	Holland et al. (2017)*		
Peru	Blackman et al. (2017)					
Uganda				Call et al. (2017)		

* Holland et al. (2017) find a positive effect on forests only when tenure security is bundled with restrictions on forest clearing and subdivision of properties.

Both Buntaine et al. (2015) and BenYishay et al. (2017) find that formalising Indigenous communities' land rights in Ecuador and Brazil, respectively, has no impact on rates of forest loss. In contrast, Blackman et al. (2017) find that titling Indigenous communities' land in Peru dramatically reduced deforestation in just three years. In order to obtain titles, the Peruvian communities had to submit sustainable management plans, which Robinson et al. (2017) argue may be necessary for tenure security to have a positive effect on forests. Blackman and Veit (2018) also find that allocation of tenure rights and management by Indigenous communities reduces deforestation in Bolivia, Brazil and Colombia (but not Ecuador).

In sum, the effect of increasing land tenure security (the perception that rights to land are recognised and will be upheld) on forests is context dependent. The existing evidence base suggests that increasing tenure security rarely leads to forest loss. However, the long-term effects are relatively understudied and hence unknown. As with direct cash transfers, there are suggestions that increasing tenure security is most likely to favour forests when accompanied by incentives or conditions that explicitly require forest conservation and sustainable management (Holland et al. 2017, Robinson et al. 2017).

1.5 Ecosystem-Based Adaptation to Climate Change

Target 1.5 is 'to build the resilience of the poor and those in vulnerable situations and reduce their exposure and vulnerability to climate-related extreme events and other economic, social and environmental shocks and disasters'. Forests have long been recognised as a safety net or form of natural insurance, providing both cash income and subsistence goods to poor rural households living on the forest margin, thus reducing their vulnerability and increasing their adaptive capacity and livelihood resilience (Agrawal et al. 2013, Byron and Arnold 1999, Pattanayak and Sills 2001). Poor and vulnerable populations tend to concentrate in remote and environmentally fragile areas (Sunderlin et al. 2008, Watmough et al. 2016), and they rely disproportionately on natural resources and ecosystem services to support their livelihoods, for both subsistence and income generation (Barbier 2010, Barrett 2005). This implies they are both more exposed (more often affected) and more vulnerable (lose more when affected relative to their income or wealth) to environmental shocks (Hallegatte et al. 2016). It also implies the effects of environmental shocks are likely to be channelled through ecosystems and moderated by the health of those ecosystems. Thus, managing for improved ecosystem health may be an effective way to reduce the exposure and vulnerability of poor populations to shocks and disasters.

There is increasing scientific and policy consensus that natural ecosystems can contribute to climate change adaptation by reducing exposure to shocks (Doswald et al. 2014, Munang et al. 2013). The role of forests in mitigating climate change itself through REDD+ is discussed in Chapter 13. Forests are often credited with reducing the sensitivity of ecosystems to extreme rainfall events, thus buffering communities from floods (Chapter 6), and mangrove forests can reduce damage from storm surges in coastal areas (Das and Vincent 2009, Chapter 14). Thus, measures to reduce exposure and vulnerability could include reforestation of slopes to prevent landslides and restoration of mangrove shelterbelts to protect coastal settlements against storms (Pramova et al. 2012). These are examples of ecosystem-based adaptation, or the conservation or restoration of natural ecosystems to reduce the vulnerability of people facing climate change threats (Vignola et al. 2009). This may be accomplished through public works programmes that jointly provide social protection and expanded forest cover (Tirivayi 2017). Ecosystem-based adaptation has been adopted in some National Adaptation Programmes of Action, as reflected in calls for afforestation and reforestation in Burkina Faso and Mali (to forestall desertification), Bangladesh (to stabilise the coast) and Haiti (to protect watersheds) (Locatelli et al. 2008). Thus, efforts to meet Target 1.5 could result in an expansion of forest area.

1.5.1 Sloping Land Conservation Program

The Sloping Land Conservation Program (SLCP) in China is another example of a national forest policy implemented to reduce exposure and vulnerability to environmental shocks. The SLCP is one of several programmes that China launched in response to a perceived 'national land-system sustainability emergency' in the late 1990s (Bryan et al. 2018). With rapid economic growth since the 1980s, China has experienced deforestation and land degradation (Liu and Diamond 2005). Deforestation and over-logging have exacerbated soil erosion, which is believed to threaten the safety of more than 100 million Chinese living in downstream sections of rivers in the eastern coastal region (Liu and Wu 2010). In particular, the massive floods of 1998, which resulted in more than 4000 deaths and serious economic damages, were blamed on soil erosion due to logging and deforestation in the Yangtze and Yellow River basins (Gutiérrez Rodríguez et al. 2016, Jin et al. 2017).

This perception that deforestation was to blame for the flooding led directly to the National Forest Protection Program, which banned logging, and the SLCP, which initially subsidised farmers to convert cropland to forest or grassland in the basins that had suffered flooding. In 2002, the Chinese government expanded the SLCP to cover most of the country (Liu et al. 2008). Under this programme, farmers with land prone to soil erosion and desertification are encouraged to convert agricultural fields to forest or grassland with subsidies from the government (Liu and Wu 2010). The subsidies, which many authors describe as PES, have been in the form of grain or cash. In their systematic review of the literature on the programme, Gutiérrez Rodríguez et al. (2016) find that most studies confirm the expected positive impact of the programme on forest cover and tree planting. Chen et al. (2015) concur that forest cover has increased, but note that some studies have raised questions about how much of the increase should be attributed to the SLCP and about the effects on ecosystem services.

In summary, forests have been recognised as potentially reducing both exposure and vulnerability to environmental shocks, including the extreme weather events that are expected to increase in frequency and severity with climate change. Particularly for the poorest and most vulnerable communities, investing in ecosystem services may be more effective, efficient and sustainable than infrastructure or technological options for adaptation (Locatelli et al. 2008). Thus, Target 1.5 provides an incentive to invest in forest protection and reforestation, especially in coastal zones, on steep slopes, in areas at risk of desertification and in critical watersheds. While these investments may be more likely to happen after a disaster (e.g. SLCP implementation in the wake of catastrophic floods), there are increasing calls to proactively implement ecosystem-based adaptation, including through afforestation and reforestation.

1.6 Conclusion and Means of Implementation

As with most SDGs, the suggested means of implementation for SDG 1 are mobilisation of resources and investment. For SDG 1, these are indicated by the proportion of government spending and international aid (grants and non-debt-creating inflows) allocated to three priorities: (1) poverty reduction; (2) education, health and social protection; and (3) sectors that disproportionately benefit women, the poor and vulnerable groups. While international aid to a country is not strictly a zero-sum game, clearly all governments operate under budget constraints. In this context, increasing the proportions of government spending and international aid on the priorities identified for SDG 1 could reduce the proportions of spending and aid allocated to forests (also a prominent concern for biodiversity; see Roe et al. 2013, Sanderson and Redford 2003). This could exacerbate the underfunding of the forest sector (Agrawal et al. 2013). Furthermore, as illustrated by the PRSPs, most governments prioritise infrastructure and agricultural development to alleviate rural poverty, with likely negative impacts on forests.

These trade-offs can be avoided if forests are understood to be fundamental to poverty reduction and hence included either as a means of poverty reduction (e.g. as part of ecosystem-based adaptation) or as a condition of poverty-reduction policies (e.g. social protection or titling policies that require commitments to forest conservation). Given the concentration of poverty in Africa, this is particularly important for the future of forests on that continent. There is some evidence that donors and governments are increasingly recognising the potential synergies between forest conservation and poverty reduction (Leisher et al. 2013). For example, Ethiopia's climate-resilient green economy strategy includes forest protection and restoration as one of the four pillars of economic development (FDRE 2011). Bilateral donors have invested resources in understanding forest–poverty relationships – for example, through the Ecosystem Services for Poverty Alleviation programme (Schreckenberg et al. 2018). This programme was funded by the UK's Department for International Development, along with the Economic and Social Research Council and Natural Environment Research Council. Likewise, CIFORs [4] Poverty and Environment Network attracted funding from international donors for rigorous research that carefully quantified the contribution of forest products to local livelihoods (Wunder et al. 2014).

While the importance of ecosystem goods and services to the rural poor has been established by this line of research, there is much less evidence on the types of interventions that can successfully achieve both forest conservation and poverty reduction (cf. Adams et al. 2004). For example, reviews of integrated conservation and development projects have generally concluded that most fail to achieve their goals (Naughton-Treves et al. 2005), and even that those goals are fundamentally contradictory (Miller et al. 2011). Likewise, the literature on PES has been cautious about its potential to simultaneously achieve forest conservation and poverty reduction, in part because these two goals may require different spatial targeting (Alix-Garcia et al. 2013, 2015; James and Sills 2018). Systematic reviews of the literature have found no evidence that PES harms recipients, but little evidence of benefits (Sills and Jones 2018).

Forest conservation is often pursued through reallocation of property rights to forests, either to government agencies to manage as protected areas or to communities to manage under sustainability requirements enforced by the government. Intuition suggests that restricting access to forest in protected areas should have a negative effect on local incomes, while decentralisation of forest management should increase incomes. However, recent research using rigorous quasi-experimental methods has found that protected areas can help alleviate poverty, with tourism as the likely mechanism (den Braber et al. 2018, Ferraro and Hanauer 2014, Pullin et al. 2013, Robalino and Villalobos-Fiatt 2015, Sims 2010). In a systematic review of impact evaluations of decentralisation, Samii et al. (2014) find three studies that report a positive effect on participants' household income (from forests or in total), suggesting a fairly thin evidence base. Further research into how impacts are moderated by institutions and other contextual factors is needed to understand the potential to achieve SDG 1 through forest initiatives (Sills and Jones 2018). Research that differentiates impacts on women, the poor and vulnerable groups may identify windows of opportunity or challenges. For example, Duchelle et al. (2018) report that REDD+ initiatives that limit deforestation have generally not negatively impacted local incomes. In sites where there have been negative impacts, they are concentrated among the highest-income households, resulting in greater equality of income. In the same sample, Larson et al. (2018) find negative impacts of REDD+ on women except

[4] Centre for International Forestry Research.

in sites where there were explicit strategies to address their priorities (also see Chapter 5). Thus, opportunities for win–wins may be fairly narrowly defined and require detailed analysis and tailored policies.

In sum, the specific targets for SDG 1 include social protection, secure land tenure and reducing exposure and vulnerability to environmental shocks. The empirical literature shows that cash transfers as well as more secure property rights – especially for community land – can be conducive to forest conservation, given the right context and conditionalities. As demonstrated by programmes to reforest hillsides and protect mangroves, initiatives to reduce vulnerability to environmental shocks can adopt an ecosystem-based adaptation approach, thereby promoting an expansion of forest cover. This approach is consistent with the scientific evidence that forests are both a mainstay of rural livelihoods and a source of natural insurance. However, there is relatively little evidence that this scientific knowledge is shaping poverty reduction and national development strategies. To the extent that those strategies are based on infrastructure and agricultural development, they are likely to remain in conflict with forest conservation and sustainable management.

References

Adams, W. M., Aveling, R., Brockington, D. et al. 2004. Biodiversity conservation and the eradication of poverty. *Science* 306(5699):1146–49.

Agrawal, A., Cashore, B., Hardin, R., et al. 2013. Economic contributions of forests. Background Paper 1: *Economic contributions of forests*. United Nations Forum on Forests. Tenth Session. 8–19 April, Istanbul. Turkey.

Alix-Garcia, J. M., McIntosh, C., Sims, K. R. E. and Welch, J. R. 2013. The ecological footprint of poverty alleviation: Evidence from Mexico's Oportunidades program. *Review of Economics and Statistics* 95(2):417–35.

Alix-Garcia, J. M., Sims, K. R. and Yañez-Pagans, P. 2015. Only one tree from each seed? Environmental effectiveness and poverty alleviation in Mexico's Payments for Ecosystem Services Program. *American Economic Journal: Economic Policy* 7(4):1–40.

Alix-Garcia, J. M. and Wolff, H. 2014. Payment for ecosystem services from forests. *Annual Review of Resource Economics* 6(1):361–80.

Alston, L. J., Libecap, G. D. and Mueller, B. 2000. Land reform policies, the sources of violent conflict, and implications for deforestation in the Brazilian Amazon. *Journal of Environmental Economics and Management* 39(2):162–88.

Angelsen, A. 2010. Policies for reduced deforestation and their impact on agricultural production. *Proceedings of the National Academy of Sciences* 107(46):19639–44.

Angelsen, A., Jagger, P., Babigumira, R. et al. 2014. Environmental income and rural livelihoods: A global-comparative analysis. *World Development* 64:12–28.

Angelsen, A. and Rudel, T. K. 2013. Designing and implementing effective REDD+ policies: A forest transition approach. *Review of Environmental Economics and Policy* 7(1):91–113.

Asfaw, S., Pickmans, R., Alfani, F. and Davis, B. 2016a. *Productive impact of Ethiopia's Social Cash Transfer Pilot Programme. A From Protection to Production report.* Rome: FAO.

Asfaw, S., Pickmans, R. and Davis, B. 2016b. *Productive impact of Malawi's Social Cash Transfer Programme – midterm report. From Protection to Production report.* Rome: FAO.

Atmadja, S. and Sills E. 2015. Identifying the causes of tropical deforestation: Meta-analysis to test and develop economic theory. In Köhl, M. and Pancel L. (eds.) *Tropical forestry handbook.* Berlin Heidelberg: Springer Verlag, pp. 1–27.

Baird, S., Ferreira, F. H., Özler, B. and Woolcock, M. 2014. Conditional, unconditional and everything in between: A systematic review of the effects of cash transfer programmes on schooling outcomes. *Journal of Development Effectiveness* 6(1):1–43.

Barbier, E. B. 2010. Poverty, development, and environment. *Environment and Development Economics* 15(4):635–60.

Barbier, E. B. and Tesfaw, A. T. 2013. Tenure constraints and carbon forestry in Africa. *American Journal of Agricultural Economics* 95(4):964–75.

Barrett, C. B. 2005. Rural poverty dynamics: Development policy implications. *Agricultural Economics* 32(s1):45–60.

Bastagli, F., Hagen-Zanker, J., Harman, L. et al. 2016. *Cash transfers: What does the evidence say. A rigorous review of programme impact and the role of design and implementation features*. London: ODI.

BenYishay, A., Heuser, S., Runfola, D. and Trichler, R. 2017. Indigenous land rights and deforestation: Evidence from the Brazilian Amazon. *Journal of Environmental Economics and Management* 86:29–47.

Blackman, A., Corral, L., Lima, E. and Asner, G. 2017. Titling Indigenous communities protects forests in the Peruvian Amazon. *Proceedings of the National Academy of Sciences* 114(16):4123–28.

Blackman, A. and Veit, P. 2018. Amazon Indigenous communities cut forest carbon emissions. *Ecological Economics* 153(C):56–67.

Bryan, B. A., Gao, L., Ye, Y. et al. 2018. China's response to a national land-system sustainability emergency. *Nature* 559(7713):193.

Buntaine, M., Hamilton, S. and Millones, M. 2015. Titling community land to prevent deforestation: An evaluation of a best-case program in Morona-Santiago, Ecuador. *Global Environmental Change* 33:32–43.

Busch, J. and Ferretti-Gallon, K. 2017. What drives deforestation and what stops it? A meta-analysis. *Review of Environmental Economics and Policy* 11(1):3–23.

Byron, N. and Arnold M. 1999. What futures for the people of tropical forests? *World Development* 27(5):789–805.

Call, M., Mayer, T., Sellers, S. et al. 2017. Socio-environmental drivers of forest change in rural Uganda. *Land Use Policy* 62:49–58.

Chazdon, R., Brancalion, P., Laestadius, L. et al. 2016. When is a forest a forest? Forest concepts and definitions in the era of forest and landscape restoration. *Ambio* 45(5):538–50.

Chen, C., König, H. J., Matzdorf, B. and Zhen, L. 2015. The institutional challenges of payment for ecosystem service program in China: A review of the effectiveness and implementation of Sloping Land Conversion Program. *Sustainability* 7(5):5564–91.

Cheng, S. H., Ahlroth, S., Onder, S. et al. 2017. What is the evidence for the contribution of forests to poverty alleviation? A systematic map protocol. *Environmental Evidence* 6(1):10.

Choumert, J., Motel, P. C. and Dakpo, H. K. 2013. Is the Environmental Kuznets Curve for deforestation a threatened theory? A meta-analysis of the literature. *Ecological Economics* 90:19–28. http://dx.doi.org/10.1016/j.ecolecon.2013.02.016

Daidone, S., Davis, B., Dewbre, J. and Covarrubias, K. 2014. *Lesotho's Child Grant Program: 24-month impact report on productive activities and labour allocation. Lesotho country case study report.* Protection to Production Project Report. Rome: FAO.

Das, S. and Vincent, J. R. 2009. Mangroves protected villages and reduced death toll during Indian super cyclone. *Proceedings of the National Academy of Sciences* 106(18):7357–60.

Daw, T., Brown, K., Rosendo, S. and Pomeroy, R. 2011. Applying the ecosystem services concept to poverty alleviation: The need to disaggregate human well-being. *Environmental Conservation* 38(4):370–79.

Day M., Gumbo, D., Moombe, K. B., Wijaya, A. and Sunderland, T. 2014. *Zambia country profile: Monitoring, reporting and verification for REDD+.* CIFOR Occasional Paper 113. Bogor, Indonesia: Center for International Research (CIFOR).

Democratic Republic of Congo. 2006. *Poverty reduction and growth strategy paper.* Available at: http://siteresources.worldbank.org/INTPRS1/Resources/Demrepofcongo_PRSP(Sept2007).pdf (Accessed 22 February 2019).

den Braber, B., Evans, K. L. and Oldekop, J. A. 2018. Impact of protected areas on poverty, extreme poverty, and inequality in Nepal. *Conservation Letters* 11(6):e12576.

Doswald, N., Munroe, R., Roe, D., et al. 2014. Effectiveness of ecosystem-based approaches for adaptation: Review of the evidence-base. *Climate and Development* 6(2):185–201.

Dorner, P. and Thiesenhusen, W. 1992. *Land tenure and deforestation: Interactions and environmental Implication. Discussion Paper.* Geneva: United Nations Research Institute for Social Development (UNRISD).

Duchelle, A. E., de Sassi, C., Sills, E. O. and Wunder, S. 2018. People and communities: Well-being impacts of REDD+ on the ground. In Angelsen, A., Martius, C., de Sy, V. et al. (eds.) *Transforming REDD+: Lessons and new directions.* Bogor, Indonesia: Center for International Forestry Research, pp. 131–43.

Elbow, K. 2014. *What is tenure security? Why does it matter?* USAID. Available at: www.land-links.org/wp-content/uploads/2017/02/USAID_Land_Tenure_2014_Haiti_Training_Module_1_Presentation_2_Elbow.pdf (Accessed 23 July 2019).

FDRE 2011. *Ethiopia's climate-resilient green economy: Green Economy Strategy. Addis Ababa,* Ethiopia: FDRE.

Ferraro, P. and Simorangkir R. 2018. *Environmental consequences of poverty alleviation programs: Evidence from conditional cash transfers in Indonesia*. Conference Paper, 6th World Congress of Environmental and Resource Economists. Gothenburg, Sweden.

Ferraro, P. J. and Hanauer, M. M. 2014. Quantifying causal mechanisms to determine how protected areas affect poverty through changes in ecosystem services and infrastructure. *Proceedings of the National Academy of Sciences*:1–6.

Ferreira, F. H. G., Chen, S., Dabalen, A. L. et al. 2015. *A global count of the extreme poor in 2012: data issues, methodology and initial results*. Policy Research Working Paper no. WPS 7432. Washington, DC: World Bank Group. Available at: http://documents.worldbank.org/curated/en/360021468187787070/A-global-count-of-the-extreme-poor-in-2012-data-issues-methodology-and-initial-results (Accessed 22 February 2019).

Gertler, P. J., Martinez, S. W. and Rubio-Codina, M. 2012. Investing cash transfers to raise long-term living standards. *American Economic Journal: Applied Economics* 4(1):164–92.

Gilliand, T. E., Sanchirico, J. N. and Taylor, J. E. 2018. *Environmental impacts of cash transfer programs: Implications for the welfare of poor communities in developing countries*. Conference Paper, 6th World Congress of Environmental and Resource Economists, Gothenburg, Sweden.

Gutiérrez Rodríguez, L., Hogarth, N. J., Zhou, W. et al. 2016. China's conversion of cropland to forest program: A systematic review of the environmental and socioeconomic effects. *Environmental Evidence* 5:21.

Hallegatte, S., Bangalore, M., Bonzanigo, L. et al. 2016. *Shock waves: Managing the impacts of climate change on poverty*. Climate Change and Development. Washington, DC: World Bank.

Handa, S., Daidone, S. Peterman, A. et al. 2017. *Myth-busting? Confronting six common perceptions about unconditional cash transfers as a poverty reduction strategy in Africa*. Innocenti Working Paper 2017–11. Florence: UNICEF Office of Research.

Handa, S., Natali, L., Seidenfeld, D., Tembo, G. and Davi, B. 2018. Can unconditional cash transfers raise long-term living standards? Evidence from Zambia. *Journal of Development Economics* 133:42–65.

Holland, M., Jones, K., Naughton-Treves, L. et al. 2017. Titling land to conserve forests: The case of Cuyabeno Reserve in Ecuador. *Global Environmental Change* 44:27–38.

Hulme, D., Hanlon, J. and Barrientos, A. 2012. *Just give money to the poor: The development revolution from the Global South*. Herndon: Kumarian Press.

IMF 2018. *Factsheet: Debt Relief under the Heavily Indebted Poor Countries (HIPC) Initiative*. Available at: www.imf.org/en/About/Factsheets/Sheets/2016/08/01/16/11/Debt-Relief-Under-the-Heavily-Indebted-Poor-Countries-Initiative (Accessed 16 February 2019).

James, N. and Sills, E. 2018. Payments for ecosystem services. *Oxford encyclopedia of environmental economics*. doi:10.1093/acrefore/9780199389414.013.580

Jin, L., Porras, I., López, A. and Kazis, P. 2017. Sloping Lands Conversion Programme, People's Republic of China. *Conditional transfers, poverty and ecosystems: National programmes highlights*. London: IIED. Available at: https://pubs.iied.org/pdfs/G04188.pdf (Accessed 28 July 2019).

Land Portal 2018. *Land and the Sustainable Development Goals (SDGs)*. Available at: https://landportal.org/book/sdgs/14/indicator-142 (Accessed 22 February 2019).

Lao PDR (People's Democratic Republic) 2006. *National Socio-Economic Development Plan (2006–2010)*. Vientiane, Lao: PDR. Available at: http://siteresources.worldbank.org/INTPRS1/Resources/LAO_PRSP2(Oct2008).pdf (Accessed 9 August 2018).

Larson, A. M., Brockhaus, M., Sunderlin, W. D. et al. 2013. Land tenure and REDD+: The good, the bad and the ugly. *Global Environmental Change* 23(3):678–89.

Larson, A. M., Solis, D., Duchelle, A. E. et al. 2018. Gender lessons for climate initiatives: A comparative study of REDD+ impacts on subjective wellbeing. *World Development* 108:86–102.

Lawlor, K. 2015. *Poverty–environment relationships under market heterogeneity: Cash transfers and producer-consumers in Zambia*. (Chapter 3 in PhD diss.: Impacts of Poverty Reduction in Remote Rural Landscapes: Evidence from Cash Transfers in Zambia. University of North Carolina, Chapel Hill, NC). Available at: https://cdr.lib.unc.edu/indexablecontent/uuid:801cb5e9-ed43-4231-89e2-be4989eb8ba9 (Accessed 16 February 2019).

Lawlor, K., Madeira, E. M., Blockhus, J. and Ganz, D. J. 2013. Community participation and benefits in REDD+: A review of initial outcomes and lessons. *Forests* 4:296–318.

Leisher, C., Sanjayan, M., Blockhus, J., Larsen, N. and Kontoleon, A. 2013. Does conserving biodiversity work to reduce poverty? A state of knowledge review. In Roe, D., Elliott, J., Sandbrook, C., Walpole, M. (eds.) *Biodiversity conservation and poverty alleviation: Exploring the evidence for a link* (1st ed.) Hoboken: John Wiley & Sons, pp. 43–59.

Lin, Y., Qu, M., Liu, C. and Yao, S. 2018. Land tenure, logging rights, and tree planting: Empirical evidence from smallholders in China. *China Economic Review*. doi:10.1016/j.chieco.2018.08.011.

Liscow, Zachary D. 2013. Do Property Rights Promote Investment but Cause Deforestation? Quasi-Experimental Evidence from Nicaragua. *Journal of Environmental Economics and Management* 65(2). Available at: https://ssrn.com/abstract=2297090 (Accessed 28 July 2019).

Liu, C. and Wu, B. 2010. *Grain for Green Programme in China: Policy making and implementation.* Nottingham: University of Nottingham China Policy Institute. Available at: www.nottingham.ac.uk/iaps/documents/cpi/briefings/briefing-60-reforestation.pdf (Accessed 16 February 2019).

Liu, J. and Diamond, J. 2005. China's environment in a globalizing world. *Nature* 435:1179–86.

Liu, J., Li, S., Ouyang, Z., Tam, C. and Chen, X. 2008. Ecological and socioeconomic effects of China's policies for ecosystem services. *Proceedings of the National Academy of Sciences* 105(28):9477–82.

Locatelli, B., Kanninen, M., Brockhaus, M. et al. 2008. *Facing an uncertain future: How forests and people can adapt to climate change*. Forest Perspectives No. 5. Bogor, Indonesia: CIFOR.

López-Feldman, A. and Chávez, E. 2017. Remittances and natural resource extraction: Evidence from Mexico. *Ecological Economics* 132:69–79.

L'Roe, J., Rausch, L., Munger, J. and Gibbs, H. 2016. Mapping properties to monitor forests: Landholder response to a large environmental registration program in the Brazilian Amazon. *Land Use Policy* 57:193–203.

Maini, J. S. 2003. International dialogue on forests: Impact on national policies and practices. In Teeter, L., Cashore, B. and Zhang, D. (eds.) *Forest Policy for Private Forestry – Global and Regional Challenges*, Wallingford: CABI Publishing, pp. 9–15.

Manley, J., Gitter, S. and Slavchevska, V. 2013. How effective are cash transfers at improving nutritional status? *World Development* 48:133–55.

Miller, T. R., Minteer, B. A. and Malan, L. C. 2011. The new conservation debate: The view from practical ethics. *Biological Conservation* 144:948–57.

Mongolia 2003. *Economic growth support and poverty reduction strategy*. Ulaanbaatar, Mongolia: Government of Mongolia. Available at: http://siteresources.worldbank.org/INTPRS1/Resources/Country-Papers-and-JSAs/Mongolia_PRSP.pdf (Accessed 22 February 2019).

Munang, R., Thiaw, I., Alverson, K. et al. 2013. Climate change and ecosystem-based adaptation: A new pragmatic approach to buffering climate change impacts. *Current Opinion in Environmental Sustainability* 5(1):67–71.

Naughton-Treves, L., Holland, M. B. and, Brandon, K. 2005. The role of protected areas in conserving biodiversity and sustaining local livelihoods. *Annual Review of Environment and Resources* 30:219–52

Pattanayak, S. K. and Sills, E. O. 2001. Do tropical forests provide natural insurance? The microeconomics of non-timber forest product collection in the Brazilian Amazon. *Land Economics* 77(4):595–612.

Pfaff, A., Robalino, J., Walker, R. et al. 2007. Road investments, spatial spillovers and deforestation in the Brazilian Amazon. *Journal of Regional Science* 47:109–23.

Pullin, A., Bangpan, M., Dalrymple, S. et al. 2013. Human well-being impacts of terrestrial protected areas. *Environmental Evidence* 2(19).

Puppim de Oliveira, J. 2008. Property rights, land conflicts and deforestation in the eastern Amazon. *Forest Policy and Economics* 10(5):303–15.

Pramova, E., Locatelli, B., Djoudi, H. and Somorin, O. A. 2012. Forests and trees for social adaptation to climate variability and change. *WIREs Climate Change* 3:581–96.

Republic of Bolivia 2001. *Poverty Reduction Strategy Paper – PRSP*. La Paz: Republic of Bolivia. Available at: http://siteresources.worldbank.org/INTPRS1/Resources/Country-Papers-and-JSAs/bolivaprsp.pdf (Accessed 9 August 2018).

Republic of Congo 2012. *Congo National Development Plan, Book 1: Growth, employment, and poverty reduction strategy paper (2012–2016 DSCERP)*. Ministry of Economy, Planning, Land Reform and Integration, Republic of Congo. Available at: http://siteresources.worldbank.org/INTPRS1/Resources/Congo-PRSP(August2012).pdf (Accessed 9 August 2018).

Republic of Guinea-Bissau 2011. *Second National Poverty Reduction Strategy Paper*. Ministry of Economy, Planning and Regional Integration, Bissau, Guinea-Bissau. Available at: http://siteresources.worldbank.org/INTPRS1/Resources/Gineau-Bisseau_PRSP(Dec2011).pdf (Accessed 9 August 2018).

Republic of Zambia 2006. *Fifth National Development Plan 2006–2010*. Republic of Zambia. http://siteresources.worldbank.org/INTPRS1/Resources/Zambia_PRSP(Dec2006).pdf (Accessed 9 August 2018).

Robalino, J. and Villalobos-Fiatt, L. 2015. Protected areas and economic welfare: An impact evaluation of national parks on local workers' wages in Costa Rica. *Environment and Development Economics* 20(3):283–310.

Robinson, B. E., Holland, M. B. and Naughton-Treves, L. 2014. Does secure land tenure save forests? A meta-analysis of the relationship between land tenure and tropical deforestation. *Global Environmental Change* 29:281–93.

Robinson, B. E., Holland, M. B. and Naughton-Treves, L. 2017. Community land titles alone will not protect forests. *Proceedings of the National Academy of Sciences* 114(29):E5764.

Rodríguez, L. C., Pascual, U., Muradian, R., Pazmino, N. and Whitten, S. 2011. Towards a unified scheme for environmental and social protection: Learning from PES and CCT experiences in developing countries. *Ecological Economics* 70(11):2163–74.

Roe, D. 2013. *Has biodiversity fallen off the development agenda*? A case study of the UK Department for International Development. *Oryx* 47(1):113–21.

Roe, D., Mohammed, E. Y., Porras, I. and Giuliani, A. 2013. Linking biodiversity conservation and poverty reduction: De-polarizing the conservation-poverty debate. *Conservation Letters* 6(3):162–71.

Royal Government of Bhutan 2004. *Poverty Reduction Strategy Paper: A cover note to the ninth plan main document*. Ministry of Finance, Royal Government of Bhutan. Available at: http://siteresources.worldbank.org/INTPRS1/Resources/Country-Papers-and-JSAs/cr04246.pdf (Accessed 9 August 2018).

Rudel, T. K., Coomes, O. T., Moran, E. et al. 2005. Forest transitions: Towards a global understanding of land use change. *Global Environmental Change* 15(1):23–31.

Samii, C., Lisiecki, M., Kulkarni, P., Paler, L. and Chavis, L. 2014. Effects of payment for environmental services (PES) on deforestation and poverty in low and middle income countries: A systematic review. *Campbell Systematic Reviews* 10(11):1–95.

Sanderson, S. E. and Redford, K. H. 2003. Contested relationships between biodiversity conservation and poverty alleviation. *Oryx* 37(4):389–90.

Schreckenberg, K., Poudyal, M. and Mace, G. 2018. *Ecosystem services and poverty alleviation: Trade-offs and governance*. Routledge. Available at: www.routledge.com/Ecosystem-Services-and-Poverty-Alleviation-OPEN-ACCESS-Trade-offs-and/Schreckenberg-Mace-Poudyal/p/book/9781138580848 (Accessed 29 July 2019)

Sen, A. 1999. *Development as freedom*. New York: Anchor Books.

Shackleton, S., Delang, C. O. and Angelsen, A. 2011. From subsistence to safety nets and cash income: Exploring the diverse values of non-timber forest products for livelihoods and poverty alleviation. In Shackleton, S., Shackleton, C. and Shanley, P. (eds.) *Non-timber forest products in the global context*. Berlin: Springer, pp. 55–81.

Sierra, R. and Russman, E. 2006. On the efficiency of environmental service payments: A forest conservation assessment in the Osa Peninsula, Costa Rica. *Ecological Economics* 59(1):131–41.

Sills, E. and Jones, K. 2018. Causal inference in environmental conservation: The role of institutions. In Dasgupta, P., Pattanayak, S. K. and Smith, V. K. (eds.) *Handbook of environmental economics*, vol. 4. Amsterdam: Elsevier, pp. 395–437.

Sims, K. R. E. 2010. Conservation and development: Evidence from Thai protected areas. *Journal of Environmental Economics and Management* 60(2):94–114.

Sjaastad, E. D. and Bromley, D. W. 2000. The prejudices of property rights: On individualism, specificity and security in property regimes. *Development Policy Review* 18:365–389.

Sunderlin, W. D., Dewi, S., Puntodewo, A. et al. 2008. Why forests are important for global poverty alleviation: A spatial explanation. *Ecology and Society* 13(2). Available at: www.ecologyandsociety.org/vol13/iss2/art24/ (Accessed 28 July 2019).

Tirivayi, N., Knowles. M. and Davis, B. 2016. The interaction between social protection and agriculture: A review of evidence. *Global Food Security* 10:52–62.

Tirivayi, N., Rodriguez, O., Juvenal, T. and Ma, Q. 2017. *Social protection for forest-dependent communities*. FAO Policy Brief. Rome: FAO. Available at: www.fao.org/3/a-i7008e.pdf (Accessed 15 February 2019).

Todd, J., Winters, P. and Hertz, T. 2010. Conditional cash transfers and agricultural production: Lessons from the Oportunidades experience in Mexico. *Journal of Development Studies* 46(1):39–67.

UNDP 2018. *Multidimensional poverty index*. http://hdr.undp.org/en/2018-MPI (Accessed 16 February 2019).

United Republic of Tanzania 2010. *National Strategy for Growth and Reduction of Poverty II*. Dar es Salaam, Tanzania: Ministry of Finance and Economic Affairs, United Republic of Tanzania. Available at: http://siteresources.worldbank.org/INTPRS1/Resources/Tanzania-PRSP-Dec2010.pdf (Accessed 9 August 2018).

Vignola, R., Locatelli, B., Martinez, C. and Imbach, P. 2009. Ecosystem-based adaptation to climate change: What role for policy-makers, society and scientists? *Mitigation and Adaptation Strategies for Global Change* 14:691–96.

Watmough, G. R., Atkinson, P. M., Saikia, A., and Hutton, C. W. 2016. Understanding the evidence base for poverty–environment relationships using remotely sensed satellite data: An example from Assam, India. *World Development* 78:188–203.

White, A. and Martin, A. 2002. *Who owns the world's forests? Forest tenure and public forests in transition*. Washington, DC: Forest Trend & Center for International Environmental Law.

World Bank 2018. *Poverty and Shared Prosperity 2018: Piecing Together the Poverty Puzzle*. Washington, DC: World Bank. License: Creative Commons Attribution CC BY 3.0 IGO

World Bank 2001. *World development report 2000/2001: Attacking poverty*. New York: Oxford University Press New York.

Wunder, S., Angelsen, A. and Belcher, B. 2014. Forests, livelihoods, and conservation: Broadening the empirical base. *World Development* 64:S1–11.

Yin, R., Zulu, L., Qi, J., Freudenberger, M. and Sommerville, M. 2016. Empirical linkages between devolved tenure systems and forest conditions: Challenges, findings, and recommendations. *Forest Policy and Economics* 73:294–9.

Chapter 2 SDG 2: Zero Hunger – Challenging the Hegemony of Monoculture Agriculture for Forests and People

Terry C. H. Sunderland*, Alida O'Connor, Giulia Muir, Lauren Nerfa, Giulia Rota Nodari, Camilla Wildmark, Nur Bahar and Amy Ickowitz

Key Points

- A 'business-as-usual' approach to food production will continue to cause mass deforestation. This is detrimental for biodiversity, consequently impacting forest-dwelling communities who depend on forests for the direct provision of food. With the loss of forests comes the loss of far-reaching ecosystem services, vital for many facets of food production relied on by the wider population.
- SDG 2 and five of its targets (Targets 2.1–2.5) are closely related to forests. These five targets underscore the reciprocity between forests and SDG 2. Forest biodiversity is integral for nutrition and the ability to grow and harvest diverse crops. In turn, investing in small-scale farming systems and sustainable farming techniques can help conserve forests.
- If we are to achieve SDG 2 sustainably, we need a reimagined food system that does not polarise agricultural production and the conservation of forest resources. This calls for land management that promotes the maintenance of biodiversity and integrated land-use planning. This is especially evident when examining the relationship between SDG 2 and the other SDGs, most of which are concomitantly contingent on each other.

2.1 Introduction and Context

For the majority of human history, we sustained ourselves by foraging edible plants and hunting animals encountered in grasslands, forests and other wild habitats. Indeed, much of our evolutionary development is based on a complex system of hunting and gathering, which provided a varied and nutritious diet (Gordon et al. 2017). All that changed around 10 000 years ago when agriculture simultaneously emerged in various parts of the world, creating a food system that is very much dominant today (Harari 2014).

* Lead author.

Agriculture dominates the global landscape. More than 40 per cent of the global land area is under some sort of cultivation, and we produce more food than ever before in human history (Ellis et al. 2010, Springmann et al. 2018). Much of this expansion has come at the expense of our forests. Agricultural expansion is also pushing other environmental boundaries. Over half of the world's freshwater is appropriated to nourish our crops, soil erosion now exceeds soil formation, chemical herbicides and pesticides result in extensive and pervasive pollution, and agriculture now accounts for around one-third of greenhouse gas emissions (Springmann et al. 2018). While this expansion has had great environmental costs, it has not necessarily resulted in better dietary and nutritional outcomes (Bahadur et al. 2018).

Our global food system is characterised by a heavy reliance on a narrow range of crops and livestock (Khoury et al. 2014). Diets across the globe have shifted from being largely plant-based with complex carbohydrates and low in fats to diets high in fats and oils, meats and refined carbohydrates (HLPE Report 2017a). As a result, almost 2 billion of our global population are over-nourished, and around the same number remain under-nourished (HLPE Report 2017a). With the latter issue, the proportion of the world's population that goes to bed hungry has actually increased in recent years (FAO et al. 2017), while almost one-third of all food produced is wasted, either post-harvest or post-purchase (FAO 2011). There are repeated calls for food production to increase by between 50 per cent and 100 per cent in order to feed an ever-growing human population – a call now echoed throughout the academic and development literature (Tomlinson 2013). In short, our global food system has failed to achieve universal food security (Vandermeer et al. 2018).

With food security and nutrition currently prominent in terms of global development priorities, we need to fully comprehend the deficiencies in our food system and the impact it has on the wider environment, including forests and other ecosystems (HLPE Report 2017b). The current global food system leaves millions of people food insecure while contributing to over-production and generating significant environmental degradation (HLPE Report 2017a). Often, however, food security is measured solely in terms of food energy (i.e. calorie production), losing sight of the fact that, by definition, food security includes secure access to the foods needed for a nutritionally balanced diet (Bahadur et al. 2018, HLPE Report 2017a, Ickowitz et al. 2019). This focus on energy production has contributed to a dichotomisation in which food production, sustainable forest management and conservation are portrayed as mutually exclusive (Brussard et al. 2010). The clear separation of biodiversity conservation and agricultural production has been an impediment in achieving optimised outcomes for either (Gordon et al. 2017). A serious reform of the current food system is clearly needed.

The question central to this chapter is: *Can we feed and nourish the growing human population without further damaging our wider environment, especially forests, in the process?* Throughout the chapter we explore this question by examining SDG 2 (Zero Hunger) in relation to forests. First, we focus on the SDG 2 targets that are impacted by or will directly impact forests. This is followed by a discussion on the relationship between SDG 2 and the other SDGs in regard to forests.

2.2 Zero Hunger and Forests

SDG 2 seeks to 'End hunger, achieve food security and nutrition and promote sustainable agriculture' (United Nations 2015). The goal aims to end hunger and all forms of malnutrition by 2030. It also commits to 'universal access to safe, nutritious and sufficient food at all times of the year' (Table 2.1). The narrative further describes how achieving SDG 2 will require sustainable food production systems and resilient agricultural practices, equitable access to land for farmers and communities, technology and markets, and international cooperation on investments in infrastructure and technology to boost agricultural productivity. Targets 2.1–2.5 (Table 2.1) are closely entwined with forests and forest-related livelihoods. Unless agriculture and forestry are designed to coexist, the possible impacts of achieving SDG 2 on forests include increasing resource use to raise production, thereby creating more pollution (e.g. phosphorus, nitrates, fossil fuels) and higher rates of deforestation (Springmann et al. 2018).

Reflecting on this chapter's guiding question – *Can we feed and nourish the growing human population without further damaging our wider environment, especially forests, in the process?* – we begin to see how Targets 2.1–2.5 address this. Targets 2.1 and 2.2 can be viewed as the desired outcomes of SDG 2: to end all hunger and ensure that everyone, especially vulnerable populations, has access to nutritious food. Target 2.4 draws our attention to the need to achieve Targets 2.1 and 2.2 in a way that will sustain rather than degrade forest ecosystems. Finally, if managed correctly, Targets 2.3 and 2.5 are key to achieving SDG 2 while maintaining the ecological integrity of forests.

2.2.1 Forests and Targets 2.1 and 2.2

Agriculture expansion is the largest cause of deforestation, responsible for approximately 80 per cent of forest loss worldwide (HLPE Report 2017b, Kissinger et al. 2012). Recent research has found that more than one-quarter of permanent forest transformation is driven by commodity expansion, notably that of cattle, soy and oil palm (Curtis et al. 2018). This has devastating

Table 2.1 SDG 2 targets and indicators

Target	Indicator(s)
2.1 End hunger and ensure access by all people to safe, nutritious and sufficient food, especially vulnerable populations	2.1.1 Prevalence of undernourishment 2.1.2 Prevalence of moderate or severe food insecurity
2.2 End all forms of malnutrition	2.2.1 Prevalence of stunting among children under 5 years of age 2.2.2 Prevalence of malnutrition among children under 5 years of age
2.3 Double the agricultural productivity and incomes of small-scale food producers	2.3.1 Volume of production per labour unit 2.3.2 Average income of small-scale food producers
2.4 Ensure sustainable food production systems and implement resilient agricultural practices	2.4.1 Proportion of agricultural area under productive and sustainable agriculture
2.5 Maintain the genetic diversity of seeds, cultivated plants and farmed and domesticated animals and their related wild species	2.5.1 Number of plant and animal genetic resources for food and agriculture secured 2.5.2 Proportion of local breeds classified as being at risk, not at risk or at unknown level of risk of extinction
2.A Increase investment in rural infrastructure, agricultural research and extension services, technology development and plant and livestock gene banks	2.A.1 The agriculture orientation index for government expenditures 2.A.2 Total official flows to the agriculture sector
2.B Correct and prevent trade restrictions and distortions in world agricultural markets	2.B.1 Agricultural export subsidies
2.C Adopt measures to ensure the proper functioning of food commodity markets and their derivatives and facilitate timely access to market information	2.C.1 Indicator of food price anomalies

Source: IAEG-SDGs 2016

consequences for both forests and people. Forests contain 80 per cent of terrestrial biomass and provide habitat for more than half of the world's known terrestrial plant and animal species (Aerts and Honnay 2011, Shvidenko et al. 2005). Forests contribute directly and indirectly to food security and nutrition in numerous ways and for various groups of people (Broegaard et al. 2017, Powell et al. 2015). All those who rely to some extent on forests and trees for their livelihood can be considered forest-dependent (HLPE Report 2017b). Byron and Arnold (1997) further this definition by making a crucial distinction between those who rely on forest use and have no alternative and those who use forest products or engage in economic activities involving forests, but do so as a matter of choice.

Communities located in remote areas in and around forests are heavily dependent on forest resources for their livelihoods, especially food (Powell et al. 2015). The types of food from forests and the ways it is harvested have cultural and traditional significance to Indigenous groups (Kuhnlein et al. 2009). These groups often live as hunter-gatherers or shifting cultivators on a subsistence basis (Padoch and Sunderland 2014). Shifting cultivation, also known as swidden agriculture, involves the intermittent clearing and burning of small patches of forest for subsistence food crop production, followed by longer periods of fallow in which the forest regenerates and restores the productivity of the land (Cramb et al. 2009). Swidden agriculture is practised in many countries in the tropical regions of Africa, Asia and Latin America (Mertz et al. 2009). In places with an abundance of land and low human population, shifting cultivation can be managed sustainably to preserve biodiversity and soil fertility while contributing to food security, and can have a long-term, often beneficial, influence on the floristics and ecology of the forests concerned (Maezumi et al. 2018). However, this is less likely to be as sustainable in places with denser populations (Peng et al. 2014, Vira et al. 2015).

People who live in proximity to forests are also somewhat dependent on forests for food security and nutrition. These people are usually involved in agricultural practices either within or outside of the forest, and use forest products partly for their own subsistence and partly for income generation (HLPE Report 2017b). For those more involved in agriculture, dietary supplements from forests are of critical importance to diet diversification for a more nutritious diet (Broegaard et al. 2017). Take bushmeat for example. Bushmeat is derived from wild terrestrial animals and is a significant source of protein extracted from the forest (Nasi et al. 2011). In tropical areas where livestock production is limited and domesticated meats are unaffordable, bushmeat is an important source of micronutrients and protein (Fa et al. 2015). Relatedly, forests act as an economic and environmental safety net, helping households and communities recover from shocks (Wunder et al. 2014). After a poor harvest or drought, food from forests often provides dietary substitutes during periods of scarcity. In times of food insecurity, maternal food deprivation can cause childhood malnutrition and ill health, effects that can linger long into adulthood, ultimately affecting life-long productivity and health (Agarwal 2018). The role of safety net that forests play is important for the most vulnerable groups and relates directly to Targets 2.1 and 2.2.

In addition to the direct provision of food, forest plants are used as feed for livestock, another source of meat and income generation (Baudron et al. 2017). Furthermore, forests contribute to food security and nutrition through the provision of energy. In places where people have no alternative energy sources, wood gathered from the forest is used as the main fuel for cooking. One-third of the global population relies on woodfuel for cooking (HLPE Report 2017b). The ability to cook food expands food options and is important for food safety and water purification (Jin et al. 2017).

The contributions of forests to Targets 2.2 and 2.3 reach far beyond communities living in and near forests. When discussing forest-dependent people, it is difficult to truly understand what this encompasses. Attempts to quantify the number of forest-dependent people worldwide have been made primarily using information on food and income generated from forests. However, these methods do not consider that most agricultural activities depend on ecosystem services provided by forests, which would drastically increase the number of forest-dependent people (HLPE Report 2017b). Forests deliver ecosystem services such as water regulation, soil protection, nutrient circulation, pest control, pollination and carbon-cycle regulation, all of which support food production at the farm, landscape and global scales and mitigate the impacts of climate change (Cumming et al. 2014).

Nevertheless, forests possess inherent trade-offs. They can harbour species that contribute to human wildlife conflict via crop or livestock damage, and pests or diseases that can transfer to crops, livestock and people. For example, in the United Kingdom badgers have been known to spread bovine tuberculosis to dairy cattle (HLPE Report 2017b). However, evidence shows that the benefits of forests to agriculture far outweigh the costs (Reed et al. 2017a). Moreover, the trade-offs mentioned here would still be a challenge, perhaps intensified by fragmentation and deforestation from agriculture. Loss of habitat leaves wildlife populations in search of food and water, resulting in livestock predation and competition for water and grazing land (HLPE Report 2017b).

To summarise, forests are vital to nutritious food production through the direct provision of diverse and nutritious food, energy for cooking and ecosystem services (Powell et al. 2015, Reed et al. 2017a). These contributions are particularly important for the nutrition and food security of vulnerable populations (Targets 2.1 and 2.2); they also impact the food security of the global population (HLPE Report 2017b). As such, the importance of conserving the ecological integrity of forests is undeniable (FAO 2019). However, the current dominant food system results from precisely the contrary: namely, a denial of this importance. This emphasises the need to bear in mind forest conservation to achieve Target 2.4's aim to 'ensure sustainable food production systems and implement resilient agricultural practices' (see Table 2.1).

2.2.2 Forests and Targets 2.3, 2.4 and 2.5

This section explores how the current food system polarises food production and forest conservation, when in fact they should and can be harmonised. We pay special attention to the role of maintaining genetic diversity (Target 2.5) and investing in small-scale food producers (Target 2.3) in sustainable food systems (Target 2.4) to reduce hunger and malnutrition (Targets 2.1 and 2.2), ultimately needed to achieve SDG 2. Throughout this discussion we draw attention to the impacts of achieving these targets on forests and people.

A POLARISING FOOD SYSTEM

Agricultural expansion, production and trade, particularly in the past 100 years, have been the greatest drivers of land conversion and habitat loss, as well as the major direct cause of deforestation (Gibbs et al. 2010). More efficient and productive agriculture has now reached increasingly into marginal lands and is a major land use worldwide (Springmann et al. 2018). With this has come trade and transport, meaning agriculture is now connected to markets and finance across the globe (Swain et al. 2018). The globalisation of agribusiness has resulted in a shift from traditional wholesale markets towards vertically coordinated supply chains that favour large-scale monocrop production (FAO 2015).

Whether rapid agricultural expansion causes deforestation or takes place on previously cleared land has been missing from the conversation on agriculture for some time. A study by Gibbs et al. (2010) reveals that the total net increase in agricultural area was more than 100 million ha across tropical regions during the 1980s and 1990s. More than 55 per cent of this new land came from intact forests and 28 per cent came from disturbed forests (forests previously affected by shifting cultivation, woodfuel collection and other forms of gradual degradation). This confirms that during those decades forests were the primary source for new agricultural land, and expansion has not come from previously cleared or degraded land (Gibbs et al. 2010). This trend persists: forest-rich tropical countries with lower production costs and fewer environmental regulations are being used to meet the continuously growing demand for agricultural land. Much of this land is used as pasture for livestock and to grow livestock feed and commodity crops such as sugarcane, soybeans and oil palm (Curtis et al. 2018).

Nutrition transitions are occurring in tandem with deforestation and environmental change. Rural communities whose land is converted to monoculture agricultural production, such as oil palm, lose not only their forests but in many cases their formerly diverse diets sourced from multi-functional landscapes (Ickowitz et al. 2016). This can equate to cultural losses, such as the loss of foods with symbolic meanings or food required for certain traditions (Cockx et al. 2018). Clearing land with no regard for conserving biodiversity has not only diminished the dietary variety of people living in or near forests, but also that of the wider population too.

Historically, the achievement of food security has focused primarily on calorie intake rather than nutrition (Ickowitz et al. 2019). There exists today a triple burden: malnutrition, consisting of deficiencies in dietary energy intake (hunger), estimated to affect more than 800 million people worldwide in 2017; nutrient deficiencies – such as a lack of iron, iodine or vitamin A – which affect some 2 billion people (2017); and the rapidly growing number of people who are overweight, estimated by the World Health Organization at 1.9 billion adults in 2016 (39 per cent of the world's adult population), of which 650 million (13 per cent) were classified as obese (HLPE Report 2017a). This is expected to intensify. As countries urbanise and incomes rise, diets tend to become high in sugar, fats, refined carbohydrates, meat and dairy (WRI 2018). Although small portions of meat and dairy can provide important micronutrients, half of the global population consume 50 per cent more protein than needed (WRI 2018). Ruminant meat (cattle, sheep, goats) consumption is expected to grow 88 per cent by 2050. Ruminant livestock uses two-thirds of global agricultural land (WRI 2018), and approximately two-thirds of all soybeans, maize and barley and one-third of all grains are used as feed for livestock (Willett et al. 2019).

The EAT–Lancet Commission describes a universal healthy reference diet that links healthy foods with improved human health and environmental sustainability. The diet consists of vegetables, fruits, whole grains, legumes, nuts, unsaturated oils and low amounts of seafood and poultry. The diet recommends low to no consumption of unhealthy foods such as red meat, processed meats, added sugar, starchy vegetables and refined grains (Willett et al. 2019). Transitioning to a diet similar to the healthy reference diet requires a reduction in global consumption of unhealthy foods by more than 50 per cent (Willett et al. 2019).

Adding to the mounting concerns of the current food system is food waste. It is estimated that a third of all food grown is wasted, either post-harvest or post-purchase (FAO 2011). Food loss occurs along the entire food chain and has negative economic, social and environmental consequences (Aschemann-Witzel et al. 2015, FAO 2011). Food waste at the beginning of the food chain is prevalent in low-income countries. These losses are largely due to technical limitations in harvesting and a lack of storage and cooling facilities, as well as packaging and marketing systems (FAO 2011). Food waste in medium- and high-income countries shows an opposite trend, with most food wasted at the consumer level. This can be attributed to poor purchase planning and best-before dates, quality standards and aesthetic expectations, enabled by consumers who can afford to waste food (Aschemann-Witzel et al. 2015, FAO 2011). When food is wasted, the resources used in its production and transportation are also squandered (Aschemann-Witzel et al. 2015).

From this discussion, it is clear the production and consumption trends driven by the current food system are problematic for both people and forests. The focus on maximising the production of select commodity crops has resulted in mass deforestation for monocrop agriculture. The logic behind our current approach to agriculture has become disconnected from what agriculture is so dependent on: nature (Gordon et al. 2017) and its biodiversity (Sunderland 2011). This is made worse by the fact that the system has failed to achieve global food security and nutrition. Rather, the current system relies on a narrow range of crops, and diets have shifted to become less diverse and nutritious, negatively impacting the health of people and forests.

The complex polarisation of the approaches needed to feed current and future populations while conserving forests and the wider environment is a fundamental development issue. It has led to the protectionist paradigm of separating nature from agriculture rather than the two operating in synergy (Harari 2014). The achievement of SDG 2 is contingent on recoupling nature and agriculture. Diverse and nutritious diets are synonymous with biodiversity. Moving forward, current production and consumption trends need to change. As much as this is for the benefit of people, it presents a challenge to our current habits. Achieving SDG 2 necessitates a behavioural change in what food we consume, as well as how we manage and produce food. The following sections explore how maintaining genetic diversity (Target 2.5) and investing in smallholder farmers (Target 2.3) will help address these needed changes for a food system that is resilient (Target 2.4) and nutritious (Targets 2.1 and 2.2).

TARGET 2.5: GENETIC DIVERSITY IN CROPS

Achieving SDG 2 can improve nutrition and positively impact the health of people and forests. This requires changing a defining characteristic of our current food system: the increased reliance on only a very few species, leading to the increased homogenisation of diets (Khoury et al. 2014). Since agriculture began some 12 000 years ago, approximately 7 000 plant species and several thousand animal species have been used for human nutrition (Burlingame et al. 2012). Today, although certain traditional and Indigenous communities continue to use a multitude of species in their diets, the general global trend has been towards diet simplification, with consequent negative impacts on human food security, nutrition and health (FAO 2019, Powell et al. 2015). It is estimated that three-quarters of the varietal genetic diversity of agricultural crops has been lost over the last century (Khoury et al. 2016). Just 12 crops and 14 animal species now provide most of the world's food (Burlingame et al. 2012, Khoury et al. 2014).

As the biodiversity used in food and agriculture declines, the food supply becomes less sustainable and more vulnerable (FAO 2019). The narrowing of the genetic basis of our food systems means there is less resilience to the consequences of climate change such as droughts, floods, fires and incidences of pest outbreaks (Deutsch et al. 2018, Schipanski et al. 2016). Less genetic diversity means fewer opportunities for the growth and innovation needed to provide food security and boost agricultural production at a time of soaring food prices and competition with production for biofuels. For example, Tigchelaar et al. (2018) estimate that the predicted 4°C temperature increase worldwide will lead to losses of up to 87 per cent in global maize production. In addition, the nutritional value of some crops could change (Smith and Myers 2018). With increased carbon dioxide in the atmosphere it is postulated that while plant growth may indeed increase, the nutritional quality of staple crops such as potatoes, barley, wheat and rice may fall due to an increase in carbohydrate production and a reduction in protein levels (Ebi and Ziska 2018). This would have a major nutritional impact on the billions of people who rely on these staples.

Not only does relying on a few, select crops leave the food system vulnerable to climatic changes, it lacks the diversity proven to have a plethora of nutritional benefits (HLPE Report 2017a). Biological diversity includes countless plants that feed and heal people, many crop varieties and aquatic species with specific nutritional characteristics, livestock species adapted to harsh environments, insects that pollinate fields and micro-organisms that regenerate agricultural soils. As discussed in Section 2.1, forests contain most of the world's terrestrial biodiversity and provide ecosystem services vital for the survival of other ecosystems (Springmann et al. 2018). An investment in conserving forest biodiversity is an investment in future food security that is diverse, nutritious and resilient (FAO 2019).

To summarise, achieving SDG 2, specifically genetic diversity (Target 2.5), requires the conservation of biodiversity. As forests are home to most of the world's terrestrial biodiversity, this means forest conservation should be at the forefront of our considerations. Further, genetic diversity in our food supply benefits people in several ways. First, genetic diversity enhances dietary nutrition and health. Second, it is integral for climate change adaptation and mitigation. This helps build resilience (Target 2.4) and lessen vulnerability to shocks, thereby benefitting the socio-economic well-being of producers and those who are supported by them. Finally, genetic diversity and maintaining biodiversity, especially that of forests, helps sustain the numerous ecosystem services on which people rely. Clearing forests to grow a minimal assortment of crops is no longer an option. Forests can and should be integrated into agriculture rather than being viewed as being 'in the way' of production. The next section discusses the importance of investing in smallholder farms and enhancing biodiversity in agriculture.

TARGET 2.3: INVESTING IN SMALLHOLDER FOOD PRODUCERS

The pervasive image of modern agriculture is of a vast swath of swaying cereals tended by industrial-scale machinery. This is certainly the case in much of the temperate world. However, in the tropics most of the food produced originates in complex multi-functional landscapes, characterised by small farms producing a wide variety of products (Ricciardi et al. 2018). This diversification is critical for livelihood strategies (don't put all your eggs in one basket) as it provides resilience against both economic and environmental shocks – the latter increasingly driven by climate-induced droughts, floods and other events. Smallholder systems are estimated to produce between 30 per cent (Ricciardi et al. 2018) and 70–80 per cent (FAO 2014) of the world's food. Even with this wide range, it still represents a significant proportion of global food production.

Unlike in temperate regions, the majority of smallholder farmers in tropical regions do not benefit from national or regional subsidies (Chirwa and Dorward 2013). Post-harvest losses are considerable in these systems, yet little to no government support is available for most farmers. They stand to lose their markets due to cheap, subsidised products being dumped on their own production range (FAO 2015, McMichael 2005). In short, smallholder farmers are a resilient and productive group that contributes to global food security in often unseen ways; they deserve more support and the opportunity to compete in fair and equitable market systems. Development support should complement existing knowledge and practices within local systems.

Unfortunately, a growing worldwide trend in the demography of farmers works against such long-term support. Many farmers support their children's education, who, in turn, tend to shun farming as an occupation. As a result, farmers are growing older and less able to manage the land. There is a general trend for small farms to consolidate into larger production units both in temperate and tropical production systems, which is problematic because large-scale commercial agriculture is the driver of approximately 40 per cent of deforestation in tropical and subtropical regions (FAO 2016).

Target 2.3 seeks to change this by doubling the agricultural productivity and income of small-scale food producers, particularly women, Indigenous peoples, family farmers, pastoralists and fishers. Smallholders would benefit from mechanisms that provide access to essential services, such as credit, electricity and transport needed to participate in agribusiness. This could be made more accessible through instruments such as group savings and loan associations, chattel mortgages and leasing (FAO 2017). Furthermore, farmers, especially youth, would benefit from opportunities to develop technical skills and entrepreneurial training. Helping smallholders build the technical capacity and access to the resources necessary to remain competitive in the food system would improve the socio-economic well-being of many.

Investing in smallholder producers is very much related to forests, as many smallholder farmers operate near forests. As discussed in Section 2.1, farmers in rural areas rely on the forest as a safety net for a bad harvest. Additionally, it is good to be near forests for the resources they provide, such as plants used to feed livestock, wood for fences and other structures, improved soil nutrients and much more. Giving smallholder farmers everywhere access to support is important for the aforementioned reasons; however, smallholder farmers near forests present a unique opportunity to conserve forest ecosystems through integrated land uses such as agroforestry (Godfray et al. 2010).

Agroforestry, as defined by Lundgren and Raintree (1982), is 'the name for land-use systems and technologies where woody perennials (trees, shrubs, palms, bamboos, etc.) are deliberately used on the same land-management units as agricultural crops and/or animals, in some form of spatial arrangement or temporal sequence. In agroforestry systems there are both ecological and economical interactions between the different components' (HLPE Report 2017b: 34). There are three classes of agroforestry systems: agrisilvicultural systems, combining agricultural crops and trees or shrubs; silvopastoral systems, combining trees and pasture for grazing livestock; and agrosilvopastoral systems, combining crops, pastures and trees (Nair 1993, Vira et al. 2015).

Trees on farms can generate an array of benefits for communities and the environment. Trees provide shade for shade-tolerant crops, which increases yields. Cocoa grown under tree shade can produce yields for 60–100 years, compared to 20 years or less without shade (Obiri et al. 2007, 2011; Ruf and Schroth 2004). Another example is the presence of fruit trees in agroforestry systems. They have been shown to help fill seasonal gaps in fruit supply (Jamnadass et al. 2011, Vinceti et al. 2013) and attract wild animals for hunting (Sylvester and Segura 2016). Essentially, agroforestry helps maintain biodiversity, genetic diversity and the associated benefits of forests – i.e. improved soil fertility resulting in increased crop yields, fodder for livestock, woodfuel for cooking, ecosystem services necessary for food production, etc. In some cases, agroforestry can be an appealing alternative to conservation agriculture. Conservation agriculture relies on reduced tillage of soils and preserving crop residues to prevent soil erosion. However, some farmers rely on crop residues to feed livestock (WRI 2018). Agroforestry allows this practice to continue and improves soil quality.

It is important to note that land rights are key to investing in land and pursuing long-term investment activities, such as planting trees. In much of the world, rural and forest dwellers lack registered or formalised rights to land. Recent work has drawn attention to the importance of recognising customary rights and shared rights to land and forests, as well as a need to reduce the bureaucracy and legal obstacles of granting community rights (WRI 2018).

Certification schemes and market-based mechanisms are one way of supporting agriculture that integrates forest conservation. These schemes encourage integrative alternatives to the polarising approach that has dominated thus far. Market-based mechanisms and certifications engage multiple stakeholders, including farmers, government, communities and private companies, incentivising sustainable management and production. There are numerous examples, including the REDD+ programme, that offer results-based payments for actions reducing forest carbon emissions, such as sustainable agriculture practices (REDD 2016).

Certifications allow an independent assessment of a defined set of management standards that promote and measure sustainable forest management (CEPI 2006, HLPE Report 2017b). Some certification schemes (e.g. The Forest Stewardship Council (FSC), and the Programme for the Endorsement of Forest Certification (PEFC)) are focused on sustainable forest management in general, while others are focused specifically on food production and forests (e.g. the Round Table on Sustainable Palm Oil (RSPO), and the Round Table on Sustainable Soy (RTRS)) (HLPE Report 2017b). These schemes are proving to be successful. Take RSPO, for example: nearly 2.5 million ha of palm oil is RSPO certified, which represents 21 per cent of global production (HLPE Report 2017b). However, forest certification is primarily focused on boreal and temperate forests, while only 6 per cent of the total certified area is in the tropics (MacDicken et al. 2015), leaving ample room for improvement.

In addition, high-income countries tend to protect their own natural resources and import from lower-income countries to sustain consumption (Mills Busa 2013). With this in mind, consumption strategies such as certification schemes can be as important a buffer to forests as protected areas (HLPE

Report 2017b). In the same vein, it is worth noting that although protected areas are undeniably important, they can be managed in a way that is restrictive to local people and the resources needed for their diets, again reinforcing the importance of systems that sustainably integrate multiple uses.

2.2.3 Integrated Landscape Management

Taking what we have learned from Targets 2.1 to 2.5, a common theme emerges: the need for management that recognises the multiple uses of landscapes and the ways they impact each other (Kremen and Merenlender 2018). As this chapter has shown through the discussion of forests and agriculture, landscapes are a mosaic of natural and human-modified ecosystems that cannot be neatly separated from one another (Reed et al. 2017b). Our failing food system and degraded forests are a testament to the need for a new approach to food production (Ickowitz et al. 2019). The landscape approach answers this call for change, as it seeks to 'provide tools and concepts for allocating and managing land to achieve social, economic, and environmental objectives in areas where agriculture, mining, and other productive land uses compete with environmental and biodiversity goals' (Sayer et al. 2013: 8349).

SDG 2 targets span sectoral and geographic boundaries and involve multiple stakeholders along the entire supply chain, including consumers, producers, policymakers and many other actors. Operationalising integrated landscape management for forests and agriculture necessitates building partnerships among states, rural communities and industry. This requires new legislation, policies and novel forms of forest governance, such as co-management or community managed forests (HLPE Report 2017b). In particular, agricultural policy should be linked to health, education and trade policies that simultaneously promote human and planetary health (Willett 2019). This can help facilitate changes in behaviour and production.

Furthermore, a landscape approach calls for enhancing stakeholder capacity and coordination. It is pertinent that stakeholders are included in decision-making processes related to land management. Stakeholder involvement is increasingly recognised as a means to manage competing interests and as a way to be explicit about potential trade-offs. More than half of the national forest policies and programmes revised since 2007 in 42 countries now include measures to enhance the involvement of traditional forest users in decision-making processes (FAO 2014, HLPE Report 2017b).

GENDER CONSIDERATIONS

Social processes are key in decisions about forest-dependent livelihoods and forest-resource management, as well as governance processes and the distribution of benefits, with strongly differentiated gender roles and impacts (HLPE Report 2017b). These differences need to be considered while managing integrated landscapes. Women tend to grow a greater diversity of products, experiment more with folk varieties and landraces (and thus agrobiodiversity) and are often reliant on broader aspects of biodiversity for herbal medicine – linking both health and nutrition (Sunderland et al. 2011). Women are primarily responsible for food preparation and allocation and, as such, are usually the 'guardians' of household food security (WFP 2002). Yet women's access and control over land and resources is generally inferior to that of men in the same household or community (Agarwal 2018). Where women do have access to land, they tend to use it for food production, and income generated from such land is more likely to be utilised for the well-being of the household, whether for nutritional, health or other benefits (Wan et al. 2011).

Women and men tend to have differing tasks and responsibilities in the production and provision of food, including wild foods (Sunderland et al. 2014). Many women face gender-specific constraints that cut their productivity and limit their income-earning potential. There are gender gaps in access to land, credit, technology, employment and markets. Even though they are often primary resource users, women usually participate much less than men in formal land management and policy decisions (Leisher et al. 2016). Cultural, socio-economic and institutional factors contribute to gender inequality. These range from the societal perceptions of women's roles and the time women have to spend on domestic responsibilities and childcare to disparities in literacy, education, physical abilities, technical skills and access to training and extension services.

Target 2.3 specifically identifies women as a group of smallholder farmers that need support in order to achieve SDG 2. This is especially timely with the 'feminisation' of agriculture due to male out-migration and moves towards off-farm sources of income (Doss 2014, FAO 2017).

Many female farmers lack access to credit and extension services despite evidence suggesting that investment aimed at women leads to the increase of both farm and non-farm incomes at the household level. Although development policymakers and agencies increasingly recognise the crucial contributions of female farmers to food security, contemporary agricultural policies and research do not often directly address the needs of female farmers, focusing instead on traditionally male-dominated cropping practices. Such 'gender blindness' in the context of sustainable agricultural development is a risk to future food security given the major contributions of women to agriculture in the Global South. This underscores the importance of gender considerations in integrated landscape management planning processes. Moreover, a better understanding of what forest resources women are using for food and agriculture is useful for knowing what forest resources are of value to local communities and how to sustain them.

2.3 Synergies and Trade-offs with Other SDGs

Nutrition is an indispensable cog without which the SDG machine cannot function smoothly (Global Nutrition Report 2017). Poor nutrition has varied causes, many of which are intimately connected to work being undertaken to accomplish other SDGs. There is huge potential for making connections among the SDGs, but there is also the potential for incoherence. These trade-offs and synergies will have varied impacts on forests. The Global Nutrition Report (2017) finds that improving nutrition can have a powerful multiplier effect across the SDGs. Indeed, it indicates that it will be a challenge to achieve any of the SDGs without addressing nutrition. The report identifies five key fields where SDG 2 interacts at a broader scale with the other SDGs. In this section we use these five fields as a backdrop to discuss the trade-offs and synergies between SDG 2 and other SDGs, and how these impact forests.

1. Sustainable food production (relevant SDGs: 13, 14, 15) is key to nutrition outcomes. Agricultural yields will decrease as temperatures increase by more than 4°C. Increased carbon dioxide will result in decreased protein, iron, zinc and other micronutrients in major crops consumed by much of the world (Ebi and Ziska 2018, Tigchelaar et al. 2018). Unsustainable fishing (SDG 14 Life below Water) threatens 17 per cent of the world's protein and a source of essential micronutrients (Golden et al. 2016). Policies and investments to maintain and increase the diversity of agricultural landscapes are needed to ensure small and medium-sized farms can continue to produce the 53–81 per cent of key micronutrients they do now (Herrero et al. 2017). As this chapter has explored, diversifying crops using sustainable agriculture practices and supporting small-scale farmers can enhance terrestrial biodiversity (SDG 15 Life on Land) and enable a food system that is more resilient in the face of climate change (SDG 13 Climate Action). While mechanisms for achieving SDG 15, such as protected areas, can benefit forest conservation, they can also restrict forest use and negatively impact the diets of forest-dependent communities. Sustainable food production approaches such as agroforestry and integrated landscape management show potential for harmonising the objectives of SDG 2 and SDG 15 (Timko et al. 2018).

2. Strong systems of infrastructure (relevant SDGs: 6, 7, 8, 9, 11, 12) play key roles in providing safe, nutritious and healthy diets (SDG 2), clean water and sanitation (SDG 6) and more resilient communities overall (SDG 11). Contamination of food from unclean water and poor sanitation is associated with 50 per cent of under-nutrition; it leads to diarrhoea and can cause death, especially among young children. Improved infrastructure (SDG 9) can help deliver resources and services to underserved areas. Improved infrastructure such as cooling systems and storage facilities can reduce food waste. Furthermore, affordable and clean energy (SDG 7) can reduce the reliance on wood from forests for cooking fuel.

The impacts of improved infrastructure on forests and SDG 2 are twofold. First, infrastructure such as roads can help smallholder farmers access previously inaccessible markets and create decent work opportunities (SDG 8). This has the potential to encourage younger generations to continue participating in agriculture and incentivise farmers to continue farming diverse crops at a small scale, which would help reduce the formation of large-scale conglomerates that put pressure on forests. Second, improved access to remote areas may be beneficial for markets and delivering services, but building roads and other infrastructure can cause environmental harm, such as deforestation. In addition, improved access to these areas makes it easier for bigger industries to move in. This is where sustainable consumption (SDG 12) is important. Market-based mechanisms and certifications, like the examples discussed in Section 2.2.2, can help regulate the impacts of new infrastructure and industry on forests.

3. Health systems (relevant SDG: 3) have an important role in promoting infant and young child feeding, supplementation, therapeutic feeding, nutrition counselling to manage overweight and underweight concerns, and screening for diet-related noncommunicable diseases in patients. Yet our analysis shows that health systems are not delivering where they should: for example, only 5 per cent of children aged 0–59 months who need zinc treatment are receiving it. As discussed in Section 2.1, forests play a substantial role in the health and well-being of people (SDG 3). Forests provide nutritious food such as bushmeat, fruits and nuts, as well as providing wood for cooking meals. However, it should be noted that food can be a health risk, as in the case of bushmeat being linked to Ebola outbreaks. Overall, nutrition from forests improves health, helping keep people out of hospitals. The importance of eating nutritious food to maintain good health is accentuated in remote areas where access to healthcare is variable.

4. Equity and inclusion (relevant SDGs: 1, 4, 5, 8, 10, 16) matter for nutrition outcomes: ignoring equity in the distribution of wealth, education and gender will make it impossible to end malnutrition in all its forms. Approximately 767 million people live in extreme poverty, and 46 per cent of all stunting falls in this group (Global Nutrition Report 2017). This group is often neglected or excluded. Reflecting on the discussion in Section 2.2.2 about lack of support for smallholder farmers and women, the relationship between SDG 2 and SDG 4 (Quality Education) shows potential to reduce inequalities (SDG 10) among men and women (SDG 5 Gender Equality) and smallholder farmers and big industry. Workshops on conservation agriculture and other knowledge-sharing opportunities would be of great benefit to farmers and forest peoples. Education relates to decent work opportunities because skill building opens up new work opportunities and stimulates economic growth (SDG 8) and reduces poverty (SDG 1). The impacts on forests depend on the type of work and how growth is managed.

Another key intersection regarding equity and inclusion is the relationship between SDG 16 (Peace, Justice, and Strong Institutions), SDG 10 (Reduced Inequality) and SDG 2. SDG 16 aims to end corruption and exploitation and develop effective, accountable and transparent institutions at all levels. Transparency and regulation is very important for industries clearing or extracting forest resources. This is heightened by the fact that there are communities that do not have recognised rights over their land. In some cases, this has resulted in the loss of land to private companies. This can lead to deforestation by the company or by communities who have lost their farmland and moved to other forested areas. SDG 16 could improve tenuous land rights, securing land and resources vital to nutritious diets, as well as regulate industry to prevent the exploitation of forests.

5. Peace and stability (relevant SDG: 16) are vital to ending malnutrition. The proportion of under-nourished people living in countries in conflict and protracted crisis is almost three times higher than in other developing countries. Long-term instability can exacerbate food insecurity in many ways. In the worst-case scenario, conflict can lead to famine. When conflict or emergencies occur, nutrition must be included in disaster risk reduction and post-conflict rebuilding. On the other hand, forests act as a safety net during periods of crises and conflicts, as they provide food substitutes during times of insecurity. Additionally, peace and stability support law and order, which fosters an environment conducive to sustainable forest management. Whether managed at the national level or the community level, sustainable management conserves biodiversity, which is necessary for food security and nutrition.

2.4 Conclusion

Under our current food system, forests are treated either as a space for agricultural expansion or a threatened resource needing protection from such expansion (HLPE Report 2017b). Breaking down this siloed thinking to realise that agriculture and forests are inextricably linked is an important step in achieving SDG 2. As this chapter highlights, our current food system is failing people and forests. Although we are producing more food than ever before, our population is characterised by both under- and over-malnutrition. Forests – exceptional sources of biodiversity and ecosystem services necessary for food and agriculture – are being destroyed for a limited selection of crops and livestock. A lack of diversity reduces nutrition and leaves our food system vulnerable to the vicissitudes of a changing climate.

SDG 2, specifically Targets 2.1–2.5, brings optimism to the future of forests and people. These targets emphasise biodiversity, sustainability and integration – all key ingredients of a resilient food system. Achieving SDG 2 should be based on the integration of food production and forests within the context of land management; this will both require and result in positive changes. For instance, major change will be required in both national and global governance systems and processes. The SDGs are an emerging opportunity in this regard, as they are all dependent on each other. Some SDGs have seamless synergies, such as SDG 2 and SDG 3, while others are bound to face trade-offs. We have reached a point where collaboration across sectors is needed more than ever. Forests can play an enormous role in facilitating this collaboration.

References

Aerts, R. and Honnay, O. 2011. Forest restoration, biodiversity and ecosystem functioning. *BMC Ecology* 11(1):29.

Agarwal, B. 2018. Gender equality, food security and the Sustainable Development Goals. *Current Opinion in Environmental Sustainability* 34:26–32.

Aschemann-Witzel, J., de Hooge, I., Amani, P., Bech-Larsen, T. and Oostindjer, M. 2015. Consumer-related food waste: Causes and potential for action. *Sustainability* 7(6):6457–77.

Bahadur, K., Dias, G. M., Veeramani, A. et al. 2018. When too much isn't enough: Does current food production meet global nutritional needs? *PLoS ONE* 13(10):e0205683.

Baudron, F., Chavarría, J. Y. D., Remans, R., Yang, K. and Sunderland, T. 2017. Indirect contributions of forests to dietary diversity in Southern Ethiopia. *Ecology and Society* 22(2):28.

Broegaard, R. B., Rasmussen, L.V ., Dawson, N. et al. 2017. Wild food collection and nutrition under commercial agriculture expansion in agriculture-forest landscapes. *Forest Policy and Economics* 84:92–101.

Brussaard, L., Caron, P., Campbell, B. et al. 2010. Reconciling biodiversity conservation and food security: Scientific challenges for a new agriculture. *Current Opinion in Environmental Sustainability* 2(1–2):34–42.

Burlingame, B., Charrondiere, U. R., Dernini, S., Stadlmayr, B. and Mondovì, S. 2012. Food biodiversity and sustainable diets: Implications of applications for food production and processing. In Boye, J. I. and Arcand, Y. (eds) *Green technologies in food production and processing*. Boston: Springer, pp. 643–57.

Byron, N. and Arnold, M. 1997. *What futures for the people of the tropical forests?* Working Paper No. 19. Bogor, Indonesia: Center for International Forestry Research.

CEPI (Confederation of European Paper Industries) 2006. *A comparison of the Forest Stewardship Council and the Programme for Endorsement of Forest Certification*. Brussels: CEPI.

Chirwa, E. and Dorward, A. 2013. *Agricultural input subsidies: The recent Malawi experience*. Oxford: Oxford University Press.

Cockx, L., Colen, L. and De Weerdt, J. 2018. From corn to popcorn? Urbanization and dietary change: Evidence from rural–urban migrants in Tanzania. *World Development* 110:140–59.

Cramb, R. A., Colfer, C. J. P, Dressler, W. and Wadley, R. L. 2009. Swidden transformations and rural livelihoods in Southeast Asia. *Human Ecology* 37(3):323–46.

Cumming, G. S., Buerkert, A., Hoffmann, E. M. et al. 2014. Implications of agricultural transitions and urbanization for ecosystem services. *Nature* 515(7525):50–7.

Curtis, P. G., Slay, C. M., Harris, N. L., Tyukavina, A. and Hansen, M. C. 2018. Classifying drivers of global forest loss. *Science* 361(6407):1108–11.

Deutsch, C. A., Tewksbury, J. J., Tigchelaar, M. et al. 2018. Increase in crop losses to insect pests in a warming climate. *Science* 361(6405):916–19.

Doss, C. 2014. Collecting sex disaggregated data to improve development policies. *Journal of African Economies* 23(suppl_1):i62–i86.

Ebi, K. L. and Ziska, L. H. 2018. Increases in atmospheric carbon dioxide: Anticipated negative effects on food quality. *PLoS Medicine* 15(7):e1002600.

Ellis, E. C., Klein Goldewijk, K., Siebert, S., Lightman, D. and Ramankutty, N. 2010. Anthropogenic transformation of the biomes, 1700 to 2000. *Global Ecology and Biogeography* 19(5)589–606.

Fa, J. E., Olivero, J., Real, R. et al. 2015. Disentangling the relative effects of bushmeat availability on human nutrition in central Africa. *Scientific Reports* 5:8168.

FAO (Food and Agriculture Organization of the United Nations) 2011. *Global food losses and food waste: Extent, causes and prevention*. Rome: FAO.

FAO 2014. *Strengthening the links between resilience and nutrition in food and agriculture. A discussion paper*. Rome. Available at: www.fao.org/3/a-i3777e.pdf (Accessed 1 November 2018).

FAO 2015. *The economic lives of smallholder farmers: An analysis based on household data from nine countries*. Rome. Available at: www.fao.org/3/a-i5251e.pdf (Accessed 1 November 2018).

FAO 2016. *State of the world's forests 2016. Forests and agriculture: Land-use challenges and opportunities*. Rome: FAO.

FAO 2017. *The future of food and agriculture: Trends and challenges*. Rome: FAO.

FAO 2019. The state of the world's biodiversity for food and agriculture. In Bélanger, J. and Pilling, D. (eds.) *Commission on Genetic Resources for Food and Agriculture Assessments*. Rome: FAO. Available at: www.fao.org/3/CA3129EN/CA3129EN.pdf (Accessed 26 February 2019).

FAO, IFAD, UNICEF, WFP and WHO 2017. *The state of food security and nutrition in the world 2017. Building resilience for peace and food security*. Rome: FAO.

Gibbs, H., Ruessch, A., Achard, F. et al. 2010. Tropical forests were the primary sources of new agricultural land in the 1980s and 1990s. *Proceedings of the National Academy of Science* 107:16732–7.

Global Nutrition Report 2017. *Nourishing the SDGs*. Bristol: Development Initiatives Poverty Research. Available at: http://globalnutritionreport.org/the-report/ (Accessed 1 November 2018).

Godfray, H. C. J., Beddington, J. R., Crute, I. R. et al. 2010. Food security: The challenge of feeding 9 billion people. *Science* 327(5967):812–18.

Golden, C., Allison, E. H., Cheung, W. W. et al. 2016. Fall in fish catch threatens human health. *Nature* 534(7607):317–20.

Gordon, I. J., Prins, H. H. and Squire, G. R. (eds.) 2017. *Food production and nature conservation: Conflicts and solutions*. London: Routledge.

Harari, Y. N. 2014. *Sapiens: A brief history of humankind*. New York: Random House.

Herrero, M., Thornton, P. K., Power, B. et al. 2017. Farming and the geography of nutrient production for human use: A transdisciplinary analysis. *The Lancet Planetary Health* 1(1):e33–42.

HLPE Report 2017a. *Nutrition and food systems*. A report by the High Level Panel of Experts on Food Security and Nutrition of the Committee on World Food Security. Rome.

HLPE Report 2017b. *Sustainable forestry for food security and nutrition*. A report by the High Level Panel of Experts on Food Security and Nutrition of the Committee on World Food Security. Rome.

IAEG-SDGs (Inter-Agency and Expert Group on SDG Indicators) 2016. United Nations Statistical Commission, 47th session. Report of the Inter-Agency and Expert Group on Sustainable Development Goal Indicators. Available at: https://sustainabledevelopment.un.org/sdg2 (Accessed 1 November 2019).

Ickowitz, A., Jones, A., Rowland, D., Powell, F. and Sunderland, T. 2019. Agricultural intensification, dietary diversity, and markets in the global food security narrative. *Global Food Security* 20:9–16.

Ickowitz, A., Rowland, D., Powell, B., Salim, M. A. and Sunderland, T. 2016. Forests, trees, and micronutrient-rich food consumption in Indonesia. *PLoS ONE* 11(5):e0154139.

Jamnadass, R. H., Dawson, I. K., Franzel, S. et al. 2011. Improving livelihoods and nutrition in sub-Saharan Africa through the promotion of indigenous and exotic fruit production in smallholders' agroforestry systems: A review. *International Forest Review* 13:338–54.

Jin, S. L., Schure, J., Ingram, V. and Byoung, I. I. 2017. *Sustainable woodfuel for food security. A smart choice: Green, renewable and affordable*. Working Paper. Rome: FAO.

Khoury, C. K., Achicanoy, H. A., Bjorkman, A. D. et al. 2016. Origins of food crops connect countries worldwide. *Proceedings of the Royal Society B* 283(1832):20160792.

Khoury, C. K., Bjorkman, A. D., Dempewolf, H. et al. 2014. Increasing homogeneity in global food supplies and the implications for food security. *Proceedings of the National Academy of Sciences* 111(11):4001–6.

Kissinger, G., Herold, M. and de Sy, V. 2012. *Drivers of deforestation and degradation: A synthesis report for REDD+ policymakers*. Vancouver: Lexeme Consulting.

Kremen, C. and Merenlender, A. M. 2018. Landscapes that work for biodiversity and people. *Science 362*(6412):eaau6020.

Kuhnlein, H. V., Erasmus, B., Spigelski, D. and Burlingame, B. (eds.) 2009. *Indigenous peoples' food systems: The many dimensions of culture, diversity and environment for nutrition and health*. Rome: FAO.

Leisher, C., Temsah, G., Booker, F. et al. 2016. Does the gender composition of forest and fishery management groups affect resource governance and conservation outcomes? A systematic map. *Environmental Evidence* 5(1):6.

Lundgren, B. O. and Raintree, J. B. 1982. Sustained agroforestry. In Nestel, B. (ed.) *Agricultural research for development: Potentials and challenges in Asia*, The Hague: International Service for National Agricultural Research (ISNAR), pp. 37–49.

MacDicken, K. G., Sola, P., Hall, J. E. et al. 2015. Global progress towards sustainable forest management. *Forest Ecology and Management* 352:47–56.

Maezumi, S. Y., Alves, D., Robinson, M. et al. 2018. The legacy of 4,500 years of polyculture agroforestry in the eastern Amazon. *Nature Plants 4* (8):40.

McMichael, P. 2005. Global development and the corporate food regime. In Buttel, F. H. and McMichael, P. (eds.) *New directions in the sociology of global development, vol. 2*. Bingley, West Yorkshire, UK: Emerald Group Publishing, pp. 265–99.

Mertz, O., Leisz, S., Heinimann, A. et al. 2009. Who counts? The demography of swidden cultivators. *Human Ecology* 37:281–9. doi:10.1007/s10745-009-9249-y.

Mills Busa, J. H. 2013. Deforestation beyond borders: Addressing the disparity between production and consumption of global resources. *Conservation Letters* 6(3):192–9.

Nair, P. K. N. 1993. *An introduction to agroforestry*. Dordrecht: Kluwer Academic Publishers.

Nasi, R., Taber, A. and van Vliet, N. 2011. Empty forests, empty stomachs? Bushmeat and livelihoods in the Congo and Amazon basins. *International Forestry Review* 13(3):355–68. Available at: www.cifor.org/publications/pdf_files/articles/ANasi1101.pdf (Accessed 1 November 2018).

Obiri, D. B., Bright, G. A., McDonald, M. A., Anglaaere, L. C. N. and Cobbina, J. 2007. Financial analysis of shaded cocoa in Ghana. *Agroforestry Systems* 71(2):139–49.

Obiri, D. B., Depinto, A. and Tetteh, F. 2011. *Cost-benefit analysis of agricultural climate change mitigation options: The case of shaded cocoa in Ghana*. Research Report. Washington, DC: International Food Policy Research Institute (IFPRI).

Padoch, C. and Sunderland, T. 2014. Managing landscapes for greater food security and improved livelihoods. *Unasylva* 64(241):3–13.

Peng, L., Zhiming, F., Luguang, J., Chenhua, L. and Jinghua, Z. 2014. A review of swidden agriculture in Southeast Asia. *Remote Sensing* 6:1654–83. doi:10.3390/rs6021654.

Powell, B., Thilsted, S. H., Ickowitz, A., Termote, C., Sunderland, T. and Herforth, A. 2015. Improving diets with wild and cultivated biodiversity from across the landscape. *Food Security* 7(3):535–54.

REDD 2016. United Nations Programme on Reducing Emissions from Deforestation and Forest Degradation. UN REDD Programme Website. www.un-redd.org/ (Accessed 1 November 2018).

Reed, J., van Vianen, J., Foli, S. et al. 2017a. Trees for life: The ecosystems service contribution for trees to food production and livelihoods in the tropics. *Forest Policy and Economics* 84:62–71.

Reed, J., van Vianen, J., Barlow, J. and Sunderland, T. 2017b. Have integrated landscape approaches reconciled societal and environmental issues in the tropics? *Land Use Policy* 63:481–92.

Ricciardi, V., Ramankutty, N., Mehrabi, Z., Jarvis, L. and Chookolingo, B. 2018. How much of the world's food do smallholders produce? *Global Food Security* 17:64–72.

Ruf, F. and Schroth, G. 2004. Chocolate forests and monocultures: A historical review of cocoa growing and its conflicting role in tropical deforestation and forest conservation. In Schroth, G., Da Fonseca, G. A. B., Harvey, C. A. et al. (eds.) *Agroforestry and biodiversity conservation in tropical landscapes*. Washington, DC: Island Press.

Sayer, J., Sunderland, T., Ghazoul, J. et al. 2013. Ten principles for a landscape approach to reconciling agriculture, conservation, and other competing land uses. *Proceedings of the National Academy of Sciences* 110(21):8349–56.

Schipanski, M. E., MacDonald, G. K., Rosenzweig, S. et al. 2016. Realizing resilient food systems. *Bioscience* 66(7):600–10.

Shvidenko, A., Barber, C. V., Persson, R., Gonzalez, P. and Hassan, R. 2005. Forest and woodland systems. In Hassan. R., Scholes, R., and Ash, N. (eds.) *Ecosystems and human well-being: Current state and trends*. Vol. 1. Millennium Ecosystem Assessment series. Washington, DC: Island Press, pp. 585–622.

Smith, M. R. and Myers, S. S. 2018. Impact of anthropogenic CO_2 emissions on global human nutrition. *Nature Climate Change* 8(9):834.

Springmann, M., Clark, M., Mason-D'Croz, D. et al. 2018. Options for keeping the food system within environmental limits. *Nature* 562:519–25.

Sunderland, T., Achdiawan, R., Angelsen, A. et al. 2014. Challenging perceptions about men, women, and forest product use: A global comparative study. *World Development* 64:S56–66.

Sunderland, T. C. H. 2011. Food security: Why is biodiversity important? *International Forestry Review* 13(3):265–74.

Swain, M., Blomqvist, L., McNamara, J. and Ripple, W. J. 2018. Reducing the environmental impact of global diets. *Science of the Total Environment* 610:1207–9.

Sylvester, O. and Segura, A. G. 2016. Landscape ethnoecology of forest food harvesting in the Talamanca Bribri Indigenous Territory, Costa Rica. *Journal of Ethnobiology* 36(1):215–33.

Tigchelaar, M., Battisti, D. S., Naylor, R. L. and Ray, D. K. 2018. Future warming increases probability of globally synchronized maize production shocks. *Proceedings of the National Academy of Sciences* 115(26):6644–9.

Timko, J., Le Billon, P., Zerriffi, H. et al. 2018. A policy nexus approach to forests and the SDGs: Tradeoffs and synergies. *Current Opinion in Environmental Sustainability* 34:7–12.

Tomlinson, I. 2013. Doubling food production to feed the 9 billion: A critical perspective on a key discourse of food security in the UK. *Journal of Rural Studies* 29:81–90.

United Nations 2015. Resolution adopted by the General Assembly on 25 September 2015.*Transformingour world: the 2030 Agenda for Sustainable Development.* UN A/RES/70/1. Available at: www.un.org/ga/search/view_doc.asp?symbol=A/RES/70/1&Lang=E (Accessed 20 July 2018).

Vandermeer, J., Aga, A., Allgeier, J. E. et al. 2018. Feeding Prometheus: An interdisciplinary approach for solving the global food crisis. *Frontiers in Sustainable Food Systems* 2:39.

Vinceti, B., Termote, C., Ickowitz , A. et al. 2013. The contribution of forests and trees to sustainable diets. *Sustainability* 5(11):4797–824. doi:10.3390/su5114797.

Vira, B., Wildburger, C. and Mansourian, S. (eds.) 2015. *Forests, trees and landscapes for food security and nutrition.* A Global Assessment Report. IUFRO World Series Vol. 33. Vienna: IUFRO.

Wan, M., Colfer, C. J. P. and Powell, B. 2011. Forests, women and health: Opportunities and challenges for conservation. *International Forestry Review* 13(3):369–87.

WFP (World Food Programme) 2002. *Gender Policy 2003–2007: Enhanced commitments to women to ensure food security.* Rome: WFP.

Willett, W., Rockström, J., Loken, B. et al. 2019. Food in the Anthropocene: The EAT–Lancet Commission on healthy diets from sustainable food systems. *The Lancet* 393(10170):447–92.

WRI (World Resources Institute) 2018. Creating a sustainable food future: A menu of solutions to feed nearly 10 billion people by 2050. *Synthesis Report.* Washington, DC: WRI.

Wunder, S., Börner, J., Shively, G. and Wyman, M. 2014. Safety nets, gap filling and forests: A global-comparative perspective. *World Development* 64:S29–42.

Chapter 3 SDG 3: Good Health and Well-Being – Framing Targets to Maximise Co-Benefits for Forests and People

Rosemary A. McFarlane*, John Barry, Guéladio Cissé, Maya Gislason, Marta Gruca, Kerryn Higgs, Pierre Horwitz, Giang Huu Nguyen, Jane O'Sullivan, Subhashis Sahu and Colin D. Butler

Key Points

- The achievement of SDG 3 depends on many other SDGs; some SDGs are logically inconsistent, especially in the attempt to increase conventionally defined GDP while preserving natural capital.
- Any short-term gains for human health from further forest conversion (e.g. food production) creates short- and long-term, direct and indirect health risks for humans, as well as for other biota.
- Failure to ensure universal access to sexual and reproductive healthcare services (including family planning) will increase pressure on forests at local, regional and global scales.
- The burning and clearing of forests cause significant harm to health via impaired quality of water, soil and air; increased exposure to infectious diseases and impacts climate regulation.
- Many infectious diseases are associated with forest disturbances and intrusions; some important infectious diseases have emerged from forests (notably HIV/AIDS).
- Greater exposure to green space, including forests, provides mental and physical health benefits for the growing global urban population.

3.1 Introduction

The third SDG is very ambitious. It includes the words 'health' and 'well-being', which both have lofty, multiple and contested meanings and aspirations. Echoing the almost-forgotten World Health Organization (WHO) slogan proclaimed at the Alma Ata conference in 1978 of 'health for all by the year 2000', SDG 3 proclaims the aspiration 'well-being for all' by 2030. However, this is a pledge for the world to promote this aspiration rather than

* Lead author.

achieve it – a task probably considered unwisely ambitious, even by the SDG framers.

In 1948, the newly formed WHO defined human health as 'a state of complete physical, mental and social well-being and not merely the absence of disease or infirmity' (WHO 2019b). Then revolutionary, this definition challenged the disease-focused medical model of health that was (and is still) the common perception of not only lay people, but also many health workers. It challenged convention by recognising the importance of the social and mental dimensions of health.

In recent decades, the WHO definition has been increasingly criticised, mainly for its use of the word 'complete'. Today, in a world with so many people with chronic illnesses and disabilities, complete health is unattainable for hundreds of millions, perhaps billions. Indeed, some argue that the pursuit of complete health is counterproductive because it promotes the re-medicalisation of the concept of health, allowing profit-seeking corporations such as pharmaceutical companies, the medical screening industry and often health workers themselves to seek opportunity and personal gain through new drugs, blurring the distinction between normal variation (including ageing) and pathology (Huber et al. 2011).

Unlike in 1948, many people today live reasonably fulfilling lives in their older years, while in the mid-twentieth century infectious diseases crippled and shortened the lives of many and were seen as the major global health problem. For young people with conditions such as undernutrition, hookworm or malaria, the potential of cure and a period of 'complete' health (lasting at least some decades following treatment) seemed a realistic aspiration for health workers in a milieu encouraged by the WHO in 1948. Today, suggesting that most 70-year-olds with diabetes or arthritis can become completely healthy is unrealistic. Nevertheless, 'rectangularising the curve' – maintaining good, though rarely (if ever) complete, health well into old age – remains an important goal. The pathways to this goal of excellent, albeit imperfect, lifelong health are increasingly understood in theory but remain out of reach for billions, as so many of the determinants of health remain fragile, damaged, endangered, unattainable and structural. Few are within the ability of individuals or communities to change.

A lesser known criticism of the WHO health definition is its lack of environmental or ecological dimension, including its relationship with forests. In 1990, the pioneer of primary health care, Maurice King, suggested that the WHO insert 'sustainable' as the second word in its health definition (King 1990). Like all other suggested amendments, this has been resisted so far. However, there is increasing understanding, including within the WHO, that there are crucial environmental underpinnings of health and that many of these determinants lie far beyond the expertise of clinicians – that is, of doctors and other health workers who encounter and treat the sick.

The Declaration of the Alma Ata conference identifies two of eight essential components of primary health care as environmental: (1) adequate nutrition, and (2) safe water and basic sanitation. The 1986 Ottawa Charter, a landmark in health promotion, pays even more attention to environmental issues. It declares that the fundamental conditions and resources for health are 'peace, shelter, education, food, income, a stable ecosystem, sustainable resources, social justice and equity' (WHO 1986). These (and others) have become known as the social (and environmental) determinants of health and are considered to determine the inequity of health outcomes among populations. Many of these conditions and resources are related to other SDGs, illustrating a difficulty not only for this chapter but also for the others. Virtually all 17 SDGs are related to health and well-being in some way, as well as to each other. One risk of the SDG approach is inadvertently reinforcing barriers among disciplines and lobby groups. On the other hand, progress with many SDGs is likely to have synergistic benefits: the WHO describes the SDGs as a blueprint for systematically addressing the social determinants of health (Government of South Australia and WHO 2017).

Although this chapter focuses on forests, health and well-being through the lens of specific SDG 3 targets (for the complete list, see Table 3.1), we stress that the public goods (the 'ends') of health and well-being have many non-environmental determinants, including caste, class, corporate, cultural, economic, educational, epigenetic, ethnic, gender, genetic, nutritional, political, social and spiritual aspects. These are listed alphabetically to stress that they are all important; prioritising any one is subjective. We acknowledge that some analysts will argue that some categories (e.g. social) may embrace subsets (e.g. political). No framework of analysis will satisfy everyone. An analogy from biology is of survival. Humans need air, water and food, but death from suffocation is fastest: this does not mean air is more important than food, considered over a longer period. While humans may survive with only air, water and food – perhaps in a windowless cell – they will certainly not thrive with those inputs alone. Other aspects, such as social connections, are also vital to foster even an imperfect state of physical, mental and social well-being. Some of these relate to forests and their services.

Well-being is also a contested, context-dependent term. The WHO defines it as part of health, whereas the Millennium Ecosystem Assessment conceptual framework considers health as one of five components of well-being, along with material sufficiency, security, good human relations, and freedom and choice (Butler et al. 2003).

This chapter discusses the impacts of achieving SDG 3 targets on forests, forest people and humans more broadly, including background on important connections between some specific SDG 3 targets and forests and their services. We argue that failure to make significant progress with Target 3.7, concerning sexual and reproductive healthcare services, will have significant adverse effects not only on forests, but on all other SDGs.

In Table 3.1 we highlight in bold those SDG 3 targets with forest connections discussed in detail in this chapter. The others are still relevant to forest-dependent populations and are referred to within the sections.

Table 3.1 SDG 3 targets, highlighting those particularly pertinent to forests and forest populations (targets in bold have forest connections discussed in detail in chapter)

Target	Description
3.1	By 2030, reduce the global maternal mortality ratio to less than 70 per 100 000 live births
3.2	By 2030, end preventable deaths of newborn and children under 5 years of age, with all countries aiming to reduce neonatal mortality to at least as low as 12 per 1000 live births and under-5 mortality to at least as low as 25 per 1000 live births
3.3	**By 2030, end the epidemics of AIDS, tuberculosis, malaria and neglected tropical diseases and combat hepatitis, water-borne diseases and other communicable diseases**
3.4	**By 2030, reduce by one-third premature mortality from non-communicable diseases through prevention and treatment and promote mental health and well-being**
3.5	Strengthen the prevention and treatment of substance abuse, including narcotic drug abuse and harmful use of alcohol
3.6	By 2020, halve the number of global deaths and injuries from road traffic accidents
3.7	**By 2030, ensure universal access to sexual and reproductive health-care services, including for family planning, information and education, and the integration of reproductive health into national strategies and programmes**
3.8	**Achieve universal health coverage, including financial risk protection, access to quality essential healthcare services and access to safe, effective, quality and affordable essential medicines and vaccines for all**

Table 3.1 (cont.)

Target	Description
3.9	**By 2030, substantially reduce the number of deaths and illnesses from hazardous chemicals and air, water and soil pollution and contamination**
3.A	Strengthen the implementation of the WHO Framework Convention on Tobacco Control in all countries, as appropriate
3.B	Support the research and development of vaccines and medicines for the communicable and non-communicable diseases that primarily affect developing countries, provide access to affordable essential medicines and vaccines, in accordance with the Doha Declaration on the TRIPS Agreement and Public Health
3.C	Substantially increase health financing and the recruitment, development, training and retention of the health workforce in developing countries, especially in least developed countries and small island developing States
3.D	Strengthen the capacity of all countries, developing countries, for early warning, risk reduction and management of national and global health risks

Source: https://sustainabledevelopment.un.org/sdg3

3.1.1 Health and Forests

We conceptualise the links between forests and population health at three scales (Figure 3.1) (World Bank 2008). We also stress that the relationship is bidirectional: forest effects on health are not all positive, and people's improved well-being can have good and bad impacts on forests.

However, forests do have important benefits for health for all people. Most proximally, about 350 million people live very close to or within dense forests and are substantially dependent on them. Of these, about 60 million (mostly Indigenous) are wholly dependent on forest ecosystem services for food, water, fuel, medicine, culture and livelihood. At a second scale is a larger population, though of uncertain size, that lives away from the forest, may never even visit one and yet depends on and, in some cases, consciously consumes services such as drinking water, firewood or bushmeat. Across these two scales are at least 13.5 million people employed formally in forestry (Garland 2018).

Most of the global population is at the third scale; many are exposed on a frequent basis to the urban forest and may visit other forests on holiday. Most

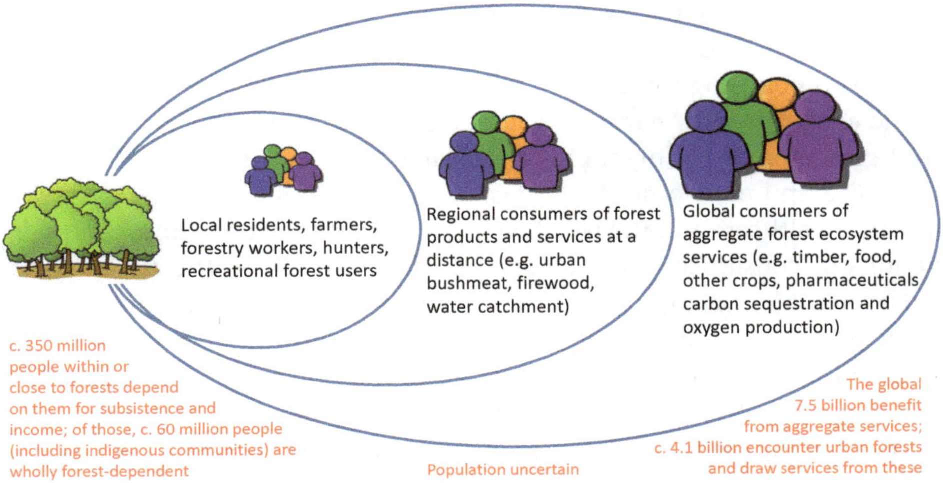

Figure 3.1 Multi-scale impacts of forests on the health and well-being of populations.

consume, often indirectly, forest products such as timber, food and pharmaceutical discoveries from around the world. This entire population benefits, whether they know it or not, from other services, especially the carbon-regulating and oxygen-providing functions of forests. A small but significant fraction of people in the second and third category consciously seek contact with forests (near and far), as well as with other aspects of nature. There is increasing appreciation that this group may experience direct health benefits.

3.2 Sustainability, Limits, Population and the 'Free Market'

The ecological impacts of achieving SDG 3 (and its specific targets), as with the other human-focused SDGs, need to be framed within the debates about ecological sustainability, population and market processes. These factors affect the pursuit of health and well-being for all and the fate of the world's forests. Importantly, decades of often-fluctuating concern about the impact of humans on natural resources have included warnings that exceeding the limits of natural resources is both possible and catastrophic for human well-being. As early as the 1970s, *The Limits to Growth* (Meadows et al. 1972) identified major aspects of future crisis brought on by accelerating industrialisation and population growth, leading to depletion of non-renewable resources and many forms of environmental decline. More recently, the planetary boundaries framework aims to establish the limits beyond which human activities fatally undermine the ecological integrity on which human life ultimately depends (Steffen et al. 2015). Forests are at the heart of three dangerous ecological trends. Biodiversity loss and biogeochemical disruption have already breached limits, while land-system change and climate disruption are approaching the danger zone.

The exploitation of natural resources is driven by economic (and population) growth and this highlights a fundamental concern: currently, achieving SDG 3 and most other human-centric SDGs requires ongoing economic growth. Healthcare and the well-being of the growing global (and local) populations of consumers is costly, and this has impacts on the environment and sometimes directly on forests. Forest transformation can provide local employment, generating cash and opportunities for financially poor and often vulnerable populations. It can also greatly increase food production. These opportunities can then be used to engage more broadly with the wider economy and to facilitate better education and healthcare. Earth system and population health concerns become lost in such discussions. Furthermore, existing market forces do little to protect nature or promote the broader social determinants of health.

Solutions to these problems require transformative thinking and alternative economic models. Impressive improvements in nutrition and health in the early twenty-first century have been achieved in Ghana, Vietnam and Brazil, where government programmes provided benefits such as cash for mothers, support for smallholders and land grants (Lappé et al. 2013). Such programmes, which may ease the pressure on forests, are in opposition to current (neoliberal) economic principles. Alternative economic models that decouple economic growth from environmental degradation (Target 8.4) are required to safeguard health. This chapter discusses some integrated approaches to poverty, population, health and environmental management that seek optimal outcomes for both forests and people (see Box 3.2).

3.3 Forests, People and SDG 3 Targets

3.3.1 Ending Epidemics and Controlling Communicable Diseases

Target 3.3 specifically calls for ending epidemics of HIV/AIDS, malaria and tuberculosis (three major 'communicable' or 'infectious' diseases that are the focus of sustained global control efforts) and of a collection of diseases grouped because of neglect. A high population burden of communicable diseases is generally linked to poverty and underdevelopment. Poverty is deepened by ill health, the costs of seeking treatment and lost livelihood. The control of communicable diseases has increased globally with good health literacy (particularly regarding sanitation, food safety and minimising interpersonal, vector and zoonotic transmission of infections), vaccination programmes, surveillance and response. In poor countries and regions, the capacity and effectiveness of health systems to provide these services (Target 3.8) is often limited – particularly for remote communities, including those in forested regions.

Many important communicable diseases have little or no association with forest ecology (such as influenza, tuberculosis, viral hepatitis and sexually transmitted infections other than HIV), although they may affect forest-associated populations. However, seeking to fulfil Target 3.3 has implications for forest management with regards to malaria and some other forest-origin or associated diseases.

Malaria affects 219 million people, with almost half a million deaths (in 2017), 90 per cent in sub-Saharan Africa (WHO 2018a). The global campaign focuses on eradicating the malarial parasite with the use of insecticide-impregnated bed nets, domestic spraying, diagnosis and treatment (Bauhoff and Busch 2018). However, the density and activity of vector mosquito populations can be affected by deforestation, particularly in Asia and Latin America. Rutted ground and roads resulting from forest clearance, forest edges and cleared patches provide free-standing water, optimum temperature and protection from desiccation for mosquito development – often transiently – and may favour vectors over non-vectors (Guerra et al. 2006). This knowledge can be used locally to reduce malaria through forest management, but is more commonly used to identify risk. Forest workers may be susceptible to local infection or exposed to new species of malaria as they penetrate new forests, as demonstrated in the emergence of monkey malaria (*Plasmodium knowlsei*) as a new human disease (Barber et al. 2017). Infected forest workers may also introduce malarial parasites to disease-free forests. In frontier settlements, children (highly susceptible), if not adults, will perpetuate and amplify malarial infection. Disease fronts can establish: 99 per cent of Brazil's malaria now occurs in the Amazon basin (Chaves et al. 2018).

The so-called Neglected Tropical Diseases with either significant prevalence in forest-based populations or with a link to forest transformation, positive or negative, are listed in Table 3.2. Many of these conditions afflict Indigenous and other forest-dwelling populations and are strongly associated with poverty, low health literacy and poor or underfunded health services (Target 3.8). Relatively simple medication, if available and affordable, can have a dramatic effect on many of these diseases. Forest management may also play a part in the prevention of some of these disease conditions.

HIV/AIDS is among a new group of infectious diseases, recognised since the 1970s, that have wildlife and environmental origins. For HIV/AIDS, its forest association is historic (Sharp and Hahn 2011). It is now a human-specific

Table 3.2 Neglected tropical diseases and forests

Neglected tropical disease	Deforestation effect					Comment
	Direct	In direct	Worsened	Improved	Mixed	
Schisto-somiasis		Y	Y			Results from post-forest land-use change, particularly involving water/irrigation; affects many tropical regions, but not South Asia
Soil transmitted helminths	Y	P	Y			Forest-to-crops conversion can change soil biodiversity; decreased soil-pathogen competition promotes threadworm/hookworm, especially with increased soil moisture following flooding; increased site-contamination for resettled (previously mobile) forest groups associated with poor sanitation, bare feet, open defecation
Buruli ulcer		Y	Y			Geographically associated with upper catchment areas; also wetlands recently converted to farming
Chagas disease	Y		Y		Y	Associated with forest workers/poor housing; deforestation favours synanthropic triatominae (e.g. *T cruzi*) and wild mammal hosts, amplified in palm oil plantations
Leishmaniasis (Kala Azar)	Y		Y		Y	Associated in Latin America with forest clearing/residential expansion and synanthropic vector/host combinations; some vector sand fly and host wildlife species persist in modified/plantation forest; associated in Sudan with forest/peri-forest exposure

Onchocerciasis (river blindness)		Y	Y		Y	Reduced disease associated with deforestation and loss of shade, but difficult to separate impact of wide scale use of DDT (1970s); in West Africa deforestation appears to have expanded the range of vector
Echinococcosis		Y	Y		Y	Land-use change in South China favours different hosts in endemic area – transient increases following deforestation, resurgence following reforestation; increases in Europe/North America due to urbanisation of foxes/landscape transformation
Snakebite	Y		Y			Risks to forest workers; some snakes have colonized suburban and urban areas, exploiting new synanthropic resources (like rodents)
Yaws		Y				Endemic in some remote forested locations; compounded by lack of healthcare access (e.g. Congo Pygmies); also serve as infection reservoir for gorillas
Lymphatic filariasis	Y	**Y** **P**				Risk to Malaysian forest workers from sub-periodic *B. malayi* vectors and wildlife reservoirs (e.g. leaf monkeys)
Sleeping sickness	Y			Y		Scrub clearing originally used to reduce tsetse fly infestation (West Africa); significant disease risk appears in SSA wet tropical forests; most cases in DR Congo
Rabies	Y	P				Deforestation impacts on host (vampire) bats increases overlap with humans (likewise for bat hosts of lyssavirus in Australia). Most human transmission via peridomestic dogs

Table 3.2 (cont.)

Neglected tropical disease	Deforestation effect					Comment
	Direct	In direct	Worsened	Improved	Mixed	
Dengue, Chikungunya **(arboviruses)**	Y	Y	Y			Associated historically with forest clearing/sylvatic cycle/peri-domestic vector; deforestation drives new risk of sylvatic strains of Dengue although urbanisation provides better vector habitat (*Ae. aegypti* most closely associated with human habitation and indoors, also *Ae. albopictus*)
Leprosy		Y/P				Wildlife reservoirs (e.g. armadillos) may be impacted in the Americas
Deep mycoses	Y	P				Risk to forest and agricultural workers from agricultural plants (e.g. tea, rubber) and forestry; highest numbers in Madagascar and Brazil
Scabies		P	Y			Wildlife reservoirs including forest spp. play minor role
Trachoma	Y	P	Y			Desertification – as sequelae to deforestation, associated with dry dusty conditions, lack of water for adequate face washing)

(P = association with poverty, poor sanitation and access to healthcare; Y = yes; SSA = sub-Saharan Africa)

virus of global significance: currently, 36.9 million people have HIV/AIDS, two-thirds of whom live in Africa (WHO 2019a).

Forest dwellers and workers are among those who continue to be impacted: it easily spreads with extractive industries, associated with ad hoc development (with transient single males or prostitution). For those afflicted it increases demands for traditional medicines, food and income (Lopez 2008). Access to adequately funded and resourced health services with appropriate education and anti-viral medication is critical to supporting at risk and infected individuals (Targets 3.8, 3.C).

Other forest-origin diseases include Ebola, Zika, Nipah and SARS corona virus. These have also caused significant outbreaks in recent decades. Increasing human density in biodiverse areas has been associated with the emergence of these diseases (Jones et al. 2008). Contact (often indirect) between wildlife (particularly bats, rats and primates) and humans through consumption, dispersal of hosts following habitat loss, amplification following loss of predators or competitors, and exposure through deforestation, road construction and wildlife farming provide opportunities for cross-species virus transmission. Relatively few diseases have gone on to be capable of sustained human-to-human transmission. Many others episodically spill over from their wildlife and environmental sources and are local health risks. For example, 39 of 187 arboviruses (transmitted by mosquitoes) identified in the Amazon basin during road construction can cause disease in humans (Vasconcelos et al. 2001).

Importantly, forest-origin diseases are not limited to the tropics, developing countries or even to deforestation. For example, Lyme disease is a significant risk to recreational forest users and residents in the USA and has expanded its range as a result of reforestation of previously cleared areas and altered host-pathogen dynamics. While bringing many benefits, the novel ecology of human-modified environments, including restored forests, plantations and urban parks, creates opportunity for new species combinations and disease emergence (McFarlane et al. 2012).

Predicting and preventing the next pandemic (i.e. multi-country epidemic) has been a focus of research and investment with consequences for forests, forest dwellers and the global population, as efforts are best employed to stop the spread, and possibly the emergence, at the source. Initiatives such as the US Agency for International Development Emerging Pandemic Threats programme have deployed scientists to remote forests to catalogue wildlife pathogens. Such programmes assist Target 3D to strengthen the capacity, particularly in developing countries, for early warning, risk reduction and management of national and global health risks. In regions where the burden of traditional infectious diseases remains significant, foreign investment in identifying pathogens that may cause future pandemics (capable of reaching developed nations) has the potential to create tensions, as well as fear. This may be sensitively and usefully navigated on a community or country basis. However, the experience of the 2013–16 Ebola pandemic in West Africa illustrates the complexity of stopping outbreaks of even identified diseases.

Ebola virus was identified in 1976 in Zaire and Sudan, with episodic outbreaks in Central Africa, associated with climatic and environmental factors, multispecies wildlife mortality and bushmeat consumption (Real et al. 2017). The virus recently spread to West Africa, potentially as habitat change and food availability impacted the ecology of the speculative bat hosts. Across West Africa in recent decades, agriculture, palm oil and other plantations have accelerated forest transformation, impacting fruit bat ecology (Wallace et al. 2014). The virus transmission that began the West African outbreak is believed to have occurred when a child played in a tree where bats roosted. However, it was the subsequent movement of infected people and the poor capacity of local health systems to perform surveillance, identification, containment or treatment of infected people that enabled the virus to spread extensively, resulting in 28 616 suspected cases and 11 310 deaths.

Disease regulation as a proposed ecosystem service of intact or pristine forests (Millennium Ecosystem Assessment 2005) is not broadly applicable. There is no doubt that ecological change can alter infectious diseases epidemiology, but there is no simple inverse relationship (Tucker Lima et al. 2017). Indeed, new zoonotic diseases from wildlife may ultimately cease emerging as a result of ongoing forest destruction and biodiversity loss. Sophisticated epidemiological understandings have provided targeted approaches to lowering risk, but these rarely promote forest conservation. For example, the 1998 outbreak of Nipah virus in Malaysia and Singapore, traced to contact between fruit bats (facing habitat loss and/or escaping haze from forest fires) and intensively farmed pigs, has not reoccurred, due in part to removing bat-attracting mango trees from pig pens (Pulliam et al. 2012). The example of Ebola illustrates the vast spatial and temporal scales over which the clearing, burning and replacement of forests may have altered the migratory patterns, population sizes and distributions of wildlife hosts. Although the capacity to address such scales is not currently available, ecosystem-based approaches to disease prevention that maximise co-benefits for people and nature are evolving (McFarlane et al. 2018).

WATERBORNE DISEASES

Waterborne diseases refer to a diverse group of pathogens including protozoa (such as giardia and cryptosporidia) and bacteria (such as typhoid, cholera and dysentery). Forests have a role to play in reducing illness and deaths from waterborne diseases and pollution through the protection of water catchments, reducing the impact of flooding, and in local climate regulation. Both the excess of water (e.g. flooding and faecal contamination) and its shortage (e.g. droughts and limited water for proper sanitation) can increase the risk of waterborne diseases, making climate change a concern for this group of diseases. The WHO estimates that waterborne diarrhoeal diseases were responsible for 2 million deaths in 2017, with most occurring in children under 5 (WHO 2018b) (see Target 3.2).

Natural (upper) catchments have reduced exposure to pollutants and waterborne pathogens associated with human and livestock activity. Additionally, forested watersheds generally offer higher-quality water than alternative land uses and do so at a lower cost than equivalent technology. For example, in 1997 New York City conserved the Catskill Mountains (the city's main water source) rather than install a new water filtration plant costing USD 4–6 billion, with USD 250 million a year in operating costs (Chichlinisky and Heal 1998).

The biophysical properties of forests also contribute to water quality. In Fiji, catchments cleared of their forest cover, or where riparian vegetation has been lost, show elevated incidence of waterborne infectious diseases such as typhoid and leptospirosis (Jenkins et al. 2016). Forested catchments also improve water discharge and protect against downstream flooding, although this is influenced by specific properties of the forest and catchment and the extremity of flooding (Chandler et al. 2018).

3.3.2 Reducing Non-Communicable Diseases and Promoting Mental Health

NON-COMMUNICABLE DISEASES

Target 3.4 calls for calls for a one-third reduction in premature mortality from non-communicable diseases (NCDs). NCDs are responsible for almost three-quarters of all deaths globally, the majority of which occur in low- and middle-income countries. Six NCDs are included in the top 10 global causes of mortality (WH0 2018c): cardiovascular diseases, stroke, chronic obstructive pulmonary disease, Alzheimer's disease, respiratory cancers and Type 2 diabetes. Unhealthy diets, physical inactivity, exposure to tobacco smoke and the harmful use of alcohol are considered the most important NCD risk factors.

The global increase in overweight and obesity and its many associated health conditions, including cardiovascular diseases, diabetes and cancer, is in part connected to excess food consumption and reduced physical exercise. Unhealthy diets, particularly with high fatty and red-meat intake, as well as nutrient-poor, energy-dense diets, are increasingly common among poor populations in low-income settings with limited health literacy. Genetic factors make some populations especially vulnerable to diabetes. Frequently, depression, social exclusion, vulnerability and a sense of being exploited contribute to these unhealthy behaviours. The importance of forest protection for exercise and mental health are discussed below. First, we draw attention to the impact of two components of unhealthy diets on the destruction of forests.

There is a strong link between high red-meat consumption (especially beef) and the risk of death from heart disease, other NCDs and several forms of cancer (Kmietowicz 2017). Worldwide, meat production has tripled over the last four decades, increasing 20 per cent in the last decade alone. Demand for red meat, historically popular in Western countries, is growing in developing economies, in part because of its perceived status. Beef production is a major driver of deforestation, woody encroachment of savannahs and desertification. Production of livestock feed crops and pastures are the major cause of deforestation of Amazonian forests (Armenteras et al. 2017). Extensive grazing drives deforestation elsewhere, including Australia's tropical savannas, the world's largest intact savanna ecosystem. Increased cattle numbers contribute to rising quantities of the potent greenhouse gas methane as well as nitrous oxide – important issues for SDG 13 (Climate Action). Excessive beef production is deeply problematic in an era striving for sustainable development, and it has profoundly adverse health consequences (Potter 2017). Awareness of these harms has not driven per capita reduction of meat production or consumption in traditional (developed) producing countries.

Another leading cause of tropical deforestation with adverse impacts on diet is the production of palm oil. Palm oil is an affordable source of cooking oil and is valuable as a replacement for polyunsaturated oils, which have the potential to form harmful trans-fatty acids. A modest ingestion of palm oil appears to be safe; however, as a widely used ingredient of calorie-dense processed foods, it can be injurious. Indonesia and Malaysia produce 86 per cent of the world's palm oil, significantly contributing to their economies (WWF 2018). Production and continued expansion in these countries comes at a very high cost to native forests (and Indigenous peoples), along with significant harms to health.

In contrast, conserving forests as a source of nutrient rich foods is important for associated Indigenous people and subsistence farmers (Ickowitz et al. 2016). Micronutrient deficiencies affect two billion people worldwide, predisposing them to disease and poor cognitive development. For forest-associated groups, there can be a cost-effective synergy of healthy people, food harvesting and stewarded forests. This should be valued against the destruction of forests for unhealthy global diets, hunger reduction (SDG 2) and the production of greenhouse gasses (SDG 13).

MENTAL HEALTH

The WHO reports that 14 per cent of the global burden of disease is attributed to mental health disorders, with 75 per cent of affected people from low-income countries (WHO 2018c). The role of forest loss in these figures is not known. However, the stimulation and stresses of urban life – more than half of the global population live in urban areas – has generated interest in the consequences of nature deprivation, a situation forewarned in the 1950s by René Dubos, one of the founders of ecological public health at the planetary scale. There are also impacts for Indigenous and other non-urban forest people. For many, the landscapes of personal and community significance have disappeared, often rapidly and recently. Solastalgia – the psychic or existential stress caused by environmental change (Albrecht et al. 2007) – can be profound and amplified by disempowerment, marginalisation, the loss of religious or cultural sites and identity.

Most of the research and practice concerning natural environments and mental (and physical) health is undertaken in urban and developed settings. That increased exposure to high-quality green and blue space (tree-lined streets, parks, gardens and water views) is beneficial to physical and mental health is intuitively attractive to all who value nature. However, as with the disease regulation theory, the idea that exposure to green space is automatically beneficial to health is simplistic. Many of the world's poorest and least healthy populations, including Indigenous and other populations living in or around forests, are exposed to considerably more green space than the average urban inhabitant, yet have poor physical health. While absence of nature contact may be harmful, abundant exposure does not fully offset other risk factors for ill health.

For most of the global population now classed as urban dwellers, there is evidence of positive effects of visiting or even having green space in one's neighbourhood, not just for the wealthier in leafy suburbs. Gains may be greatest for groups otherwise deprived of access (Taylor et al. 2015). There is growing evidence that biological diversity is responsible for some of the reported positive effect (Aerts et al. 2018). Biological mechanisms account for some of the reported health benefits. For example, in addition to the physiological contributions to cardiovascular health from increased exercise, cleaner air and less noise (Donovan et al. 2015), there is growing evidence that the human microbiome may be enhanced by exposure to biodiverse environments, including forests (Prescott et al. 2016). Immune function and other objective biomarkers for health, such as cortisol levels and blood pressure, may also be improved (Rook 2013). Some studies have found that even brief interactions with nature can produce marked increases in cognitive function (Berman et al. 2008).

The distribution of green space across cities is now considered a source of health inequity, recognised broadly, for example, by the European Environment Agency and as a specific target within SDG 11 (Sustainable Cities and Communities). Box 3.1 discusses this in more detail.

Box 3.1 Urban Forests and Health

The term 'urban forest' has been in use since the 1970s and includes all trees within a metropolitan boundary. There is growing recognition of urban forests' importance to health and, more recently, also its ecological value.

Health benefits include reduced flows and nutrients in storm water, pollution control, shade and urban heat island reduction, by 4–5°C in some settings (Livesley et al. 2016) – of growing importance under climate change. Trees reduce air pollution due to cars, industry and coal burning. All tree species capture PM10 (particulate matter up to 10 microns in diameter), but some are much more efficient than others (Manes et al. 2016, Yang et al. 2005). Trees can store and remove carbon and, depending on type and form, reduce particulate matter by 7–24 per cent. The urban forest also provides opportunity for residents to have some contact with nature and to garner its benefits, such as reducing developmental issues in children and behavioural issues in young adults, improving mental health more generally, and as a backdrop for exercise (see Target 3.4). Improved recovery time in patients with natural views or direct exposure to nature has led to the purposeful planting of trees around hospitals and medical centres (CSH 2018). Interestingly, exposure to virtual forests (and nature) is reported to have significant impact on cognitive function in people with dementia (Moyle et al. 2017).

However, some trees (such as poplars) emit volatile organic compounds that interact with car exhausts and increase the concentration of ground-level ozone, particularly during heatwaves (Willis and Petrokofsky 2017). Trees provide habitat for urban animals, including birds, squirrels, possums, monkeys and bats, providing many valuable benefits, though some carry infectious diseases, e.g. West Nile virus, Lyme disease, Hendra virus (McFarlane et al. 2012). Another drawback is hay fever, which can be debilitating, from allergenic plants, including trees with high pollen counts. Increased risks of fires in towns and cities due to climate change can be aggravated by urban forests.

Nevertheless, urban forest design, sometimes referred to as green infrastructure, is potentially a significant factor – and is increasingly recognised – in human health and ecosystem service protection and conservation (Kowarik and von der Lippe 2018).

Outside cities, there is further evidence of positive effects (Maller et al. 2006). In South Korea – a highly industrialised, urbanised nation with a high rate of suicide – the therapeutic exposure to national parks is now being vigorously promoted. In Japan, the term *shinrin-yoku* refers to the practice of 'forest bathing' for well-being, with growing evidence that this is beneficial. The positive benefits of spending time in wilderness to deal with death, including one's own, have been investigated in Canada. New Zealand has pioneered 'green prescriptions' that recommend physical activity, a concept that has grown to include the added benefits of exercising in nature. The Healthy Parks Healthy People (HPHP) movement, particularly active in Australia and the USA, promotes the benefits of park use to increase social well-being and reduce NCDs. HPHP has also pioneered some significant cross-sectoral relationships between government health and environment sectors.

In summary, there is persuasive evidence of benefits to mental and physical health from increased exposure to forests, at least for those for whom it is not a daily event. These benefits are likely greatest for those living in urban areas and whose basic health needs (nutrition, housing and an income allowing dignity and physical security) are largely met (Tomita et al. 2017).

MENTAL AND PHYSICAL WELL-BEING BEYOND THE URBAN FOREST

The suffering, especially mental, when Indigenous and other peoples lose their forests has been relentless for centuries and continues today, although poorly documented and almost universally overlooked by the colonising groups. An authentic commitment to reduce mental health suffering resulting from the forest-conversion actions of others requires protecting forests and the rights of their traditional custodians (e.g. West Papuans and the Congo Pygmies; Ohenjo et al. 2006). Prohibition of forest use, including for culturally important products such as bushmeat or medicine, regardless of the abundance of non-traditional alternatives, may cause psychological unrest and affect well-being (e.g. various Congo basin forest peoples; Dounias and Ichikawa 2017).

Protecting the eudemonic well-being of many groups and populations not resident in forests is still deeply grounded in forest protection. Collective well-being is reflected, for example, in respect for sacred sites or ancestors and the opportunity to pass on biodiverse natural resources and customary tenure rights to future generations, in turn protecting well-being, identity and kinship (Fritz-Vietta 2016). An estimated 5–8 per cent of global forests are considered to be sacred. Protecting these forests has profound consequences for people as well as conservation. For example, monk-led community conservation of 18 000 ha of rare lowland evergreen forest in Northern Cambodia, motivated by reverence for the example and teaching of Buddha, has been focal in post-Khmer Rouge community recovery (ARC 2010). In summary, forest conservation promotes mental well-being in diverse ways.

FORESTRY ACCIDENTS

Statistics on forestry accidents are difficult to obtain and are sparsely reported outside developed nations. Available reliable data suggest that forestry-related work is extremely hazardous. Occupational health and safety for many who work in forestry is poorly regulated, particularly for those who work informally. Tree felling is the deadliest occupation in forestry; in developing countries, chainsaws may be involved in nearly half of all forestry accidents. Other reported issues include chemical exposure (e.g. pesticides), hearing loss, heat- and cold-related diseases, repetitive stress syndromes and musculoskeletal trauma (Garland 2018). Sedentary machine operators have increased risk of diabetes and obesity. Addressing SDG 3 targets would benefit these groups and those exposed to hazardous chemicals. Halving road traffic accidents globally (Target 3.6) would also benefit forestry workers.

3.3.3 Reproductive Health and Family Planning

Target 3.7 calls for universal access to sexual and reproductive healthcare services, including family planning, information and education, and the integration of reproductive health into national strategies and programmes. Although this target encompasses a range of important issues, its most relevant aspect to forests (and the other SDGs) is to promote access to contraception and thus to slow population growth (Starbird et al. 2016).

There is abundant evidence that rapid population growth hinders economic development, intensifies resource insecurities and environmental damage, and fuels conflicts (Bongaarts 2016, Butler and Higgs 2018, Husain et al. 2016, Population Institute 2015). No country has been able to advance from least-developed status while fertility remains above four children per woman, unless (and only for as long as) it has vast oil or other natural-resource income. Generally, economic development has only taken off after fertility falls well below three children, being highest in countries with below-replacement fertility (O'Sullivan 2017). Successful voluntary family planning programmes preceded the economic growth of the East Asian tiger economies.

Forest-dependent people, including Indigenous minority groups, tend to have higher fertility than their national averages and are disadvantaged in access to family planning. This impacts family finances, resource security and the health and well-being of women and children, as well as demands on forests. Smaller families and wider child spacing mean more investment per child. Universal access to family planning would help close the development gap between forest dwellers and urbanised communities.

Even in developed countries, more than 40 per cent of pregnancies are unintended. Access to contraception and the outcomes of unwanted pregnancies are problematic for many (Foster et al. 2018), and each birth draws more heavily on Earth's natural resources (Wynes and Nicholas 2017).

Population growth is a major driver of forest loss. The Food and Agriculture Organisation of the UN (FAO 2016) reports a strong correlation across regions between forest loss and increase in rural population. Traditional swidden agriculture relies on long forest fallows, but as fallow periods shorten under population pressure, forest remnants shrink and become degraded before being permanently cleared. Commodity-driven deforestation is reported as the largest category of forest loss (Curtis et al. 2018) but small-holder plots are also cleared for 'commodity agriculture', often enabled by nearby large-scale commercial plantations or new roads through forests. In Africa, subsistence agriculture is the dominant cause of forest conversion (Curtis et al. 2018). Burgeoning population and affluence in emerging economies increases commodity demand; this is also driven by affluent populations globally.

The IPCC-led[1] modelling of future climate change scenarios found that limiting warming to 2°C is only feasible with population growth much lower than current UN projections. A major stumbling block is agricultural demand, making further deforestation unavoidable (Riahi et al. 2017). The World Resources Institute estimates that achieving replacement-level fertility (about 2.1 children per woman) by 2050 could save an area of forest the size of Germany, seeing this as 'a multi-win solution to humanitarian, economic and environmental challenges, and an important item on the menu for a sustainable food future' (Searchinger et al. 2013: 2).

How much population-growth reduction may be achieved by the SDGs is complex and uncertain. Since the 1980s, the 'demographic-economic' rationale for slowing population growth, once a central pillar of the development agenda that influenced many norms, has been diluted, largely replaced by reproductive health services, emphasising only the 'reproductive health and rights' rationale for family planning, undertaken in many developing countries by poorly funded health ministries and some NGOs (Bongaarts 2016). The reproductive health and rights framing is today missing two vital ingredients: political will, stemming from the conviction that high population growth threatens economic development, and a focus on motivating people to want smaller families and to use contraception. Together, these elements can reduce the fatalistic acceptance of large families and gain support for family planning even in patriarchal societies.

[1] Intergovernmental Panel on Climate Change.

There are tentative signs of rekindled interest in the demographic rationale. Integrated development projects, under the 'population, health and environment' (PHE) model (Oglethorpe et al. 2008), are gaining recognition for achieving behavioural changes more rapidly than single-sector interventions, e.g. in environmental management, health and sanitation practices and diversified livelihoods, as well as those embracing smaller families and women's access to education and employment (see Box 3.2).

Box 3.2 Conservation, Community Health, Family Planning and Livelihoods

Conservation initiatives have recognised the importance of supporting the health and livelihoods of local populations in high-conservation-value areas for some time (Ancrenaz et al. 2007). This is strongly supported within organisations such as the International Union for Conservation of Nature (IUCN), the UN Development Programme (UNDP) and the Word Wide Fund for Nature (WWF), and it takes many forms.

In North Kayong, Kalimantan, Indonesia, the Alam Sehat Lestari (ASRI) clinic provides villagers with the most extensive healthcare services in the area and incentives to stop them from logging in the adjacent Gunung Palung National Park, such as 70 per cent discounts on medical fees. The clinic represents an alternative to health services provided by forestry companies. Patients who cannot afford medical fees, and so might otherwise resort to illegal logging, can choose to pay with various non-cash options, including native seedlings or labour. ASRI replants forests and trains ex-loggers to farm and run alternative businesses through a chainsaw buy-back scheme. Since inception, the number of logging households has decreased by 89 per cent, primary forest loss has stabilised and infant mortality has declined from 3.4 to 1.1 deaths per 100 households. This model is now being replicated elsewhere on the island of Borneo (Webb et al. 2018).

The Bwindi Impenetrable National Park in Uganda is home to about half of the world's remaining mountain gorillas. Conservation Through Public Health (CTPH) was initially established to address the transmission of diseases between gorillas, livestock and human populations. As it quickly became apparent that diversified livelihood assistance was needed to reduce park incursions, the project added livestock and microfinance programmes. It soon added family planning, realising that many parents were having more children than they wanted, while population growth countered conservation efforts. The project demonstrated strong synergies in cross-sectoral work, as the trust built through one area made communities more receptive to information in other sectors. Health and livelihood activities built support for conservation goals; ecological

Box 3.2 (cont.)

understanding generated enthusiasm for family planning. Contraception use increased twelvefold, to more than 60 per cent of women. Gorilla numbers have subsequently increased (Wilson Center 2013).

The model adopted by CTPH is the PHE approach (Oglethorpe et al. 2008). PHE projects recognise that tackling population growth is crucial for the long-term sustainability of environment and development interventions. By engaging with communities on their own priorities and enabling them to draw linkages between their livelihoods, resource base, family size and ability to educate and provide for children, they are motivated and empowered to overcome cultural barriers to change. PHE projects have particularly built male support for family planning and female participation in natural resource management. Since the early 2000s, PHE projects have gained increasing recognition. Many established projects, such as Ethiopia's Ethio Wetlands and Natural Resources Association and Papua New Guinea's Tree Kangaroo Conservation Programme, have adopted PHE approaches to enhance their impact. CTPH is a role model for several other PHE projects and a successful advocate for PHE to be recognised and scaled up through government agencies.

3.3.4 Universal Health Coverage and Affordable Essential Medicines

Target 3.8 strives to achieve universal health coverage, including financial risk protection, access to quality essential healthcare services and access to safe, effective, good-quality and affordable essential medicines and vaccines. Seeking this goal has potential beneficial effects for forests and people. Universal health coverage would benefit forest dwellers and forests since few options exist in remote forest locations to access modern healthcare or pay for it – other than through illegal timber felling, hunting bushmeat or illegal pet or medicinal plant trade. Many conservation groups recognise this relationship (Box 3.2). Additionally, through enhanced protection of genetic resources, as proposed by the Nagoya Protocol (Convention on Biological Diversity), the vast traditionally used and potential medical resources provided by forests may receive better protection.

PROTECTING THE NATURAL PHARMACOPOEIA

Traditional knowledge derived from a close relationship to nature is extremely important for pharmacological resources locally and globally (Fabricant and Farnsworth 2001). Many drugs are derived from compounds found in plants, often identified via their traditional use (Chivian and Bernstein 2008).

Conservative estimates of flowering plant species worldwide is 250 000; there is likely an abundance of drugs yet to be discovered.

Preserving and maintaining biodiversity and associated traditional knowledge is extremely important for the cultural well-being of local communities. The WHO estimates that up to 80 per cent of developing country populations rely, in part, on traditional medicine for their primary healthcare needs (WHO 2015). In many settings, traditional health systems are culturally preferred, and often based on complex social and spiritual relationships and directly dependent on natural resources. However, traditional knowledge and associated pharmacological knowledge is vanishing very quickly (Reyes-García et al. 2013). So too are many therapeutic plant and fungal species, although intensive efforts are being made to identify species with potential therapeutic value.

Fulfilling Target 3.8 is likely to protect forest ecosystem services in two ways. The first benefit is the protection of unidentified 'natural pharmacopeia': ethically exploring, cataloguing and protecting traditional knowledge about natural pharmaceutical properties can help identify useful remedies. In parallel, protecting as many species as possible, particularly in their traditional settings (as stated in the 2010 Nagoya Protocol, also linked to SDG 15, Life on Land), can ensure this knowledge can be tested and applied if found beneficial. This can be summarised as protecting the unidentified 'natural pharmacopeia'.

The second benefit will flow from a more thorough investigation of the possible 'false pharmacopeia', referring particularly to animal parts, but also some plants and fungi, that have zero, marginal or even adverse health benefits, yet are harvested from the wild (especially for consumption by large urban markets) and have significant harmful ecological effects despite their cultural importance. Numerous wildlife species, some inhabiting forests, have been pursued to (near) extinction for their alleged pharmaceutical benefits, including charismatic mammals such as the rhinoceros, snow leopard and tiger, as well as the humble pangolin (Byard 2016). Evidence of therapeutic benefits for many traditional remedies involving animal parts is extremely limited.

3.3.5 Improving the Quality of Air, Water and Soil

Target 3.9 calls for a *substantial* reduction in the number of deaths and illnesses from hazardous chemicals and air, water and soil pollution and contamination. There is broad recognition of the health burden of air pollution, particularly from fossil fuels and petrochemicals, as well as its disproportionate impact on the poor (Landrigan et al. 2018). This, and to a lesser extent water and soil pollution control, has several direct implications for forests. Exposure to hazardous chemicals, particularly pesticides, is a recognised risk for forestry workers (Garland 2018).

AIR POLLUTION

Air pollution is chiefly from the combustion either of fossil fuels such as coal and petroleum or of biomass. The latter include forest fires, a burn following clearing of forests, and/or the burning of organic-rich forest soils (peats). The frequency of peat fires appears to be increasing due to climate change (Seidl et al. 2017), via intensifying drought, heat, stronger winds and increased dry lightning storms. Other forms of biomass combustion relate to burning as an agricultural practice, as a form of land management and as a household fuel.

HOUSEHOLD-ASSOCIATED AIR POLLUTION

Biomass, including dung, crop residues and wood, provide the main cooking fuel for at least 2.8 billion people (Bruce et al. 2015). Much of this is combusted inside dwellings with poor ventilation, contributing to high rates of respiratory and other diseases. Increasingly referred to as household-associated air pollution (Goldemberg et al. 2018), smoke-induced diseases are responsible for the premature death of 4.3 million people annually, with women and young children most affected (Bruce et al. 2015). An estimated 500 000 children under 5 die each year of preventable and treatable respiratory conditions, worsened by household air pollution, undernutrition and inadequate health care (Langbein 2017). Considerable effort over decades has been directed to developing low-cost, safer cooking stoves, but with variable success (Goldemberg et al. 2018). Furthermore, gathering fuelwood can have significant environmental impact, in some areas driving desertification (Masera et al. 2015).

Any successful attempt to attain SDG 3.9 needs to promote the replacement of forest products with forms of energy less damaging to health and the environment. This may be possible through large-scale electrification using wind and solar power (or with hydroelectricity, as in the case of Ecuador); yet, in most settings the cost of clean electrical power (e.g. solar, wind or hydro), though increasingly plausible for lighting, is still too high to make this a likely prospect for populations living away from centralised, reliable energy supplies. Gas is being increasingly used, especially in Brazil and India; though far better for health, it still results in significant carbon emissions and is unaffordable for most (Goldemberg et al. 2018). The flooding of forests and farmland to produce hydroelectricity has been a source of controversy in many areas, particularly when benefits are realised at a distance from the land lost.

RESPIRATORY DISEASE ASSOCIATED WITH FORESTRY

Specific hazards associated with wood processing and manufacturing industries (e.g. inhaled sawdust, pulp and mould) can lead to a range of bacterial, fungal and airborne endotoxin infections, generating respiratory disorders such as wood workers' lung and hypersensitivity pneumonitis (Adhikari et al. 2015, Sforza and Marinou 2017). These conditions can affect not only workers but also people living in the vicinity of these industries.

FOREST FIRES AND TRANSNATIONAL AIR POLLUTION

In some parts of the world, especially Southeast Asia, the Amazon and sub-Saharan Africa, the deliberate, seasonal burning of forests, mainly to promote agriculture and plantations, creates a substantial health hazard (Johnston et al. 2012). For example, forest burning during the dry season in Kalimantan and Sumatra (Indonesia) contributes regularly to dangerous levels of air pollution. In 2015, as climatic conditions resulted in drought and greatly intensified fire activity in the region, persistent, hazardous levels of smoke pollution (haze) resulted in an estimated 100 000 deaths across Indonesia, Malaysia and Singapore – more than double those from previous reported events in 1997 and 2006 (Koplitz et al. 2016). Such hazardous conditions can cause schools and many workplaces to close, grounding of air traffic and residents being encouraged to stay indoors. Not all residents can gain respite indoors or have means to evacuate. In 2002, several ASEAN[2] nations passed a Transboundary Haze Pollution Act that financially penalises companies for smoke-haze activities beyond the borders of individual countries. Since 2017, sustainability certification of forest industries in Indonesia has significantly reduced deforestation and associated fires (Carlson et al. 2017). Similar concerns over the quartet of forest fires, particulate matter pollution, respiratory disease and carbon emissions exist in the Amazon basin.

FORESTS, WATER AND SOIL

All measures that protect agricultural productivity and ensure food and water safety have significant benefits for health. Forests enhance soil biodiversity and organic matter recycling; limit desiccation, erosion and dryland salinity; and promote pest control by providing shelter for pest-predators. The capacity of forests to reduce soil contamination from pollutants may be significant, yet it remains under-researched. Phytoremediation of contaminated

[2] Association of South East Asian Nations

land (including heavy metals and radioactive material) does utilise some tree species, as well as grassland and forbs. Large, uncontrolled forest fires can result in water pollution that reduces the quality of water emerging from forests.

CLIMATE CHANGE, FORESTS AND HEALTH

Climate change primarily results from the accumulation of heat-trapping gases in the atmosphere. It can thus be conceptualised as a form of air pollution, worsened not only by the transfer of carbon in fossil fuels but also from biomass (including in the soil) and, increasingly, from tundra and peat to the atmosphere (and ocean). Its health effects are protean, but still largely in the future. Heat stress for outdoor labourers, already significant where conditions and underlying health are poor, will amplify in impact, affecting many people in and near forests engaged in such labour. Further impacts, including direct trauma, are anticipated from heatwaves from other forms of extreme events, including rising sea levels, storm surges, droughts, flooding, fires and high winds.

Addressing Target 3.4 by reducing forest destruction, including fires, would have the co-benefit of reducing carbon emissions (not only of overlying vegetation but in some cases underlying peat), which can be of global significance during El Niño years (Page et al. 2002). Emissions from peat oxidation resulting from water-table lowering add to the carbon burden. In Southeast Asia, forest and swamp conversion for plantations and other agriculture means that a major carbon sink is now a carbon source (Miettinen et al. 2017). Reducing emissions from deforestation and forest degradation, and reducing health impacts of climate change, are already important elements of UN climate policy. Additionally, the potential for coastal mangrove forests to provide protection from coastal storm surges is now contributing to their conservation (Feller et al. 2017), and the ability of the urban forest to reduce a city's temperature is increasing recognition of the value of trees (Box 3.1). Action to avoid tipping globally important forest areas into other, post-clearance ecological states additionally protects future carbon sequestration (Miettinen et al. 2017).

Other impacts on health through climate change are manifold: global food price increases, local crop failures (to which subsistence populations are particularly vulnerable), reduced labour productivity and alterations in the epidemiology of vector-borne and other infectious diseases. In the long run, perhaps most importantly, the highly indirect, politically mediated 'tertiary' effects from famine, economic disruption, population displacement and conflict will prove most harmful (Butler 2014).

3.4 Summary and Recommendations

Table 3.3 summarises the impacts of implementing SDG 3 targets on forests and forest-dwelling people. Benefits to forest people assume targets will be pursued through a social justice lens (poor-preferencing) or at least neutrality, so that, at the minimum, forest people are not disadvantaged by development.

In theory, most SDG 3 targets can be improved in ways that protect forests, as discussed in this chapter. In reality, the pathways are predicated on increased economic growth (and possibly population) and that is likely to cause harm to forests, rather than be beneficial or neutral. Finally, we conceptualised the links between forests and population health at three scales (Figure 3.1) and summarise the findings of this chapter in this way.

Table 3.3 Summary of impacts of implementing SDG 3 targets on forests and forest people: benefits (green), context-dependent harms or benefits (yellow)

SDG 3 targets		Impacts	
		On forests	On people
1	Reduced maternal mortality		
2	Reduced neonatal and under-5 mortality		
3	Communicable disease control		
4	Reduction of non-communicable diseases and mental health problems		
5	Prevent substance abuse		
6	Road traffic accidents		
7	Reproductive health		
8	Universal health coverage		
9	Pollution and hazardous chemical control		
A	Tobacco control		
B	Improved vaccines and medicine access		
C	Health financing and recruitment		
D	Early risk warning		

HOW ATTAINING SDG 3 COULD AFFECT FORESTS AND POPULATIONS DEPENDENT ON FORESTS

- Improving health and well-being of Indigenous and other forest-adjacent communities can positively impact forests where these groups play a crucial role in forest stewardship, including sustainable management of natural resources.
- Economic development of forested areas that ignores harm to Indigenous and local people and the ecosystem services on which they rely will almost inevitably decrease their health and well-being.
- Access to family planning, health education, investment in clean water and sanitation, alternatives to wood biomassfuel and control of large forest fires have co-benefits to people and forests (Targets 3.7, 3.8, 3.9, 3.C and other SDGs).
- Achieving universal health coverage has an important role to play in making healthcare affordable and reducing pressure on forests from (catastrophic) health expenditure (Target 3.8). Health workers should be aware of potential negative consequences of development (such as disempowerment and its mental health consequences; sedentary lifestyles; nutrient-poor and unhealthy food; tobacco, alcohol and other substances; commercial sex trade) (Targets 3.3, 3.4, 3.5, 3.7, 3.8, 3.C).
- Traditional medical systems should be integrated into contemporary healthcare to ensure the most culturally appropriate treatment for Indigenous peoples. Medical and other healthcare personnel should collaborate with traditional healers to provide more efficient services and gain better understanding of traditional practices necessary for appropriate healthcare. Preservation of cultural and ecological knowledge is also valued by pharmacological research (Target 3.8).
- Research and surveillance for emerging diseases and health risks at the forest interface is best coupled with support to address existing disease burdens, reducing risks and improving health literacy and capacity (Targets 3.3, 3.D).

HOW ATTAINING SDG 3 COULD AFFECT FORESTS AND THOSE WHOSE COMMERCIAL LIVELIHOODS ARE DEPENDENT ON FORESTS

- Reduction of hazardous chemicals and air, water and soil pollution and contamination will improve the health of forestry workers and adjacent communities (Target 3.9).
- The work-related health problems of Indigenous and traditional people and other people engaged in unorganised sectors (e.g. leaf plate-making,

handicraft) are largely unreported and not specifically identified within SDG 3. Work-related injuries and even mortality resulting from forestry is important but also largely unreported.

- Control of vast forest and peat fires and the resulting haze and transnational air pollution and associated loss of life is relevant to local as well as distant populations; efforts to strengthen the capacity to reduce and manage global health risks include such fire control (Target 3.D).
- Infectious diseases emerging from disturbed forests or from the hunting or dispersal of vectors and wildlife hosts may be a particular risk to forestry workers (e.g. malaria) or plantation farmers (e.g. Chagas disease). Surveillance and management of forests to limit infectious disease risks requires further research, but includes positive and negative outcomes for forests and forest people (Targets 3.3, 3.D).
- The target to reduce non-communicable diseases, where linked to dietary commodities associated with health risks such as red meat or palm oil, may challenge extensive forest clearing for their production (Target 3.4).

HOW ATTAINING SDG 3 COULD AFFECT FORESTS AND REGIONALLY DEPENDENT POPULATIONS

- Recognition that natural forest catchments are cost-effective in addressing water pollution and quality should encourage their protection and re-establishment (Target 3.9).
- Urban demand for bushmeat, bush medicines and some timbers is driving unsustainable pressure on forests. Alternatives need to be identified and promoted.

HOW ATTAINING SDG 3 COULD AFFECT FOREST AND GLOBAL AND URBAN POPULATION

- Climate change is not singled out as a global health risk in SDG 3; however, we note not only that this exists, but also that reducing forest destruction has the co-benefit of maintaining carbon sequestration and local climate regulation. International cooperation is required to address this risk (Target 3.9).
- As with all scales described here, universal access to family planning has an important role in reducing human pressure on forests (and natural resources). This is relevant not only for populations in forested areas, but also for those at any distance where consumption drives demand for forest products (Target 3.7).
- Cultural and spiritual ecosystem services of forests contribute to the well-being of many people who may rarely (if ever) visit them. Recreational

forest users may gain additional mental, physical and immunological benefits from forests. This is an area that warrants further research (Target 3.4).

- The urban forest is increasingly valued and developed for the mental, social and physical well-being it can provide urban dwellers. Linking improvements in health with urban-forest use and proximity could contribute to forest maintenance and expansion more generally (Target 3.4).

3.5 Conclusions

This chapter has reviewed many ways in which genuine attempts to attain the targets associated with SDG 3 can protect forests, forest ecosystem services and the people who rely on them. It has discussed the close relationship between many aspects of health and forests, not only for Indigenous and other peoples directly dependent on forest benefits but, less directly, for the global population. Forests play an important part in maintaining earth systems, and their erosion has potentially negative and catastrophic consequences for the health and well-being of the global human population, particularly those already vulnerable. Adjustments to our definitions of health, protection of its social and ecological determinants and recognition of planetary limits will contribute to global health and, in so doing, will safeguard forests.

This chapter has identified the cognitive dissonance evident in the SDGs that seek to expand economic growth (as conventionally defined) yet protect natural capital, including forests. The chapter has also discussed how a failure to improve reproductive health, especially by inadequate provision of family planning services and other influences on fertility, threatens forests and forest populations and will also threaten the achievement of many other SDGs. A way forward may be to foster the understanding, among those with more political and economic power, that their health and well-being will be promoted by more biosensitive activities, such as a diet less reliant on animal products, less wasteful consumption and more contact with nature. These principles also apply for many people in the global aspirational class, and give homage to those Indigenous and traditional groups that still live by these values.

References

Adhikari, A., Sahu, S., Bandyopadhyay, A., Blanc, P. D. and Moitra, S. 2015. Fungal contamination of the respiratory tract and associated respiratory impairment among sawmill workers in India. *ERJ Open Research* 1:00023–2015.

Aerts, R., Honnay, O. and Van Nieuwenhuyse, A. 2018. Biodiversity and human health: Mechanisms and evidence of the positive health effects of diversity in nature and green spaces. *British Medical Bulletin* 127:5–22.

Albrecht, G., Sartore, G. M., Connor, L. et al. 2007. Solastalgia: The distress caused by environmental change. *Australasian Psychiatry* 15:S95–S98.

Ancrenaz, M., Dabek, L. and O'Neil, S. 2007. The costs of exclusion: Recognising a role for local communities in biodiversity conservation. *PLOS Biology* 5:e289.

ARC 2018. *Monks' community forest in Cambodia wins prestigious Equator Prize.* UK: Alliance of Religions and Conservation. Available at: www.arcworld.org/news.asp?pageID=412 (Accessed 4 February 2019).

Armenteras, D., Espelta, J., Rodríguez, N. and Retana, J. 2017. Deforestation dynamics and drivers in different forest types in Latin America: Three decades of studies (1980–2010). *Global Environmental Change* 46:139–147.

Barber, B. E., Rajahram, G. S., Grigg, M. J., William, T. and Anstey, N. M. 2017. World malaria report: Time to acknowledge *Plasmodium knowlesi* malaria. *Malaria Journal* 16:135.

Bauhoff, S. and Busch, J. 2018. *Does deforestation increase malaria prevalence? Evidence from satellite data and health surveys*. Working Paper 480. Washington, DC: Center for Global Development.

Berman, M., Jonides, J. and Kaplan, S. 2008. The cognitive benefits of interacting with nature. *Psychological Science* 19:1207–12.

Bongaarts, J. 2016. Development: Slow down population growth. *Nature* 530:409–12.

Bruce, N., Pope, D., Rehfuess, E. et al. 2015. WHO indoor air quality guidelines on household fuel combustion: Strategy implications of new evidence on interventions and exposure–risk functions. *Atmospheric Environment* 106:451–457.

Butler, C., Chambers, R., Chopra, K. et al. 2003. Ecosystems and human well-being. In *Ecosystems and human well-being. A framework for assessment.* Millennium Ecosystem Assessment. Washington DC: Island Press, pp. 71–84.

Butler, C. D. (ed.) 2014. *Climate change and global health*, Wallingford: CABI.

Butler, C. and Higgs, K. 2018. Health, population, limits and the decline of nature. In Marsden, T. (ed.) *The Sage Handbook of Nature.* London: Sage Publications, pp. 1142–49.

Byard, R. W. 2016. Traditional medicines and species extinction: Another side to forensic wildlife investigation. *Forensic Science, Medicine, and Pathology* 12:125.

Carlson, K. M., Heilmayr, R., Gibbs, H. K. et al. 2017. Effect of oil palm sustainability certification on deforestation and fire in Indonesia. *Proceedings of the National Academy of Sciences* 115(1):121–6.

Chandler, K., Stevens, C., Binley, A. and Keith, A. 2018. Influence of tree species and forest land use on soil hydraulic conductivity and implications for surface runoff generation. *Geoderma* 310:120–7.

Chaves, L., Conn, J. E., López, R. V. and Sallum, M. A. 2018. Abundance of impacted forest patches less than 5 km^2 is a key driver of the incidence of malaria in Amazonian Brazil. *Scientific Reports* 8:7077.

Chichlinisky, G. and Heal, G. 1998. Economic returns from the biosphere. *Nature* 391:629–30.

Chivian, E. and Bernstein, A. (eds.) 2008. *Sustaining life. How human health depends on biodiversity*. Oxford: Oxford University Press.

CSH (Centre for Sustainable Healthcare) 2018. *The NHS Forest*. Oxford. Available at: https://sustainablehealthcare.org.uk/what-we-do/green-space/nhs-forest.

Curtis, P. G., Slay, C. M., Harris, N. L., Tyukavina, A. and Hansen, M. C. 2018. Classifying drivers of global forest loss. *Science* 361:1108–11.

Donovan, G. H., Michael, Y. L., Gatziolis, D., Prestemon, J. P. and Whitsel, E. A. 2015. Is tree loss associated with cardiovascular-disease risk in the Women's Health Initiative? A natural experiment. *Health & Place* 36:1–7.

Dounias, E. and Ichikawa, M. 2017. Seasonal bushmeat hunger in Congo basin. *EcoHealth* 14:575–90.

Fabricant, D. S. and Farnsworth, N. R. 2001. The value of plants used in traditional medicine for drug discovery. *Environmental Health Perspectives* 109:69.

FAO 2016. *State of the world's forests 2016. Forests and agriculture: Land-use challenges and opportunities*. Rome. Available at: www.fao.org/publications/sofo/en/ (Accessed 4 February 2019).

Feller, I. C., Friess, D. A., Krauss, K. W. and Lewis, R. R. 2017. The state of the world's mangroves in the 21st century under climate change. *Hydrobiologia* 803:1–12.

Foster, D. G., Biggs, M. A., Ralph, L. et al. 2018. Socioeconomic outcomes of women who receive and women who are denied wanted abortions in the United States. *American Journal of Public Health* 108:407–13.

Fritz-Vietta, N. V. M. 2016. What can forest values tell us about human well-being? Insights from two biosphere reserves in Madagascar. *Landscape and Planning* 147:28–37.

Garland, J. J. 2018. Accident reporting and analysis in forestry: guidance on increasing the safety of forest work. *Forestry Working Paper No. 2*. Rome: FAO.

Goldemberg, J., Martinez-Gomez, J., Sagar, A. and Smith, K. R. 2018. Household air pollution, health, and climate change: Cleaning the air. *Environmental Research Letters* 13:030201.

Government of South Australia and WHO 2017. *Progressing the Sustainable Development Goals through Health in All Policies: Case studies from around the world*. Adelaide: Government of South Australia.

Guerra, C. A., Snow, R. W. and Hay, S. I. 2006. A global assessment of closed forests, deforestation and malaria risk. *Annals of Tropical Medicine and Parasitology* 100:189–204.

Huber, M., Knottnerus, J. A., Green, L. et al. 2011. How should we define health? *BMJ*:343:d4163.

Husain, I., Patierno, K., Zosa-Feranil, I. and Smith, R. 2016. *Fostering economic growth equity and resilience in sub-Saharan Africa: The role of family planning*. Washington DC: US Agency for International Development.

Ickowitz, A., Rowland, D., Powell, B., Agus Salim, M. and Sunderland, T. 2016. Forests, trees and micronutrient rich food consumption in Indonesia. *PLoS ONE* 11:e0154139.

Jenkins, A. P., Jupiter, S., Mueller, U. et al. 2016. Health at the sub-catchment scale: typhoid and its environmental determinants in Central Division, Fiji. *EcoHealth* 13:633–51.

Johnston, F. H., Henderson, S. B., Chen, Y. et al. 2012. Estimated global mortality attributable to smoke from landscape fires. *Environmental Health Perspectives* 120:695–701.

Jones, K. E., Patel, N. G., Levy, M. A. et al. 2008. Global trends in emerging infectious diseases. *Nature* 451:990.

King, M. 1990. Health is a sustainable state. *The Lancet* 336:664–67.

Kmietowicz, Z. 2017. Red meat consumption is linked to higher risk of death from most major causes. *BMJ* 357:j2241.

Koplitz, S. N., Mickley, L. J., Marlier, M. E. et al. 2016. Public health impacts of the severe haze in Equatorial Asia in September–October 2015: Demonstration of a new framework for informing fire management strategies to reduce downwind smoke exposure. *Environmental Research Letters* 11:094023.

Kowarik, I. and Von Der Lippe, M. 2018. Plant population success across urban ecosystems: A framework to inform biodiversity conservation in cities. *Journal of Applied Ecology* 55:2354–61

Landrigan, P. J., Fuller, R., Acosta, N. J. R. et al. 2018. The Lancet Commission on Pollution and Health. *The Lancet* 391:462–512.

Langbein, J. 2017. Firewood, smoke and respiratory diseases in developing countries – the neglected role of outdoor cooking. *PLoS ONE* 12(6):e0178631.

Lappé, F., Clapp, J., Anderson, M. and Al, E. 2013. *Framing Hunger: A Response to The State of Food Insecurity in the World 2012*. Cambridge, MA: Small Planet Institute. Available at: www.iatp.org/documents/framing-hunger-a-response-to-the-state-of-food-insecurity-in-the-world-2012 (Accessed 4 January 2019).

Livesley, S., Mcpherson, E. and Calfapietra, C. 2016. The urban forest and ecosystem services: Impacts on urban water, heat, and pollution cycles at the tree, street, and city scale. *Journal of Environmental Quality* 45:119–24.

Lopez, P. 2008. The subversive links between HIV/AIDS and the forest sector. In Colfer, C. J. P. (ed.) *Human health and forests: A global overview of issues, practice and policy*. London: Earthscan, pp. 221–38.

Maller, C., Townsend, M., Pryor, A., Brown, P. and St Leger, L. 2006. Healthy nature healthy people: 'Contact with nature' as an upstream health promotion intervention for populations. *Health Promotion International* 21:45–54.

Manes, F., Marando, F., Capotorti, G. et al. 2016. Regulating ecosystem services of forests in ten Italian metropolitan cities: Air quality improvement by PM10 and O_3 removal. *Ecological indicators* 67:425–40.

Masera, O. R., Bailis, R., Drigo, R., Ghilardi, A. and Ruiz-Mercado, I. 2015. Environmental burden of traditional bioenergy use. *Annual Review of Environment and Resources* 40:121–50.

McFarlane, R. A., Butler, C. D., Maynard, S., Cork, S. and Weinstein, P. 2018. Ecosystem-based translation of health research: Expanding frameworks for environmental health. *Australian and New Zealand Journal of Public Health* 42:437–40.

McFarlane, R. A., Sleigh, A. C. and Mc Michael, A. J. 2012. Synanthropy of wild mammals as a determinant of emerging infectious diseases in the Asian-Australasian region. *EcoHealth* 9:24–35.

Meadows, D., Meadows, D., Randers., J. and Behrens III, W. 1972. *The limits to growth*. New York: Universe Books.

Miettinen, J., Hooijer, A., Vernimmen, R., Liew, S. C. and Page, S. E. 2017. From carbon sink to carbon source: Extensive peat oxidation in insular Southeast Asia since 1990. *Environmental Research Letters* 12(2):024014.

Millennium Ecosystem Assessment 2005. *Ecosystems and human well-being: Biodiversity synthesis*, Washington, DC: World Resources Institute.

Moyle, W., Jones, C., Dwan, T. and Petrovich, T. 2017. Effectiveness of a virtual reality forest on people with dementia: A mixed methods pilot study. *The Gerontologist* 58:478–87.

Oglethorpe, J., Honzak, C. and Margoluis, C. 2008. *Healthy people, healthy ecosystems: A manual for integrating health and family planning into conservation projects*. Washington DC: WWF.

Ohenjo, N. O., Willis, R., Jackson, D., Nettleton, C., Good, K. and Mugarura, B. 2006. Health of indigenous people in Africa. *The Lancet* 367:1937–46.

O'Sullivan, J. N. 2017. Synergy between population policy, climate adaptation and mitigation. In Hossain, M., Hales, R. and Sarker, T. (eds.) *Pathways to a sustainable economy: Bridging the gap between Paris Climate Change Commitments and Net Zero Emission*. Springer International Publishing, pp. 103–25

Page, S. E., Siegert, F., Rieley, J. O. et al. 2002. The amount of carbon released from peat and forest fires in Indonesia during 1997. *Nature* 420:61.

Population Institute 2015. *Demographic vulnerability: Where population growth poses the greatest challenges*. Available at: www.populationinstitute.org/resources/reports/dvi/ (Accessed 4 February 2019).

Potter, J. D. 2017. Red and processed meat, and human and planetary health. *BMJ* 357:j2190.

Prescott, S. L., Millstein, R. A., Katzman, M. A. and Logan, A. C. 2016. Biodiversity, the human microbiome and mental health: Moving toward a new clinical ecology for the 21st century? *International Journal of Biodiversity* 2016:1–18.

Pulliam, J. R. C., Epstein, J. H., Dushoff, J. et al. 2012. Agricultural intensification, priming for persistence and the emergence of Nipah virus: A lethal bat-borne zoonosis. *Journal of the Royal Society Interface* 9:89–101.

Real, R., Olivero, J., Fa, J. E. et al. 2017. *The biogeographic basis of Ebola-virus disease outbreaks: A model for other zoonotic diseases?* Bogor: CIFOR.

Reyes-García, V., Guèze, M., Luz, A. C. et al. 2013. Evidence of traditional knowledge loss among a contemporary indigenous society. *Evolution and Human Behavior* 34:249–57.

Riahi, K., van Vuuren, D. P. and Kriegler, E. 2017. The shared socioeconomic pathways and their energy, land use and greenhouse gas emissions implications: An overview. *Global Environmental Change* 42:153–168.

Rook, G. A. 2013. Regulation of the immune system by biodiversity from the natural environment: An ecosystem service essential to health. *Proceedings of the National Academy of Sciences of the United States of America* 110:18360–67.

Searchinger, T., Hanson, C., Waite, R. et al. 2013. *Achieving replacement level fertility*. Working Paper, Installment 3 of Creating a sustainable food future. Washington, DC: World Resources Institute.

Seidl, R., Thom, D., Kautz, M. et al. 2017. Forest disturbances under climate change. *Nature Climate Change* 7:395.

Sforza, G. G. R. and Marinou, A. 2017. Hypersensitivity pneumonitis: A complex lung disease. *Clinical and Molecular Allergy* 15:6.

Sharp, P. M. and Hahn, B. H. 2011. Origins of HIV and the AIDS pandemic. *Cold Spring Harbor Perspectives in Medicine* 1(1):a006841.

Starbird, E., Norton, M. and Marcus, R. 2016. Investing in family planning: Key to achieving the Sustainable Development Goals. *Global Health: Science and Practice* 4:191–210.

Steffen, W., Richardson, K., Rockström, J. et al. 2015. Planetary boundaries: Guiding human development on a changing planet. *Science* 347:736–46.

Taylor, M. S., Wheeler, B. W., White, M. P., Economou, T. and Osborne, N. J. 2015. Research note: Urban street tree density and antidepressant prescription rates–a cross-sectional study in London, UK. *Landscape and Urban Planning* 136:174–79.

Tomita, A., Vandormael, A. M., Cuadros, D. et al. 2017. Green environment and incident depression in South Africa: A geospatial analysis and mental health implications in a resource-limited setting. *The Lancet Planetary Health* 1:e152–e162.

Tucker Lima, J. M., Vittor, A., Rifai, S. and Valle, D. 2017. Does deforestation promote or inhibit malaria transmission in the Amazon? A systematic literature review and critical appraisal of current evidence. *Philosophical Transactions of the Royal Society B: Biological Sciences* 372(1722):20160125.

Vasconcelos, P. F., Travassos Da Rosa, A., Rodrigues, S. G. et al. 2001. Inadequate management of natural ecosystem in the Brazilian Amazon region results in the emergence and reemergence of arboviruses. *Cadernos de Saúde Pública* 17:S155–S164.

Wallace, R. G., Gilbert, M., Wallace, R. et al. 2014. Did Ebola emerge in West Africa by a policy-driven phase change in agroecology? Ebola's social context. *Environment and Planning A* 46:2533–42.

Webb, K., Jennings, J. and Minovi, D. 2018. A community-based approach integrating conservation, livelihoods, and health care in Indonesian Borneo. *The Lancet Planetary Health* 2:S26.

WHO 2015. *Connecting global priorities: biodiversity and human health*. World Health Organization and Secretariat of the Convention on Biological Diversity. Geneva: WHO.

WHO 2018a. *World Malaria Report 2018*. Geneva: WHO.

WHO 2018b. *Waterborne disease related to unsafe water and sanitation*. Available at: www.who.int/sustainable-development/housing/health-risks/waterborne-disease/en/ (Accessed 5 December 2018).

WHO 2018c. *Global Health Estimates: Disease burden by cause, age, sex, by country and by region, 2000–2016*. Geneva, WHO. Available at: www.who.int/healthinfo/global_burden_disease/estimates/en/index1.html (Accessed 5 January 2018).

WHO 2019a. *Global Health Observatory data (HIV/AIDs)*. Available at: www.who.int/gho/hiv/en/ (Accessed 4 January 2019).

WHO 2019b. *WHO remains firmly committed to the principles set out in the preamble to the Constitution*. Available at: www.who.int/about/who-we-are/constitution (Accessed 28 July 2019).

Willis, K. J. and Petrokofsky, G. 2017. The natural capital of city trees. *Science* 356:374–76.

Wilson Center 2013. *Gorillas and Family Planning: At the Crossroads of Community Development and Conservation. Environmental Change and Security Program*. Available at: www.wilsoncenter.org/event/gorillas-and-family-planning-the-crossroads-community-development-and-conservation (Accessed 5 December 2018).

World Bank 2008. *Forests sourcebook: Practical guidance for sustaining forests in development cooperation* (English). Washington, DC: World Bank. Available at: http://documents.worldbank.org/curated/en/356731468155739082/Forests-sourcebook-practical-guidance-for-sustaining-forests-in-development-cooperation (Accessed 4 February 2019).

World Health Organisation 1986. *Ottawa Charter for Health Promotion: First International Conference on Health Promotion Ottawa, 21 November 1986*. Available at: www.healthpromotion.org.au/images/ottawa_charter_hp.pdf (Accessed 28 July 2019).

WWF 2018. *Palm oil*. Available at: www.worldwildlife.org/industries/palm-oil (Accessed 4 February 2019).

Wynes, S. and Nicholas, K. A. 2017. The climate mitigation gap: Education and government recommendations miss the most effective individual actions. *Environmental Research Letters* 12:074024.

Yang, J., Mcbride, J., Zhou, J. and Sun, Z. 2005. The urban forest in Beijing and its role in air pollution reduction. *Urban Forestry & Urban Greening* 3:65–78.

Chapter 4 SDG 4: Quality Education and Forests – 'The Golden Thread'

Peter Kanowski*, Dollie Yao* and Stephen Wyatt

Key Points

- Education is argued to be at the heart of sustainable development. SDG 4 aims to broaden and deepen education to people of all ages and expand its scope to a lifelong process spanning formal, non-formal and informal settings. SDG 4 emphasises quality of educational access, particularly for girls, women and marginalised groups.
- Education plays a foundational role in developing the knowledge, competencies and attitudes that foster pro-environment behaviour, yet this relationship is not simple or direct. Individual and community attitudes to the environment, their competencies in managing it and their sense of connectedness to nature are key factors in fostering pro-environmental behaviour.
- Pro-forest behaviours are those intended to benefit forests, or the components of forest ecosystems, in some way. There are many manifestations of and pathways to these behaviours.
- Encouraging and enabling pro-forest behaviours, in all their forms and contexts, is the basis of positive linkages between SDG 4 and forests.
- The formal, non-formal and informal elements of education systems have complementary and synergistic roles in facilitating pro-forest behaviours and outcomes.
- In these contexts, progress towards SDG 4 will benefit forests if education:

  1. Informs, encourages and enables pro-forest behaviour;
  2. Respects, nurtures and enables Indigenous and traditional knowledge;
  3. Promotes forest-related environment and sustainability education in each of formal, non-formal and informal settings;

* Lead authors.

4. Strengthens forest-related professional, technical and vocational education and training, and capacity development;
5. Capitalises on the power of both established and new media.

4.1 Introduction

This chapter explores the relationships between *SDG 4 Ensure inclusive and equitable quality education and promote lifelong learning opportunities for all*[1] and forests – specifically forest ecosystem services, forest-related livelihoods and human well-being. The 2030 Agenda for Sustainable Development understands education to be 'at the heart' of sustainable development (UNESCO et al. 2016: 24) and as 'the golden thread that runs through all 17 [SDGs]' (Thomson 2017). This is in part because SDG 4 conceives of education in very broad terms, encompassing formal, non-formal and informal elements over a person's lifetime (UNESCO 2016). The Incheon Declaration (UNESCO et al. 2016: 27), which articulates the rationale for SDG 4, argues that:

> Evidence of education's unmatched power to improve lives, particularly for girls and women, continues to accumulate. Education has a key role in eradicating poverty: it helps people obtain decent work, raises their incomes and generates productivity gains that fuel economic development. Education is the most powerful means of achieving gender equality, of enabling girls and women to fully participate socially and politically, and of empowering them economically.

The ambition articulated by SDG 4 builds on both the Millennium Development Goals and the *UN Decade of Education for Sustainable Development 2005–2014* (UNDESD)[2] (UNESCO 2016). The UNDESD drew from precursor initiatives and experiences in both environmental and sustainability education (Thomas 2017, UNESCO 2016 Table 1.2, Wals and Benavot 2017), including initiatives addressing forest-related topics such as biodiversity conservation, climate change and the green economy.

However, little of the research exploring the relationships between education and sustainable development focuses explicitly on forests; rather, as in the SDGs, forests are present as part of wider cultural, social and terrestrial landscapes (Buckler and Creech 2014; Introduction (this volume)). Nevertheless, inferences can be drawn for forests because many of the challenges to and

[1] Commonly abbreviated to 'Quality education'.

[2] Education for Sustainable Development (ESD) is also characterised as Education for Sustainability (EfS); see Buckler and Creech (2014).

opportunities for sustainable development are manifest in and for forests (UNEP 2011), and because experience in environmental and forest-related education informs education for sustainable development, and vice-versa (Gilless 2015, NEEF 2015).

We first overview education as conceived under SDG 4 (Section 4.2) and discuss how key contexts frame the relationships between SDG 4 and forests (Section 4.3). We then explore how progress towards SDG 4 targets might have impacts on forests and interact with other SDGs (Section 4.4), and ways to develop elements of SDG 4 to the benefit of forests (Section 4.5). Drawing on pro-environment behaviour concepts, we propose *pro-forest* behaviour as foundational to SDG 4 progress benefitting forests. Finally, we briefly note synergies between SDG 4 and other SDGs (Section 4.6) and offer concluding observations (Section 4.7).

4.2 SDG 4: Quality Education

SDG 4 is avowedly 'comprehensive, holistic, aspirational, ambitious and universal' (UNESCO et al. 2016: 24). It focuses on broadening and deepening education, to reach people of all backgrounds and all ages with effective and relevant learning (UNESCO et al. 2016); it expands the scope of education beyond the traditional focus of the formal schooling environment and years, to a lifelong process in a wide range of formal, informal and non-formal settings.

SDG 4 characterises *formal education* as education delivered in an organised system, occurring in institutions and leading to a recognised award. *Non-formal education* occurs in planned learning settings outside of formal systems, such as professional and capacity development. *Informal education,* which includes Indigenous knowledge, happens outside of organised programmes. It includes learnings from everyday activities and is increasingly facilitated by new technologies (UNESCO 2016, Figure 4.1).

The breadth of SDG 4 (Table 4.1) is reflected in its targets. Each is supported by specific indicators; the UN reports annual evaluations of progress towards targets (UN SDG Knowledge Platform 2019).

Figure 4.2 presents a stylised representation of SDG 4 targets: how they are situated and interact along axes represents the type of education and the stage of life. Some targets, such as those directed at gender equality and inclusivity, apply across the full spectrum of educational settings and stages; others, such as access to early education, are specific to stages. While only Target 4.7 ('Sustainable Development and Global Citizenship') of SDG 4 explicitly addresses sustainability, it is argued that progress towards other SDG 4 targets also underpins progress towards sustainability, and towards other SDGs (UNESCO 2016, Wals and Benavot 2017). This assertion is necessarily qualified, as 'education *can* make a critically important contribution to progress towards the SDGs, but this is by no means inevitable' (Sterling 2016: 211).

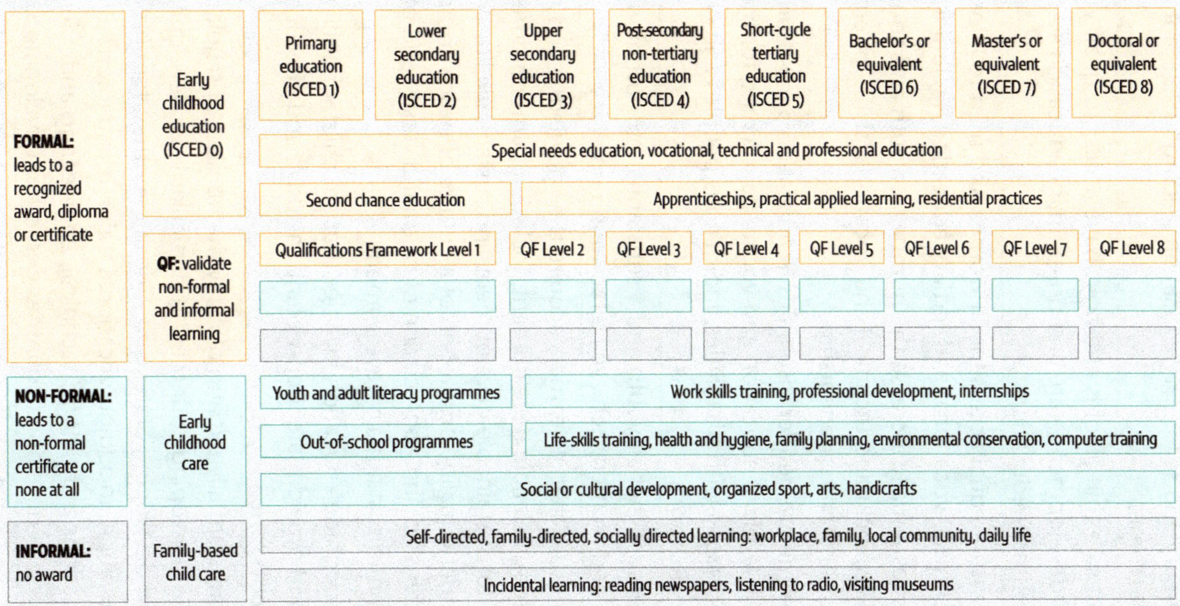

Source: GEM Report team.

Figure 4.1 General form, structure and elements of lifelong education, as conceived by the post-2015 development agenda. Source: UNESCO 2016: Figure 0.1. CC BY-SA 3.0 IGO [5077].

Table 4.1 SDG 4 targets

SDG 4 Targets	
4.1	By 2030, ensure that all girls and boys complete free, equitable and quality primary and secondary education leading to relevant and effective learning outcomes
4.2	By 2030, ensure that all girls and boys have access to quality early childhood development, care and pre-primary education so that they are ready for primary education
4.3	By 2030, ensure equal access for all women and men to affordable and quality technical, vocational and tertiary education, including university
4.4	By 2030, substantially increase the number of youth and adults who have relevant skills, including technical and vocational skills, for employment, decent jobs and entrepreneurship
4.5	By 2030, eliminate gender disparities in education and ensure equal access to all levels of education and vocational training for the vulnerable, including persons with disabilities, Indigenous peoples and children in vulnerable situations
4.6	By 2030, ensure that all youth and a substantial proportion of adults, both men and women, achieve literacy and numeracy
4.7	By 2030, ensure that all learners acquire the knowledge and skills needed to promote sustainable development, including, among others, through education for sustainable development and sustainable lifestyles, human rights, gender equality, promotion of a culture of peace and nonviolence, global citizenship and appreciation of cultural diversity and of culture's contribution to sustainable development
4.A	Build and upgrade education facilities that are child, disability and gender sensitive and provide safe, non-violent, inclusive and effective learning environments for all
4.B	By 2020, substantially expand globally the number of scholarships available to developing countries, in particular least developed countries, small island developing States and African countries, for enrolment in higher education, including vocational training and information and communications technology, technical, engineering and scientific programmes, in developed countries and other developing countries
4.C	By 2030, substantially increase the supply of qualified teachers, including through international cooperation for teacher training in developing countries, especially least developed countries and small island developing states

Source: UN SDG Knowledge Platform 2019

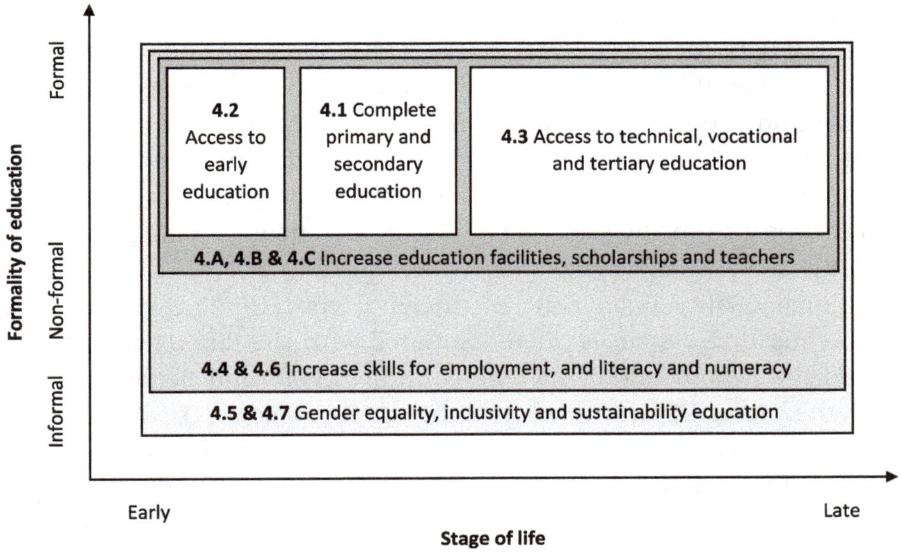

Figure 4.2 Stylised representation of the coverage of SDG 4 targets (numbered), in relation to formality of education and stage of life. (Source: Inspired by UNESCO 2016, Figure 0.1).

As we discuss further, this caveat applies to the impacts of SDG 4 on forests as much as it does to the SDGs more generally.

4.3 Contextual Conditions

We identify four sets of contexts for the adoption of SDG 4 and its impacts on forests: the quality and reach of education (Section 4.3.1); Environment and Sustainability Education (Section 4.3.2); the relationship between education and behavioural change (Section 4.3.3); and the nature of relationships between people, forests and pro-forest behaviour (Section 4.3.4).

4.3.1 The Education System

The characteristics of national education systems – often comprising subnational, both public and private components – provide a foundational context for SDG 4, particularly levels of access at different stages and quality at all stages. The education system encompasses all formal, non-formal and informal elements of education, and their 'life-wide contexts (family, school, community, workplace and so on)' (UNESCO et al. 2016: 30). It therefore includes the various forms of adult learning and education (UNESCO et al. 2016) and capacity building (Bloomfield et al. 2018) related to forests.

A central focus of SDG 4 is to improve access to education, particularly for school-aged children. Despite substantial progress over the past 50 years

(World Bank 2018), some 263 million children worldwide aged 6–17 do not attend school (UIS 2016). Currently, only 2 of 8 world regions have achieved the goal of universal lower-secondary education, and 3 are projected to not even achieve universal primary education by 2030 (UNESCO 2016). There are significant gender dimensions to access: worldwide, girls are twice as likely as boys to not start school, and rates of completing primary school are as low as 25 per cent for girls in the poorest families in low-income countries (World Bank 2018).

Educational quality is an issue of universal concern. The quality of a country's education system is often associated with the difference between richer and lower- and middle-income countries (Wals and Benavot 2017), although there is significant variation within these categories (UNESCO 2017, Figure 20.1). Richer countries are characterised as having well-developed and relatively well-funded formal education systems, with high rates of participation and effective learning through to post-secondary level; the situation in lower- and middle-income countries is typically the converse (WEF 2016a).

Consequently, the situation in many poorer countries' school systems has been described as a learning crisis, characterised by inadequate educational systems and schools (World Bank 2018). Unless these are addressed (for proposed actions WEF 2016a, World Bank 2018), neither the ambitions of SDG 4 in those countries nor the potential positive impacts on forests we discuss herein are likely to be realised.

4.3.2 Environment and Sustainability Education

The second context is that of environment and sustainability education (ESE; Sterling et al. 2017).[3] ESE was founded on promoting environmental literacy, which extends beyond simply knowledge of the environment to adoption and promotion of pro-environment behaviours (Leicht et al. 2018); it does so particularly by fostering relevant competencies and a sense of connectedness to the environment through experiential learning (NEEF 2015). ESE programmes specifically focused on forests have been developed to complement school curricula in many countries (e.g. Australia: Forest Education Foundation 2018; Scotland: OWL Scotland 2018; the USA: Project Learning Tree 2018), often beginning at the pre-school level, e.g. European forest kindergartens (Gregory 2017).

The UNDESD extended the environmental literacy concept to sustainability more broadly, seeking – in the SDG context – to integrate education

[3] ESE includes 'Education for Sustainable Development' (ESD) and 'Education for Sustainability' (EfS) (UNESCO 2016).

into sustainable development, and vice-versa (Leicht et al. 2018), and enable transformative societal change (UNESCO 2014). There was global progress in developing and implementing education for sustainable development (ESD) during the UNDESD (Buckler and Creech 2014), but there is significant variation in SDG 4 indicators among otherwise comparable countries (UNESCO 2017).

4.3.3 Education and Pro-Environment Behavioural Change

Quality education plays a fundamental role in achieving sustainability globally by fostering pro-environment behaviour (UNESCO 2016) – i.e. 'behaviour that is undertaken with the intention to [positively] change the environment' (Stern 2000: 408). For example, holistic pedagogical practices that complement immersive environment experiences with pre-experience preparation and post-experience follow-up are more likely to foster pro-environment attitudes than less holistic approaches (Stern et al. 2014). The pathways through which education exerts influence are not simple, linear or direct. Behaviour is determined by a suite of complex and interconnected elements that vary contextually; it is easy to oversimplify these elements and overestimate their causality (Heimlich 2010, Steg and Vlek 2009). With these caveats, we summarise the most influential elements in terms of Kollmuss and Agyeman's (2002) categorisation of demographic, external and internal factors in Figure 4.3, and discuss them below.

Education and gender are the most influential demographic factors (Kollmuss and Agyeman 2002). Increasing the duration, intensity or quality of education increases pro-environment behaviour (Zsóka et al. 2013). In many (but not all) contexts (Villamor et al. 2014), women are more likely

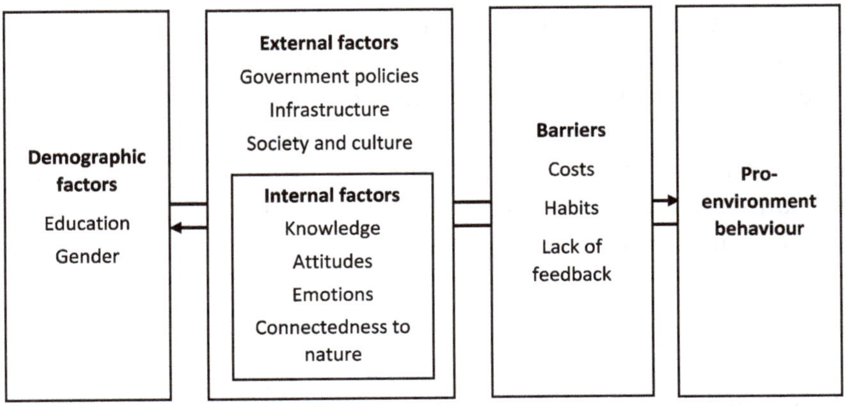

Figure 4.3 Simplified model of factors shaping pro-environment behaviour. Source: Adapted from Kollmuss and Agyeman 2002, Figure 7.

than men to empathise with environmental causes and behave accordingly (Hunter et al. 2004).

External factors including infrastructure, policies and social and cultural factors form the context in which behavioural decisions are made. Infrastructure (e.g. the accessibility of recycling bins) enables or hinders pro-environment behaviour (Freed 2018). Government policies (e.g. taxes) can successfully deter certain behaviours, such as plastic bag usage (Convery et al. 2007). Social and cultural norms are particularly powerful because they set standards, e.g. in relation to energy and water-conserving behaviour (Reese et al. 2013).

Internal factors comprise various psychological factors, notably knowledge, attitudes, emotions and habits (Kollmuss and Agyeman 2002). These are often the target of education-based interventions (Stern et al. 2014). Knowledge, including of behavioural options to achieve environmental outcomes (Frick et al. 2004), is foundational but not solely influential. Positive attitudes and emotions towards the environment are relatively strong determinants of pro-environment behaviour (Roczen et al. 2014), particularly a sense of *connectedness to nature* (Otto and Pensini 2017).

Pro-environment behaviour is inhibited by various barriers, conceptualised by Diekmann and Preisendörfer (2003) in terms of cost. Low-cost behaviours (i.e. relatively easy or inexpensive, such as using a recycling bin) are more likely to be performed than high-cost behaviours, such as using public transport instead of a car (Boyes and Stanisstreet 2012). Many behaviours are habits – learned routines performed without conscious intention – and are challenging to change (Steg and Vlek 2009). Moreover, pro-environment behaviours can wane without positive feedback, such as a sense of satisfaction or social approval (Kollmuss and Agyeman 2002).

Furthermore, the benefits pro-environment behaviours convey can be overshadowed by the overall impact of higher-consumption lifestyles. Behaviours are therefore unlikely to be transformative in isolation. For example, environmentally conscious people who recycle can have a similar overall ecological footprint to their less environmentally conscious counterparts who do not (Csutora 2012).

In summary, education is an important foundation for pro-environment behaviour, but such behaviour depends on a suite of complex, interconnected and contextual factors. Fostering behavioural change requires strategies developed thoughtfully in this light.

4.3.4 Relationships between People, Forests and Pro-Forest Behaviours

The fourth context is the diverse relationships between people and forests. Broadly, we characterise these at individual, household and community levels; we distinguish those dependent directly or indirectly on forests for livelihoods (e.g. forest-dwelling people or forestry sector employment, respectively) from those with less-dependent relationships (e.g. most urban residents). There are also socially and culturally constructed relationships, which differ, for example, between Indigenous and non-Indigenous peoples and their environments (Tengö et al. 2017); over time and between actors in a particular country (Dargavel 1995, Hull 2011); or between societies in forest-rich compared to forest-poor countries (Sands 2013).

These different relationships are recognised in various ways: for example, through major groups in international intergovernmental processes (e.g. the UN Forum on Forests); as stakeholder groups in international or national multi-stakeholder platforms (e.g. The Forest Dialogue and Brazilian Diálogo Florestal, respectively); or in relevant principles and criteria under mechanisms promoting sustainable forest management (SFM), such as forest certification systems (e.g. FSC and PEFC[4]) or SFM processes (e.g. the Montreal Process).

Attitudes and behaviours towards forests are shaped and mediated by a range of internal and external factors. We suggest it is helpful to focus on pro-forest behaviours, which we define by adapting Stern's (2000) definition of pro-environment behaviours as those that are intended to benefit forests, or the components of forest ecosystems, in some way. We recognise that there are many pathways to and manifestations of pro-forest behaviour (Beery and Wolf-Watz 2014).

We suggest that pro-forest behaviours are evident and can be fostered across the full spectrum of people–forest relationships for natural and planted forests in urban and rural landscapes. They may manifest in forest protection and conservation activities undertaken by individuals and groups, ranging from Indigenous peoples to environmental and forestry agencies and corporations; in SFM implementation by Indigenous and local communities, private landowners and public forest managers; in various forms of forest and landscape restoration; and in product choices made by consumers. We argue that education has a key (albeit complex) role in fostering pro-forest behaviours.

4.4 Possible Impacts of Progress towards SDG 4 on Forests

SDG 4 is anticipated to have a range of societal benefits, as discussed in Section 4.1. Progress towards SDG 4 may affect forests in various ways, which we categorise (from general to specific) under the following overlapping outcomes:

---

[4] Forest Stewardship Council (FSC), The Programme for the Endorsement of Forest Certification (PEFC).

1. improved education, in the broad sense intended by SDG 4, for individuals, communities and societies (Targets 4.1, 2, 3 and 6);

2. greater equality and inclusiveness, for women and vulnerable people, including Indigenous peoples (Target 4.5);

3. greater knowledge about and skills for sustainable development (Target 4.7);

4. employment associated with forests and the forest-based economy (Target 4.4);

5. post-secondary education relevant to the environment and sustainability, and professional, technical and vocational education and training specifically relevant to forests (Target 4.3).

4.4.1 Improved Education

Progress towards SDG 4, at levels from the most foundational and general (e.g. improved literacy and numeracy) to the more specific and targeted (e.g. increased numbers of qualified teachers), is expected to lead to benefits at a range of scales, from those of the individual and family to those of community and society (Table 4.2). Multinational surveys of representative

Table 4.2 Generalised examples of benefits of education

	Individual/family	Community/society
Monetary	Higher probability of employment Greater productivity Higher earnings Reduced poverty	Higher productivity More rapid economic growth Poverty reduction Long-run development
Non-monetary	Better health Improved education and health of children/family Greater resilience and adaptability More engaged citizenship Better choices Greater life satisfaction	Increased social mobility Better-functioning institutions/service delivery Higher levels of civic engagement Greater social cohesion Reduced negative externalities

Source: Adapted from World Bank 2018, Table 1.1.

national adult populations since 1993 demonstrate both that the aggregate level of environmental concern increases with national wealth (as measured by GDP), and that people with higher levels of formal education are more likely to express concern for the environment, regardless of personal wealth, political preference or individual characteristics (Franzen and Vogl 2013). While there are obvious caveats to these results – they are limited to middle- and high-income countries (Franzen and Vogl 2013) and are unlikely to adequately sample the views of groups for whom forests have particular significance, such as Indigenous peoples – they nevertheless suggest a strong role for education in raising environmental awareness. However, to adapt Sterling's (2016) caution: while education can contribute to pro-forest behaviour, this not guaranteed.

As discussed in Sections 4.3.3 and 4.3.4, such awareness and concerns may foster pro-forest actions – e.g. landowners supporting biodiversity conservation (Drescher et al. 2017) or individual awareness, mitigation and adaptation regarding climate change (Wamsler et al. 2012). More educated individuals are more likely to follow up environmental concerns with activism to advance a pro-environment political agenda (Clery and Rhead 2013). However, specific outcomes for forests from educational improvements envisaged by SDG 4 depend on complex interactions, across and within levels of social organisation and individual and group values, worldviews, norms and behaviours (Drescher et al. 2017).

4.4.2 Greater Equality and Inclusiveness

Improving equality of access to and inclusivity in education has significant benefits for disadvantaged groups, and potentially for forests.

ADDRESSING GENDER DISPARITY

Gender disparity is manifest in most societies, but is most marked in terms of educational access and participation in low-income countries and regions, where the out-of-school population is disproportionately high (UIS 2017). Correspondingly, the general consequences for forests of addressing this disparity differ between lower-income and higher-income societies.

LOWER-INCOME SOCIETIES

Improving participation by women and girls in education is central to the goal of improving their lives, the lives of the families and communities of which they are members, and educational outcomes generally:

> Better educated women tend to be healthier, participate more in the formal labor market, earn higher incomes, have fewer children, marry at a later age, and enable better health care and education

> for their children, should they choose to become mothers. All these factors combined can help lift households, communities, and nations out of poverty. (World Bank 2017)

Women with fewer children have more time to engage in productive work or education, which reduces their preferred family size and helps normalise educational attainment for women (Colfer et al. 2008). While population growth, particularly in poorer countries, usually increases direct pressures on forests, this pressure can be mediated by greater human development (Jha and Bawa 2006), to which education is a fundamental contributing factor (UNDP 2018).

In general, 'increases in women's incomes have greater impacts on food, health and education expenditure and therefore on overall household well-being than increases in men's incomes' (FAO 2013: 9). As an additional year of schooling can increase a woman's earnings by 10–20 per cent (UN Women 2012), women's education offers a more direct pathway to improving household well-being, and also diminishes – at least in principle – the need for household members, typically men (Sunderland 2014), to access forests for commercial products at unsustainable rates.

Improved literacy, education and practical skills related to income generation or employment increase women's social status and self-confidence, thereby increasing the effectiveness of their participation in forest management through organisations such as community forest user groups (Agarwal 2010, Coleman and Mwangi 2013, FAO 2013). Women's participation in decision-making can reduce gender-based conflict because it leads to more equitable access to forests (Coleman and Mwangi 2013). Furthermore, women's participation can lead to greater forest conservation and restoration through a range of direct and indirect pathways (Agarwal 2009).

The importance of empowering women in relation to forest and tree management is amplified by the *feminisation* of rural communities and economies globally, as men migrate for or in search of employment elsewhere (Alston et al. 2018, Mukhamedova and Wegerich 2018, Tamang et al. 2014).

HIGHER-INCOME SOCIETIES

Gender disparity also remains significant in most rich countries. As the World Economic Forum (WEF 2016b: 1) notes, 'Female talent remains one of the most underutilised resources, so in addition to the moral case for gender equality, which has mostly been won, there is a business case'. In nearly 100 countries, women make up most university enrolments, but overarching cultural and societal factors result in skews against women in Science, Technology, Engineering and Medicine (STEM) fields, where women comprise only 32 per cent of graduates (WEF 2016b). This impacts on forest-related professions, as well as others.

In forest-sector contexts specifically, gender gaps persist (Brown et al. 2010, Eriksen et al. 2016, Hansen et al. 2016). However, as Lawrence et al. (2017: 113–14) note:

> Female leadership potential has been recently emphasised as a source of untapped potential in forest industry. ... Higher diversity is also associated with better sector image, retention of much required talent pool, innovation and better reflection of customer and stakeholder needs, all of which are significant sources of market and financial benefits over the longer run.

Consequently, addressing the educational, employment and societal constraints that limit women's participation in the forest-sector workforce can be expected to deliver a range of positive outcomes: for individuals and organisations, for innovation and workforce capacity in forest management and forest-based value chains, and for the rural and regional economies on which these value chains are typically embedded.

ADDRESSING INDIGENOUS RIGHTS, INTERESTS AND DISADVANTAGES

The importance of access to appropriate education for Indigenous peoples is now well-established internationally (e.g. UNCED Forest Principles 5a and 12d, UN 1992), but implementation remains challenging. Article 14 of the UN Declaration on the Rights of Indigenous Peoples (UNDRIP 2007) asserts that Indigenous people have a right to control education systems so they are culturally appropriate and in their own language. The relevance of Indigenous knowledge (IK)[5] is increasingly recognised in contemporary forest management (Parrotta and Trosper 2012) for the benefits it delivers to both Indigenous and wider communities, and for SFM (Ens et al. 2012, Lyver et al. 2017).

IK is typically rooted in distinct ontologies, incorporating cultural values and norms:

> Knowledge is not secular. It is a process derived from creation, and as such, it has a sacred purpose. It is inherent in and connected to all of nature, to its creatures, and to human existence ... Traditions, ceremonies, and daily observations are all integral parts of the learning process. They are spirit-connecting processes that enable the gifts, visions, and spirits to emerge in each person. (Battiste 2002: 14–15)

Consequently, IK is inherently place- and context-specific; it is often privileged, with restrictions on knowledge sharing and learning (e.g. to elders,

[5] Also referred to as traditional forest-related knowledge, TFRK, and other terms.

men or women), typically intended to ensure that those who hold knowledge appreciate how this knowledge may and should be used. While the foundations and perspectives of IK and modern Western science differ (Fenstad et al. 2002), it is important to appreciate the complementary relationship between various tenets of traditional knowledge and those of ecological sciences, and the value of learning from both realms (Parrotta and Trosper 2012). Education systems for both Indigenous and non-Indigenous peoples can capitalise on this complementarity and its synergies, while respecting differences.

The importance of culture and values, and of oral and experiential teaching and learning, can mean that IK education is seen principally in terms of informal and non-formal approaches, in a family or a community setting. However, Indigenous education also occurs in other modes, including the most formal and advanced (Allen and Krogman 2013, Dockry et al. 2016, Hoagland et al. 2017), and practice-based co-learning through co-management (Ens et al. 2012). Many such examples demonstrate how awareness, understanding and respect for IK can be integrated into both formal and non-formal education about forests; and how forest-related IK can contribute to enhancing forest management.

More broadly, recognition of the validity and utility of IK and of Indigenous education systems can empower Indigenous communities seeking an enhanced role in forest management, or the recognition of their traditional rights, e.g. in post-colonial societies such as Canada, Australia and New Zealand (Wyatt et al. 2010). Thus, an important outcome of advancing SDG 4 ambitions in terms that are respectful and inclusive of IK can be the greater empowerment of Indigenous peoples in relation to their rights and interests in forests (Bulkan 2017).

4.4.3 Greater Knowledge and Skills

ESD begins from the premise that 'certain knowledge and skills promote sustainability more than others' and aims to create empowered and responsible global citizens (UNESCO 2016: 11). Education generally, and that for sustainability specifically (EfS), support sustainable development in two ways: through knowledge and skills that foster values and behavioural change; and through building 'greater agency to address complex sustainability challenges' (UNESCO 2016: 11). The former is most relevant in addressing issues about which there is a high level of agreement, and the latter where there is uncertainty and contingency (UNESCO 2016). Forest-related examples of such issues might be, respectively, the significance of biodiversity loss and the best means to address trade-offs between conservation and development.

In the EfS context specifically, Wals and Benavot (2017) characterise these approaches as instrumental and emancipatory, respectively. Instrumental education communicates the knowledge and skills that foster sustainable behaviours; emancipatory education operates on a deeper, value-based level to foster independent, reflective, responsible behaviours (Wals and Benavot 2017). They suggest three general ways education supports sustainable development: recognising and drawing on diverse viewpoints, including IK; emphasising learning across disciplinary and societal boundaries; and helping learners acquire new life skills and competencies, and interpret and apply them in more holistic and systemic understandings of complex realities (Wals and Benavot 2017).

Enhancing environmental and sustainability literacy fosters commitment and action, enables the identification of environmental issues and the capability to respond and provides the agency required to tackle wicked problems and facilitate transformative change (UNESCO 2016). Developing such literacy is the premise of established environmental education programmes (NEEF 2015), including those focused specifically on forests, which provide platforms for greater pro-forest thinking and decision-making at all levels of social organisation, from individual to international.

4.4.4 Employment and the Forest-Based Economy

Forest management, production systems and value chains are an important source of employment, particularly for forest-dependent and other rural communities, employing some 54 million people formally and informally worldwide (World Bank 2016). In a world in which the importance and value of the green economy (UNEP 2011) and bioeconomy (Lawrence et al. 2017) are growing, forest-related employment should expand far beyond traditional roles associated with management of forests and harvesting and processing of wood and non-wood products. A much wider knowledge and skill base will be required for the sustainable management of forests and trees, including those on farms and in cities, for the full range of ecosystem goods and services, and for the continuing development and success of innovative and sustainable forest industries, on both small and large scales (Macqueen et al. 2018, Panwar et al. 2016, Sanchez Badini et al. 2018).

While there remain some forest-related roles that require little formal education, including those for which high levels of informal Indigenous and local knowledge are particularly valuable, the knowledge and skill requirements for forest-sector employment continue to evolve (Brandth and Haugen 2000, Lawrence et al. 2017): away from simply labour-based and towards more knowledge-based skills requiring post-secondary education, including advanced mechanical and information technology skills, entrepreneurship and business skills (e.g. ecotourism) and high-level communication, organisation and people management skills (Lawrence et al. 2017). As noted in Section 4.4.2, many of these skills are associated with women, reiterating the importance of gender equality in access to relevant education and training.

The evolution of forest-related employment needs to be supported and facilitated by equitable and affordable access to relevant education and training, particularly in formal and non-formal contexts. Consistent with the breadth of SDG 4, such education and training needs to be broadly conceived and accessible, to reach diverse groups:

- Members of Indigenous communities acquiring higher levels of technical, specialist and business knowledge to complement their traditional knowledge, to better participate in green economy roles, such as those created by ecosystem services markets or ecotourism (Altman and Kerins 2012, Ens et al. 2012, Russell-Smith et al. 2013, UNEP 2011);
- Members of rural communities acquiring higher levels of technical, specialist and business knowledge, to allow them to better participate in or capitalise on forestry sector employment in various ways (Hiedanpää and Salo 2017, Mayett-Moreno et al. 2017, Sanchez Badini et al. 2018);
- Public, private and community organisations relying on technically and vocationally skilled staff to respond to the diversification of the forest-based economy, which is creating the need for new knowledge and skill sets in a wider array of organisations (UNEP 2011);
- Tertiary students in a wide range of forest-sector-related programmes, whose participation reflects a growing interest in enhanced degree programmes and leads to a stronger and more diverse professional workforce (Gilless 2015).

4.4.5 Professional, Technical and Vocational Education and Training

The evolution of scientific professional and technical forestry education has been described elsewhere (Innes and Ward 2010, Kanowski 2001). Notwithstanding its strengths in developing cadres of well-educated and trained professional and technical foresters, the limitations of this model are apparent – e.g. in terms of its privileging of particular interests (Ojha et al. 2009), or its focus on only some elements of forested landscapes and on only some of the diverse skills required to manage them in dynamic social and landscape contexts (Gilless 2015, Hull 2011).

Consequently, both professional and technical education and training relevant to forests have changed significantly in the late twentieth and early twenty-first centuries (AP-FECM 2018, Rekola et al. 2017, Temu and Kiwia 2008). University forestry curricula have been broadened, strengthening the social sciences, humanities and interdisciplinarity; programmes have become more inclusive, and more networked and internationalised; and student communities have become more diverse (Gilless 2015). Topic areas that were once marginal, such as agroforestry or community forestry, are now mainstream, and the focus of specific institutions and programmes as well as elements of broader curricula (RECOFTC 2018, Yayé et al. 2015). International collaboration seeks to strengthen forest-related education networks, student mobility and curricula (Kanowski 2015, Rekola et al. 2017, Temu and Kiwia 2008, Yayé et al. 2015), as forestry education continues to evolve and adapt to ensure its relevance. However, challenges remain in aligning curricula and skills sought by employers, particularly in terms of the balance and relevance of generic and technical skills (Ramcilovic-Suominen et al. 2016).

These challenges are paralleled at the technical and vocational levels, which are historically underdeveloped in many lower-income countries and for the natural resource sectors (Robinson-Pant 2016, UNEP 2017), and which must contend with perceptions, particularly among youth, that rural-based occupations and work are those of last resort (Robinson-Pant 2016). However, as Robinson-Pant (2016) and Lawrence et al. (2017) note for the agriculture and forestry sectors, respectively, there are significant opportunities to improve household livelihoods, rural communities' resilience, and environmental outcomes from more effective technical and vocational education that is also more inclusive of women, who now comprise a much greater proportion of farmers and rural workers. While green knowledge and skills are foundational in technical and vocational education and training (TVET) for rural work (INRULED 2012), they should also be embedded in TVET more widely (UNESCO-UNEVOC 2017).

Such challenges also extend into the arena of non-formal education, in its many forms relevant to forests. The continuing decline of traditional public extension services in many countries (Mogues et al. 2015) has fostered new approaches, including those capitalising on the rapid development and reach of information and communication technologies (Sagor et al. 2014), and more community-based approaches (Catacutan et al. 2015, Reid 2017). These activities are increasingly seen in the context of broader knowledge and innovation systems (Lubell et al. 2014), based on capacity development for co-production of useable knowledge (Clark et al. 2016), in which boundary workers may play critical roles. Forest-related examples illustrative of the diversity of actors and approaches include the UK Sylva Foundation's *myForest* and Forest Schools initiatives (Sylva Foundation 2018), which facilitate forest information and knowledge exchange for landowners and schools, respectively; structured multi-stakeholder dialogue processes, such as Brazil's Forest

Dialogue (Diálogo Florestal 2018); and the research partnerships and outputs of international public good research centres such as the World Agroforestry Centre (World Agroforestry Centre 2018) and international forest-related initiatives such as forest and landscape restoration (Chazdon et al. 2017).

4.5 Advancing the Ambitions of SDG 4 Relevant to Forests

While the ambitions of SDG 4 are global, transcending countries and sectors, many forest-related actors have particular interest in fostering synergies between SDG 4 and forests. We propose five priorities that forest-related actors might seek to advance in this context.

4.5.1 Encouraging and Enabling Pro-Forest Behaviour

It is evident that pro-forest behaviour at various levels of social organisation, from the individual to the international, derives from a complex combination of factors that are both internal and external to the individual and the community. Education that builds and reinforces understanding and knowledge of forests and competencies in forest management, and that helps individuals and communities to feel or stay connected to forests, has a foundational role in fostering or sustaining pro-forest attitudes and behaviours. The formal, non-formal and informal elements of education systems have

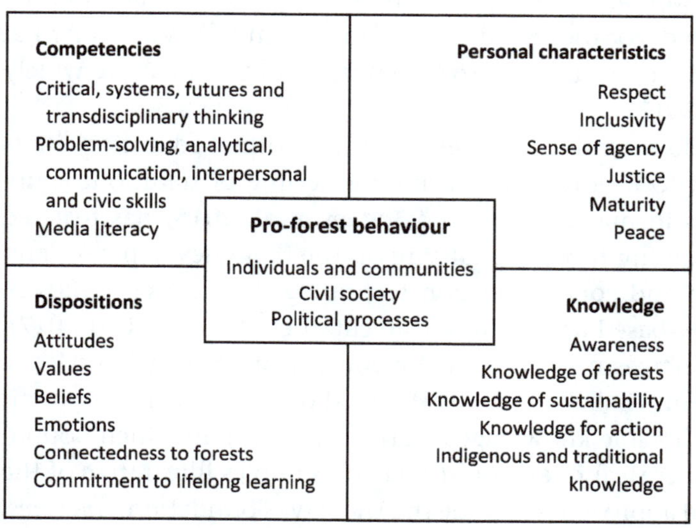

Figure 4.4 Outcomes of forest-related education contributing to pro-forest behaviour.
Source: Adapted from Ardoin et al. 2017, Lozano et al. 2017 and University of Florida 2017.

complementary and synergistic roles in facilitating these outcomes. Figure 4.4 draws from generic models of learning outcomes from environmental education (Ardoin et al. 2017), the ESD literature (Lozano et al. 2017) and sustainable development curricula (University of Florida 2017) to present a stylised model of educational outcomes that contribute to pro-forest behaviour.

A central goal of education about forests should be to provide opportunities and enable experiences that help individuals develop a sense of connectedness to forests, or that sustain and enrich connectedness that already exists. Strategies to achieve this need to be diverse, to reflect the diversity of ways in which people learn and the diversity of their backgrounds and circumstances (Collins and Bilge 2016), and will obviously differ between, for example, Indigenous communities in which individuals have intimate cultural and material connections to their forests, rural communities in which connectedness to the forested environment is part of daily life, and urban communities for whom the most common experience of forests is of urban and peri-urban settings. They will differ in their form and elements between higher-income and lower-income countries. At their core, these strategies share the common purpose of fostering a personal sense of connectedness to forests, as the basis for fostering pro-forest attitudes and behaviour.

It is evident that these behaviours are most likely to be expressed when external actors and factors enable and support pro-forest actions. Such enabling and support measures are embedded or implicit in concepts such as a landscape approach (Sayer et al. 2013), forest and landscape restoration (Chazdon et al. 2017), locally controlled forestry (Elson 2012) or biophilic cities (Beatley and Newman 2013). These principles need to be translated into policies, processes and outcomes that recognise and respect different forms of knowledge and enable partnerships for its use: e.g. between state management agencies, researchers and Indigenous and local communities (Fisher et al. 2017); between investors and traditional forest owners (Elson 2012); or between local authorities and communities in urban environments (Mattijsssen et al. 2017).

Encouraging and enabling pro-forest behaviour, in whatever context and form, is the basis of connecting SDG 4 and forests. It underpins each of the following areas of activity.

4.5.2 Respecting, Nurturing and Enabling Indigenous and Traditional Knowledge

The standing and value of Indigenous and other forms of traditional knowledge for forests and their management are now well-recognised, as are both the epistemological differences and potential complementarities with modern scientific knowledge (Mistry et al. 2016). Forest management that draws on both Indigenous and scientific knowledge can explore a wider range of options than that limited to either paradigm (Mistry et al. 2016, Parrotta et al. 2009), and can be an important element of empowering Indigenous communities (Altman and Kerins 2012, Bulkan 2017, Tengö et al. 2017).

Capitalising on Indigenous and other forms of traditional knowledge to the benefit of Indigenous and local communities, and of society more widely, faces a range of challenges. These include the privileging of scientific knowledge in environmental governance and management, the restrictions on access to some elements of IK to specific knowledge holders, the loss of Indigenous and traditional knowledge due to loss of agency and to a range of societal forces, and challenges of integrating elements of traditional and scientific knowledge in contemporary policy and management contexts (Mistry et al. 2016, Tengö et al. 2017). Nevertheless, a diverse range of examples (Bulkan 2017, Parrotta and Trosper 2012) and policy development at international and national levels (Tengö et al. 2017) illustrate how these challenges can be addressed.

The common theme that underlies these examples is one of respect by other parties for Indigenous and traditional knowledge, and of a range of measures to nurture this knowledge and enable its use. Fundamentally, governments and other actors have to create the space in knowledge systems and in policy and decision processes for IK (Hill et al. 2012, Tengö et al. 2017); and, where Indigenous people have lost agency and standing, as in many settler societies, foster and support the engagement of Indigenous peoples in those processes. Non-governmental and community-based organisations and forestry businesses can play significant enabling roles in these diverse contexts (Chhetri et al. 2013, Nikolakis and Nelson 2015, Waller and Reo 2018).

Commitment by non-Indigenous actors to respecting, nurturing and enabling Indigenous and traditional knowledge benefits both Indigenous and non-Indigenous communities and the relations between them, and should lead to more adapted and sustainable forest management.

4.5.3 Promoting Forest-Related ESE in Formal, Non-Formal and Informal Settings

FORMAL

Forest-related ESE is already well-established in many formal education systems, at pre-school, primary and secondary levels. While the UNDESD fostered progress for ESE curriculum integration globally, including the institutionalisation of ESE in many countries, teacher capacity and curriculum implementation remain limited in others (UNESCO 2014). For example, in some countries, ESE has been de-emphasised due to a focus on content and skills relevant to economic growth, and greater emphasis on standardised curricula and testing (McBeath et al. 2016, Witoszek 2018).

Immersive and experiential forest-related ESE is especially effective in fostering pro-forest behaviour and delivers a range of wider learning and behavioural benefits (Project Learning Tree 2018), and so complements and extends classroom-based learning. Therefore, programmes that connect children to forests from the outset of their formal education (we have noted a small number of the many examples in preceding sections), and those that engage tertiary students similarly in a variety of settings (Hill et al. 2008, van Wynsberghe and Moore 2015), are most likely to enable pro-forest behaviours. ESE principles are reinforced and demonstrated by whole-of-institution approaches that embed sustainability into the facilities and operations of the learning environment (UNESCO-UNEVOC 2017) – a goal to which many institutions are already committed (University Alliance for Sustainability 2018).

NON-FORMAL

Non-formal forest-related ESE is an essential complement to formal approaches in fostering pro-forest behaviour. For example, businesses are seeking training and professional development through a range of actors to improve their sustainability performance (UNESCO 2014). Non-formal modes of education can be more effective than formal modes in reaching marginalised groups, such as women forest owners who have little agency in a traditionally male domain (Redmore and Tynon 2011). Experiential co-learning approaches (e.g. farmer field schools) can be effective in many contexts, particularly for those who are resource-poor, such as smallholder farmers and tree growers, and can facilitate both scaling up and fostering local adaptation (FAO 2017). In contrast, eco-tourists – a resource-rich group – are demonstrably willing to pay for non-formal ESE (Walter 2009). In urban environments, community engagement programmes offer non-formal ESE that foster and support pro-forest behaviour: e.g. Chicago's long-established Treekeepers (Dwyer and Schroeder 1994) or Singapore's Community in Bloom and Community in Nature (Er 2018). Non-formal education can also be an effective and targeted way to reach groups on the margins of society; e.g. ESE delivered through the USA's Sustainability in Prisons Project reduced recidivism (LeRoy et al. 2012).

INFORMAL

Informal education is widely encompassing and ubiquitous, and therefore also important for forest-related ESE, as the following examples illustrate. Informal learning frequently occurs in social settings when knowledge is transferred through social networks; for example, children learn pro-environment behaviour directly and indirectly from their parents (Ando et al. 2015), and, conversely, environment-related learning from school can reach parents through their children (Eilam and Trop 2012). Children also learn from role models, especially adults, whose behaviour instils environmental literacy and responsibility and helps develop relevant character and leadership traits (Stern et al. 2018).

NGOs and community groups are active informal educators and can promote pro-forest knowledge and behaviours through awareness campaigns, such as those directed at reducing deforestation or responsible consumption, or engagement programmes such as those for community-based forest restoration (Boyer-Rechlin 2010). Online communities can be effective means of improving people's scientific literacy and increasing pro-environment behaviour (Robelia et al. 2011). Researchers and knowledge institutions can engage, educate and learn from the public through citizen–science projects (Bonney et al. 2014).

In urban environments, parks and green spaces are important sites for learning about trees, particularly for children, whose play and interaction with nature not only develops appreciation for the environment but also improves their cognitive abilities and physical growth (Clements 2004). However, they can be equally important for adults, especially those with low levels of environmental knowledge. Similarly, community-based activities, such as community gardens or environment groups, are an important vehicle for informal knowledge and skills development and exchange (Krasny and Tidball 2009). The increasing body of evidence of positive relationships between people's physical and mental health and various forms of experience of trees and forests (Dzhambov et al. 2018), and of feelings of well-being associated with exposure to wood in buildings compared to harder materials (Strobel et al. 2017), also offer potentially powerful means of informal learning about the value of forests and forest products, as the basis for pro-forest behaviours.

MUTUALLY REINFORCING FORMAL, NON-FORMAL AND INFORMAL EDUCATION ABOUT FORESTS

There is strong circumstantial evidence that learning about and experiencing forests – in informal, non-formal and formal settings – forms the foundations of pro-forest behaviour. This suggests that, from a forest perspective, SDG 4 implementation should focus on promoting forest-related content and opportunities to experience trees, forests and forest products. New technologies can assist this in a variety of ways, complementing established structures and modes. For example, social media can support self-regulated, on-demand learning through personal learning environments (PLEs); these are personalised learner-driven platforms to aggregate, create and share knowledge using digital tools, and so help to bridge formal and informal learning (Dabbagh and Kitsantas 2012). The highly autonomous nature of PLEs synergises well with other forms of learning, such as lifelong and workplace learning, and so this approach is widely applicable (Attwell 2007) as well as increasingly available.

4.5.4 Strengthening Professional, Technical and Vocational Education and Training

Tertiary forestry education has evolved (see Section 4.4.5), and frequently in the context of significant changes to national higher education systems (Kanowski 2015). In conjunction with shifts in student preferences and employment opportunities that parallel the emergence of a wider landscape approach (Sayer et al. 2013) to forests and forestry, these changes challenge tertiary educators and institutions to deliver both a broader curriculum and specific elements that address the need for increasing specialist knowledge across the natural and social sciences and their intersection, and in relevant generic knowledge and skills such as those in business and communication.

These challenges suggest a range of responses, which themselves demand new or more effective partnerships within and between tertiary education institutions and other actors, notably employers and professional associations. These partnerships should support:

- New modes of teaching and learning, including online learning using a variety of platforms and mechanisms, ranging from mass participation Massive Open Online Courses (MOOCs) to personalised micro-credentials (Carey and Stefaniak 2018, Carrera and Ramírez-Hernández 2018);
- More interdisciplinary and integrated curricula and programmes, providing students with a more diverse and individually relevant portfolio of knowledge and skills, which in turn allows them the wider suite of employment opportunities and career pathways necessary in contemporary and future employment markets (WEF 2016c);
- Shifting the locus of professional forestry education to Masters-level programmes (Innes 2015), and strengthening learning and knowledge partnerships with industry at all stages of professional and technical education (Sagor et al. 2014, Yayé et al. 2015);
- Further internationalising programmes by enabling international participation of students in a variety of ways (e.g. exchanges, joint degrees or degree elements, placements), explicitly internationalising curricula, and supporting complementary activities such as students' active participation in international processes (Yunita et al. 2017);

- Through all of these means, broadening access to and inclusivity of programmes, and enhancing the diversity of those studying forest and forestry-related courses and programmes. These goals are being actively pursued by many institutions and networks (e.g. AP-FECM 2018, Gilless 2015, Rekola et al. 2017).

These challenges are perhaps more marked in many countries for TVET than professional education, because TVET systems for rural sectors are often less well-developed. Strengthening 'skills related to the quality of life, productivity skills and skills related to organization, attitudes and values', and 'providing business and entrepreneurial skills training to improve understanding of market opportunities and improve managerial expertise', should underpin future TVET, in recognition that rural people deriving their livelihoods from farms and forests are also likely to want or need to derive income from other, non-farm or forest activities (Robinson-Pant 2016: 19–20).

4.5.5 Capitalising on the Power of the Media

Both old (print, radio and television) and new (online, social) media are near-ubiquitous and influential forces in disseminating environmental information and messaging, reflecting and changing attitudes and norms, encouraging or discouraging pro-environment behaviour, and enhancing or subverting educational experiences. Digital disruption is changing the ways in which people access information, and is challenging established models of reporting and programming (Newman et al. 2017). Media literacy is arguably now more important than ever, in an era of post-truth news and of social media that can facilitate the propagation of misinformation (Williams et al. 2015).

Despite the media transition, traditional means of communication continue to be important. Television is still the primary news source in many countries (Newman et al. 2017) and remains influential in shaping viewers' understanding of environmental issues (Hofman and Hughes 2018, Huang 2016). For example, nature documentaries supported by post-viewing material have been demonstrated to instigate long-term behavioural change (Hofman and Hughes 2018).

The power of social media has been harnessed by many actors – government, business, NGOs and community groups – to promote their perspectives on pro-forest behaviour. One of the strengths of social media is its interactivity, which enables the strategic building of communities and relationships through two-way communication and networking (Lovejoy and Saxton 2012). These online communities can create engaging informal learning environments, especially when users continue to generate and post content (Mason and Rennie 2007). However, both old and new media can work against pro-forest efforts. Journalists may compromise the accuracy of scientific information to increase entertainment value, thereby misrepresenting a story (Frank 2014). Environmental issues can also be framed through narrow perspectives, reinforcing perspectives that prioritise economic growth, or disseminating misconceptions such as that deforestation is confined to the Global South (Lewis 2000). Media can disseminate information unsupported by science, as is evident in reporting of climate scepticism (Painter 2011). In this context, Boykoff and Boykoff (2007) note the role of journalistic norms such as *personalisation* (to focus on human-interest perspectives) and *balance* (to present both sides of a story).

Media literacy is therefore an increasingly essential component of education to foster pro-forest behaviour. It enables people to critically analyse the accuracy and credibility of media content, to identify intents, and to effectively access and create media (Koltay 2011). Education can also help bridge knowledge inequality gaps and empower people to learn through media, as educated people are more likely to use media for personal information gain (Wei and Hindman 2011). In summary, both old and new media can facilitate or constrain the ambitions of SDG 4 in relation to forests.

4.6 Synergies

Education is at the heart of sustainable development, underpinning progress towards all other SDGs through various direct and indirect pathways. Core competencies, such as literacy and numeracy, are the basis for fostering individual agency to participate in society in terms more likely to realise their potential. It is this human potential that other SDGs variously seek to nurture or capitalise on. Education catalyses virtuous circles: those who receive early education are more likely to continue learning formally and non-formally (OECD 2014); educated parents are more likely to invest in their children's education (Pufall et al. 2016); education provides the platform for knowledge generation and capacity building to support SDG implementation; and education, in conjunction with experience of forests, fosters pro-forest behaviours across the domains of other SDGs. However, as Rieckmann et al. (2017: 7) warn, 'not all kinds of education support sustainable development. Education that promotes economic growth alone may well also lead to an increase in unsustainable consumption patterns'. This caution emphasises the rationale and need for education to be embedded in an environmental and sustainability context, as discussed in Section 4.3.2.

In these terms, education is pivotal to improving well-being and livelihoods, particularly through securing income from decent employment (SDG 8, Hanushek and Wößmann 2007), enabling the alleviation of poverty (SDG

1) and hunger (SDG 2), and access to clean water (SDG 6) and clean energy (SDG 7). Education, especially maternal education, improves child health and reduces family sizes (SDG 3, Colfer et al. 2008). Education empowers women (SDG 5) and marginalised groups (SDG 10) to participate fully in society by instilling values of inclusion and challenging the socio-cultural norms that contribute to inequality. Education is also core to climate action (SDG 13) as it fosters concern and capacity for action, particularly for those vulnerable to climate-related disasters (Wamsler et al. 2012).

Economic development (SDG 8) is strongly linked to education quality (Hanushek and Wößmann 2007) and, similarly, underpins multiple facets of development, including sustainable built environments (SDG 11) where knowledge institutions can cluster and collaborate. Universities and other knowledge sector actors are key to generating and applying knowledge to drive sustainable development, generally through partnerships (SDG 17, Charles 2011) and inclusion and diffusion mechanisms such as international scholarships to build capacity at a global scale (SDG 17). However, as cities grow, education inequalities may widen without adequate education infrastructure (SDG 9), particularly in poorer and peri-urban areas (UNESCO 2016).

As discussed in Sections 4.4 and 4.5, targeted education can foster pro-forest behaviour more directly, through research and training to build capacity for SFM (SDGs 14, 15); education campaigns to encourage responsible consumer choices and production that minimise consumption and waste (SDG 12) and conserve energy (SDG 7); corporate education to discourage unsustainable business models (SDG 12); farmer education to discourage deforestation (SDGs 14, 15; Sills and Caviglia-Harris 2015); technical training to enable forest-conserving technologies and their applications (SDG 9); and civic education that empowers people to participate in public policy processes and challenge elite interests (SDG 16) or support planning decisions that better protect forests (SDG 9).

4.7 Conclusions

There is a persuasive case that progress towards SDG 4 is a foundation for progress towards the other SDGs. However, it is also the case that progress towards SDG 4 will not necessarily benefit forests, or the livelihoods of those who depend on forests, unless the *inclusive and equitable quality education and lifelong learning for all* envisaged by SDG 4 fosters pro-forest behaviour by individuals, communities and societies. Pro-forest behaviour is supported by education – formal, informal and non-formal – that shapes pro-forest attitudes and builds and enriches relevant competencies and a sense of connectedness between people and forests. As in other arenas of forest knowledge and management, non-Indigenous people and those not dependent on forests have much to learn in this realm from Indigenous peoples and other holders of traditional and local forest knowledge; there are both synergies and power in partnerships between these and scientific forms of knowledge. There is compelling evidence that engagement with nature from an early age fosters connectedness between people and forests across diverse societies, in both rural and urban contexts, providing the basis for the formation of pro-forest attitudes and behaviours. These can be further amplified, with relevant capacities and skills developed, by subsequent formal, non-formal and informal education.

There are many examples globally of approaches to developing knowledge about forests, and of fostering pro-forest attitudes and behaviours. These can be part of formal curricula from pre-school to tertiary levels, of non-formal education such as capacity development, and of informal learning among families, peers and communities. Educational systems that recognise the significance of each of these modes, and the ways in which they reinforce each other over an individual's lifetime and within their societal contexts, will be most effective in encouraging pro-forest behaviour. However, access to education and the quality of education remain major constraints for many of the world's poorer people, for girls and women in many societies, and for marginalised groups such as Indigenous and forest-dependent peoples. Addressing such disadvantage, as SDG 4 seeks to do, has the potential to realise significant benefits for forests as well as for these people, many of whom depend directly or closely on forests. Correspondingly, fostering a greater sense of connectedness to forests among those in the world who are advantaged – typically those in richer countries, and in cities – can be expected to benefit forests; such connectedness also benefits the well-being of people whose day-to-day lives are more physically distant from forests. There are both great opportunities and considerable challenges for all involved in formal, non-formal and informal education, if the ambitions of SDG 4 are to be realised in ways that benefit forests and our many forms of dependency on them.

Acknowledgements

We thank the editors for their invitation to participate in this work, and their support for it; other chapter leads for stimulating workshop discussions; Riley Schnurr and Kate Sherren of Dalhousie University for preliminary research and ideas that greatly assisted us to begin this chapter; Natalie Cheong of the Singapore National Parks Board and Jakob Terwitte for their timely contributions; and those editors and reviewers whose helpful comments much improved earlier drafts.

References

Agarwal, B. 2009. Gender and forest conservation: The impact of women's participation in community forest governance. *Ecological Economics* 68(11):2785–99.

Agarwal, B. 2010. Does women's proportional strength affect their participation? Governing local forests in South Asia. *World Development* 38(1):98–112.

Allen, T. and Krogman, N. 2013. Unheard voices: Aboriginal content in professional forestry curriculum. In Tindall, D. B., Trosper, R. and Perreault, P. (eds.) *Aboriginal peoples and forest lands in Canada*. Vancouver: University of British Columbia Press, pp. 279–98.

Alston, M., Clarke, J. and Whittenbury, K. 2018. Contemporary feminist analysis of Australian farm women in the context of climate changes. *Social Sciences* 7(2):16.

Altman, J. and Kerins, S. (eds.) 2012, *People on country: Vital landscapes, Indigenous futures*. Sydney: The Federation Press.

Ando, K., Yorifuji, K., Ohnuma, S., Matthies, E. and Kanbara, A. 2015. Transmitting pro-environmental behaviours to the next generation: A comparison between Germany and Japan. *Asian Journal of Social Psychology* 18(2):134–44.

AP-FECM 2018. *Growing higher forestry education in a changing world*. Beijing: China Forestry Publishing House.

Ardoin, N. M., Bowers, A. W., Wyman Roth, N. and Holthuis, N. 2017. Environmental education and K-12 student outcomes: A review and analysis of research. *Journal of Environmental Education* 49(1):1–17.

Attwell, G. 2007. Personal Learning Environments – the future of eLearning? *eLearning Papers* 2(1):1–8.

Battiste, M. 2002. *Indigenous knowledge and pedagogy in First Nations education: A literature review with recommendations*. Ottawa: National Working Group on Education.

Beatley, T. and Newman, P. 2013. Biophilic cities are sustainable, resilient cities. *Sustainability* 5(8):3328–45.

Beery, T. H. and Wolf-Watz, D. 2014. Nature to place: Rethinking the environmental connectedness perspective. *Journal of Environmental Psychology* 40:198–205.

Bloomfield, G., Bucht, K., Martínez-Hernández, J. C. et al. 2018. Capacity building to advance the United Nations sustainable development goals: An overview of tools and approaches related to sustainable land management. *Journal of Sustainable Forestry* 37(2):157–77.

Bonney, R., Shirk, J. L., Phillips, T. B. et al. 2014. Next steps for citizen science. *Science* 343(6178):1436–7.

Boyer-Rechlin, B. 2010. Women in forestry: A study of Kenya's Green Belt Movement and Nepal's Community Forestry Program. *Scandinavian Journal of Forest Research* 25(Suppl 9):69–72.

Boyes, E. and Stanisstreet, M. 2012. Environmental education for behaviour change: Which actions should be targeted? *International Journal of Science Education* 34(10):1591–614.

Boykoff, M. T. and Boykoff, J. M. 2007. Climate change and journalistic norms: A case-study of US mass-media coverage. *Geoforum* 38(6):1190–204.

Brandth, B. and Haugen, M. S. 2000. From lumberjack to business manager: masculinity in the Norwegian forestry press. *Journal of Rural Studies* 16(3):343–55.

Brown, G., Harris, C. and Squirrell, T. 2010. Gender diversification in the US Forest Service: Does it still matter? *Review of Public Personnel Administration* 30(3):268–300.

Buckler, C. and Creech, H. 2014. *Shaping the future we want: UN Decade of Education for Sustainable Development (final report)*. Paris: UNESCO.

Bulkan, J. 2017. Indigenous forest management. *CAB Reviews: Perspectives in Agriculture, Veterinary Science, Nutrition and Natural Resources* 12(4):1–16.

Carey, K. L. and Stefaniak, J. E. 2018. An exploration of the utility of digital badging in higher education settings. *Educational Technology Research and Development* 66(5):1211–29.

Carrera, J. and Ramírez-Hernández, D. 2018. Innovative education in MOOC for sustainability: Learnings and motivations. *Sustainability* 10(9):2990.

Catacutan, D., Muller C., Johnsson, M. and Garrity, D. 2015. Landcare – a landscape approach at scale. In Minang, P., van Noordwijk, M., Freeman, et al. (eds.) *Climate-smart landscapes: Multifunctionality in practice*. Nairobi: World Agroforestry Centre, pp. 151–62.

Charles, D. 2011. The role of universities in building knowledge cities in Australia. *Built Environment (1978-)* 37(3):281–98.

Chazdon, R. L., Brancalion, P. H. S., Lamb, D., Laestadius, L., Calmon, M. and Kumar, C. 2017. A policy-driven knowledge agenda for global forest and landscape restoration. *Conservation Letters* 10(1):125–32.

Chhetri, B. B. K., Johnsen, F. H., Konoshima, M. and Yoshimoto, A. 2013. Community forestry in the hills of Nepal: Determinants of user participation in forest management. *Forest Policy and Economics* 30:6–13.

Clark, W. C., Tomich, T. P., van Noordwijk, M. et al. 2016. Boundary work for sustainable development: Natural resource management at the Consultative Group on International Agricultural Research (CGIAR). *Proceedings of the National Academy of Science of the United States of America* 113(17):4615–22.

Clements, R. 2004. An investigation of the status of outdoor play. *Contemporary Issues in Early Childhood* 5(1):68–80.

Clery, E. and Rhead, R. 2013. *Education and attitudes towards the environment*. Paris: UNESCO.

Coleman, E. A. and Mwangi, E. 2013. Women's participation in forest management: A cross-country analysis. *Global Environmental Change* 23(1):193–205.

Colfer, C. J. P., Dudley, R. G. and Gardner, R. 2008. Forest women, health and childbearing. In Colfer, C. J. P. (ed.) *Human health and forests: A global overview of issues, practice and policy*. London: Earthscan, pp. 113–33.

Collins, P. H. and Bilge, S. 2016. *Intersectionality*. Cambridge: Polity Press.

Convery, F., McDonnell, S. and Ferreira, S. 2007. The most popular tax in Europe? Lessons from the Irish plastic bags levy. *Environmental and Resource Economics* 38(1):1–11.

Csutora, M. 2012. One more awareness gap? The behaviour–impact gap problem. *Journal of Consumer Policy* 35(1):145–63.

Dabbagh, N. and Kitsantas, A. 2012. Personal Learning Environments, social media, and self-regulated learning: A natural formula for connecting formal and informal learning. *The Internet and Higher Education* 15(1):3–8.

Dargavel, J. 1995. *Fashioning Australia's forests*. Oxford: Oxford University Press.

Diálogo Florestal 2018. *About Diálogo Florestal*. Available at: http://dialogoflorestal.org.br (Accessed 20 September 2018).

Diekmann, A. and Preisendörfer, P. 2003. Green and greenback: The behavioural effects of environmental attitudes in low-cost and high-cost situations. *Rationality and Society* 15(4):441–72.

Dockry, M. J., Hall, K., Van Lopik, W. and Caldwell, C. M. 2016. Sustainable development education, practice, and research: An indigenous model of sustainable development at the College of Menominee Nation, Keshena, WI, USA. *Sustainability Science* 11(1):127–38.

Drescher, M., Warriner, G. K., Farmer, J. R. and Larson, B. M. H. 2017. Private landowners and environmental conservation: a case study of social-psychological determinants of conservation program participation in Ontario. *Ecology and Society* 22(1):44.

Dwyer, J. F. and Schroeder, H. W. 1994. The human dimensions of urban forestry. *Journal of Forestry* 92(10):12–15.

Dzhambov, A. M., Markevych, I., Hartig, T. et al. 2018. Multiple pathways link urban green- and bluespace to mental health in young adults. *Environmental Research* 166:223–33.

Eilam, E. and Trop, T. 2012. Factors influencing adults' environmental attitudes and behaviors and the role of environmental schools in influencing their communities. *Education and Urban Society* 46(2):234–63.

Elson, D. 2012. *Guide to investing in locally controlled forestry*. London: Growing Forest Partnerships.

Ens, E. J., Finlayson, M., Preuss, K., Jackson, S. and Holcombe, S. 2012. Australian approaches for managing 'country' using Indigenous and non-Indigenous knowledge. *Ecological Management & Restoration* 13(1):100–7.

Er, K. 2018. *Growing a biophilic city in a garden*. Available at: www.csc.gov.sg/articles/growing-a-biophilic-city-in-a-garden (Accessed 14 September 2018).

Eriksen, C., Waitt, G. and Wilkinson, C. 2016. Gendered dynamics of wildland firefighting in Australia. *Society & Natural Resources* 29(11):1296–310.

FAO 2013. *Forests, food security and gender: linkages, disparities and priorities for action*. Rome: FAO.

FAO 2017. *Discovery-based learning in land and water management: A practical guide for farmer field schools*. Rome: FAO.

Fenstad, E. J, Hoyningen-Huene, P., Hu, Q. et al. 2002. *Science and traditional knowledge: Report from the ICSU Study Group on Science and Traditional Knowledge*. Paris: International Council for Science.

Fisher, M. R., Workman, T., Mulyana, A. et al. 2017. Striving for PAR excellence in land use planning: Multi-stakeholder collaboration on customary forest recognition in Bulukumba, South Sulawesi. *Land Use Policy*. doi:10.1016/j.landusepol.2017.09.057.

Forest Education Foundation 2018. *Forest Education Foundation*. Available at: www.forest-education.com/ (Accessed 15 June 2018).

Frank, A. K. 2014. Writing about sustainability science for the media: How to be both true-to-fact and tell a good story. *Applied Environmental Education & Communication* 13(3):203–11.

Franzen, A. and Vogl, D. 2013. Two decades of measuring environmental attitudes: a comparative analysis of 33 countries. *Global Environmental Change* 23(5):1001–8.

Freed, A. 2018. The relationship between university students' environmental identity, decision-making process, and behavior. *Environmental Education Research* 24(3):474–5.

Frick, J., Kaiser, F. and Wilson, M. 2004. Environmental knowledge and conservation behavior: Exploring prevalence and structure in a representative sample. *Personality and Individual Differences* 37(8):1597–613.

Gilless, J. K. 2015. The Berkeley Summit – Looking to the future for forestry education. *Journal of Forestry* 113(6):587–91.

Gregory, A. 2017. *Running free in Germany's outdoor preschools. The New York Times.* 18 May 2017. Available at: www.nytimes.com/2017/05/18/t-magazine/germany-forest-kindergarten-outdoor-preschool-waldkitas.html (Accessed 20 September 2018).

Hansen, E., Conroy, K., Toppinen, A. et al. 2016. Does gender diversity in forest sector companies matter? *Canadian Journal of Forest Research* 46(11):1255–63.

Hanushek, E. A. and Wößmann, L. 2007. *Education quality and economic growth.* Washington, DC: The World Bank.

Heimlich, J. E. 2010. Environmental education evaluation: Reinterpreting education as a strategy for meeting mission. *Evaluation and Program Planning* 33(2):180–5.

Hiedanpää, J. and Salo, M. 2017. Emerging forest ecosystem service entrepreneurship in Finland and Peru. *International Forestry Review* 19(1):113–24.

Hill, R., Grant, C., George, M. et al. 2012. A typology of Indigenous engagement in Australian environmental management: Implications for knowledge integration and social-ecological system sustainability. *Ecology and Society* 17(1):23.

Hill, T. R., Birch-Thomsen, T., Traynor, C. H., de Neergaard, A. and Bob, U. 2008. Problem-based, interdisciplinary field-based courses: reflections from South African experiences. *South African Geographical Journal* 90(2):122–33.

Hoagland, S. J., Miller, R., Waring, K. M. and Carroll, O. 2017. Tribal lands provide forest management laboratory for mainstream university students. *Journal of Forestry* 115(5): 484–90.

Hofman, K. and Hughes, K. 2018. Protecting the Great Barrier Reef: analysing the impact of a conservation documentary and post-viewing strategies on long-term conservation behaviour. *Environmental Education Research* 24(4):521–36.

Huang, H. 2016. Media use, environmental beliefs, self-efficacy, and pro-environmental behavior. *Journal of Business Research* 69(6):2206–12.

Hull, R. B. 2011. Forestry's conundrum: High value, low relevance. *Journal of Forestry* 109(1): 50–6.

Hunter, L. M., Hatch, A. and Johnson, A. 2004. Cross-national gender variation in environmental behaviors. *Social Science Quarterly* 85(3):677–94.

Innes, J. L. 2015. Master's degrees and other postgraduate education options for foresters. *Journal of Forestry* 113(6):561–5.

Innes, J. and Ward, D. 2010. Professional education in forestry. In *Commonwealth forests 2010: An overview of the forests and forestry sectors of the countries of the Commonwealth*. Shropshire: Commonwealth Forestry Association, pp. 76–95.

INRULED 2012. *Education and training for rural transformation: skills, jobs, food and green future to combat poverty*. Beijing: INRULED.

Jha, S. and Bawa, K. S. 2006. Population growth, human development, and deforestation in biodiversity hotspots. *Conservation Biology* 20(3):906–12.

Kanowski, P. J. 2001. Forestry education in a changing landscape. *International Forestry Review* 3(3):175–83.

Kanowski, P. J. 2015. Internationalizing forestry education. *Journal of Forestry* 113(6):574–8.

Kollmuss, A. and Agyeman, J. 2002. Mind the gap: Why do people act environmentally and what are the barriers to pro-environmental behavior? *Environmental Education Research* 8(3):239–60.

Koltay, T. 2011. The media and the literacies: media literacy, information literacy, digital literacy. *Media, Culture & Society* 33(2):211–21.

Krasny, M. E. and Tidball, K. G. 2009. Community gardens as contexts for science, stewardship, and civic action learning. *Cities and the Environment* 2(1):8.

Lawrence, A., Spinelli, R., Toppinen, A. and Salo, E. 2017. What are the implications of the bioeconomy for forest-related jobs? In Winkel, G. (ed.) *Towards a sustainable European forest-based bioeconomy: Assessment and the way forward. What science can tell us 8*. Joensuu: European Forest Institute, pp. 108–17.

Leicht, A., Heiss, J. and Byun, W. J. 2018. Introduction. Chapter 1. In Leicht, A., Heiss, J. and Byun, W. J. (eds.) *Issues and trends in education for sustainable development: Education on the move*. Paris: UNESCO, pp. 7–16.

LeRoy, C. J., Bush, K., Trivett, J. and Gallagher, B. 2012. *The sustainability in prisons project: an overview 2004–2012*. Olympia: Sustainability in Prisons Project.

Lewis, T. L. 2000. Media representations of 'sustainable development': Sustaining the status quo? *Science Communication* 21(3):244–73.

Lovejoy, K. and Saxton, G. D. 2012. Information, community, and action: How nonprofit organizations use social media. *Journal of Computer-Mediated Communication* 17(3):337–53.

Lozano, R., Merrill, M. Y., Sammalisto, K., Ceulemans, K. and Lozano, F. J. 2017. Connecting competences and pedagogical approaches for sustainable development in higher education: A literature review and framework proposal. *Sustainability* 9(10):1889.

Lubell, M., Niles, M. and Hoffman, M. 2014. Extension 3.0: managing agricultural knowledge systems in the network age. *Society & Natural Resources* 27(10):1089–103.

Lyver, P. O., Timoti, P., Gormley, A. M. et al. 2017. Key Māori values strengthen the mapping of forest ecosystem services. *Ecosystem Services* 27(Part A):92–102.

Macqueen, D., Bolin. A., Greijmans, M. and Grouwels, S. 2018. Innovations towards prosperity emerging in locally controlled forest business models and prospects for scaling up. *World Development*. doi:10.1016/j.worlddev.2018.08.004.

Mason, R. and Rennie, F. 2007. Using Web 2.0 for learning in the community. *The Internet and Higher Education* 10(3):196–203.

Mattijssen, T. J. M., van der Jagt, A. P. N., Buijs, A. E. et al. 2017. The long-term prospects of citizens managing urban green space: From place making to place-keeping? *Urban Forestry & Urban Greening* 26:78–84.

Mayett-Moreno, Y., Villarraga-Flórez, L. F. and Rodríguez-Piñeros, S. 2017. Young farmers' perceptions about forest management for ecotourism as an alternative for development, in Puebla, Mexico. *Sustainability* 9(7):1134.

McBeath, J., Huang McBeath, J., Qing, T. and Huang, Y. 2016. *Environmental education in China*. New York: Elgar.

Mistry, J., Bilbao, B. A. and Berardi, A. 2016. Community owned solutions for fire management in tropical ecosystems: case studies from Indigenous communities of South America. *Philosophical Transactions of the Royal Society B: Biological Sciences* 371(1696):20150174.

Mogues, T., Fan, S. and Benin, S. 2015. Public investments in and for agriculture. *The European Journal of Development Research* 27(3):337–52.

Mukhamedova, N. and Wegerich, K. 2018. The feminization of agriculture in post-Soviet Tajikistan. *Journal of Rural Studies* 57:128–39.

NEEF 2015. *Environmental Literacy in the United States: An agenda for leadership in the 21st century*. Washington, DC: National Environmental Education Foundation.

Newman, N., Fletcher, R., Kalogeropoulos, A., Levy, D. A. and Nielsen, R. K. 2017. *Reuters Institute digital news report 2017*. Oxford: Reuters Institute for the Study of Journalism.

Nikolakis, W. and Nelson, H. 2015. To log or not to log? How forestry fits with the goals of First Nations in British Columbia. *Canadian Journal of Forest Research* 45(6):639–46.

OECD 2014. Indicator C6: How many adults participate in education and learning? In *Education at a Glance 2014: OECD Indicators*. Paris, OECD Publishing.

Ojha, H. R., Cameron, J. and Kumar, C. 2009. Deliberation or symbolic violence? The governance of community forestry in Nepal. *Forest Policy and Economics* 11(5–6):1–10.

Otto, S. and Pensini, P. 2017. Nature-based environmental education of children: Environmental knowledge and connectedness to nature, together, are related to ecological behaviour. *Global Environmental Change* 47:88–94.

OWL Scotland (Outdoor and Woodland Learning Scotland) 2018. *About us*. Available at: www.owlscotland.org/about-us/ (Accessed 1 February 2019).

Painter, J. 2011. *Poles apart: The international reporting of climate scepticism*. Oxford: Reuters Institute for the Study of Journalism.

Panwar, R., Kozak, R. A. and Hansen, E. (eds.) 2016. *Forests, business and sustainability*. England: Routledge.

Parrotta, J. A., Hin Fui, L., Jinlong, L., Ramakrishnan, P. S. and Yeo-Chang, Y. 2009. Traditional forest-related knowledge and sustainable forest management in Asia. *Forest Ecology and Management* 257(10):1987–8.

Parrotta, J. A. and Trosper, R. (eds.) 2012. *Traditional forest-related knowledge: Sustaining communities, ecosystems and biocultural diversity*. New York: Springer.

Project Learning Tree 2018. *Why environmental education is important*. Available at: www.plt.org/about-us/why-environmental-education-is-important/ (Accessed 20 September 2018).

Pufall, E., Eaton, J. W., Nyamukapa, C. et al. 2016. The relationship between parental education and children's schooling in a time of economic turmoil: The case of east Zimbabwe, 2001 to 2011. *International Journal of Educational Development* 51:125–34.

Ramcilovic-Suominen, S. Puentes Rodriguez, Y., Kirongo, B. and Pitkänen, S. 2016. Higher forestry education in Kenya: bridging the gap between educational training and job market competencies. *International Forestry Review* 18(1):56–67.

RECOFTC 2018. *Our vision and mission.* Available at: www.recoftc.org/about/our-vision-and-mission (Accessed 1 February 2019).

Redmore, L. E. and Tynon, J. F. 2011. Women owning woodlands: Understanding women's roles in forest ownership and management. *Journal of Forestry* 109(5):255–9.

Reese, G., Loew, K. and Steffgen, G. 2013. A towel less: Social norms enhance pro-environmental behavior in hotels. *Journal of Social Psychology* 154(2):97–100.

Reid, R. 2017. Developing farmer and community capacity in Agroforestry: is the Australian Master TreeGrower program transferable to other countries? *Agroforestry Systems* 91(5):847–65.

Rekola, M., Abbas, D., Bal, T. et al. 2017. *Global Outlook on Forest Education (GOFE): A pilot study report.* Vienna: IUFRO.

Rieckmann, M., Mindt, L. and Gardiner, S. 2017. *Education for Sustainable Development Goals: Learning objectives.* Paris: UNESCO.

Robelia, B. A., Greenhow, C. and Burton, L. 2011. Environmental learning in online social networks: adopting environmentally responsible behaviors. *Environmental Education Research* 17(4):553–75.

Robinson-Pant, A. 2016. *Learning knowledge and skills for agriculture to improve rural livelihoods.* Paris: UNESCO.

Roczen, N., Kaiser, F. G., Bogner, F. X. and Wilson, M. 2014. A competence model for environmental education. *Environment and Behavior* 46(8):972–92.

Russell-Smith, J. Cook, G. D., Cooke, P. M. et al. 2013. Managing fire regimes in north Australian savannas: applying Aboriginal approaches to contemporary global problems. *Frontiers in Ecology and the Environment* 11(s1):e55–e63.

Sagor, E. S., Kueper, A. M., Blinn, C. R. and Becker, D. R. 2014. Extension forestry in the United States: A national review of state-level programs. *Journal of Forestry* 112(1):15–22.

Sanchez Badini, O., Hajjar, R. and Kozak, R. 2018. Critical success factors for small and medium forest enterprises: A review. *Forest Policy and Economics* 94:35–45.

Sands, R. 2013. *Forestry in a global context.* 2nd ed. Oxfordshire: CABI.

Sayer, J. Sunderland, T., Ghazoul, J. et al. 2013. Ten principles for a landscape approach to reconciling agriculture, conservation, and other competing land uses. *Proceedings of the National Academy of Sciences of the United States of America* 110(21):8349–56.

Sills, E. O. and Caviglia-Harris, J. L. 2015. Evaluating the long-term impacts of promoting 'green' agriculture in the Amazon. *Agricultural Economics* 46(S1):83–102.

Steg, L. and Vlek, C. 2009. Encouraging pro-environmental behaviour: An integrative review and research agenda. *Journal of Environmental Psychology* 29(3):309–17.

Sterling, S. 2016. A commentary on education and Sustainable Development Goals. *Journal of Education for Sustainable Development* 10(2):208–13.

Sterling, S., Glasser, H., Rieckmann, M. and Warwick, P. 2017. 10. 'More than scaling up': a critical and practical inquiry into operationalizing sustainability competencies. In Blaze Corcoran, P., Weakland, J. P. and Wals, A. E. J. (eds.) *Envisioning futures for environmental and sustainability education*. Wageningen: Wageningen Academic Publishers, pp. 153–68.

Stern, M. J., Frensley, B. T., Powell, R. B. and Ardoin, N. M. 2018. What difference do role models make? Investigating outcomes at a residential environmental education center. *Environmental Education Research* 24(6):818–30.

Stern, M. J., Powell, R. B. and Hill, D. 2014. Environmental education program evaluation in the new millennium: what do we measure and what have we learned? *Environmental Education Research* 20(5):581–611.

Stern, P. C. 2000. Toward a coherent theory of environmentally significant behavior. *Journal of Social Issues* 56(3):407–24.

Strobel, K., Nyrud, A. Q. and Bysheim, K. 2017. Interior wood use: linking user perceptions to physical properties. *Scandinavian Journal of Forest Research* 32(8):798–806.

Sunderland, T., Achdiawan, R., Angelsen, A. et al. 2014. Challenging perceptions about men, women, and forest product use: A global comparative study. *World Development* 64(1):S56–S66.

Sylva Foundation 2018. *Sylva Foundation*. Available at: https://sylva.org.uk/home (Accessed 1 March 2018).

Tamang, S., Paudel, K. P. and Shrestha, K. K. 2014. Feminization of agriculture and its implications for food security in rural Nepal. *Journal of Forest and Livelihood* 12(1):20–32.

Temu A. B. and Kiwia A. 2008. *Future forestry education: Responding to expanding societal needs*. Nairobi: ICRAF.

Tengö, M., Hill, R., Malmer, P. et al. 2017. Weaving knowledge systems in IPBES, CBD and beyond – lessons learned for sustainability. *Current Opinion in Environmental Sustainability* 26–7:17–25.

Thomas, I. 2017. Post-sustainability and environmental education: remaking the future for education. *Environmental Education Research* 24(6):1–3.

Thomson, P. 2017. *Uniting to put education at heart of sustainable development*. Available at: www.unesco.org/new/en/media-services/single-view/news/uniting_to_put_education_at_heart_of_sustainable_development/ (Accessed 1 March 2018).

UIS 2016. *Leaving no one behind: How far on the way to universal primary and secondary education?* Paris: UNESCO.

UIS 2017. *Reducing global poverty through universal primary and secondary education. Policy Paper 32/Fact Sheet 44*. Paris: UNESCO.

UN 1992. *Non-legally binding authoritative statement of principles for a global consensus on the management, conservation and sustainable development of all types of forests*. Available at: www.un.org/documents/ga/conf151/aconf15126-3annex3.htm (Accessed 15 June 2018).

UN SDGs Knowledge Platform 2019. *Sustainable Development Goal 4*. Available at: https://sustainabledevelopment.un.org/sdg4 (Accessed 30 January 2019).

UN Women 2012. *The future women want: A vision of sustainable development for all*. Available at: www.unwomen.org/-/media/headquarters/media/publications/en/thefuturewomenwant.pdf?la=en&vs=947 (Accessed 28 July 2019).

UNDP 2018. *Human Development Index (HDI)*. Available at: http://hdr.undp.org/en/content/human-development-index-hdi (Accessed 1 February 2018).

UNDRIP 2007. *United Nations Declaration on the Rights of Indigenous Peoples*. Available at: www.un.org/esa/socdev/unpfii/documents/DRIPS_en.pdf (Accessed 1 February 2018).

UNEP 2011. *Forests in a green economy: A synthesis*. Nairobi: UNEP.

UNEP 2017. *Africa Environmental Education and Training Action Plan 2015–2024: Strengthening sustainable development in Africa*. Nairobi: UNEP.

UNESCO 2014. *Roadmap for implementing the Global Action Programme on Education for Sustainable Development*. Paris: UNESCO.

UNESCO 2016. *Education for people and planet: Creating sustainable futures for all*. Paris: UNESCO.

UNESCO 2017. *Accountability in education: meeting our commitments*. Paris: UNESCO.

UNESCO, UNDP, UNPFA, UNHCR, UNICEF and UN Women 2016. *Incheon Declaration and Framework for Action for the implementation of Sustainable Development Goal 4*. Available at: http://uis.unesco.org/sites/default/files/documents/education-2030-incheon-framework-for-action-implementation-of-sdg4-2016-en_2.pdf (Accessed 20 January 2018).

UNESCO-UNEVOC 2017. *Greening technical and vocational education and training: A practical guide for institutions*. Paris: UNESCO.

University Alliance for Sustainability 2018. *University Alliance for Sustainability*. Available at: www.fu-berlin.de/en/sites/uas/index.html (Accessed 20 September 2018).

University of Florida 2017. *Master of Sustainable Development Practice program, manual 2017–18*. Available at: http://sites.clas.ufl.edu/africa-mdp/files/MDP-Program-Manual-2017.pdf (Accessed 23 October 2018).

van Wynsberghe, R. and Moore, J. L. 2015. UN decade on education for sustainable development (UNDESD): enabling sustainability in higher education. *Environment, Development and Sustainability* 17(2):315–30.

Villamor, G. B., Desrianti, F., Akiefnawati, R., Amaruzaman S. and van Noordwijk, M. 2014. Gender influences decisions to change land use practices in the tropical forest margins of Jambi, Indonesia. *Mitigation and Adaptation Strategies for Global Change* 19(6):733–55.

Waller, D. M. and Reo, N. J. 2018. First stewards: ecological outcomes of forest and wildlife stewardship by Indigenous peoples of Wisconsin, USA. *Ecology and Society* 23(1):45.

Wals, A. E. J. and Benavot, A. 2017. Can we meet the sustainability challenges? The role of education and lifelong learning. *European Journal of Education* 6(5):404–13.

Walter, P. 2009. Local knowledge and adult learning in environmental adult education: community–based ecotourism in southern Thailand. *International Journal of Lifelong Education* 28(4):513–32.

Wamsler, C., Brink, E. and Rantala, O. 2012. Climate change, adaptation, and formal education: The role of schooling for increasing societies' adaptive capacities in El Salvador and Brazil. *Ecology and Society* 17(2):2.

WEF 2016a. *Quality basic education.* Available at: https://toplink.weforum.org/knowledge/insight/a1Gb0000000LPFfEAO/explore/dimension/a1Gb00000016UZWEA2/summary (Accessed 23 October 2018).

WEF 2016b. *The industry gender gap: Women and work in the fourth industrial revolution (executive summary).* Geneva: World Economic Forum.

WEF 2016c. *The future of jobs: Employment, skills and workforce strategy for the fourth industrial revolution.* Geneva: World Economic Forum.

Wei, L. and Hindman, D. B. 2011. Does the digital divide matter more? Comparing the effects of new media and old media use on the education-based knowledge gap. *Mass Communication and Society* 14(2):216–35.

Williams, H. T. P., McMurray, J. R., Kurz, T. and Hugo Lambert, F. 2015. Network analysis reveals open forums and echo chambers in social media discussions of climate change. *Global Environmental Change* 32:126–38.

Witoszek, N. 2018. Teaching sustainability in Norway, China and Ghana: challenges to the UN programme. *Environmental Education Research* 24(6):831–44.

World Agroforestry Centre 2018. *Outputs.* Available at: www.worldagroforestry.org/output (Accessed 1 March 2018).

World Bank 2016. *Forests generate jobs and income.* Available at: www.worldbank.org/en/topic/forests/brief/forests-generate-jobs-and-incomes (Accessed 1 March 2018).

World Bank 2017. *Girls' education.* Available at: www.worldbank.org/en/topic/girlseducation (Accessed 1 March 2018).

World Bank 2018. *World Development Report 2018: Learning to realize education's promise.* Washington, DC: World Bank.

Wyatt, S., Natcher, D. C., Smith, P. and Fortier, J.-F. 2010. Aboriginal land use mapping: What have we learned from 30 years of experience? In Stevenson, M. and Natcher, D. (eds.) *Planning co-existence: Aboriginal issues in forest and land use planning.* Edmonton: CCI Press, pp. 185–98.

Yayé, A. D., Ochola, A. O., Chakeredza, S. and Aucha, J. 2015. Strengthening capacity for agribusiness in agroforestry and natural resources in tertiary agricultural education in Africa: African Network for Agriculture, Agroforestry and Natural Resources Education (ANAFE). *Agroforestry Systems* 91(5):835–45.

Yunita, S. A. W., Soraya, E. and Maryudi, A. 2017. 'We are just cheerleaders': Youth's views on their participation in international forest-related decision-making fora. *Forest Policy and Economics* 88:52–8.

Zsóka, Á., Szerényi, Z. M., Széchy, A. and Kocsis, T. 2013. Greening due to environmental education? Environmental knowledge, attitudes, consumer behavior and everyday pro-environmental activities of Hungarian high school and university students. *Journal of Cleaner Production* 48:126–38.

Chapter 5 SDG 5: Gender Equality – A Precondition for Sustainable Forestry

Seema Arora-Jonsson*, Shruti Agarwal, Carol J. Pierce Colfer, Stephanie Keene, Priya Kurian and Anne M. Larson

Key Points

- Forestry cannot be thought of in isolation from its relations with other sectors and other parts of people's lives – for both the health of the forests and the well-being of forest peoples.
- Forest governance and everyday management are upheld by a superstructure of gendered forest relations – invisible to mainstream forestry – that often disadvantages women as a social group.
- Well-intentioned gender programmes can backfire, causing adverse effects on forests and forest peoples, if the efforts are not cognisant of context and power relations.
- Constant awareness of differences among various social groups – men, women, different classes, ethnicities – and how their interests intersect differently in various forest contexts is needed for everyone's energy, creativity and motivation to contribute to sustainable forest management.
- Research suggests that greater democratic governance of forests leads to better environmental outcomes.
- The gender-neutral framing of some SDG goals undermines efforts towards achieving the outcomes called for in SDG 5.

5.1 Introduction

SDG 5's ambition to 'achieve gender equality and empower all women and girls' is extremely important in forestry contexts. It brings attention to aspects that make forest livelihoods possible but often get subsumed in conventional forestry definitions, associating forests only with timber, woody biomass or biodiversity conservation. Taking SDG 5 seriously in relation to forests brings to the forefront what is usually taken for granted or backgrounded in forest

* Lead author.

debates: people, and their relationships to one another and to the forests, which determine forest outcomes.

We first analyse the context for SDG 5 in relation to forests; second, we consider how taking SDG 5 seriously might impact forests and people's livelihoods. So far, little progress has been made in implementing these targets within the forestry sector. Our analysis is built around SDG 5's nine targets (Table 5.1). Though defined separately in Goal 5, the issues the targets raise are inextricably linked to one another in the everyday lives of women and men. Bringing change to one would affect other aspects. We demonstrate the need to understand large systemic connections from a broad perspective. To do so, we turn to compelling feminist research on gender and forest livelihoods. We also go beyond forestry research to cutting-edge gender research on themes and targets where research in relation to forest contexts is scarce. We then theorise on insights from this research to what we know about forest contexts.

Taken together, the subsections titled 'Implications' in this chapter build a framework for analysis and raise serious questions in relation to interventions

Table 5.1 SDG 5 targets

5.1 End all forms of discrimination against all women and girls everywhere
5.2 Eliminate all forms of violence against all women and girls in the public and private spheres
5.3 Eliminate all harmful practices, such as child, early and forced marriage and female genital mutilation
5.4 Recognize and value unpaid care and domestic work
5.5 Ensure women's full and effective participation and equal opportunities for leadership at all levels in political, economic and public life
5.6 Ensure universal access to sexual and reproductive health and reproductive rights
5.A Undertake reforms to give women equal rights to economic resources, and access to ownership and control over land and other forms of property, financial services, inheritance and natural resources, in accordance with national laws
5.B Enhance the use of enabling technology, in particular information and communications technology, to promote the empowerment of women
5.C Adopt and strengthen sound policies and enforceable legislation for the promotion of gender equality and the empowerment of all women and girls
Source: Adapted from https://sustainabledevelopment.un.org/SDG5

in forestry if equality and sustainability are to guide forest action. The analysis, while focused on forest contexts, is also relevant for other sectors (e.g. agriculture), as is evident from the research we draw on and the ways these issues are intertwined in people's everyday lives.

Gender, as it is discussed here, is not synonymous with women – a common mainstream assumption. Rather, gender, as a category of analysis, studies relationships of power based on sex/sexuality and the ways in which relationships may be organised. We therefore analyse not only how this organisation in different contexts may discriminate against certain social groups based on their sex, but also how men's and women's positions in society are always cross-cut by intersecting dimensions of power such as class, caste, age, ethnicity and sexual orientation.[1] Much of the research on forestry and gender comes from the Global South, but research on this topic is gaining ground in the Global North. There are differences in these contexts, but also striking parallels and connections, as we show in the following sections.

We begin with Target 5.1: the elimination of discrimination against women and girls. This section establishes the context of gender and forest relations and the potential for SDG 5 targets to be achieved in forest contexts. Extensive research on forestry shows that what is taken by mainstream forestry to be 'work' often ignores women's work in forests. We reflect on the implications of this research in relation to forestry and the anticipated impact that achieving the target might have for forests and forest livelihoods.

5.2 Elimination of Discrimination: What Does It Look Like in Forestry Today? – Target 5.1

Women are central to the work done in and around forests, yet forests have long been a male public domain. Most recently, FAO's *The State of the World's Forests* (2018) report, building on global data, states that women's forest-related work often surpasses that of men. Research from both the Global South and North shows how what has been considered work or valuable in forests has commonly involved activities associated with men: activities related to the commodification of forest products whose trade is often male-dominated. This is especially problematic as, in many places (if not most), men and women have different divisions of labour and differences in their ability to act and make decisions regarding forests and forest resources. Not actively addressing discrimination in this sector is not only

[1] Most human rights law continues to deploy 'gender' as a synonym for 'women'. This failure to truly grapple with gendered systems of power means that these policy instruments are unable to effectively tackle gendered inequalities.

a setback for an equitable society, but also a huge obstacle for sustainable forest management.

Although research since the 1970s has highlighted gendered differences, gender-neutral approaches have coloured forestry policies and programmes, both in the Global North and South. The view of institutions as gender-neutral and meant for all is an important factor in discrimination against women. These institutions tend to take the interests of certain men (of a particular class, age, ethnicity and race or caste) as the norm – as is evident in past forestry programmes in the Global South, such as social forestry, joint forest management or community forestry in the 1970s–90s (Agarwal 2010). This is equally true of the Global North (Reed 2008) and in relation to women from Indigenous communities (Mills 2006). Current programmes such as REDD+ and large land-investment schemes seem only to be repeating past errors. A study of 23 early-stage REDD+ projects found that none listed women as a stakeholder group, although five initiatives listed fair benefits to women as an equity goal. A follow-up study three years later found that women's well-being in REDD+ sites had fared worse than the villages' as a whole, and that being in a REDD+ site was significantly associated with a drop in women's well-being when compared to a control group over the same period (Larson et al. 2018). Similarly, studies in Indonesia (Li 2015) bring to light negative economic consequences shouldered by women and their larger communities when forest-based roles are ignored during negotiation processes with investors proposing large-scale land acquisitions. Women in particular were absent from informational meetings and had little or no knowledge of what would happen to their lands.

The insecurity of women's forest rights under national law continues to be an obstacle. A recent global assessment of the legally recognised rights of Indigenous and rural women to community forests in 30 low- and middle-income countries (RRI 2017) concludes that none of the assessed countries adequately recognise women's rights. Rights to inheritance, community membership, community-level governance (voting and leadership) and community-level dispute resolution are wanting, despite constitutional commitments to protect women's rights.

Research also shows that the personal, spiritual, emotional and non-economic aspects in women's and men's everyday lives cannot be separated from decision-making about forests. In both the Global North (Arora-Jonsson 2013) and South (Agarwal and Saxena 2018), women have often chosen to forego economic benefits in favour of other forest outcomes they see as benefitting their communities, families and themselves (see Box 5.1). These elements central to the lives of forest-dependent peoples are often disregarded in academic and political discourses that prioritise the economic value of forests.

Box 5.1 Beyond Economic Benefits

India's Forest Rights Act (FRA) of 2006 recognises the 'rights of ownership, access to collect, use, and dispose of minor forest produce' of forest-dwelling communities.[2] Tendu (*Diospyros melanoxylon*) leaves, an important minor forest product in Central India, are collected primarily by tribal women and constitute a crucial source of cash. As tendu is a nationalised minor forest product (MFP), the forest department had a monopoly on its trade. This changed with the enactment of the FRA, which empowers forest-dwelling communities to sell such MFPs to traders of their choice.

In 2017, when a private trader offered a better rate for tendu leaves than the forest department, tribal women from six Odisha villages contracted a deal with the trader. The forest department strongly opposed the deal and insisted that the leaves could only be sold to the department. Tendu leaves are a significant source of revenue for the state government (estimated at more than USD 70 million).

The choice available to these women was to sell the leaves to the forest department and earn whatever the department had to offer, or not to sell the leaves at all. Time was a crucial factor in the women's decision, as tendu leaf quality deteriorates quickly without proper storage, which was unavailable to the villages. The women agreed unanimously to trade on their own terms and forego the potential revenue from tendu if the forest department did not come around. The women wanted the department to respect their rights under the FRA. In a major victory for the women, the forest department acknowledged – albeit after six months of sustained protests – that communities have the right to engage in the private trade of nationalised MFPs.

Source: Agarwal and Saxena 2018.

Violence by men (and sometimes by female relatives), lack of access to birth control or decisions regarding childbearing, domestic work and lack of access to information or education prevent many women from participating, owning or managing forests and resources in and beyond the household (Colfer 2011). As a woman from a forest in Odisha remarked, 'What is the point of protecting the forests when we cannot protect ourselves?' (Arora-Jonsson 2013: 204). In forest livelihoods, as elsewhere, the feminist slogan remains clearly relevant: the personal is political. Matters around bodily integrity, domestic partnerships and household-level power dynamics are deeply intertwined with what takes place in the public sphere of management, conservation and business.

[2] https://forestrights.nic.in/pdf/FRAAct.pdf

Studies shows that women are consistently at a disadvantage in relation to institutional support in extension, information, technical support and other services (Lambrou and Nelson 2010). In Sweden, a governmental inquiry demonstrated that male-dominated forestry networks and greater links to economic resources for men than women have contributed to the slow progress of gender equality within the forestry sector (DS 2004: 39). A technical study of formalisation procedures on forest tenure across four countries – Indonesia, Uganda, Peru and Nepal – shows that most government officials managing these processes in each country were men. Only 18 per cent of the officials were women, and only 17 per cent of officials believed that strengthening the rights of special groups such as women and Indigenous peoples was a formalisation objective (Herawati et al. 2017).

The lack of female extension agents and officers is especially troubling in light of research showing that women often prefer female extension agents in order to discuss their interests regarding agriculture. A study in Tanzania shows that men too prefer female extension agents as they feel women are more inclined to listen to them than the male extension agents (Due et al. 1997). Another such example (from Arora-Jonsson 2013) is an assessment report of community forestry groups in Odisha by male authors with little direct contact with the village women that reported the women as being oppressed and lacking agency in forest contexts. In contrast, ethnographic research by a woman at the same time and in the same place presents a different picture, pointing to the many ways in which women's groups were taking action both for themselves and for the forests, showing ways in which they could be supported for forest health and themselves. Research in Senegal (Moore et al. 2001) shows that contact with women officers was a strong predictor of the level of women's knowledge about natural resource management (NRM) and adoption of management practices, also contributing to the level of men's knowledge. Mechanisms established specifically to have contact with women, such as employing women agents, are important for women.

These examples have implications for forestry since it is clear that the lack of networks, good extension and sensitivity to the experiences of different groups is likely to hamper forest production and health. They highlight the need for officers and researchers who might have better access to women. More importantly, research stresses that the main difference lies not in the sex of extension workers or forest officers, but in their ability and training to listen to the contextual needs of different groups and the importance of making an effort to reach out to them (Due et al. 1997, Jafrey and Sulaiman 2013, Quisumbinq and Pandolfelli 2010).

5.2.1 Implications

This research shows overwhelmingly that forestry cannot be considered in isolation. As the previously given quote by the Odisha woman highlights, forestry cannot only be about trees but needs to link to other parts of people's lives. Questions of safety in the forests and the home, the spiritual and cultural significance that forests represent for many Indigenous and local communities worldwide, and the ability of women and other vulnerable groups to participate in forest-related decision-making are equally important for forestry agencies to consider.

Past lessons are being ignored. Discrimination against women often results from the institutional make-up of official bureaucracies and other outsiders and discriminatory legislation as much as it derives from customs within communities. The role of forestry officials on the ground, as well as other natural resource officials, is extremely important. The ratio of men to women working in forestry is significant: it is important that forest departments hire more women. Yet this is rarely the case. It is even more important that forest officers are trained to listen to concerns of different groups and to concerns that may differ from mainstream forestry as usually defined.

As the SDGs are implemented we need to confront the gendered nature of institutions, particularly in local areas, but also across the scale to the national and international levels responsible for the policymaking and projects that also shape conditions on the ground. Policymaking needs to be sensitive to these nuances when promulgating reforms intended to be gender-inclusive.

5.3 Women's Rights over Their Own Bodies – Targets 5.2, 5.3 and 5.6

Forests are particularly linked to violence or the equally debilitating fear of violence. In studies from the Global South, women speak of their fear of violence at the hands of contractors engaged in forest investments (Arora-Jonsson 2013), often associated with big multinational companies and other forest outsiders (Zamora and Monterroso 2017). Forests are seeing increased violence against environmental defenders in ongoing conflicts over territory and resources, and sometimes violence perpetrated by government authorities in their zeal for conservation (RRI 2018). In 2016, at least 200 forest defenders were murdered (almost 10 per cent more than in 2015) in different conflicts over land and resources; 40 per cent of the victims were Indigenous (Global Witness 2017). Female land and human rights defenders are murdered less often, but are more often subject to sexual violence and are less likely to be able to denounce these abuses. Nevertheless, there are emblematic cases such as that of Bertha Caceres in Honduras, assassinated in 2016.[3] Caceres was an Indigenous Lenca woman and a Goldman Environmental Prize recipient in 2015,[4] and was killed for her opposition to a hydroelectric dam. Women and transgender forest defenders are especially vulnerable to sexual assault by fellow activists due to remoteness to the outside world (Mallory 2006).

Insidious everyday violence takes place within the walls of the home, including in forested areas. Studies show that poverty and alcohol abuse as well as climate-induced socio-economic changes in Kenya (Scheffran et al. 2014), or drought- and income-related stress in Australia (Whittenbury 2013) have led to an increase in violence against women. While these factors may provoke some men into violence against women in forested areas, incontrovertible evidence demonstrates that men's violence against women is widespread everywhere (for a review of 34 countries in North America and Europe, see Alhabib 2010). In India, lower-caste working women are subjected to routine violence and sexual abuse by the upper castes (Jayal 2003, Kumari 2017). In Sweden, considered to be one of the most progressive countries in terms of gender equality indices, Lundgren et al. (2001) found that nearly every second woman, at some point since her 15th birthday, has experienced violence at the hands of a man, regardless of ethnicity or social class. In Aotearoa, New Zealand, more than one-third of women (35.4 per cent) reported physical and/or intimate partner violence in her lifetime (Simon-Kumar et al. 2017). Research from South Africa shows the increasingly negative effects of violence on the mental health of victims (Lagdon et al. 2014). Data for 87 countries from 2005 to 2016 shows that 19 per cent of girls and women aged 15–49 experienced physical and/or sexual violence from an intimate partner in the previous year (HLPF 2017: 1).

Strategies to counter violence against women have had varying success. It is clear that male violence against women, and some men, is enabled by constructions of masculinity and the desire for control that go far beyond catalytic factors such as alcohol consumption. Scholars argue, however, that in some cases reducing alcohol consumption (Jewkes 2002) and access to arms – as, for example, in Sudan and Kenya (Budlender and Alma 2011, Scheffran et al. 2014) – can contribute to curtailing the frequency of violence. Education is important: women who are highly educated have an edge, although the relation between empowerment and the risk of violence is non-linear and education does not always mitigate the risk of violence (Jewkes 2002, Mabsout and van Staveren 2010).

---

[3] www.theguardian.com/world/2017/feb/28/berta-caceres-honduras-military-intelligence-us-trained-special-forces www.greenpeace.org/international/en/news/Blogs/makingwaves/revealed-investigation-uncovers-the-plot-to-m/blog/60633/

[4] www.goldmanprize.org/recipient/berta-caceres

Kusuma and Babu (2017) argue for the need to coordinate efforts among a range of relevant areas and groups: health, education, police, judiciary and community groups. This is especially relevant for forest contexts, which are often remote and far from such services. A review study of 142 documents on violence against women, mainly in high-income settings, indicates that education, youth projects and interventions in school-based dating violence can be successful in counteracting intimate partner and sexual violence among adolescents (Lundgren and Amin 2015). Research from Turkey shows that the involvement of medical professionals such as nurses and midwives has a key role in counselling women, including about the legal rights that protect them from the risk of violence (Özcan et al. 2016). Access to reproductive health services, including family planning, has helped reduce poverty, contributed to improved nutrition and educational outcomes, and saved mothers' lives, benefitting not only women but society more generally. Supporting girls' and women's education and the prevention of HIV infections by providing contraceptives has been shown to reduce the number of births, and that in turn mitigates deforestation effects as less land is taken over for agriculture (Starbird et al. 2016).

Context-based factors as well as resources and household assets can protect women from violence, as shown in 30 sub-Saharan African countries (Cools and Kotsadam 2017), in Nicaragua (Grabe 2010) and in different parts of India (Bhattacharya et al. 2011, Panda and Agarwal 2005). Ownership of resources also correlates with health. For example, propertied women in South Africa are better able to protect themselves from HIV/AIDS infections (Swaminathan et al. 2008). Income equality or men's economic subordination, not only in relation to women, is seen as a threat to a certain hegemonic masculinity. In such cases violence may be used to resolve a crisis of male identity (Jewkes 2002, Lwambo 2013). Behrman et al. (2014) in South Asia and sub-Saharan Africa and Mabsout and van Staveren (2010) in Ethiopia note a perception among some that increasing women's access to resources can increase violence against them. However, no linear relation between the two has been noted in practice.

Violence should be dealt with not only as a women's issue, but also as a gendered and systemic one. Researchers link increasing violence (warfare, domestic abuse) and criminality in sub-Saharan Africa with changing sex roles that no longer allow men to perform an idealised form of masculinity and act as breadwinners (Barker and Ricardo 2005, Lwambo 2013; also noted in Nicaragua, Evans et al. 2017). The Responsible Men's Club in Vietnam (Hoang et al. 2013) was one attempt to work with men that yielded important insights about men's relationships to their wives and also how these relationships were interconnected across scale, extending far beyond the household.

Many link violence to the weakness of the state (Dolan 2002, Schroeder 1999) and its inability to address caste and class violence, as in India (Jayal 2003, Kumari 2017). Violence against women must be tackled in a variety of ways and across various scales. Forest actors are vital to this in forest areas.

5.3.1 Implications

In contexts where forest issues are being debated, there is a need to work actively against gender-based violence and the lack of access to sexual and reproductive health. Legal rights are important, as is the work of state authorities, including forest authorities. At the same time, violence is not only an individual action but is tied to a larger question of gender and power, and contextual factors are extremely important.

A weak state can lead to greater gender-based violence. Efforts to counter violence have often resorted to the criminalisation of perpetrators, disregarding larger structural contexts and minimising success in reducing violence. The increasing focus on criminalisation that has emerged in both international law and the international-security domain risks obfuscating and downsizing the collective and public dimension of state responsibilities to reduce violence. Indeed, criminalisation strategies allow states to circumvent their duty to address the social, political and economic structural dimensions at the root of this severe form of violation of women's human rights (Pividori and Degani 2018).

Parental support and peer networks (Chandra-Mouli et al. 2015, Faxon et al. 2015), along with safe spaces for women, are important in giving women agency and safety from violence (Eduards 2002). Again, forest authorities have an important role to play: for example, in providing training within forest departments, helping to provide safe spaces for women and spaces for their participation in decision-making on their own terms. The need for greater attention to these issues in the forest context is pressing since the official realm of forestry the world over is male-dominated. By not actively taking up questions that concern SDG 5 in other arenas because they appear unrelated to their forestry work, forest actors help to ensure that issues surrounding violence against women remain barriers to an equal, democratic and sustainable forestry.

5.4 Recognition of Unpaid Work through Social Policy and Public Provisions – Target 5.4

Women are often overburdened with care work in the home and perform much of the unpaid domestic labour everywhere. Research in Nicaragua shows that women identified their domestic work, and men's failure to share it, as a main reason why they did not participate in community decision-making regarding forests (Mairena et al. 2012). Unpaid care work adversely affects both women and men, though it lowers women's income more (Qi and Dong 2016, on China). Even in a context of increasingly egalitarian policies such as in Canada, the gendered care gap is widening, with women taking on the bulk of this work (Proulx 2016). Time-use studies have led to a recognition of women's unpaid work in both the Global North and South. As research shows, however, recognition is insufficient. We need to rethink the androcentric socio-economic institutions and narratives that lead to such disparities. As a case from Malta indicates, pressure to conform to gender norms, combined with poor government policies (no paid parental leave, higher male salaries), coincides with an overall resistance among spouses to reallocate responsibilities (Camilleri-Cassar 2017).

This is particularly so in the forest sector. Most of women's forest-based labour is unpaid, and forest products that women are responsible for are often less economically valuable. Research in Scandinavia (Kaldal 2000) shows that women's forest work not associated with timber tends not to be regarded as work. Even where women do not do forest-based labour themselves, their tremendous care duties make forest-based labour possible for other family members. Especially in lower-income countries, a whole regime of unacknowledged care work upholds the forest sector. Exacerbating this hidden work is migration, especially in the Global South. Islam and Shamsuddoha (2017) show that in Bangladesh women left behind are burdened with additional work, including securing food for the family while the men are away. In cases where men migrate and are not able to or do not send money home, women must generate income to sustain their families. This is not always culturally acceptable, raising concerns for trafficking and exploitation. This is confounded by development-induced displacement and resettlement where women usually receive fewer benefits than men (Lin 2008). Michocha (2015) argues women act as shock absorbers when families are displaced, taking responsibility for sourcing food, cooking and child care.

In some countries, social provisions such as help with childcare or child allowances have been central to women entering the labour force and achieving economic independence. These provisions have counteracted women's weak social positions vis-á-vis public decision-making and enabled them to enter public arenas. Yet, the forestry sector remains an aberration. In Sweden some argue that the increasing number of women forest owners is likely to presage a change in forest management (Nordlund and Westin 2011). So far, practice has yet to bear this out. The forest sector continues to be dominated by a narrow group of actors – mostly men – despite increasing numbers of women forest owners (Holmgren and Arora-Jonsson 2015).

In cases where inequitable norms that privilege men remain intact, social provisions may be inadequate. Regardless of incentives, the participation of men in care work is lower than that of women in both the Global North and South. In Sweden, despite generous paternity benefits, fewer men than women take childcare leave (Duvander et al. 2010), due in part to higher wages for men. In Korea (Peng 2011) women are encouraged to enter the labour market, but their jobs are more often in non-regular and service sectors, impeding the attainment of gender equality. Yoon (2014) shows that in Korea the state tends to omit the role of family-provided care services, overestimating its own role and suggesting that much more in-depth understanding of the contributions to care, both qualitative and quantitative, is needed. Furthermore, policies that focus on flexible work hours – intended as an alternative to the male model of 'working hours'– disregard the gendered pay gap and social norms (Lewis and Plomien 2009) and can reproduce rather than reduce gender inequalities (Figart and Mutari 2000). Neither do they reduce the job deficit (Estévez-Abe and Hobson 2015). The increasing commodification of care work is gendered female worldwide, and tends to be low-paid.

Scholars argue that state provision of 'merit goods' – goods and services deemed valuable for every citizen, such as child allowance, unemployment insurance, schooling and healthcare – accompanied by targeted cash payments are much more likely to increase productivity and reduce inequalities (Bergmann 2004). They call for valuing care work on a par with other activities (Baker 2008). Research from Chile and Mexico shows that childcare services can enhance gender equality if the state commits to such goals, gives sufficient financial resources and develops effective regulations (Staab and Gerhard 2011). Moreover, findings from Argentina, Bolivia, Brazil and Chile indicate that non-contributory pension funds can be a key measure to reduce gender gaps in benefits at old age (Arza 2017). Drawing on findings from a multi-country cash-transfer programme in sub-Saharan Africa, Asfaw (2016) concludes that promoting cash programmes can positively affect livelihoods as well as agricultural productivity, provided there is coordination with other sectoral development programmes and attention to local contexts.

5.4.1 Implications

The research cited herein is thus important to consider in the forestry context as much as in other walks of life. Meeting Target 5.4 in the forestry world requires recognising that women's care work, often invisible, underpins the recognised and overtly valued commercial work, widely considered as men's work.

Informal employment characterises the lives of most working women, especially in the Global South: see Ghosh (2015) for India and Lopez-Ruiz et al. (2017) for Central America. Women's need to control their participation in local non-timber forest products (NTFPs) and other forest-related markets needs to be addressed (Section 5.6 discusses women's participation in markets). Merit goods such as those described here have been shown to be beneficial not only for men and women, but also for agricultural productivity (Asfaw 2016). Similar outcomes could be expected within forestry if merit goods provide resources and time for men and women.

The thrust of this research indicates the need for a systemic and contextual understanding, including addressing paid and unpaid forest-related labour in one frame. This entails acknowledging care work in the home and the subsistence work that men or women carry out in the forest. The need for interconnected policies and programmes is urgent (Peng 2011). As the examples demonstrate, the role of authorities and other official actors is vital in these efforts.

5.5 Women's Right to Decision-Making – Target 5.5

In contrast to the themes discussed earlier, there is considerable research on the participation of marginalised social groups, including women, in forest decision-making. Few women participate in forestry decision-making forums in the Global North and South (Agarwal 2010, Reed 2008). The RRI study (2017) across 30 countries in Asia, Africa and Latin America found that women's right to participate in community-level forest governance processes was the most inadequately protected community-level right analysed in the study.

The devolution of forest decision-making has been an important global trend in recent decades. In settings related to rural development and forestry in both the Global North and South, decision-making power is often captured by male elites. Studies from South Asia report intense conflict in local groups managing high-value forests. Older and elite men become more active in forest-user groups managing high-value forests, making women's effective participation harder to achieve (Adhikari and Di Falco 2008, Lama et al. 2017). Similarly, studies in Nicaragua (Evans et al. 2017) and Colombia (Sandoval-Ferro 2013) show that women in some Indigenous communities, with equal rights on paper to manage the forest, forego their own interests. They are pressured to accept decisions guided by male interests and men's groups, defined as having more benefit to the wider community.

A similar pattern of elite capture can be seen in Europe in a forest management context (Arora-Jonsson 2013), as well as in EU programmes that stipulate the participation of local groups in development. In a process of devolution of local governance activities in the UK, for example, Tickell and Peck (1996) observed that male power was naturalised as the legitimate conduit for effective local governance. The political process in unelected bodies privileged pre-existing male networks, while their modus operandi marginalised and excluded women and their interests.

In such cases, quotas for women have been useful, though sometimes quotas have also resulted in token rather than substantive and comprehensive participation among women (RRI 2017). According to studies in rural Andhra Pradesh, India (Afridi et al. 2017), women village council heads in reserved seats were initially seen to be less effective at administration and tackling corruption, though in several districts they caught up after one year. The findings suggest that learning speeds among female Executive Committee members may depend on their starting point (e.g. level of education, intra-group dynamics and reciprocal support among the women members) as well as the complexity of their administrative tasks and responsibilities. Programmes to make women's representation more effective from the start require a better and more detailed understanding of hurdles to this effectiveness and its variation with individual, resource and community characteristics (Afridi et al. 2017). Importantly, this entails working actively to reduce disparities with men in access to technology and information (Mwangi et al. 2011).

Research across South Asia indicates that women's groups tend to be more rigorous in forest conservation (Agarwal 2010). It remains unclear whether this leads to better conditions for them.

Gender balance in forest-user groups in East Africa and Latin America was shown to be associated with more participation and enhanced forest sustainability (Mwangi et al. 2011). From her research in Northern India, Minocha (2015) concludes that a perception among women interviewed was that more active participation by women in council meetings and similar decision-making bodies would result in more resistance to big infrastructure projects that adversely affect livelihoods or cause displacement.

Adding a few women to committees will not ensure that all women's interests are addressed or that governance necessarily becomes more effective. Real positive change requires a willingness to accommodate the interests and issues taken up by marginalised groups and openness to changing decision-making forums and structures (Arora-Jonsson 2013). In forestry forums, women often raise issues not considered central to forests or forestry, such as violence, lack of decision-making power and inattention to other community issues they consider related to forests. Such issues that touch upon spaces other than forests, such as the home or village, while not directly related to timber and forest products, impinge on people's relations to the forests and to each other. They play an important part in what actually happens in forests. This is the 'space-off'[5] of forestry, i.e. the invisible relations and spaces often disregarded in decision-making on forests (frequently focused on certain economic interests) that are nonetheless vital in supporting forestry and the well-being of forest peoples (Arora-Jonsson 2013).

Separate spaces or networks supported from outside mainstream decision-making forums have been important for women to press their demands in forest contexts, suggesting a need for new thinking about forest governance. These demands have been expressed when women have organised across loyalties such as ethnicity, indigeneity, class or caste (Arora-Jonsson 2013). In such cases, support (not necessarily monetary) from the outside (NGOs, government agencies) has been important for women's groups to make claims and be heard (Arora-Jonsson 2013, Schroeder 1999, Sundar 1998). Similarly, support for collectives may also need to be combined with individualised support, as shown by microfinance interventions such as an HIV-prevention measure for transgender and cisgender[6] women using drugs (Lall et al. 2017). It is clear that no one solution fits all contexts. Women's groups are not the solution in all contexts. Even in the same place, different strategies might be needed at different times, including both individual and collective approaches.[7]

5.5.1 Implications

Better, more equitable management is required. The empirical research makes clear that decision-making would represent a wider diversity of interests related to forests if marginalised people were included. Research also suggests that greater democratic management leads to better environmental outcomes.

To achieve equitable management, individualised support is important, but so is a wider systemic approach. Hurdles to women's actual participation exist even when there are quotas or legally recognised governance rights. Contextual factors need to be addressed in each case to ensure the realisation

[5] Arora-Jonsson borrows the term from de Lauretis, who explains 'space-off' as the 'spaces in the margins of hegemonic discourses, social spaces carved in the interstices of institutions and in the chinks and cracks of the power-knowledge apparati. It is there that the terms of a different construction of gender can be posed … in the micropolitical practices of daily life and daily resistances that afford both agency and sources of power' (1989: 25).

[6] Cis women identify with the gender they are assigned at birth, unlike transgender women, who do not.

[7] For example, in Odisha, some women's microcredit groups were groups in name only. Others carried out microcredit activities, but were controlled by a few men from village committees. Still others were able to use the space provided to make demands and bring about change (Arora-Jonsson, 2013).

of women's governance rights. A willingness to address systemic obstacles and undertake structural change is required, rather than merely adding women and other marginalised groups to existing structures. Practically, this would entail recognising the 'space off' of forestry in mainstream discussions and a disposition to consider dynamic and flexible organisational forms. If we are serious about involving women as a group in decision-making on forests, we need to allow possibilities for women to take part in decision-making in various ways – in concert with each other, from within their own spaces and also as individuals from within formal institutions (Arora-Jonsson 2013).

5.6 Rights to Economic Resources and Control over Land and Resources – Target 5.A

This section discusses women's rights to resources including: (1) rights, access and control over land, and (2) policies and projects on income generation designed to give women increased economic resources.

5.6.1 Tenure Rights

Strong legislation and accessible mechanisms to implement women's forest rights are essential, but the enabling conditions needed for such achievements must be understood within local contexts. Depending on context, law can work as perceived authority preventing violence and enabling women's rights in ways that village leadership may be unable to do. Rao (2007) advocates hybridity in the pursuit of gender equality, wherein socially validated rights are addressed alongside formal instances of law. Rankin (2003) calls for recognition of the limits to undertaking change within the boundaries of households or villages, cautioning against strengthening place-based institutions such as the local civil society built on hierarchical premises. Specific components of secure tenure[8] must be considered carefully before and during actions designed to secure women's tenure rights and legal entitlements.

The question of collective as opposed to individual rights is extremely pertinent in the forest context since a substantial percentage of forest-dependent communities, especially in the Global South, rely on community-based tenure systems (formal and/or informal) in order to control and access land. Under many such collective tenure systems, the community rather than any individual community member is understood as 'owning' land. This can make the situation for women more complex to understand and to guarantee

[8] Tenure refers to the bundle of legal entitlements that comprise a landholder's ability to control, use, access and benefit from land and natural resources.

rights due to various layers of governance, especially in customary systems. In several instances in South Asia, women's customary rights of access were undermined by the creation of forest management committees that created new rules of access disregarding women's previous customary access (Agarwal 1995). Interestingly, a recent study across the 80 community-based legal frameworks analysed within 30 low- and middle-income countries found that the frameworks that provided the strongest legal protection for communities as a whole also provided the most robust protection for women community members (RRI 2017).

Projects with little understanding of the context often create new inequalities. For instance, Chung (2017) examined a large-scale land deal for industrial sugar-cane production in the coastal region of Tanzania. The forms for land valuation and compensation claims distributed by the government only included the names of male household heads (unless the households were headed by females) 'It was assumed that husbands and fathers were the de facto owners of the land and that they were the ones that would collect compensation payments on behalf of their families' (Chung 2017: 115). Even though wives of landowners saw this as common practice, nearly all women interviewed speculated that their husbands would use the money and/or run away with it to pay bride-wealth for younger women, leaving them and their children destitute. Women of the Barabaig tribe (among whom polygyny is actively encouraged) noted that some men took on more wives or mistresses, using the prospect of cash compensation as credit.

5.6.2 Income Generation

The ubiquitous adoption of income-generation programmes and policies advocating the opening of markets for women (connections with SDG 9 – Industry, Innovation and Infrastructure) are also relevant for women's rights to resources. In forestry contexts, these have included the establishment of markets for NTFPs, making handicrafts, being involved in forest work, etc. These initiatives are meant as a key to empowerment, improved family income and nutrition, and children's education.

Access to markets is important for women, but the benefits depend on the kind of control they have over their own involvement and its implications for forest sustainability. For example, in Burkina Faso's plans for REDD+ programmes, engaging the women-dominated shea trade is an important strategy. Related decisions were made, however, without consulting the women or women's groups involved (Westholm and Arora-Jonsson 2015). An example from Odisha, India, shows how women may work to circumvent such disempowerment: women from the lower castes in some areas opted not to sell their bamboo goods in the new local markets especially for NTFPs because men made all the decisions on pricing and location. Instead, they chose to sell in kind only to other women in order to have control over their trade (Arora-Jonsson 2013). This was an attempt to address their own needs and to avoid situations where dependency on markets led to further poverty for their communities. In Burkina Faso, the linking up to international markets has led to increased pressure to provide shea nuts and consequent environmental degradation (Elias and Arora-Jonsson 2017). These instances provide important insights to consider when addressing SDG 5 and SDG 1 (No Poverty). They also challenge targets proposed by SDG 9 (Industry, Innovation and Infrastructure).

Many income-generation programmes meant ostensibly to empower women have in fact become projects where women are expected to raise incomes for domestic use and take part in projects that serve other development and market agendas (Chant 2016). Indigenous women's income-generating activities with respect to large-scale oil palm projects in Indonesia provided women with sorely needed income. In some communities this income was accompanied by a more equal distribution of household work among partners. In others, women's workload simply increased in ways that were fundamentally unsustainable since domestic tasks did not decrease. Furthermore, working conditions for women were part-time, insecure and lower paid (Li 2015).

Indeed, a significant body of research suggests that women's agency and well-being cannot be secured through an increase in income alone. A study of intra-household decision-making in 3000+ households in Ethiopia (Mabsout and van Staveren 2010) found that increases in women's incomes do not necessarily result in increased decision-making power within households. Instead, many women compensated for having taken on some of men's ideal responsibilities by more assiduously performing women's traditional roles, including submission. The research on microcredit enterprises, extremely popular in the last two decades, reports similar findings. There are no automatic benefits since the programmes insufficiently explore the dynamics of women's social networks (Maclean 2010) and ignore the role of men and of gender power dynamics (Chant 2014).

5.6.3 Implications

Tenure rights: Given the cross-cutting nature of women's tenure rights, legal reforms that strengthen these rights must extend beyond the realm of forestry as traditionally conceived. They must address women's underlying needs with respect to family law, access to justice and a broad array of obstacles to women's economic agency (i.e. their capacity for choice and action) that render them vulnerable to both economic and physical forms of violence. Women and men need support to be able to negotiate changing norms within the community.

Work is needed to devise constructive approaches to bolstering women's forest rights and titling, which can have negative effects without sufficient attention to particular contexts. This is especially so for women's livelihoods in cases where women are dependent on commonly held and managed resources. Strong norms that 'good women do not inherit land' (Rao 2008) put pressure on women not to claim a share in their inheritance even if there are no brothers. Mainstream neoliberal policies that advocate individual ownership can backfire for women, who may be exploited and divested of their titles (Ramdas 2009, Ahlers and Zwarteveen 2009).

How government agencies act in such contexts is of vital importance. Institutions directly involved in the formalisation processes granting forest and land rights often prioritise men. More creativity is needed to develop titling processes that positively affect women's lives in different social and cultural contexts.

Income generation: Income generation and involvement in projects can be counterproductive and can become a 'feminisation of responsibility' (Chant 2016). But they can also have surprising outcomes, not necessarily related to the income they generate but to the space they provide for women to take up their particular concerns (Arora-Jonsson 2013). In such cases, outside support for women is important. For example, in a Dominican Republic project, support from NGOs enabling women to control their money gave them an edge in negotiating HIV protective behaviour vis-à-vis their partners (Ashburn et al. 2008) – linked also to SDG 3 (Good Health and Well-Being). Likewise, in Bangladesh participation in microcredit-based productive activity (SDG 1 – No Poverty) protected women from poor communities against marital violence (Hadi 2005), and in South Africa women participating in women's HIV support groups received both practical and emotional support (Dageid and Duckert 2008). In such cases, the role of outsider help, such as from NGOs or officials, is vital.

Strategic and practical interests: These debates touch on the question of women's strategic and practical interests in relation to forests and how those are intertwined. Feminists have long been critical of development interventions such as income-generation programmes that might fulfil individual women's practical interests but fail to deliver on larger structural changes and their strategic interests. It is believed that strategic interests are those derived from an analysis of women's subordination and the formulation of strategic objectives to overcome it. Practical gender interests, on the other hand, arise from the concrete conditions of women's lives and are usually a response to an immediate perceived need. They do not generally entail a strategic goal such as women's emancipation or gender equality (Molyneux 1985).

Arora-Jonsson (2013) draws on her studies of women's grassroots activism in forest contexts to argue that though useful to understand some aspects of discrimination, this division of women's lives into practical and strategic interests can be counterproductive. The articulation of practical needs and claim-making in relation to forestry have in themselves challenged structural disadvantage. On the other hand, strategic considerations such as tenure or a conscious involvement of women in organisations for forest governance could well become a tool to rubber stamp existing inequalities.

These studies suggest that there is no clear division between the two. Depending on the process and the context, income-generation activities or short-term practical help for individual women in forestry contexts *can* lead to larger changes, as can structural change initiatives. What is clear is the need for supporting structures in the home and in public that enable women and other marginalised or vulnerable groups to participate in maintaining forests sustainably and to voice their particular interests.

5.7 Enhance the Use of Enabling Technology for Women's Empowerment – Target 5.B

Information and communication technologies (ICTs) are enabling the participation of women in social, political and economic processes at greater rates than in the past (Alves and Steiner 2017). However, structural and socio-cultural factors, including poverty, illiteracy and gender norms and practices, may limit women's access to ICTs and other technologies, with resulting impacts on women's empowerment and agricultural productivity (Mogues et al. 2009, O'Brien et al. 2016). Agricultural extension work and access to technology packages focus more on men, tending to view women as less important to development initiatives (Mogues et al. 2009). O'Brien et al. (2016) show that involving men and women as couples in technology-related training events can improve women's access to new and emerging agricultural information.

Gender biases in technology access and dissemination and disparities in information access have consequences even when women are part of the decision-making process (Mwangi et al. 2011). A study on female ICT intermediaries in rural China and India found that although access to ICTs could improve women's status in the short term, such gains were reversed in the absence of broader changes in society. The choices they made while using ICTs always took place in the context of 'societal expectations (and their own) [which] continued to be structured by patriarchal values' (Oreglia and Srinivasan 2016: 506).

Socio-cultural values and gender norms are, however, not static or unchallenged. Masika and Bailur (2015) argue that ICTs should be understood not as automatic sources of women's empowerment, but as a site of contestation where women carefully calibrate gender relations in complex ways. Clearly, more attention should be paid to women's socio-cultural contexts and the bargains they make. In comparison to men, they are less likely to have wide informal networks they can tap into to fulfil their needs.

5.7.1 Implications

The research on ICTs and gender draws attention to the significance of power structures and societal norms in shaping women's access to technologies and their impacts on gender relations. This has implications for facilitating women's agency and empowerment. Given this, addressing structural factors that shape forestry management is more critical than focusing exclusively on individual constraints.

Additionally, technologies should be targeted in ways that facilitate women's technology adoption. This could be accomplished, for example, by ensuring that extension visits to user groups be undertaken by both men and women, or by raising awareness and providing requisite skills among male agents of the sustainability benefits of proactively engaging women resource users (Mwangi et al. 2011). Authors suggest that (1) technologies should be designed to take into account women's time constraints, and (2) extension should be accompanied by training in the skills necessary for ensuring and sustaining technology adoption (Mwangi et al. 2011). More structured and formal access for women and other marginalised groups is also necessary. ICT can play an important role in that process.

5.8 Policies and Legislation for Gender Equality and Empowerment – Target 5.C

Not all gender-sensitive policies are necessarily implemented. Indeed, practice often strays far from the progressive language of policymaking. Passing gender-sensitive laws and policies is nevertheless a key component of gender equality. The manner in which gender-sensitive laws are crafted, implemented and legislated have significant consequences on their enforceability.

Emphasis on women or gender in policies often implies an increase in responsibility. Ecofeminist arguments about women's closeness to nature were mobilised by bureaucrats to enrol women in conservation and soil improvement programmes, primarily increasing women's workload without much desirable change in their everyday lives (Leach 2007). In other cases, assumptions about women's vulnerability to climate change have led to policymaking that has increased their responsibilities rather than addressing their disadvantages (Arora-Jonsson 2011). Holmes and Jones (2013) observe that policies often reproduce unequal gender norms. Social policy has made positive inroads, but it needs to move beyond reproducing harmful stereotypes that define women only as mothers and men as oppressors.

Critics have argued that gender mainstreaming policies have served merely to bureaucratise gender and that adding women to existing programmes merely underwrites their previous invisibility by reducing them to a check mark on required forms. This has absolved agencies from doing anything substantive about gender discrimination. Some have in fact argued for doing away with the idea of gender mainstreaming altogether (see Arora-Jonsson, 2014, for an overview of this debate).

5.8.1 Implications

An individual-based approach can disregard systemic gender-based discrimination. For example, Sweden's forest-sector policy on gender equality encourages women to take an active part in the forest sector and focuses on their role as economic agents benefitting the sector. While these overtures to individual women are important, the approach ignores the systemic reasons for women's absence from the sector (Holmgren and Arora-Jonsson 2015).

On the other hand, a gender-neutral approach or the absence of policies on gender can make it difficult for officials and others wanting to create space for change. For example, while Swedish environmental policy has promoted gender equality as an important cornerstone of its work in development aid, there has been little attention to gender in domestic environmental policies. This lack of policy support within the country has made it more difficult for forestry and environmental officials to challenge discrimination related to gender and power relations. On the other hand, its policies on gender in development aid have forced NGOs and others to pay attention to gender-based discrimination, which has sometimes helped women to organise themselves (see Case Study 16.4 in Chapter 16).

The discourse on gender has made space for intersectional approaches within international organisations in a way that was previously much more difficult (Arora-Jonsson and Sijapati 2018). This indicates that there is a need for policies on gender, but they must allow for hybridity (Rao 2007) and flexibility (Arora-Jonsson 2013) and enable a careful analysis of gender and power relations in each context. Blanket statements about the poverty or vulnerability of all women or women's closeness to the environment often become counterproductive to the interests of women.

5.9 Conclusions

One of the major conclusions from the literature overview is the importance of understanding the contextual and systemic nature of inequalities if we want to act for greater justice and sustainable forestry. There are no automatic gains in gender equality from greater development, expansion of markets for women, inclusion in forestry forums or poverty alleviation programmes. They might bring economic benefits to some, but for others they can exacerbate adverse conditions. As is clear from the instances cited in the chapter, concern for the dignity and welfare of forest-based peoples requires contextual responses that go beyond these measures. They need institutional support and structural change from 'business as usual'. As is clear from the research discussed here, forest governance and everyday management are upheld by a superstructure of gendered forest relations (invisible to mainstream forestry) that often disadvantages women as a social group. Paying close attention to this 'space-off' of forestry is vital if we are to reach towards sustainable and equitable forest relations promoted by the SDGs.

Forests are a key site where the goal of sustainable development and its linkages with gender equality play out. Yet, there are significant challenges and barriers to the implementation of SDG 5 across the North and South. While the contexts in these places differ greatly, similar features recur in forestry contexts across the world.

Decision-making on forests at all levels is dominated by groups of men from certain castes, class or age groups. Women often have less access to the information needed for decision-making. Men are also overwhelmingly the targets for forestry interventions – reflective of current tenure systems wherein more men than women own forest land. However, beyond ownership, perception biases as well as gender norms and values tend to position forestry as a male domain. Poverty and the lack of supportive infrastructure in countries in the Global South do correlate with discrimination, but it is also clear that welfare and development do not automatically lead to greater gender equality, and inequitable relations of power in forestry stretch across the Global North and South (Arora-Jonsson 2013). This is true in universities and international organisations where the legacy of purely technical approaches to forestry education is still entrenched. Thinking through how SDG 5 targets may be applied in various forestry contexts provides a space for new ideas to emerge and to challenge convention at a time when new directions are sorely needed.

Taking SDG 5 seriously implies a fundamental change in approaches to forests and the environment – one that incorporates systemic and contextual factors as well as people's relations outside of forestry. This change entails learning from the past. New forest-related initiatives have yet to take up these gender lessons (e.g. REDD+).

Progress will entail taking into account connections between the Global North and South. Forestry as a profession and field of work has interconnected features in its organisation and the ideologies that drive it. The responses to challenges within it must also recognise these interconnections. For example, violence in the forests in the Global South often results from struggles with multinational companies based in the Global North, which derive their influence from their work and trade there. Concepts that travel between the North and South have different implications in different places. For instance, women in (Indigenous) communities in some areas of the Global South risk backlash when using the term 'gender'. It has been associated with taking power away from men and with Northern or external ideologies that threaten custom (Geetha 2002). In other cases women have taken up the English term 'women's rights' (even if they do not speak English) to forward their claims as a group. The likelihood of success with this strategy has been enhanced with support from NGOs and others that have also engaged men's help. In contrast, questions of gender in forest contexts in the Global North have been ignored by practitioners, with the argument that gender equality has already been reached (Arora-Jonsson 2013). Taking account of SDG 5 in different contexts requires different strategies, as well as paying attention to the various connections between contexts and scales.

Closer attention to SDG 5 highlights the invisible labour and relations so crucial to good forest management, and helps to develop democratic and sustainable strategies so key to forest relations and of benefit to forest people (with close connections to most other SDGs – especially SDGs 1–4, 6, 7, 10, 13–16). Such attention could promote voice and a focus on dignity and rights. It could demand compliance on the part of institutions that perpetrate injustice, sometimes unconsciously, by carrying on with business as usual (in relation to SDGs 8, 9, 17) – such as the current focus on business within forestry and agriculture and the assumption of gender neutrality, as in the Indonesian palm oil case (Section 5.6), in the ways education is gendered in content and the student body, the provision of information and so on.

Forest management can improve with the involvement of heterogeneous groups, and especially women (Agarwal 2010, Mwangi et al. 2011). Increasing women's access to funds and social provisions such as childcare allowances benefits their families and larger communities (e.g. Bergmann 2008, Butler et al. 2012). It provides some recompense for their 'invisible work' in forests that remains otherwise unacknowledged. Valuing such care work as well as women's work in the forests on a par with other conventional forest activities is likely to increase productivity and reduce inequalities (Baker 2008). As women's household-level bargaining power increases, rates of HIV and undernutrition and violence decrease (Ashburn et al. 2008, Mabsout and van Staveren 2010), and additional energy, creativity and motivation to sustain forests could be brought to bear.[9] These aspects are enhanced by more secure forest rights (RRI 2017).

The gender-neutral framing of the other SDG goals could undermine efforts towards rights called for in SDG 5. This requires serious attention to the norms that form the basis of many of the other SDGs. For example, SDG 8's focus on economic growth could lead to serious disadvantages for marginalised groups. The call for decent work for women, without acknowledging the underlying discrimination within the current system, could merely exacerbate gender inequalities. The focus on trade and women's involvement in markets called for in SDGs 8 and 9 can undermine both the environment (Elias and Arora-Jonsson 2017) and gender equality (Li 2015).

From the point of view of SDG 5, political will is needed to transform unequal relationships, challenge privilege based on sex, class, ethnicity or caste, and destabilise inequitable micro- and macroeconomic structures (based on notions of private property, commodification etc.). Forest agencies and other actors need to interact seriously with other governmental agencies, NGOs and others that provide services and pay attention to community concerns beyond the forests. Greater democratic governance of forests leads to better forest management, but the institutional forms need to be flexible and responsive to the context. Companies need to be more accountable, and forest-sector education needs to expand to include the social and the cultural. The International Union of Forest Research Organizations (IUFRO) has an important role to play in this context – in questioning business as usual in what IUFRO does, and also in undertaking gender research and analysis of forestry policies and programmes to better understand gender dynamics in forestry. The welfare and dignity that achieving SDG 5 would bring to forest peoples and livelihoods is essential to ensuring better managed and sustainable forests.

[9] The Center for International Forestry Research's Adaptive Collaborative Management programme, which worked closely with rural forest women in 11 countries in the early 2000s, saw these benefits accrue (see www.cifor.org/acm/ and Colfer 2005).

References

Adhikari, B. and Di Falco, S. 2008. *Social inequality and collective action: An empirical study of forest commons*. IFRI Working Paper W081-5. Ann Arbor, MI, USA: University of Michigan.

Afridi, F., Iversen, V. and Sharan, M. R. 2017. Women political leaders, corruption, and learning: Evidence from a large public programme in India. *Economic Development and Cultural Change* 66:1–30.

Agarwal, B. 1995. *A field of one's own: Gender and land rights in South Asia*. Cambridge: Cambridge University Press.

Agarwal, B. 2010. *Gender and green governance: The political economy of women's presence within and beyond community forestry*. Oxford: Oxford University Press.

Agarwal, S. and Saxena, A. K. 2018. *People's forests: Is community forest resource governance the future of India's jungles?* New Delhi: Centre for Science and Environment.

Ahlers, R. and Zwarteveen, M. 2009. The water question in feminism. *Gender, Place & Culture* 16:409–26.

Alhabib, S., Nur, U. and Jones, R. 2010. Domestic violence against women: Systematic review of prevalence studies. *Journal of Family Violence* 25:369–82.

Alves, E. E. C. and Steiner, A. Q. 2017. Globalization, technology and female empowerment: Breaking rights or connecting opportunities? *Journal for Quality-of-Life Measurement* 133: 859–77.

Arora-Jonsson, S. 2011. Virtue and vulnerability: Discourses on women, gender and climate change. *Global Environmental Change* 21:744–51.

Arora-Jonsson, S. 2013. *Gender, development and environmental governance*. New York: Routledge.

Arora-Jonsson, S. 2014. Forty years of gender research and environmental policy: Where do we stand? *Women's Studies International Forum* 47:295–308.

Arora-Jonsson, S. and Sijapati, B. B. 2018. Disciplining gender in environmental organizations: The texts and practices of gender mainstreaming. *Gender, Work & Organization* 25:309–25.

Arza, C. 2017. Non-contributory benefits, pension re-reforms and the social protection of older women in Latin America. *Social Policy and Society* 16(3):361–75.

Asfaw, S. 2016. 'From protection to production': Do social cash transfer programmes promote agricultural activities and livelihoods? *Global Social Policy* 16:205–8.

Ashburn, K., Kerrigan, D. and Sweat, M. 2008. Micro-credit, women's groups, control of own money. *AIDS and Behavior* 12:396–403.

Baker, J. 2008. All things considered, should feminists embrace basic income? *Basic Income Studies* 3(3). https://doi.org/10.2202/1932-0183.1129

Barker, G. and Ricardo, C. 2005. Young men and the construction of masculinity in sub-Saharan Africa. In Bannon, I. and Correia, M. C. (eds.) *The other half of gender: Men's issues in development*. Washington, DC: The World Bank, pp. 159–94.

Behrman, J. Karelina, Z., Peterman, A., Roy, S. and Amelia Goh, A. (eds.) 2014. *A toolkit on collecting gender & assets data in qualitative & quantitative program evaluations*. Washington, DC: International Food Policy Research Institute (IFPRI).

Bergmann, B. R. 2004. A Swedish-style welfare state or basic income: Which should have priority? *Politics & Society* 32:107–18.

Bergmann, B. R. 2008. Basic income grants or the welfare state: Which better promotes gender equality? *Basic Income Studies* 3(3). https://doi.org/10.2202/1932-0183.1128.

Bhattacharya, M., Bedi, A. S. and Chhachhi, A. 2011. Marital violence and women's employment and property status: Evidence from North Indian villages. *World Development* 39:1676–89.

Budlender, D. and Alma, E. 2011. *Women and land: Securing rights for better lives*. Ottawa, Canada: International Development Research Centre.

Butler, L. GY, K., NA. A. et al. 2012. Microcredit-nutrition education link: A case study analysis of Ghanaian women's experiences in income generation and family care. *African Journal of Food, Agriculture, Nutrition and Development* 12.

Camilleri-Cassar, F. 2017. About time: Gender equality in Malta's working-time regime? *Social Policy and Society* 16:561–75.

Chandra-Mouli, V., Svanemyr, J., Amin, A. et al. 2015. Twenty years after International Conference on Population and Development: Where are we with adolescent sexual and reproductive health and rights? *Journal of Adolescent Health* 56:S1–S6.

Chant, S. 2014. Exploring the 'feminisation of poverty' in relation to women's work and home-based enterprise in slums of the Global South. *International Journal of Gender and Entrepreneurship* 6:296–316.

Chant, S. 2016. Women, girls and world poverty: Empowerment, equality or essentialism? *International Development Planning Review* 38:1–24.

Chung, Y. B. 2017. Engendering the new enclosures: Development, involuntary resettlement and the struggles for social reproduction in coastal Tanzania. *Development and Change* 48:98–120.

Colfer, C. J. P. (ed.) 2005. *The equitable forest: Diversity, community and resource management.* Washington, DC: Resources for the Future and CIFOR.

Colfer, C. J. P. 2011. Marginalized forest peoples' perceptions of the legitimacy of governance: An exploration. *World Development* 39(12):2147–64. doi:10.1016/j.worlddev.2011.04.012.

Cools, S. and Kotsadam, A. 2017. Resources and intimate partner violence in sub-Saharan Africa. *World Development* 95:211–30.

Dageid, W. and Duckert, F. 2008. Balancing between normality and social death: Black, rural, South African women coping with HIV/AIDS. *Qualitative Health Research* 18:182–95.

De Lauretis, T. 1989. *Technologies of Gender: Essays on Theory, Film, and Fiction.* Basingstoke: Macmillan.

Dolan, C. 2002. Collapsing masculinities and weak states. In Cleave, F. (ed.) *Masculinities matter! Men, gender and development.* New York: Zed Books, pp. 57–83.

DS (Departmentsserien) 2004. *Det går långsamt fram: Jämställdheten inom jord – och skogsbrukssektorn.* Stockholm: Jordbruksdepartementet Regeringsdepartementet.

Due, J. M., Magayane, F. and Temu, A. 1997. Gender again – views of female agricultural extension officers by smallholder farmers in Tanzania. *World Development* 25(5):713–25.

Duvander, A.-Z., Lappegård, T. and Andersson, G. 2010. Family policy and fertility: Fathers' and mothers' use of parental leave and continued childbearing in Norway and Sweden. *Journal of European Social Policy* 20:45–57.

Eduards, M. 2002. *Förbjuden handling*. Malmö: Liber ekonomi.

Elias, M. and Arora-Jonsson, S. 2017. Negotiating across difference: Gendered exclusions and cooperation in the shea value chain. *Environment and Planning D: Society and Space* 35:107–25.

Estévez-Abe, M. and Hobson, B. 2015. Outsourcing domestic (care) work. *Social Politics: International Studies in Gender, State & Society* 22:133–46.

Evans, K., Flores, S., Larson, A. M. et al. 2017. Challenges for women's participation in communal forests: Experience from Nicaragua's Indigenous territories. *Women's Studies International Forum* 65:37–46.

FAO 2018. *The State of the World's Forests 2018: Forest pathways to sustainable development*. Rome: FAO. Available at: www.fao.org/3/ca0188en/ca0188en.pdf (Accessed 5 February 2019).

Faxon, H., Furlong, R. and Sabe Phyu, M. 2015. Reinvigorating resilience: violence against women, land rights, and the women's peace movement in Myanmar. *Gender & Development* 23: 463–79.

Figart, D. and Mutari, E. 2000. Work time regimes in Europe: Can flexibility and gender equity coexist? *Journal of Economic Issues* 34(4):847–71.

Geetha, V. 2002. *Gender (theorizing feminism)*. Calcutta: Stree.

Ghosh, J. 2015. Growth, industrialisation and inequality in India. *Journal of the Asia Pacific Economy* 20:42–56.

Global Witness 2017. *Defenders of the Earth: Global killings of land and environmental defenders 2016*. London: Global Witness.

Grabe, S. 2010. Promoting gender equality: The role of ideology, power, and control in the link between land ownership and violence in Nicaragua. *Analyses of Social Issues and Public Policy* 10:146–70.

Hadi, A. 2005. Women's productive role and marital violence in Bangladesh. *Journal of Family Violence* 20:181–9.

Herawati, T. H., Mwangi, E., Larson, A. et al. 2017. *Forest tenure reform implementation:* Presented at the XVI Biennial IASC Conference 'Practicing the commons: Self-governance, cooperation, and institutional change', in Utrecht, The Netherlands, 11 July 2017.

HLPF 2017. *Thematic review of SDG 5: Achieve gender equality and empower all women and girls*. Available at: https://sustainabledevelopment.un.org/content/documents/14383SDG5format-revOD.pdf (Accessed 3 September 2018).

Hoang, T.-A., Quach, T. T. and Tran, T. T. 2013. 'Because I am a man, I should be gentle to my wife and my children': Positive masculinity to stop gender-based violence in a coastal district in Vietnam. *Gender & Development* 21:81–96.

Holmes, R. and Jones, N. 2013. *Gender and social protection in the developing world: Beyond mothers and safety nets*. London: Zed Books.

Holmgren, S. and Arora-Jonsson, S. 2015. The Forest Kingdom – with what values for the world? Climate change and gender equality in a contested forest policy context. *Scandinavian Journal of Forest Research* 30(3):235–45.

Islam, M. R. and Shamsuddoha, M. 2017. Socioeconomic consequences of climate induced human displacement and migration in Bangladesh. *International Sociology* 32:277–98.

Jafry, T. and Sulaiman V. R. 2013. Gender-sensitive approaches to extension programme design. *The Journal of Agricultural Education and Extension* 19:469–85.

Jayal, N. G. 2003. Locating gender in the governance discourse. In Nussbaum, M., Basu, A., Tambiah Y. and Jayal, N. G. (eds.) *Essays on gender and governance*. New Delhi: United Nations Development Program, pp. 96–142.

Jewkes, R. 2002. Intimate partner violence: Causes and prevention. *Lancet* 359:1423–29.

Kaldal, I. 2000. Skog, Arbeid og daglivliv i Kvinners og Mens Fortellninger fra Trysil og Nord-Värmland Etter 1930. In Kandal, I., Johansson, E. and Fritzböger, B. (eds.) *Skogsliv: Kulturella processer i skogsbygden*. Lund: Historisk Media, pp. 85–117.

Kumari, N. 2017. A grassroot picture of untouchability practices against Dalit women in Haryana. *International Journal of Multidisciplinary Approach and Studies* 4(4):76–85.

Kusuma, Y. S. and Babu, B. V. 2017. Elimination of violence against women and girls as a global action agenda. *Journal of Injury and Violence Research* 9:117–21.

Lagdon, S., Armour, C. and Stringer, M. 2014. Adult experience of mental health outcomes as a result of intimate partner violence victimisation: A systematic review. *European Journal of Psychotraumatology* 5:24794.

Lall, P., Shaw, S. A., Saifi, R. et al. 2017. Acceptability of a microfinance-based empowerment intervention for transgender and cisgender women sex workers in Greater Kuala Lumpur, Malaysia. *Journal of the International AIDS Society* 20:21723.

Lama, A. S., Kharel, S. and Ghale, T. 2017. When the men are away: Migration and women's participation in Nepal's community forestry. *Mountain Research and Development* 37:263–70.

Lambrou, Y. and Nelson, S. 2010. *Farmers in a changing climate: Does gender matter? Food security in Andhra Pradesh, India*. Rome: FAO.

Larson, A. M., Solis, D., Duchelle, A. E. et al. 2018. Gender lessons for climate initiatives. *World Development* 108:86–102.

Leach, M. 2007. Earth mother myths and other ecofeminist fables: How a strategic notion rose and fell. *Development and Change* 38:67–85.

Lewis, J. and Plomien, A. 2009. 'Flexicurity' as a policy strategy: The implications for gender equality. *Economy and Society* 38:433–59.

Li, T. M. 2015. *Social impacts of oil palm in Indonesia: A gendered perspective from West Kalimantan*. Bogor, Indonesia: CIFOR.

Lin, C. Y. O. 2008. Autonomy reconstituted: Social and Gender Implications of Resettlement on the Orang Asli of Peninsular Malaysia. In Resurreccion, P. B. and Elmhirst, R. (eds.) *Gender and natural resource management: Livelihoods, mobility and interventions*. London: Eartchscan, pp. 109–126.

Lopez-Ruiz, M., Benavides, F. G., Vives. A. and Artazcoz, L. 2017. Informal employment, unpaid care work, and health status in Spanish-speaking Central American countries. *International Journal of Public Health* 62:209–18.

Lundgren, E., Heimer, G., Westerstrand, J. and Kalliokoski, A. 2001. *Slagen dam.* Brottsoffermyndigheten/Uppsala Universitet, Umeå/Uppsala.

Lundgren, R. and Amin, A. 2015. Addressing intimate partner violence and sexual violence among adolescents: Emerging evidence of effectiveness. *Journal of Adolescent Health* 56:S42–S50.

Lwambo, D. 2013. 'Before the war, I was a man': Men and masculinities in the Eastern Democratic Republic of Congo. *Gender & Development* 21:47–66.

Mabsout, R. and van Staveren, I. 2010. Disentangling bargaining power from individual and household level to institutions: Evidence on women's position in Ethiopia. *World Development* 38:783–96.

Maclean, K. 2010. Capitalizing on women's social capital? Women-targeted microfinance in Bolivia. *Development and Change* 41:495–515.

Mairena, E., Lorio, G., Hernández, X., Wilson, C., Müller, P. and Larson, A. M. 2012. *Gender and forests in Nicaragua's Indigenous territories: From national policy to local practice.* CIFOR Working Paper No. 95. Bogor, Indonesia: CIFOR.

Mallory, C. 2006. Ecofeminism and forest defense in Cascadia: Gender, theory and radical activism. *Capitalism Nature Socialism* 17:32–49.

Masika, R. and Bailur, S. 2015. Negotiating women's agency through ICTs: A comparative study of Uganda and India. *Gender, Technology and Development* 19:43–69.

Mills, S. 2006. Segregation of women and Aboriginal people within Canada's forest sector by industry and occupation. *The Canadian Journal of Native Studies* 26:147–71.

Minocha, R. 2015. Gender, environment and social transformation: A study of selected villages in Himachal Pradesh. *Indian Journal of Gender Studies* 22:335–57.

Mogues, T., Cohen, M. and Birner, R. 2009. *Agricultural extension in Ethiopia through a gender and governance lens.* Washington, DC: IFPRI.

Molyneux, M. 1985. Mobilization without emancipation? Women's interests, the state, and revolution in Nicaragua. *Feminist Studies* 11:227–54.

Moore, K. M., Hamilton, S., Sarr, P. and Thiongane, S. 2001. Access to technical information and gendered NRM practices. *Agriculture and Human Values* 18:95–105.

Mwangi, E., Meinzen-Dick, R. and Sun, Y. 2011. Gender and sustainable forest management in East Africa and Latin America. *Ecology and Society* 16. Available at: www.ecologyandsociety.org/vol16/iss1/art17/ (Accessed 28 July 2019).

Nordlund, A. and Westin, K. 2011. Forest values and forest management attitudes among private forest owners in Sweden. *Forests* 2:30.

O'Brien, C., Gunaratna, N. S., Gebreselassie, K. et al. 2016. Gender as a cross-cutting issue in food security. *World Medical & Health Policy* 8:263–86.

Oreglia, E. and Srinivasan, J. 2016. ICT, intermediaries, and the transformation of gendered power structures. *MIS Quarterly* 40:201–10.

Özcan, N. K., Günaydın, S. and Çitil, E. T. 2016. Domestic violence against women in Turkey. *Archives of Psychiatric Nursing* 30:620–29.

Panda, P. and Agarwal, B. 2005. Marital violence, human development and women's property status in India. *World Development* 33:823–50.

Peng, I. 2011. The good, the bad and the confusing: the political economy of social care expansion in South Korea. *Development and Change* 42:905–23.

Pividori, C. and Degani, P. 2018. Reflecting on criminalizing male violence against women under human rights and human security discourses: A feminist legal and policy analysis. *Global Jurist* 10. https://doi.org/10.1515/gj-2017-0028.

Proulx, C. 2016. The provision of unpaid care across cohorts and gender: A research note. *Canadian Studies in Population* 43. https://doi.org/10.25336/P6SW37.

Qi, L. and Dong, X.-Y. 2016. Unpaid care work's interference with paid work and the gender earnings gap in China. *Feminist Economics* 22:143–67.

Quisumbing, A. R. and Pandolfelli, L. 2010. Promising approaches to address the needs of poor female farmers: Resources, constraints and interventions. *World Development* 38:581–92.

Ramdas, S. 2009. Women, forestspaces and the law: Transgressing the boundaries. *Economic and Political Weekly* XLIV:65–74.

Rankin, K. N. 2003. Cultures of economies: Gender and socio-spatial change in Nepal. *Gender, Place & Culture* 10:111–29.

Rao, N. 2007. Custom and the courts: Ensuring women's rights to land, Jharkhand, India. *Development and Change* 38:299–319.

Rao, N. 2008. *Good women do not inherit land: Politics of land and gender in India*. New Delhi: Social Science Press.

Ray, B., Mukherjee, P. and Bhattacharya, R. N. 2017. Attitudes and cooperation: Does gender matter in community-based forest management? *Environment and Development Economics* 22:594–623.

Reed, M. G. 2008. Reproducing the gender order in Canadian forestry: The role of statistical representation. *Scandinavian Journal of Forest Research* 23:78–91.

RRI 2017. *Power and Potential: A comparative analysis of national laws and regulations concerning women's rights to community forests*. Washington, DC: RRI.

RRI 2018. *Cornered by protected areas*. Washington, DC: RRI.

Sandoval-Ferro, B. 2013. *Overcoming inequalities without challenging women's loyalty to the Indigenous community* (master's thesis). Swedish University of Agricultural Sciences, Uppsala.

Scheffran, J., Ide, T. and Schilling, J. 2014. Violent climate or climate of violence? *The International Journal of Human Rights* 18:369–90.

Schroeder, R. 1999. *Shady Practices: Agroforestry and gender politics in The Gambia*, Berkeley: University of California Press.

Simon-Kumar, R., Kurian, P., Silcock, F. and Narasimhan, N. 2017. Mobilising culture against domestic violence in migrant and ethnic communities: Practitioner perspectives from Aotearoa/New Zealand. *Health and Social Care in the Community* 25(4):1387–95.

Staab, S. and Gerhard, R. 2011. Putting two and two together? Early childhood education, mothers' employment and care service expansion in Chile and Mexico. *Development and Change* 42:1079107.

Starbird, E., Norton, M. and Marcus, R. 2016. Investing in family planning: Key to achieving the Sustainable Development Goals. *Global Health: Science and Practice* 4:191–210.

Sundar, N. 1998. Asna women: Empowered or merely enlisted? In Kalland, A. and Persson, G. (eds.) *Environmental Movements in Asia*. Richmond: Curzon Press, pp. 227–49.

Swaminathan, H. Ashburn, K., Kes, A. et al. 2008. *Women's property rights, HIV and AIDS, and domestic violence. Research findings from two rural districts in South Africa and Uganda*. Cape Town: Human Sciences Research Council.

Tickell, A. and Peck, J. 1996. The return of the Manchester men: Men's words and men's deeds in the remaking of the local state. *Transactions of the Institute of British Geographers* 21:595–616.

Westholm, L. and Arora-Jonsson, S. 2015. Defining solutions, finding problems: Deforestation, gender, and REDD+ in Burkina Faso. *Conservation and Society* 13:189–99.

Whittenbury, K. 2013. Climate change, women's health, wellbeing and experiences of gender based violence in Australia. In Alston, M. and Whittenbury, K. (eds.) *Research, action and policy: Addressing the gendered impacts of climate change*. Springer, pp. 207–21.

Yoon, J. 2014. Counting care work in social policy: Valuing unpaid child- and eldercare in Korea. *Feminist Economics* 20:65–89.

Zamora, A. and Monterroso, I. 2017. *Una visión regional y local sobre la seguridad de tenencia comunal de la tierra y el bosque en Loreto*. Bogor, Indonesia: CIFOR.

Chapter 6 SDG 6: Clean Water and Sanitation – Forest-Related Targets and Their Impacts on Forests and People

Jaime Amezaga*, James Bathurst*, Andrés Iroumé, Julia Jones, Rajan Kotru, Laxmi Dutt Bhatta and Elaine Springgay

Key Points

- SDG 6 seems unlikely to exert a major influence on forest cover and indeed is unlikely to be pursued with forests in the forefront of consideration.
- Full implementation of Targets 6.1 and 6.2 could positively impact forest people, yet this is not an implementation priority.
- SDG 6 may focus attention on the role of forests (as providers of hydrological ecosystems services) in protecting clean drinking water resources; the exact role of forests here requires careful consideration.
- Particular attention should be given to reforestation strategies to improve water availability in areas with soil degradation and reduced infiltration. Target 6.4 may restrict the spread of new plantations in semi-arid areas and Target 6.5 may drive a more integrated view of catchments and their management.
- It is necessary to consider forest–water interactions at the catchment, regional and continental scales; actions with a beneficial impact at one scale may have an adverse impact at another.

6.1 Introduction

SDG 6 is designed to 'ensure the availability and sustainable management of water and sanitation for all' (UN 2018). It defines clean, accessible water as an essential part of the world we want to live in, one that should be universally and easily accessible across the globe. As we shall see in this chapter, the anthropocentric orientation of the SDG title is later modulated by a more integrative view of some of the specific targets. However, it already shows the potential for conflict in the competition for water under conditions of scarcity. While not a central consideration in the development of SDG 6, there

* Lead authors.

are intrinsic links between forests and water. Trees, as living organisms, need water to exist and thrive; as critical landscape components, they strongly influence water availability at local and continental scales. The exact nature of this influence is still a point of scientific debate, although in recent decades we have witnessed the slow emergence of a more nuanced picture of forest–water relationships. Understanding the potential impacts of SDG 6 on forests and people requires a balanced appraisal of these relationships.

The chapter first briefly summarises the current understanding of forest–water interactions in order to identify the critical SDG 6 targets for forestry. An in-depth discussion of target impacts on forest cover focuses on two areas: South America and South Asia. These two areas have diverse conditions where forests play an important role in upstream/downstream and inter-catchment interactions, where achieving SDG 6 will require active interventions. The final section takes a wider perspective to discuss the key considerations for improving SDG 6 and forestry interactions at the global scale.

6.2 Forest and Water Interactions

The traditional understanding of how forests and water interact is influenced by long-standing beliefs regarding the role of forests in the water cycle, which are not always supported by science (Calder 2005). The current scientific understanding is much more nuanced and needs to be stated clearly in order to understand the links with the new requirements on water management emerging from SDG 6, enabling us to distinguish between positive synergies and potential misconceptions. Important initial considerations for this discussion are the big regional differences worldwide in forest cover, climate zones and land-use changes. The following summary considers both the traditional catchment-scale water balance and recent interest in the recycling of evaporated moisture at much larger, inter-catchment scales.

1. At the catchment scale, decades of research with paired catchments and process studies have shown that, relative to shorter vegetation, forest cover *reduces* catchment run-off at the annual scale because trees have higher rainfall interception rates and higher transpiration rates during dry periods (Andréassian 2004, Bosch and Hewlett 1982, Zhang et al. 2017). Dry season flows are particularly likely to be reduced in forest catchments as tree roots can extract soil water from greater depths than shorter vegetation. Reductions in annual run-off for the most extreme change from 100 per cent grass cover to 100 per cent forest cover can be substantial, ranging from 15 per cent to at least 50 per cent (Fahey and Payne 2017, Marc and Robinson 2007). Run-off reduction has been found at catchment scales as large as thousands of square kilometres (Iroumé and Palacios 2013, Silveira and Alonso 2009, Zhang et al. 2017).

2. Run-off reduction is greatest for young, growing forests. The reduction may be smaller for old mature forests with low leaf-area indices. Different tree species take up water at different rates (Huber et al. 2010).

3. In certain cases, by increasing soil infiltration and thus groundwater recharge, forests may allow a temporally more even redistribution of run-off, thus increasing dry season flows (but still with reduced annual flow) (Calder 2005). Most evidence, however, points to a reduction in dry season flows following afforestation, although the pattern in areas with seriously degraded soils is less clear (Bruijnzeel 2004) and the overall effect is likely to vary with tree density (Ilstedt et al. 2016).

4. The special case of cloud forests, which intercept fog and cloud droplets, may possibly increase annual yields in the very specific (typically mountain) locations in which they occur (Bruijnzeel 2001, 2011). However, the fog formation may itself depend in part on recycling of evaporated moisture from upwind forests.

5. At very large (subcontinental) scales, recycling of forest evapotranspiration potentially increases the downwind rainfall (Ellison et al. 2017, Sheil 2018, van der Ent et al. 2010) and thus run-off (after any interception losses in the recipient catchment).

6. The impact of forest cover on flood peaks, as opposed to run-off, is more controversial, both because the effect on extreme flows is uncertain and because the means of quantifying the impact is disputed (see discussion in Alila and Green 2014).

7. Because of the greater evapotranspiration and consequently lower (on average) soil moisture content in forested catchments, the generally higher infiltration capacity of forest soils and the greater carbon sequestration (which aids water storage), forests can absorb more of the rain and so reduce flood peaks for given low to moderate rainfall events. This effect does not occur if the soil is already saturated, for example from a previous rain event or from soil water accumulated over a wet season (Bathurst et al. 2011).

8. Forests may not be effective in reducing flood peaks produced by extreme (but rare) rainfall events as the above absorption effect is overwhelmed by the amount of rain (Bathurst et al. 2011).

9. Forests can reduce the *frequency* with which a given flood peak occurs for all (not just low to moderate) flood sizes (Kuraś et al. 2012).

10. Forest cover often (but not always) reduces sediment yield compared with other land covers, especially those involving soil disturbance. The annual specific sediment yield in a logged catchment may exceed that in an undisturbed forested catchment by up to one order of magnitude under conditions of best management practice or two orders of magnitude in cases of severe ground disturbance or extreme events coinciding with the logged condition (Bathurst and Iroumé 2014). However, in areas of high natural sediment yields or during certain extreme events such as tropical hurricanes, the vegetation cover may have relatively little impact on overall sediment yield (Calder 2005).

11. By excluding other management approaches (e.g. fertiliser application) and limiting soil erosion, forests usually imply less-polluted water. Deposition of most atmospheric pollutants are generally higher to forests and, in regions of high (industrial) pollution, this has historically caused acidification of catchments and run-off, especially with coniferous forest cover (Calder 2005). However, this threat is reducing in the advanced economies as industrial emissions are controlled and energy production moves away from coal and other fossil fuels. In certain areas (e.g. the southwest of Western Australia), the rise in the water table following the removal of forest cover has resulted in a redistribution of soluble salts, causing severe soil salinisation and loss of crop-growing capacity (Peck and Hatton 2003). Lowering of the water table following afforestation of grasslands has also been associated with soil salinisation (e.g. in Brazil and Hungary) (Jobbágy and Jackson 2004, Tóth et al. 2013).

12. Forest–water interactions have traditionally been studied at the river catchment scale, emphasising the impact of forest cover or its absence on downstream water users and communities. The increasing interest in the recycling of forest evapotranspiration at the subcontinental scale, though, highlights the impacts that may be felt downwind over large distances and across catchment and national boundaries (Ellison et al. 2017). Additionally, forestry activities (including forest loss) often take place on a patchwork basis, rather than uniformly across an entire catchment, especially at the larger catchment scale. Moreover, the phenomenon of deforestation and its impacts on landscape and water availability are evident at scales larger than the catchment. It is therefore necessary to consider forest–water interactions at the catchment, landscape and subcontinental scales; actions with a beneficial impact at one scale may have an adverse impact at another.

6.3. Relevance of SDG 6 to Forests and Forest People

SDG 6 is one of the new goals that emerged in 2015. Indeed, water was a notable omission from the SDG predecessors, the Millennium Development Goals (MDGs). In spite of its importance for achieving many of the MDG targets (WWAP 2009), it was hidden within MDG 7 Ensure Environmental Sustainability under the dual Target 7.C: 'To halve the proportion of the universal population without sustainable access to clean and safe drinking water and basic sanitation by 2015'. This has important implications for SDG 6's policy context. While most of SDG 6 is new, the goal starts with an inherited focus on the water, sanitation and hygiene (WASH) agenda, which has developed a strong momentum after nearly two decades of MDG work. The drinking-water target was considered a big success as it was met five years before the deadline, but the sanitation target was never achieved. Although the proportion of the global rural population without access to improved sanitation has declined by nearly a quarter, half of people living in rural areas, including forested areas, do not have access to these facilities (UN 2015). Given this failure and the momentum behind the WASH agenda, it seems likely that a large proportion of the resources allocated to SDG 6 will be focused on sanitation. The reality is that the impact of MDG Target 7.C on the forestry sector and related policies was minimal; the same may be expected of the WASH impact. Two exceptions are an increasing interest in (1) the role of forests – as providers of hydrological ecosystem services – in protecting clean drinking water resources (Brauman et al. 2007), and (2) the use of wastewater in forestry (FAO 2018a). In Section 6.5.2 we discuss interest in nature-based solutions, forests' role therein and the wider role of forests in precipitation recycling.

Pointers towards an increased interaction between global water goals and forestry appear in the final MDG Report (UN 2015). It devotes attention to the 663 million people still using unimproved drinking water sources, mainly in sub-Saharan Africa but also in South Asia, and the shocking 2.4 billion still using unimproved sanitation. It highlights how 30 per cent of the planet's land area is covered by forests that not only support 1.6 billion people but also help 'provide additional benefits ... such as clean air and water' (UN 2015: 52) and support river catchments yielding three-quarters of the globally available freshwater. It implicitly assumes that changes in deforestation, afforestation and reforestation rates affect water resources. There are big regional differences in the way these interactions take place. South America and Africa have experienced the larger net losses of forest area, while large-scale afforestation programmes in China have offset continued rates of net loss in Southern and Southeastern Asia, all with corresponding impacts, positive or negative, on water balances. The exact nature of these processes is very important because one of the other key identified global environmental drivers is water scarcity, which affects more than 40 per cent of the global population – a figure that is projected to rise. Although the main problems are in the dry areas of Northern Africa and Western Asia, scarcity affects every continent. Major sectors that compete for water are agriculture (for irrigation, livestock and aquaculture), industries and municipalities. Agriculture, mainly through irrigation, takes nearly 70 per cent of freshwater withdrawals. Forests are not mentioned in the section about scarcity of the MDG 7 report, but their role in determining total water quantity and quality in catchments is critical.

After intense water-sector lobbying and proven interest from the public and governments in the consultations after Rio+20 (UNESCO-IHP 2014), SDG 6 has gone much further than the MDGs, with a set of completely new targets covering the whole gamut of integrated water resources management, as the water sector wanted (Table 6.1). This substantially increases the potential impact of SDG 6 on forests and forestry, as four of the new targets (6.3–6.6) are focused on water resources and not just on WASH. The drinking water and sanitation targets (6.1 and 6.2) are maintained and indeed enhanced with an ambitious 'for all' specification, which substantially increases their difficulty and cost. The means of implementation targets (6.A and 6.B) are neutral for the forest sector, although Target 6.B could have implications for hydrological ecosystems services involving forests. The UN SDG 6 synthesis report (UN 2018) reinforces the message that water management is critical. Water scarcity, flooding and quality are identified as the key determinants in social and economic development, and water efficiency is identified as the main factor to balance growing competing demands. The new SDG 6 targets incorporate all these aspects and, consequently, define the areas where SDG 6 implementation potentially impacts upon forests significantly.

Progress towards achieving each SDG 6 target is quantified by at least one indicator (UN Water 2018a). Section 6.4 assesses target impacts on forests through the actions that will be needed to ensure a positive direction for the respective indicator(s). Indicators 6.3.1–6.6.1 are the main focus of the analysis. Indicators 6.1.1 and 6.2.1 are the proportion of the population using, respectively, safely managed drinking water and sanitation services.

Table 6.1 SDG 6 targets and monitoring indicators for Targets 6.3–6.6

6.1 By 2030, achieve universal and equitable access to safe and affordable drinking water for all
6.2 By 2030, achieve access to adequate and equitable sanitation and hygiene for all and end open defecation, paying special attention to the needs of women and girls and those in vulnerable situations
6.3 By 2030, improve water quality by reducing pollution, eliminating dumping and minimizing release of hazardous chemicals and materials, halving the proportion of untreated wastewater and substantially increasing recycling and safe reuse globally *Indicators: 6.3.1 Proportion of wastewater safely treated* *6.3.2 Proportion of bodies of water with good ambient water quality*
6.4 By 2030, substantially increase water-use efficiency across all sectors and ensure sustainable withdrawals and supply of freshwater to address water scarcity and substantially reduce the number of people suffering from water scarcity *Indicators: 6.4.1 Change in water-use efficiency over time* *6.4.2 Level of water stress: freshwater withdrawal as a proportion of available freshwater resources*
6.5 By 2030, implement integrated water resources management at all levels, including through transboundary cooperation as appropriate *Indicators: 6.5.1 Degree of integrated water resources management implementation (0–100)* *6.5.2 Proportion of transboundary basin area with an operational arrangement for water cooperation*
6.6 By 2020, protect and restore water-related ecosystems, including mountains, forests, wetlands, rivers, aquifers and lakes *Indicator: 6.6.1 Change in the extent of water-related ecosystems over time*
6.A By 2030, expand international cooperation and capacity-building support to developing countries in water- and sanitation-related activities and programmes, including water harvesting, desalination, water efficiency, wastewater treatment, recycling and reuse technologies
6.B Support and strengthen the participation of local communities in improving water and sanitation management
Source: https://sustainabledevelopment.un.org/sdg6 (orange = WASH targets; yellow = targets and indicators with potential forest impacts; white background = means of implementation targets)

6.4 SDG 6 and Forests: Key Links

This section examines the potential impacts of SDG 6 on forests (summarised in Table 6.2).

Table 6.2 Potential impacts of SDG 6 indicators on forests

Indicator	Response for favourable indicator score
6.1.1 Proportion of population using safely managed drinking water services	Maintain forest cover to ensure good water quality in water supply catchments. Establish forested riparian buffer strips to maintain stream water quality.
6.2.1 Proportion of population using safely managed sanitation services, including a hand-washing facility with soap and water	No direct link but target generally favourable for forest people.
6.3.1 Proportion of wastewater safely treated	Encourage agroforestry schemes using treated wastewater. Maintain forest cover for treating wastewater (e.g. in schemes for induced precipitation recycling).
6.3.2 Proportion of bodies of water with good ambient water quality	Maintain or increase forest cover to enhance water quality. Change plantation tree species to enhance water quality and quantity.
6.4.1 Change in water-use efficiency over time	Require increased water efficiency from forests as forestry is combined with (relatively inefficient) agriculture in allocating available water between economic activities. Change plantation tree species, density and location to improve water-use efficiency.
6.4.2 Level of water stress: freshwater withdrawal as a proportion of available freshwater resources	Restrict establishment and continuation of plantation forest in water-limited areas (e.g. semi-arid regions) to maximise water availability. Maintain forest cover in upwind areas to safeguard downwind water resources dependent on recycled evapotranspiration.

Table 6.2 (cont.)

Indicator	Response for favourable indicator score
6.5.1 Degree of integrated water resources management implementation (0–100)	Integrate forest management with water resources management. Change plantation tree species, plantation characteristics and riparian buffer strips to optimise water availability and quality.
6.5.2 Proportion of transboundary basin area with an operational arrangement for water cooperation	Integrated consideration of forest management, including downstream impacts (within catchment) and downwind impacts (between catchments).
6.6.1 Change in the extent of water-related ecosystems over time	Reforest agricultural land, replace exotic tree plantations with native forests and implement other scenarios to maintain water availability and quality, with potential impacts on forest people.

6.4.1 Access to Safe and Affordable Drinking Water and Adequate and Equitable Sanitation and Hygiene – Targets 6.1 and 6.2

As noted in Section 6.3, Target 6.1 is somewhat neutral regarding the forest sector. Forests do not, of themselves, provide safe drinking water or sanitation services. Nevertheless, to the extent that stream waters in forested catchments tend to be of higher quality than in agricultural or urban environments, the treatment costs to bring them to a safe potable level may be lower, with a beneficial effect on affordability. It is common to find catchments maintained with a forest cover to form a source of clean water for a nearby city (e.g. Valdivia, Chile, receives water from a 12.7 km^2 catchment hosting evergreen native forest). Also, direct extraction of drinking water from streams without treatment is also generally safer in forested catchments than elsewhere (e.g. important for Indigenous populations of tropical forests). The performance of Indicator 6.1.1 (Proportion of population using safely managed drinking water services) could therefore be enhanced by a greater forest cover in water supply catchments. In some regions, forest cover is increasing through natural regeneration following abandonment of agricultural land. In other cases it may be worth deliberately afforesting catchments to provide purer water for water treatment plants, as the cost of treating lower-quality water in the absence of forest cover can be high (hundreds of thousands to millions of US dollars per year for individual cities; Ashagre et al. 2018). That cost would have to be compared with the costs of afforestation (including the potential removal of people from the land and the loss of agriculture). Reduction in run-off, and thus water availability, resulting from the afforestation would also have to be considered (e.g. Target 6.4). A more feasible and cheaper option may be the introduction of forested buffer strips along riparian zones, to reduce or interrupt nutrient fluxes to streams in agricultural catchments and sediment fluxes from both agricultural catchments and forested catchments undergoing logging. In the absence of other pressures, Indicator 6.1.1 is likely to favour maintaining existing forest covers. Given the multiple pressures on land resources, however, it would not be surprising if the forests' (high) worth to drinking-water quality was simply ignored, leaving the successful achievement of Target 6.1 increasingly dependent on artificial water treatment. The target would then be irrelevant to forests. Target 6.2, with its emphasis on sanitation and hygiene *for all*, should have a positive impact on forest people – that is, those who live in forests and whose lives and livelihoods depend directly on the forest environment and forest resources. Difficulties of accessibility, though, are likely to mean that forest people in remote areas will be among the last to benefit from this target (although perhaps being among those least in need of it).

6.4.2 Improving Water Quality – Target 6.3

Indicator 6.3.1 (Proportion of wastewater safely treated) is not closely linked to forests. However, to the extent that there is an interest in using treated wastewater for forestry, this indicator may drive an increase in, for example, agroforestry schemes. Forests have themselves been proposed as treatment areas for wastewater as part of wider schemes for induced precipitation recycling (Layton and Ellison 2016). Indicator 6.3.2 (Proportion of bodies of water with good ambient water quality) is more relevant in view of the potential for land use to affect water quality in rivers, reservoirs, estuaries and downstream wetlands. For example, in streams in native forests in Chile, nitrate (NO_3-N) and ammonium concentrations are very low and nitrogen (N) export is very low (0.2–3.5 kg N ha^{-1} yr^{-1}) (Perakis and Hedin 2001, 2002). The conversion of native forests to other land uses may therefore be expected to lead to increased NO_3-N and total N export. Conversion of native forest to pasture in southern Chile is associated with increased N fluxes and increased dominance of NO_3-N (exports up to 11 kg ha^{-1} yr^{-1}), although some of this increase may be attributable to pasture fertilisation (Oyarzún and Huber 2003). Conversely, afforestation may change soil pH and alter nutrient cycles (Hong et al. 2018). Nevertheless, because fertiliser use in plantation forestry – in terms of total use – involves a lower application frequency and smaller land areas compared with agricultural systems (May et al. 2009), the conversion of crop or pastureland to plantations (or even reverting to natural forest) is expected to improve water quality.

In most cases, logging of planted forests may produce elevated sediment export, mainly because of mechanised site preparation, road building and clear-cutting operations in steep terrain, rather than exposure of soil to rainfall (Bathurst and Iroumé 2014). Research in forest plantations in central Chile found that forest roads produce more sediment than hill slopes; after clear-cutting, the relative contributions increased from 16 per cent to 25 per cent for hill slopes and from 37 per cent to 45 per cent from forest roads (Schuller et al. 2013). Sediment delivery to streams increases if logging operations take place in rainy periods (with adverse impacts on drinking water and aquatic habitats), but the application of contemporary best management practices, which include guidelines for logging during dry periods, can limit logging-related sedimentation (Bathurst and Iroumé 2014). Although forest certification has enhanced the adoption of best management practices in Chile, forestry operations are still associated with increased sediment transport and decreases in water quality relative to unmanaged forests.

Because of forests' generally positive impact on water quality, Indicator 6.3.2 favours maintaining forest cover and increasing cover in protected or degraded catchments. Specifically, achieving good indicator scores for plantation forests will require careful consideration of tree species and the development and implementation of best management practices.

6.4.3 Water Efficiency and Improved Availability – Target 6.4

Indicator 6.4.1 (Change in water-use efficiency over time) tracks the value added (in US dollars) per volume of water withdrawn (cubic metres), by a given economic activity over time (UN Water 2018a). The UN International Standard Industrial Classification of All Economic Activities, Revision 4 code combines forestry with agriculture and fishing. This means that forestry as an economic activity will be considered jointly with agriculture when comparing water use with other sectors. It is already acknowledged that irrigation, as the largest consumer of water by volume, should be one of the big targets for water efficiency (HLPW 2018). There will therefore be increased attention to the efficient use of water in the forestry sector, which may eventually constrain the establishment and continued presence of plantation forests in the water-stressed areas highlighted by Indicator 6.4.2 (Level of water stress: freshwater withdrawal as a proportion of available freshwater resources). This indicator is demand-driven and measures the ratio between withdrawals and the difference between total renewable water resources and the environmental flows (Vanham et al. 2018). Strictly, it considers only blue water: the liquid water in rivers, lakes, wetlands and aquifers. However, the amount of blue water is determined by the upstream flows of green water – rainwater held in the unsaturated zone of the soil and available to plants – which is determined by terrestrial ecosystem functions or natural land use (e.g. forests or natural grasslands) and by consumptive water use in rain-fed agriculture and forest plantations. Therefore, analyses of water-stressed environments may lead to closer examinations of the consumptive use of water by forests.

In terms of biomass production per litre of water, trees are considered efficient users of water. Nevertheless, Soto-Schönherr and Iroumé (2014) found in Chile that water-use efficiency (i.e. kilograms of biomass produced per unit of water consumed) differs only a little between forests and grasslands: forests produce 0.1–4 kg of biomass per cubic metre of water, while grasslands produce 0.5–1.3 kg. However, because trees use more water than shorter vegetation, there is a central inconsistency (at least at the catchment scale) between the aims of maintaining forest cover (desirable for many reasons, including lower soil erosion and higher water quality) and of increasing water availability (which implies reducing forest cover). This is less concerning in high-rainfall areas (where there is enough water for all activities) but could be critical in semi-arid areas. This means that forestry as an economic activity will be compared directly with agriculture and other activities when deciding on use of limited water resources. Replacing forest by agriculture could increase annual run-off (and food supply), but at the expense of the forest ecosystem and timber supply. For example, replacement of natural vegetation with agricultural cover in a 175 360 km^2 catchment in South America produced a significant increase in annual mean discharge and high-flow season discharges because of reduced infiltration and evapotranspiration rates (Costa et al. 2003). Conversion of forest cover may ultimately lead to destruction of the land resource itself (Contreras et al. 2013). Thinning of forests considered unnaturally overgrown as a result of fire-suppression programmes has been proposed to increase water supply, e.g. in North America (Poulos 2018), but this may ignore the many other changes produced by forest management (Jones et al. 2009, NRC 2008). For example, forests play important roles in regulating the world's temperatures and freshwater flows, storing carbon and providing a broad range of important but less recognised benefits (Ellison et al. 2017). At the subcontinental scale, replacement of forests by shorter vegetation could imply less rainfall in downwind regions (Creed and van Noordwijk 2018) and possibly therefore less run-off, although the magnitude of this effect remains to be quantified.

Overall, forest plantations with fast-growing species use more water than native forests, although many of the comparisons are limited to old-growth native forests versus young plantations (Soto-Schönherr and Iroumé 2016). Results from southern Chile, for a wide range of latitudes and forest compositions and ages, showed that annual interception accounts for approximately 21 per cent of incoming precipitation in the mean, albeit with some margin of variation (Soto-Schönherr and Iroumé 2016). Within the range of variation, broad-leaved forests (including native broad-leaved and eucalyptus forests) generally exhibit higher interception losses than conifers. Indicative of the level of uncertainty, Huber et al. (2010) found that interception is lower, and water use is higher, in eucalyptus compared with pine plantations in southern Chile. Because of the relatively limited difference in canopy interception loss between native forests and forest plantations (Huber and Iroumé 2001, Soto-Schönherr and Iroumé 2016), the observed differences in water yield between the two must be explained, at least in part, by different transpiration losses.

Despite the above, forest plantations may not use more water than native forests at all stages of the forest rotation. As expected, water use is highest (and yield lowest) in the late stages of plantation growth, especially in short-rotation plantations with a high tree density, but water yield (especially summer water yield) increases just after clear-cutting, in the early phases of replanting (Iroumé et al. 2005, 2006). Again, variations between plantation species may exist (e.g. pine versus eucalyptus). Thus, water consumption by forests could be moderated by a small amount through careful choice of tree species (with an eye towards suitability to future climates), maintaining a mix of old and new growth (i.e. avoiding large-scale plantation of new growth), regulating tree density and choosing plantation location carefully, possibly implying longer growth periods or reduced timber yield. Nevertheless, such moderation is likely to be small compared with the effect of forest removal. Overall, the demands of Target 6.4 are likely to be inimical towards forest cover in many parts of the world.

6.4.4 Integrated Water Resources Management – Target 6.5

Integrated water resources management (IWRM) is defined as 'a process which promotes the coordinated development and management of water, land and related resources in order to maximise economic and social welfare in an equitable manner without compromising the sustainability of vital ecosystems' (Global Water Partnership 2017). It acknowledges the interconnected nature of hydrological resources and the interdependence of different water uses. Within the wider requirements of the definition, IWRM implementation implies the integrated management of water supply, water quality, flood management, navigation, hydroelectric power generation and other water provisions and services within a river basin. Within this balancing act, forest cover reduces water supply relative to other vegetation covers (at the annual scale) but has the potential to improve water quality. From points 8 to 10 in Section 6.2 it can be concluded that forest cover could probably be reduced without necessarily increasing damaging flood magnitudes, but flood frequencies might increase, as would soil erosion. These points refer to the catchment scale. At the subcontinental scale, forest evapotranspiration is not lost from the system but may be recycled as precipitation downwind. Water resources management needs to be integrated with forest management, an end likely encouraged by Indicator 6.5.1 (Degree of integrated water resources management implementation). Such integration is most required in regions where there are multiple and conflicting demands for water, where water supplies are restricted and forests account for significant water use (e.g. semi-arid regions) and where soil erosion is a significant problem (e.g. semi-arid regions, degraded lands and areas of unregulated agricultural expansion). The exact impact of Indicator 6.5.1 on forest cover will vary from basin to basin depending on economic, social, political and other circumstances. It seems more likely, though, that IWRM is implemented as a function of whatever level of forest cover happens to exist, and therefore reacts to rather than drives forest cover change.

In South America conversion of forest to agriculture is the major reason for changes in water availability to rivers and streams (Jones et al. 2017). The concept of virtual water (Yang and Zehnder 2007) assesses water-use efficiency based on water used to grow products traded globally. From 1986 to 2007, South America increased its annual use of water from 42 km^3 to 178 km^3 and became the continent using the greatest amount of water in food products traded globally, with significant increases in soy exports to China – which in turn has contributed to deforestation in Amazonia (Dalin et al. 2012).

In the Federal District of Brazil, river basins with substantial expansion of agriculture since the end of the 1970s show a dramatic decrease of base flow discharge by 40–70 per cent – presumably the effect of irrigation extractions more than compensating for the increase in run-off otherwise expected from reduced forest evapotranspiration. Additionally, the run-off ratio is significantly positively related to the cover of natural vegetation (Lorz et al. 2012). In south-central Chile, the run-off ratios in four large catchments were positively related to the area of native forest and negatively related to the area of eucalyptus and pine plantations (Lara et al. 2009).

Careful choice of tree species and plantation characteristics could moderate water consumption by forests. Reducing the area of forest plantations can potentially increase water availability at the catchment scale, as might replacing exotic fast-growing trees with native forest species, although it has yet to be proven that this would allow adequate timber yields to be produced with less water use. Native forest riparian buffers may increase water yield and improve water quality in forest plantations of eucalyptus in south-central Chile (Little et al. 2015) and along rivers in degraded native forest in south-east Brazil (de Souza et al. 2013). Forest thinning (Poulos 2018) should be approached with caution: thinned plantations may initially increase water yield (Forrester et al. 2012), but subsequent forest growth might take up the additional water, depending on the timing and degree of thinning (Perry and Jones 2017).

At the international level, Indicator 6.5.2 (Proportion of transboundary basin area with an operational arrangement for water cooperation) may drive a more integrated consideration of the forest environment, especially where a downstream or downwind state feels adversely affected by the forestry activities of an upstream or upwind state. At the subcontinental scales typical of many politically sensitive transboundary basins, the downstream impacts of upstream forestry activities may not always be significant. For example, there is no clear evidence for the so-called Himalayan environmental degradation theory, which proposes that loss of forest cover caused by a rapidly growing population in the Himalayan headwaters of the Ganges and Brahmaputra Rivers increases soil erosion and flood run-off, thereby generating increased flooding and siltation in the delta regions of Bangladesh (Hofer 1993). Downwind impacts at subcontinental scales are increasingly thought to be important but have yet to be proven conclusively (Creed and van Noordwijk 2018). Large amounts of moisture from evapotranspiration may be recycled within Amazonia and transferred to other parts of South America. However, the extent to which forest conversion has affected this precipitation recycling so far is yet to be clearly quantified (Bagley et al. 2014, Khanna et al. 2017, Spera et al. 2016, Swann et al. 2015). The impact of Target 6.5 is most likely to be a more integrated consideration of forest management, especially with respect to water resources management, both within and among river basins.

6.4.5 Protecting and Restoring Water-Related Ecosystems – Target 6.6

Indicator 6.6.1 (Change in the extent of water-related ecosystems over time) will drive careful consideration of the relationship between forest cover and the health and sustainability of water-related ecosystems. Although the title of Target 6.6 includes forests along with mountains, wetlands, rivers, aquifers and lakes, Indicator 6.6.1 considers only mangroves. The absence of other types of forests severely limits this indicator's degree of protection.

Despite the restricted nature of the indicator, forest-related water ecosystems services form an important underlying concept linked to the essence of Target 6.6. They are indeed one of the most proffered links between SDG 6 and forests, but they are not always properly understood and are difficult to quantify. For instance, not all forested areas safeguard water quality: rapid and aggressive afforestation and reforestation with exotic species reduces water availability, affecting its quality, especially in dry seasons (Filoso et al. 2017). Achieving good indicator scores requires maintenance of water flows – and of good water quality – to wetlands. This could imply reforesting agricultural areas, replacing plantations of exotic tree species and monocultures with a wider range of native trees (which may be slower growing), removing existing plantations or avoiding new plantations (e.g. in peatlands, páramo soils or semi-arid regions) and other scenarios. Such land management could have adverse consequences for livelihoods based on the existing agricultural and plantation activity (and for the production of the associated crops) but might open new employment prospects in forestry activities. Also, there is a high potential for payments for ecosystem services if those services are clearly proven. In the Hindu Kush Himalaya region, research is attempting to show the tangible value of water-related ecosystem services; for example, purification of the downstream water supply via upstream forests is an important ecosystem function that payments for ecosystem services schemes have attempted to quantify and that can contribute to achieving Target 6.3 by natural means (Ashagre et al. 2018). However, there is still a gap in our understanding of how individual attributes (such as changes in land-use patterns) impact ecosystem service flows, including water (Polasky et al. 2011, Nelson et al. 2013, Su and Fu 2013). In particular, it is difficult to correlate change in a land unit with change in the volume of ecosystem services that this unit provides (Bhatta et al. 2017).

Despite these uncertainties, payments are already being made to promote ecosystem services. Recently, the Chilean government, acting according to the UN Framework Convention on Climate Change (UNFCCC), gave a commitment to afforest (mainly with native tree species) 100 000 ha of degraded soils as a contribution to reducing greenhouse gases (CONAF 2016). To this is added some 470 000 ha that were burnt during the 2016–2017 (southern) summer season. Of this, the government will finance the afforestation of 100 000 ha on lands belonging to small- and medium-size owners, while the remaining burnt lands owned by larger forest companies are already being afforested. The aims of these afforestations are mainly to restore and improve the ecosystem services of the degraded areas, except for the afforestation of lands owned by larger private forest companies whose purpose remains timber production. Not all those forested areas would safeguard water quantity;

potentially, there could be resistance to forestry plantations in water-stressed regions, as highlighted by Indicator 6.4.2.

In conclusion, SDG 6 seems unlikely to exert a major influence on forest cover, and indeed is unlikely to be pursued with forests at the forefront of consideration (Table 6.2). Possible exceptions are that Target 6.4 may restrict the spread of new plantations in semi-arid areas and Target 6.5 may drive a more integrated view of catchments and their management. Plantations may be developed with more careful consideration of tree species and plantation characteristics. It seems much more likely, however, that native forest cover in much of the world will continue to decline in the face of pressures greater than SDG 6: (1) to convert forest for food production, driven by population growth and increasing aspirations for living standards worldwide; and (2) to exploit timber and other forest resources, driven first by those seeking a profit but ultimately responding to individual demand globally, with little consideration for the resulting impacts. The decline is exacerbated by the inability or unwillingness of governments in many countries to control such developments, and possibly by climate change reducing or shifting the areas of the world suitable for sustaining the current forests (Guardian 2019, WWF 2019). The extent of monoculture plantations, on the other hand, could increase or decrease: demand for plantation products (e.g. palm oil and timber) is likely to increase, but water efficiency considerations may curtail the spread of plantations in water-stressed areas.

6.5 Future Policy Considerations

6.5.1 Contextual Factors for SDG 6: The Hindu Kush Himalayas

Trying to understand the real impacts that SDG 6 may have on forests and forest people requires a careful evaluation of the context of water–forest interactions, in particular physical and social settings and the interactions with other SDGs. While context is always affected by local conditions, some situations do recur. The Hindu Kush Himalayas case illustrates some of the contextual factors that must be considered in implementing SDG 6 and highlights the upstream/downstream relationships, inherent where forested mountain areas feed major river systems, which can be found on all continents.

The Hindu Kush Himalayas harbour major river systems providing services, particularly in the form of recharge, to a mountain population of 240 million and a downstream population of 1.9 billion. Indirectly, 3 billion people are dependent on numerous ecosystem services, including climate and hydrological services provided at regional and global scales, and harvested commodities traded at multiple economic scales (Kotru et al. 2015). The observed overall increase in forest cover in India does not mean that forest degradation is controlled. An alarming rate of deforestation in parts of the Himalayas, primarily for agricultural land and fuel supply, threatens the sustained flow of forest ecosystem services. As is generally the case in South Asia, multiple sectors and actors influence forests and forest management; it is not only through forest management that the forest–water relationship can be improved for sustained water yield.

Multiple water-related objectives across a portfolio of SDGs present new challenges for policymakers and managers of forests and landscapes with partial tree cover. Hence, SDG 6 cannot be seen in isolation from other key challenges in the Himalayas, such as SDGs 1 (No Poverty), 2 (Zero Hunger) and 5 (Gender Equality). Thus, investments (e.g. in mass tourism) made upstream for addressing other SDGs are not necessarily environmentally friendly or complemented by good governance, making it potentially more difficult to achieve the SDG 6 targets (6.1, 6.3, 6.4 and 6.6). On the other hand, the policies, practices and investments necessary to achieve SDG 6 may not be coordinated with those for other SDGs, so positive outcomes for people are not ensured (Singh and Kotru 2018). The transboundary nature of hydrological resources – overlaying local, regional and national boundaries – make the challenges to safe water access more complex. New institutional responses are needed to tackle multiple water-related objectives across the full portfolio of SDGs, taking a multiple benefits approach (Creed and van Noordwijk 2018). A distinction may be made between a first group of SDGs (SDGs 1, 2, 6, 7) implying an increased demand for clean, reliably flowing water, and a second group of SDGs (5, 10, 12, 16) that stresses a change in power-sharing that allows multi-stakeholder involvement, thus increasing the need for transparency and equity in decision-making.

Several socio-economic and governance realities challenge forest regimes in fulfilling their socio-ecological role (as envisaged under SDG 6) in the Hindu Kush Himalayas (Kotru et al. 2017):

- poverty and inequity are still prevalent in South Asia, a water-deficit area;
- institutional capacities and existing policies are inadequate to meet the future challenge of forest management for sustained water yields;
- research on the forest–water relationship is essentially very limited, with no long-term monitoring data or studies available;
- there are very limited cross-sector policy interfaces (e.g. water policy and forest policy interface) that focus on a forest or landscape approach aimed at sustaining water services;
- sustainable forest management is seriously disadvantaged by a lack of proactive management, itself arising from policy deficits;

- data deficits and a lack of harmonised methodologies and data sharing mean that the planning and application of conservation and development strategies contributing to SDG 6 have only a limited foundation on firm data.

Future progress towards achieving Targets 6.3–6.6 in the Hindu Kush Himalayas will require improved upstream–downstream integration, improved transboundary cooperation and greater coordination and simultaneous progress in the implementation of different SDGs: for example, SDGs 1, 2 and 5 (already mentioned) and SDGs 13 (Climate Action), 15 (Life on Land) and 17 (Partnerships for the Goals). Adoption of a landscape approach would allow stakeholder priority interventions to be matched with public and private investments but, equally, there is a need for an improved understanding of the role of forests in influencing ecosystems services at the larger landscape scale. Greater efforts are required to make the communities struggling on the frontline of sustainable forest management more climate resilient.

6.5.2 Implementation of SDG 6

The final impact of SDG 6 on forests and forest people will be determined by the extent of its implementation. There are four important considerations here: finances, institutions, data and other SDGs. First, implementation costs are increasing: the estimated cost of achieving the WASH targets is USD 1.7 trillion (Hutton and Varughese 2016). While there are no reliable estimates of the whole cost of achieving SDG 6, it is clear that the required threefold increase for Targets 6.1 and 6.2 alone indicates a huge increase in water targets expenditure. As aid is decreasing, it is not at all clear where this money will come from. The UN calls for more technology transfer and new financing mechanisms, with some based on the recognition of the economic value of water and freshwater ecosystems (UN 2018). Forests may eventually benefit from the growing interest in nature-based solutions, which use or mimic natural processes to enhance water availability and water quality and to reduce risks associated with water-related disasters and climate change. The UN High Level Panel on Water specifically mentions that natural capital solutions, including the 'water-retaining abilities of forest', can be used at a fraction of the cost of engineering solutions (HLPW 2018). It labels forests as 'natural infrastructure' required to assure future supplies of water, calling for a better alignment of incentives to recognise the value of these services. It is of the utmost importance that natural capital solutions recognise the nuanced role of forests currently accepted as best practice and take into account local conditions. They should be particularly considered in the context of deforestation and forest degradation while recognising the need to understand water quantity effects at catchment, regional and continental scales. It is important to acknowledge that not all water-poor locations have forests to use as improvement tools.

Second, the success of SDG 6 depends on the existence of national and global institutions able and willing to implement the goal. While the WASH sector has spent nearly 20 years trying to achieve global targets, the level of institutional readiness for the new water resources targets is frequently low or non-existent at the country level. Even with the apparently successful MDG 7.C drinking water target, 53 countries were seriously off-target and 19 could not produce data. Good water governance depends on strong formal and informal institutions and the accompanying human resources. There is an acute lack of capacity across most developing countries, particularly in sub-Saharan Africa and South and Southeast Asia (UN 2018). With low institutional capacity, we can expect a slow articulation of the new targets within SDG 6 and, subsequently, low impact on the forest sector. However, since the creation of UN Water in 2003 as a focal point for coordinating efforts of UN entities and international organisations working on water and sanitation issues, the alignment of global water initiatives has increased (UN Water 2018b). Eventually this will lead to actions on the ground. The existence of SDG 6 in itself is a clear proof of the strength of these efforts. The HLPW (High Level Panel on Water 2018) has identified a number of initiatives especially relevant for states trying to implement SDG 6: the World Water Data Initiative; the OECD Water Governance Initiative; the Delta Coalition; High-level Experts and Leaders Panel on Water and Disasters, including an Alliance of Alliances on disaster risk-reduction researches; the initiative on Financing Water Infrastructure convened by the OECD; and the Water Innovation Engine. Whether these will accelerate SDG 6 implementation is yet to be seen, but they may support natural infrastructure projects with a role for forests.

The third consideration is the challenge of having enough good-quality data for monitoring SDG 6. It took a serious global effort from 1990 onwards to develop the WHO/UNICEF Joint Monitoring Programme, now the custodian of global WASH data for Targets 6.1 and 6.2. The situation is more complicated with the other targets as many countries lack the financial, institutional and human resources to acquire and analyse the required data. Fewer than half of UN member states have comparable data available on progress towards meeting the SDG 6 targets (UN 2018). This is important because countries will focus on being able to report to the Integrated Monitoring Initiative (UN Water 2018a). Accordingly, the factors monitored for each target are likely to become the focus of public policy. It is therefore important to understand how forests relate to this monitoring programme.

The final consideration is the dynamic interdependence between SDG 6 and the other SDGs (UN Water 2018c). The majority of these interlinkages are positive and mutually reinforcing. Since the MDG programme, the WASH targets have been identified as critical to: reducing poverty (SDG 1), malnutrition (SDG 2) and diseases (SDG 3); supporting education (SDG 4); and addressing gender (SDG 5) and other inequalities (SDG 10). Moreover, SDG 6 highlights how water of sufficient quality and quantity is required for food production (Target 2.4) and sustainable consumption and production (SDG 12). However, there are some targets – such as doubled agricultural productivity (Target 2.3), energy for all (Target 7.1) and sustained economic growth (Target 8.1) – that potentially could impact negatively on water resources and water ecosystems and, as such, on forests and forest people. IWRM (Target 6.5) is the appropriate framework to balance all these competing needs, for water and forests.

6.5.3 Integration of SDG 6 and Forests

The discussion of the contextual challenges in South Asia clearly shows both the difficulty and the necessity of aligning SDG 6 and forest policies. For this alignment to succeed in an IWRM framework, we need to follow a landscape approach at all levels. As larger forms of vegetation, trees use great amounts of water to produce biomass and for the process of evapotranspiration, more so than many other vegetation types, including crops and grasses. However, when considering Target 6.4 on increasing water-use efficiency and ensuring sustainable withdrawals and supply of freshwater and how it may impact forests and forest management, several issues need to be considered.

Firstly, most forests or tree-based landscapes are naturally occurring, rainfed systems. Globally, only 7 per cent of forests are planted forests, predominantly found in temperate zones (FAO 2015). Moreover, most of these planted forests are native species: only 20 per cent of planted forests – 1.4 per cent of forests globally – are exotic, and these are located mainly in the southern hemisphere (FAO 2015). It is assumed that natural forests will not be removed for the purposes of achieving Target 6.4 as they provide a wealth of other goods and services, including water-related ecosystem services. Consequently, discussion of water-use efficiency will be limited to planted forests, despite their representing only a small proportion of global forests.

Trees are highly resilient and adaptive organisms that optimise their water use. In other words, they drastically reduce their water consumption in periods of drought and use what they can when water is available (Chaves et al. 2002). This means that during periods without rain trees can use water stored in the soil; they generally have higher annual rates of water use than shallow-rooted, annual cropping and pasture systems. In high-rainfall areas (> 1500 mm per annum), planted forests can use up to 200 mm more water than pastures, but only if the water use is not energy limited. In low rainfall areas (< 600 mm per annum), forest plantations use an amount of water similar to annual crops and pastures. In intermediate rainfall areas, planted forests potentially use more water than annual crops and pastures. If this is in conflict with other demands for water, policies are required to regulate allocation of water to plantations among other uses. Policy instruments to regulate plantation water can be direct (e.g. a moratorium on land concessions) or indirect (e.g. a market for allocable water). For example, South Africa, Australia and India have implemented policies to regulate or limit plantation establishment (Brown et al. 2005, Dye and Versfeld 2007, Farley et al. 2005, van Dijk and Keenan 2007, Whitehead and Beadle 2004).

Even if trees are able to optimise their water use, it is important to note that management can be improved in order to further optimise water-use efficiency – including tailoring species selection, as well as thinning and harvesting techniques – to the environmental conditions such as slope, soil type and condition. More importantly, as planted forests are managed at the stand scale and water is managed at a catchment or basin scale, it is critical when planning to look at planted forests, and even tree-based systems more broadly, at the landscape level, taking into consideration the mosaic of land uses and their effects on water. This requires a cross-sectoral approach to land and water planning and management.

Integrated planning and management may require reframing our approach, taking into account integrated solutions such as agroforestry and the use of recycled wastewater in planted forests. Studies show that agroforestry increases water-use efficiency (Bai et al. 2016, Droppelmann et al. 2000). The recycling of treated wastewater for planted forests can reduce competition for water use (particularly in semi-arid and arid areas where water is scarce), reduce the costs associated with water treatment and reduce downstream contamination (FAO 2018a). Planted forests irrigated with treated wastewater in turn improve soil water-storage capacity, reduce soil degradation and erosion, combat desertification in arid areas and provide essential goods that support livelihoods, such as timber, pulpwood and fuelwood (FAO 2018a). According to FAO's Aquastat database (FAO 2018b), only 52 per cent of the municipal wastewater produced globally is recycled, so there is ample opportunity to explore such options. Egypt, Jordan, Mexico and Spain, among others, are exploring the use of treated wastewater for agroforestry and planted forests. In Jacksonville, North Carolina, USA, upstream forests are being irrigated with treated wastewater, with the forests acting as the final stage of the filtering process and returning water back into the catchment for use downstream (Tew 2016).

Much can be gained from a deeper integration of SDG 6 and forest policies. However, this integration must be guided by a shared understanding of the complex relationships between water and forests and their impacts on both forest people and the communities downstream, and possibly downwind.

Acknowledgements

The authors thank the internal and external reviewers, all anonymous, for suggestions that have helped to improve the chapter.

References

Alila, Y. and Green, K. C. 2014. Reply to comment by Bathurst on 'A paradigm shift in understanding and quantifying the effects of forest harvesting on floods in snow environments'. *Water Resources Research* 50:2759–64. doi:10.1002/2013WR014334.

Andréassian, V. 2004. Waters and forests: From historical controversy to scientific debate. *Journal of Hydrology* 291:1–27.

Ashagre, B. B., Platts, P. J., Njana, M. et al. 2018. Integrated modelling for economic valuation of the role of forests and woodlands in drinking water provision to two African cities. *Ecosystem Services* 32:50–61.

Bagley, J. E., Desai, A. R., Harding, K. J., Snyder, P. K. and Foley, J. A. 2014. Drought and deforestation: Has land cover change influenced recent precipitation extremes in the Amazon? *Journal of Climate* 27(1):345–61.

Bai, W., Sun, Z., Zheng, J. et al. 2016. Mixing trees and crops increases land and water use efficiencies in a semi-arid area. *Agricultural Water Management* 178(C):281–90.

Bathurst, J. C., Birkinshaw, S. J., Cisneros, F. et al. 2011. Forest impact on floods due to extreme rainfall and snowmelt in four Latin American environments 2: Model analysis. *Journal of Hydrology* 400:292–304.

Bathurst, J. C. and Iroumé, A. 2014. Quantitative generalizations for catchment sediment yield following forest logging. *Water Resources Research* 50(11):8383–402. doi:10.1002/2014WR015711.

Bhatta, L. D., Khadgi, A., Rai, R. K. et al. 2017. Designing community-based payment scheme for ecosystem services: A case from Koshi Hills, Nepal. *Environment, Development and Sustainability* 20(4):1831–48.

Bosch, J. M. and Hewlett, J. D. 1982. A review of catchment experiments to determine the effect of vegetation changes on water yield and evapotranspiration. *Journal of Hydrology* 55:3–23.

Brauman, K. A., Daily, G. C., Duarte, T. K. and Mooney, H. A. 2007. The nature and value of ecosystem services: An overview highlighting hydrologic services. *Annual Review of Environment and Resources* 32(1):67–98.

Brown, A. E., Zhang, L., McMahon, T. A., Western, A. W. and Vertessy, R. A. 2005. A review of paired catchment studies for determining changes in water yield resulting from alterations in vegetation. *Journal of Hydrology* 310:28–61.

Bruijnzeel, L. A. 2001. Hydrology of tropical montane cloud forests: A reassessment. *Land Use and Water Resources Research* 1:1–18.

Bruijnzeel, L. A. 2004. Hydrological functions of tropical forests: Not seeing the soil for the trees? *Agriculture, Ecosystems and Environment* 104(1):185–228.

Bruijnzeel L. A., Mulligan, M. and Scatena, F. N. 2011. Hydrometeorology of tropical montane cloud forests: Emerging patterns. *Hydrological Processes* 25(3):465–98.

Calder, I. R. 2005. *Blue revolution: Integrated land and water resource management*, 2nd ed. London: Earthscan.

Chaves, M. M., Pereira, J. S., Maroco, J. et al. 2002. How plants cope with water stress in the field: Photosynthesis and growth. *Annals of Botany* 89(7):907–16. doi:10.1093/aob/mcf105.

CONAF (Corporación Nacional Forestal) 2016. *Estrategia nacional de cambio climático y recursos vegetacionales 2017–2025, Chile*. Santiago, Chile: Ministerio de Agricultura, CONAF. Available at: http://portal.mma.gob.cl/wp-content/doc/ENCCRV-2017-2025-web.pdf (Accessed 6 March 2019).

Contreras, S., Santoni, C. S. and Jobbágy, E. G. 2013. Abrupt watercourse formation in a semiarid sedimentary landscape of central Argentina: The roles of forest clearing, rainfall variability and seismic activity. *Ecohydrology* 6(5):794–805.

Costa, M. H., Botta, A. and Cardille, J. A. 2003. Effects of large-scale changes in land cover on the discharge of the Tocantins River, southeastern Amazonia. *Journal of Hydrology* 283:206–17.

Creed, I. F. and van Noordwijk, M. (eds.) 2018. *Forest and water on a changing planet: Vulnerability, adaptation and governance opportunities. A global assessment report.* IUFRO World Series Vol. 38. Vienna: International Union of Forest Research Organizations (IUFRO).

Dalin, C., Konar, M., Hanasaki, N., Rinaldo, A. and Rodriguez-Iturbe, I. 2012. Evolution of the global virtual water trade network. *Proceedings of the National Academy of Sciences* 109 (16):5989–94.

de Souza, A. L., Fonseca, D. G., Liborio, R. A. and Tanaka, M. O. 2013. Influence of riparian vegetation and forest structure on the water quality of rural low-order streams in SE Brazil. *Forest Ecology and Management* 298:12–18.

Droppelmann, K. J., Lehmann, J., Ephrath, J. E. and Berliner, P. R. 2000. Water use efficiency and uptake patterns in a runoff agroforestry system in an arid environment. *Agroforestry Systems* 49(3):223–42. doi.org/10.1023/A:1006352623333.

Dye, P. and Versfeld, D. 2007. Managing the hydrological impacts of South African plantation forests: An overview. *Forest Ecology and Management* 251:121–8.

Ellison, D., Morris, C. E., Locatelli, B. et al. 2017. Trees, forests and water: Cool insights for a hot world. *Global Environmental Change* 43:51–61.

Fahey, B. and Payne, J. 2017. The Glendhu experimental catchment study, upland east Otago, New Zealand: 34 years of hydrological observations on the afforestation of tussock grasslands. *Hydrological Processes* 31:2921–34. doi:10.1002/hyp.11234.

FAO 2015. *Global Forest Resources Assessment 2015*. Rome: FAO.

FAO 2018a. *SFM Toolbox – Using treated wastewater in forestry and agroforestry in drylands.* Available at: www.fao.org/sustainable-forest-management/toolbox/modules/use-of-treated-water-in-forestry-and-agroforestry/basic-knowledge/en/ (Accessed 6 March 2019).

FAO 2018b. *Aquastat website.* Available at: www.fao.org/nr/aquastat/ (Accessed 6 March 2019).

Farley, K. A., Jobbágy, E. G. and Jackson, R. B. 2005. Effects of afforestation on water yield: A global synthesis with implications for policy. *Global Change Biology* 11:1565–76.

Filoso, S., Bezerra, M. O., Weiss, K. C. B. and Palmer, M. A. 2017. Impacts of forest restoration on water yield: A systematic review. *PLoS ONE* 12(8):0183210. doi:org/10.1371/journal.pone.0183210.

Forrester, D. I., Collopy, J. J., Beadle, C. L., Warren, C. R. and Baker, T. G. 2012. Effect of thinning, pruning and nitrogen fertiliser application on transpiration, photosynthesis and water-use efficiency in a young Eucalyptus nitens plantation. *Forest Ecology and Management* 266:286–300.

Global Water Partnership 2017. *The need for an integrated approach.* Available at: www.gwp.org/en/About/why/the-need-for-an-integrated-approach/ (Accessed 13 February 2019).

Guardian 2019. Jair Bolsonaro launches assault on Amazon rainforest protections. *The Guardian*, 2 January 2019. Available at: www.theguardian.com/world/2019/jan/02/brazil-jair-bolsonaro-amazon-rainforest-protections (Accessed 13 February 2019).

HLPW 2018. *Making every drop count: An agenda for water action.* High-Level Panel on Water Outcome Document. Panel convened by United Nations and World Bank Group. Available at: https://sustainabledevelopment.un.org/HLPWater (Accessed 13 February 2019).

Hofer, T. 1993. Himalayan deforestation, changing river discharge, and increasing floods: Myth or reality? *Mountain Research and Development* 13(3):213–33.

Hong, S., Piao, S., Chen, A. et al. 2018. Afforestation neutralizes soil pH. *Nature Communications* 9(520). doi:10.1038/s41467-018–02970-1.

Huber, A. and Iroumé, A. 2001. Variability of annual rainfall partitioning for different sites and forest covers in Chile. *Journal of Hydrology* 248(1–4):78–92.

Huber, A., Iroumé, A., Mohr, C. and Frene, C. 2010. Effect of *Pinus radiata* and *Eucalyptus globulus* plantations on water resource in the Coastal Range of Biobío region, Chile. *Bosque* 31(3):219–30.

Hutton, G. and Varughese, M. 2016. *The costs of meeting the 2030 Sustainable Development Goal targets on drinking water, sanitation, and hygiene.* Water and Sanitation Program Technical Paper, Washington, DC: World Bank.

Ilstedt, U., Bargués Tobella, A., Bazié, H. R. et al. 2016. Intermediate tree cover can maximize groundwater recharge in the seasonally dry tropics. *Scientific Reports* 6: 21930.

Iroumé, A., Huber, A. and Schulz, K. 2005. Summer flows in experimental catchments with different forest covers, Chile. *Journal of Hydrology* 300(1–4):300–13.

Iroumé, A., Mayen, O. and Huber, A. 2006. Runoff and peak flow responses to timber harvest and forest age in southern Chile. *Hydrological Processes* 20(1):37–50.

Iroumé, A. and Palacios, H. 2013. Afforestation and changes in forest composition affect runoff in large river basins with pluvial regime and Mediterranean climate, Chile. *Journal of Hydrology* 505:113–25. doi:10.1016/j.hydrol.2013.09.031.

Jobbágy, E. G. and Jackson, R. B. 2004. Groundwater use and salinization with grassland afforestation. *Global Change Biology* 10:1299–312. doi:10.1111/j.1365-2486.2004.00806.x.

Jones, J. A., Achterman, G. L., Augustine, L. A. et al. 2009. Hydrologic effects of a changing forested landscape – challenges for the hydrological sciences. *Hydrological Processes* 23(18):2699–2704.

Jones, J., Almeida, A., Cisneros, F. et al. 2017. Forests and water in South America. *Hydrological Processes* 31(5):972–80. doi:10.1002/hyp.11035.

Khanna, J., Medvigy, D., Fueglistaler, S. and Walko, R. 2017. Regional dry-season climate changes due to three decades of Amazonian deforestation. *Nature Climate Change* 7(3):200.

Kotru, R., Rathore, B. M. S., Pradhan, N. et al. 2015. *Transforming mountain forestry in the Hindu Kush Himalayas: Toward a third-generation forest management paradigm*. ICIMOD Working Paper 2015/9. Kathmandu, Nepal: International Centre for Integrated Mountain Development (ICIMOD).

Kotru, R., Sharma, S., Sharma, E. and Hofer, T. 2017. *Everybody lives upstream: The watershed approach for the changing climate of Hindu Kush Himalaya*. ICIMOD Working Paper 2017/11. Kathmandu, Nepal: ICIMOD.

Kuraś, P. K., Alila, Y. and Weiler, M. 2012. Forest harvesting effects on the magnitude and frequency of peak flows can increase with return period. *Water Resources Research* 48:W01544. doi:10.1029/2011WR010705.

Lara, A., Little, C., Urrutia, R. et al. 2009. Assessment of ecosystem services as an opportunity for the conservation and management of native forests in Chile. *Forest Ecology and Management* 258(4):415–24.

Layton, K. and Ellison, D. 2016. Induced precipitation recycling (IPR): A proposed concept for increasing precipitation through natural vegetation feedback mechanisms. *Ecological Engineering* 91:553–65.

Little, C., Cuevas, J. G., Lara, A., Pino, M. and Schoenholtz, S. 2015. Buffer effects of streamside native forests on water provision in watersheds dominated by exotic forest plantations. *Ecohydrology* 8(7):1205–17.

Lorz, C., Abbt-Braun, G., Bakker, F. et al. 2012. Challenges of an integrated water resource management for the Distrito Federal, western central Brazil: Climate, land-use and water resources. *Environmental Earth Sciences* 65(5):1575–86.

Marc, V. and Robinson, M. 2007. The long-term water balance (1972–2004) of upland forestry and grassland at Plynlimon, mid-Wales. *Hydrology and Earth System Science* 11(1):44–60.

May, B., Smethurst, P., Carlyle, C. et al. 2009. *Review of fertiliser use in Australian forestry*. Report prepared for Forest & Wood Products Australia, Melbourne, Australia: Forest & Wood Products Australia. Available at: www.fwpa.com.au/ (Accessed 13 February 2019).

Nelson, E. J., Kareiva, P., Ruckelshaus, M. et al. 2013. Climate change's impact on key ecosystem services and the human well-being they support in the US. *Frontiers in Ecology and the Environment* 11(9):483–93.

NRC (National Research Council) 2008. *Hydrologic effects of a changing forest landscape*. Washington, DC: National Academies Press. Available at: http://nationalacademies.org/wstb (Accessed 13 February 2019).

Oyarzún, C. E. and Huber, A. 2003. Nitrogen export from forested and agricultural watersheds of southern Chile (Exportación de nitrógeno en cuencas boscosas y agrícolas en el sur de Chile). *Gayana Botánica* 60(1):63–8.

Peck, A. J. and Hatton, T. J. 2003. Salinity and the discharge of salts from catchments in Australia. *Journal of Hydrology* 272(1–4):191–202.

Perakis, S. S. and Hedin, L. O. 2001. Fluxes and fates of nitrogen in soil of an unpolluted old-growth temperate forest, southern Chile. *Ecology* 82(8):2245–60.

Perakis, S. S. and Hedin, L. O. 2002. Nitrogen loss from unpolluted South American forests mainly via dissolved organic compounds. *Nature* 415(6870):416.

Perry, T. D. and Jones, J. A. 2017. Summer streamflow deficits from regenerating Douglas-fir forest in the Pacific Northwest, USA. *Ecohydrology* 10(2):e1790.

Polasky, S., Nelson, E., Pennington, D. and Johnson, K. 2011. The impact of land-use change on ecosystem services, biodiversity and returns to landowners: A case study in the state of Minnesota. *Environmental and Resource Economics* 48(2):219–42.

Poulos, H. 2018. Why North American cities must thin overgrown forest to improve water supplies. *The Source*. 5 January [online]. International Water Association. Available at: www.thesourcemagazine.org/trees-became-enemy/ (Accessed 13 February 2019).

Schuller, P., Walling, D. E., Iroumé, A. et al. 2013. Using 137Cs and 210Pbex and other sediment source fingerprints to document suspended sediment sources in small forested catchments in south-central Chile. *Journal of Environmental Radioactivity* 124:147–59.

Sheil, D. 2018. Forests, atmospheric water and an uncertain future: The new biology of the global water cycle. *Forest Ecosystems* 5(19). doi:10.1186/s40663-018–0138-y.

Silveira, L. and Alonso, J. 2009. Runoff modifications due to the conversion of natural grasslands to forests in a large basin in Uruguay. *Hydrological Processes* 23:320–9. doi:10.1002/hyp.7156.

Singh, V. and Kotru, R. 2018. *Sustainable tourism in the Indian Himalayan region*. Report of Working Group II. New Delhi: NITI Aayog. Available at: http://niti.gov.in (Accessed 13 February 2019).

Soto-Schönherr, S. and Iroumé, A. 2014. *Eficiencia de uso de agua (EUA) de plantaciones forestales y cultivos de secano*. Technical report for Bioforest SA (Unpublished).

Soto-Schönherr, S. and Iroumé, A. 2016. How much water do Chilean forests use? A review of interception losses in forest plot studies. *Hydrological Processes* 30:4674–86. doi:10.1002/hyp.10946.

Spera, S. A., Galford, G. L., Coe, M. T., Macedo, M. N. and Mustard, J. F. 2016. Land-use change affects water recycling in Brazil's last agricultural frontier. *Global Change Biology* 22(10):3405–13.

Su, C. H. and Fu, B. J. 2013. Evolution of ecosystem services in the Chinese Loess Plateau under climatic and land use changes. *Global and Planetary Change* 101:119–28.

Swann, A. L., Longo, M., Knox, R. G., Lee, E. and Moorcroft, P. R. 2015. Future deforestation in the Amazon and consequences for South American climate. *Agricultural and Forest Meteorology* 214:12–24.

Tew, E. 2016. Jacksonville implements new way to clean wastewater, preserve forest, WNCT, 2 August. Available at: www.wnct.com/news/jacksonville-implements-new-way-to-clean-wastewater-preserve-forest/1091517118 (Accessed 13 February 2019).

Tóth, T., Balog, K., Szabó, A. et al. 2013. Influence of lowland forests on subsurface salt accumulation in shallow groundwater areas. *AoB PLANTS* 6. doi:10.1093/aobpla/plu054.

UN 2015. *The Millennium Development Goals Report*. New York: United Nations.

UN 2018. *Sustainable Development Goal 6 Synthesis Report 2018 on Water and Sanitation*. New York: United Nations.

UNESCO-IHP (International Hydrological Programme) 2014. *Water in the post-2015 development agenda and Sustainable Development Goals*. Discussion paper prepared for the 21st session of the Intergovernmental Council of the International Hydrological Programme of UNESCO, 18–20 June 2014, UNESCO Headquarters, Paris.

UN Water 2018a. *Monitoring Sustainable Development Goal #6*. Available at: www.sdg6monitoring.org/ (Accessed 7 July 2018).

UN Water 2018b. *About United Nations Water*. Available at: www.unwater.org/about-unwater/ (Accessed 7 July 2018).

UN Water 2018c. *Water and sanitation interlinkages across the 2030 Agenda for Sustainable Development*. Geneva.

van der Ent, R. J., Savenije, H. H. G., Schaefli, B. and Steele-Dunne, S. C. 2010. Origin and fate of atmospheric moisture over continents. *Water Resources Research* 46(9). doi:10.1029/2010WR009127.

van Dijk, A. and Keenan, R. J. 2007. Planted forests and water in perspective. *Forest Ecology and Management* 251:1–9.

Vanham D., Hoekstra A., Wada Y. et al. 2018. Physical water scarcity metrics for monitoring progress towards SDG Target 6.4: An evaluation of Indicator 6.4.2 'Level of water stress'. *Science of the Total Environment* 613–14:218–32.

Whitehead, D. and Beadle, C. 2004. Physiological regulation of productivity and water use in Eucalyptus: A review. *Forest Ecology and Management* 193:113–40.

WWAP (World Water Assessment Programme) 2009. *The United Nations World Water Development Report 3: Water in a changing world*. Paris: UNESCO and London: Earthscan.

WWF (World Wide Fund for Nature) 2019. *Tackling threats that impact the Earth: Deforestation*. Available at: www.worldwildlife.org/threats/deforestation (Accessed 13 February 2019).

Yang, H. and Zehnder, A. 2007. 'Virtual water': An unfolding concept in integrated water resources management. *Water Resources Research* 43(12).

Zhang, M., Liu, N., Harper, R. et al. 2017. A global review on hydrological responses to forest change across multiple spatial scales: Importance of scale, climate, forest type and hydrological regime. *Journal of Hydrology* 546:44–59.

Chapter 7 SDG 7: Affordable and Clean Energy – How Access to Affordable and Clean Energy Affects Forests and Forest-Based Livelihoods

Pamela Jagger*, Robert Bailis, Ahmad Dermawan, Noah Kittner and Ryan McCord

Key Points

- The role of traditional woodfuels in energy service provision will decline, though energy stacking that includes traditional woodfuels is likely to persist low- and middle-income countries.
- The role of processed woodfuels, forest-derived liquid biofuels, and biopower in achieving SDG 7 will depend on relative costs and innovation in storage capacity of renewables including solar, wind and micro-hydro.
- Transitions to modern fuels (including electricity generated with large-scale hydropower and heavy reliance on agriculture-derived liquid biofuels) threatens forests and forest-based livelihoods.
- Energy transitions involving decreased reliance on traditional woodfuels and increased use of forest-derived modern fuels (e.g. pellets, biofuel) are generally synergistic with achieving other SDGs.

7.1 Introduction

Throughout the world, forests play a significant role in the supply of energy services. The role of forests in ensuring access to affordable, reliable and sustainable energy for all – the overarching objective for SDG 7 – varies widely. In the developing world, an estimated 3–4 billion people rely on solid fuels, primarily traditional woodfuels (e.g. firewood and charcoal) harvested from natural forests and woodlots, for cooking and heating (WHO 2016). For people in low- and middle-income countries where traditional woodfuels dominate the energy portfolio, reliance on biomass for household energy will decline overall in the coming decades, though the absolute number of traditional woodfuel users in sub-Saharan Africa and South and Southeast Asia will grow (Bonjour et al. 2013). To date, evidence suggests that traditional woodfuel harvesting affects deforestation and forest degradation in only a few hotspot

* Lead author.

locations (Bailis et al. 2015). However, rapid urbanisation in Africa and South and Southeast Asia, signalling a potential shift from firewood to charcoal for cooking and heating, raises concerns about the associated impacts on forests in the absence of introduction of clean fuels.

Many middle- and high-income countries are diversifying their domestic energy portfolios. Processed woodfuels and liquid biofuels are an increasingly important component alongside wind, solar, hydro and geothermal energy sources to increase the share of renewable energy in the global energy mix (Ellabban et al. 2014). The majority of liquid biofuels are produced from agricultural crops and residues that have negative impacts on forests when they are cleared to establish plantations. Sustainable uses of bioenergy are important pathways to ensure diversified renewable energy service provision and can broaden livelihood strategies in a wide range of settings. However, in the USA and Europe, renewable energy portfolios for electricity and heat increasingly demand industrially produced pellets, raising concerns about sustainability and high costs of transportation when pellets are not locally produced (Hanssen et al. 2017, Searchinger et al. 2018). New and more efficient technologies for producing electricity from biopower have increased attention and interest in South-eastern Europe, Japan and elsewhere (UNESCAP 2017). Notably, strategies to meet SDG 7 indicators that involve large-scale hydro projects, which frequently inundate forests, lead to deforestation and loss of livelihoods.

This chapter provides an analysis of the implications of achieving SDG 7 (Table 7.1) for forests and for people whose livelihoods depend on forests.

Table 7.1 SDG 7 targets for 2030
7.1 Ensure universal access to affordable, reliable and modern energy services
7.2 Increase substantially the share of renewable energy in the global energy mix
7.3 Double the global rate of improvement in energy efficiency
7.4 Enhance international cooperation to facilitate access to clean energy research and technology, including renewable energy, energy efficiency and advanced and cleaner fossil-fuel technology, and promote investment in energy infrastructure and clean energy technology
7.5 Expand infrastructure and upgrade technology for supplying modern and sustainable energy services for all in developing countries, particularly least developed countries, small island developing states and land-locked developing countries, in accordance with their respective programmes of support
Source: Adapted from United Nations 2015

To frame our analysis, we identify four forest energy pathways: (1) sustainable use of traditional woodfuels, (2) processed woodfuels, (3) liquid biofuels and (4) biopower and cogeneration.[1] We discuss their potential to address SDG 7 as well as their hypothesised effects on forest and forest-based livelihoods in the near to medium term.[2] We highlight that in the context of energy service provision at the household level, the major role of forest-based energy is for cooking (developing countries) and heating (globally), and that liquid biofuels primarily support transitions in the transportation and industrial sectors. Cogeneration of electricity or biomass gasification using forest products are the main pathways towards addressing electricity access using wood-based fuels. This study fills an important gap, given that most recent peer-reviewed articles about SDG 7 explicitly focus on energy for lighting and do not address energy for heating and cooking despite the fact that these are the main uses of forest-based energy (Baptista and Plananska 2017, Mentis et al. 2017, Yang and Yang 2017).

We first review theories related to energy transitions and consider the role that forest-based energy plays in both the energy ladder and energy stacking transitions. We then describe the four forest energy pathways we have identified and their implications for supporting both sustainable forest management and forest-based livelihoods. We connect each pathway to its potential contribution to the energy ladder and stacking transitions and the realisation of SDG 7. We also discuss energy transitions that have a large potential impact on forests and forest-reliant peoples, such as large-scale hydro development, but that do not include forest-derived fuels. We provide several case studies that highlight different ways in which forest products influence SDG 7 and how, in turn, progress towards SDG 7 targets impacts forests and people. The cases examined include a small-scale woodfuel (e.g. pellets) and improved cookstove enterprise in Rwanda, global experience with *Jatropha curcas*, and heating and electricity biopower from forest products in Southeastern Europe. We discuss palm-derived liquid biofuels as an example of how an energy transition to modern fuels contributes to deforestation and loss of forest-based livelihoods. The cases intentionally highlight the diverse range of impacts forests have on energy provision and the potential ways that meeting SDG 7 could affect forests – for better and for worse. Finally, we consider

[1] For this analysis, we consider biomass from forests and woodlands, and their contributions to energy production. Forests are land spanning more than 0.5 hectares with trees higher than 5 metres and a canopy cover of more than 10 per cent, or trees able to reach these thresholds in situ. Woodlands include trees able to meet the forest definition and land with a combined cover of shrubs, bushes and trees above 10 per cent. It does not include land that is predominantly under agricultural or urban land use (FRA 2015).

[2] We do not discuss in any depth non-forest-based energy pathways, which include fossil fuels and renewables other than bioenergy (i.e. solar, wind, geothermal, nuclear).

how the realisation of SDG 7 through forest-based energy pathways will influence other SDGs.

7.2 Energy Ladder and Energy Stacking Theories and SDG 7

Two competing theories posit a relationship between increases in income and energy consumption. The energy ladder theory (Leach and Mearns 1988) differentiates energy use into traditional (animal dung, crop residues and woodfuels), transitional (kerosene and coal) and modern (liquefied petroleum gas [LPG], electricity and other renewables). Under the energy ladder theory, household energy choice moves from traditional to transitional to modern fuels as incomes increase. Traditional fuels are more polluting and less efficient but cheaper, while modern fuels are more energy efficient and cleaner, but more expensive. Energy ladder transitions are linear and assume that as transitional and modern fuels are adopted traditional fuels are no longer used. A competing theory suggests that households will instead stack fuels and technologies as incomes increase. Energy stacking involves the use of multiple fuels by the same household, taking advantage of the benefits each fuel provides (Gupta and Köhlin 2006, Masera et al. 2000, Masera and Navia 1997, Nansaior et al. 2011). Under the energy stacking theory, modern fuel users continue to use traditional and/or transitional fuels irrespective of income level, and assume a gradually partial or full transition to modern fuels, including stacking of multiple fuels and technologies in diverse ways.

An important caveat of both the energy ladder and energy stacking hypotheses is that they place emphasis on household income as the major driver of energy transitions. Indeed, in several studies income is the most important factor in determining fuel choices (Arnold et al. 2006, Cooke et al. 2008, Foster et al. 2000, Heltberg 2005, Hiemstra-van der Horst and Hovorka 2008). We note that few studies have explored supply-side factors affecting fuel choice (Jagger and Shively 2014, Lewis and Pattanayak 2012, Rehfuess et al. 2010). Global estimates (GEA 2012, UN DESA 2015) predict the absolute number of people dependent on biomass fuels will increase through 2030, suggesting that policy-makers should be attentive to factors that influence the supply, demand, spatial distribution and governance dimensions of biomass fuels, including traditional woodfuels. Several studies have noted the lack of information available about fuelwood harvesting practices, geography and dynamics, specifically with respect to woody biomass availability within different land uses (Foley 2005, Hiemstra-van der Horst and Havorka 2009, Smeets and Faaij 2007). Insights into the combined spatial and behavioural dynamics of woodfuel supply and demand support a broader understanding of the role of forest products in sustainable energy transitions (Masera et al. 2006, Rehfuess et al. 2010).

7.3 Forest-Based Energy Pathways

Forest products play a range of different roles in energy service provision and sustainable energy transitions depending on a variety of contextual factors. We identify four energy service pathways for forest products: (1) traditional woodfuels, (2) processed woodfuels, (3) liquid biofuels and (4) biopower and cogeneration (Table 7.2). For each pathway we discuss: (1) sustainability or extent to which they contribute to renewable energy targets, (2) socio-economic dimensions of the pathway and (3) how the pathway connects to theories of energy ladders/stacking. We also briefly touch on the regional setting where the pathway is most common, actors engaged in the pathway, the scale of operation and governance. We acknowledge the challenge of differentiating forest-based energy from the broader category of bioenergy. The term 'bioenergy' refers to energy derived from any organic matter available on a renewable basis, including forest and mill residues, agricultural crops (including field and processing residues), wood and wood waste, animal dung, fast-growing trees and herbaceous crops, etc. In practice, it is quite difficult to disaggregate the relative contribution of forests to bioenergy in most contexts due to how data are reported and depending on the definition of forest. In energy terms, the most common use of biomass after traditional cooking/heating is for industrial heat and space heating (REN21 2018). The biomass feedstocks for electricity cogeneration are predominantly forest residues (including black liquor), bagasse and other agricultural residues.

7.4 Traditional Woodfuels

7.4.1 Context

Traditional woodfuels, which include both firewood and charcoal, represent more than half of the global wood harvest and nearly 8 per cent of the primary global energy supply (FAOSTAT 2015, REN21 2018). Roughly 2.8 billion people worldwide (Bonjour et al. 2013), including the world's poorest and most marginalised, burn traditional woodfuels to satisfy their basic energy needs, with cooking and heating being the major services provided. Globally, the absolute number of traditional woodfuel users will increase at least through to 2030 (Riahi et al. 2012). The traditional woodfuel sector is typically comprised of large numbers of small to medium-scale actors. Many traditional woodfuel consumers collect or produce their own woodfuels for subsistence consumption, though there is a rapidly growing trade in charcoal, particularly in sub-Saharan Africa and the Middle East. Traditional woodfuel markets often lack regulatory frameworks or operate in environments where rules related to the production, transport and sale of woodfuels are not enforced.

Table 7.2 Forest-based energy pathways

	Traditional woodfuels	Processed woodfuels	Liquid biofuels	Biopower and cogeneration
Products	Fuelwood and charcoal	Pellets, torrefied biomass; Other compressed wood products	Transportation biofuels; biodiesel	Pellets and other biomass converted to electricity; Co-firing with coal or fossil fuels
Regional focus	Low- and middle-income countries in the Global South	North America and Europe; China; Small-scale examples throughout the Global South	Central America; tropical areas of Africa and Asia; Europe	Europe; Japan; South Korea
Actors	Large number of small-scale producers and consumers	Small number of producers at various scales; Small- and medium-scale consumers	Medium- and large-scale producers; Small- and medium-scale consumers	Industrial and government sectors
Scale	Local with some regional trade	Local, regional or global	Regional or global	Regional or global
Governance	Unregulated; Informal sector	Regulated; Formal sector	Regulated; Formal sector	Regulated; Formal sector
Sustainability	Conditionally renewable but sometimes associated with forest degradation	Pressure to manage forest resources sustainably; Feedstock a supply issue in some settings	Pressure to manage forest resources sustainably; GHG, energy, water, land intensive	Pressure to manage forest resources sustainably; GHG, energy, water, land intensive
Livelihoods/ services/final energy use	Cooking and heating	Cooking and heating	Transportation sector; Electricity; Industrial development	Electricity; Heating, Cooking, Industrial development
Energy ladder or energy stacking	Stacking	Stacking	Ladder and stacking	Stacking

7.4.2 Sustainability

Woodfuel demand is frequently associated with deforestation and forest degradation (de Montalembert and Clement 1983, Eckholm 1975, Eckholm et al. 1984). Concerns about the impacts of firewood and charcoal consumption on forests have motivated interventions to reduce woodfuel consumption several decades. Often implemented by development agencies or other outside actors, interventions have tried to enhance fuel supply through tree nurseries and community woodlots, production of briquettes and promoting fuel-saving cook stoves. Despite decades of attempts, few interventions have achieved widespread success.

Researchers have quantified traditional woodfuel sustainability in different locations (Drigo et al. 2015, Ghilardi et al. 2016). One pantropical assessment estimated that roughly 30 per cent of the global wood harvest is unsustainable, leading to localised degradation, with hotspots concentrated in South Asia and East Africa (Bailis et al. 2015). The loss of terrestrial carbon resulting from woodfuel-consumption-driven land-cover change is equivalent to 1–2 per cent of global CO_2 emissions, and roughly 20 per cent of global land-use change emissions. There is now consensus that, in the absence of other drivers of land-cover change, woodfuel demand rarely results in long-term deforestation. However, under many circumstances it can cause forest degradation.[3] We understand that geographically specific biophysical and socio-economic factors play a critical role in woodfuel sustainability (Hosonuma et al. 2012 Hansfort and Mertz 2011, Mayaux et al. 2013, Singh et al. 2010, Smith et al. 2014b). Biophysical factors include land cover, species distribution, climatic conditions and topography, among others. Socio-economic factors include population distribution, growth and urbanisation rates, wood energy demand and other drivers of land-cover change. Policies affecting land use, forest management and energy preferences are also important to protecting forests and people in areas with populations dependent on traditional woodfuels.

A noteworthy exception of the impact of traditional woodfuel consumption on forest sustainability is the impact of woodfuel demand in humanitarian settings. Sudden influxes of people, and their need for cooking energy in particular, can place pressure on forest resources, as observed in large refugee camps in Kenya, Sudan, Somalia, Nigeria, Myanmar, Bangladesh and

[3] For this discussion, it is important to distinguish between deforestation, defined as direct human-induced conversion of forested land to non-forested land, and forest degradation, defined as long-term reduction of the overall potential supply of benefits from the forest, which includes carbon, wood, biodiversity and other goods and services.

elsewhere (Caniato et al. 2017, Thulstrup et al. 2018). Many of these camps are located in already fragile ecological settings.

7.4.3 Livelihoods

Woodfuels play an important role in the livelihoods of billions of people in the Global South. In a study of forest reliance among rural populations in 25 countries throughout the Global South, Angelsen et al. (2014) estimate that traditional woodfuels account for 35 per cent of forest income (subsistence plus cash). Despite the important role that woodfuels play in income generation and diversification, woodfuel reliance has negative consequences for human health (Bruce et al. 2000, Smith et al. 2014a). Household air pollution (HAP) related to burning solid fuels (including firewood and charcoal) was responsible for 2.6 million deaths in 2016 (Health Effects Institute 2018). Exposure to HAP related to burning biomass as fuel is the thirteenth-largest risk factor overall, and the second-largest environmental risk factor (in low and middle-income countries) for global burden of disease after ambient air pollution (Forouzanfar et al. 2015). Other health effects associated with firewood collection include risk of physical assault (O'Brien 2006), musculoskeletal injuries from carrying fuelwood bundles and insect and snake bites (Haile 1991). In addition to health burdens, reliance on traditional woodfuels has implications for allocation of productive time, particularly for women and children. Where forest resources are scarce, people dedicate more time to wood collection and frequently involve children in the activity. When young girls spend more time collecting wood, they spend less time in school and do not progress to higher levels of education (Oluwafemi et al. 2012).

7.4.4 Link to Energy Transition Theories and SDG 7

Traditional woodfuels have a limited role to play in the way that SDG 7 is articulated. Firewood and charcoal are at the lowest rung on the energy ladder. Munro et al. (2017), in a study in Sierra Leone, express concern for both the lack of attention to energy poverty in SDG 7 discussions and for the flawed dismissal of the role of sustainably sourced traditional woodfuels in supporting the realisation of SDG 7. They cite an overemphasis on 'modern energy', much of which is out of reach for lower-income groups, advocating for an approach that allows for the promotion of multiple energy sources, including traditional woodfuels. This view supports energy stacking as the theory of change required to move towards achieving SDG 7. However, ensuring modern and affordable access to clean energy will likely involve significant reductions in traditional woodfuels.

7.5 Processed Woodfuels and Liquid Biofuels

7.5.1 Context

We distinguish between two main types of forest-based bioenergy: processed woodfuels (densified or torrefied solid fuels), and production of liquid biofuels from forest and woodland products (e.g. *Jatropha curcas*). Processed woodfuels are widely used for home heating in Northern Europe and China; in a small but growing number of countries in the Global South they are used in tandem with micro-gasification cook stoves (Case Study 7.1). Liquid biofuels help society respond to the increased demand for renewable energy sources to meet EU climate policy and renewable energy targets and comply with international agreements on climate change. The transition to renewable fuels in countries addressing the SDG 7 framework may create demand for new forest-based fuel products (Case Study 7.2).

Case Study 7.1 Densified Pellets (Processed Wood Fuel) in Rwanda

Founded in 2012, *Inyenyeri*, a for-profit social benefit company in Rwanda, is a private sector firm marketing processed woodfuels (e.g. pellets, briquettes) and micro-gasification stoves. Other than *Supamoto*, a firm in Zambia, no other pellet/cook-stove promoter in sub-Saharan Africa operates at the same scale. The experience of *Inyenyeri* provides insights into the challenges related to pellet production, improved cook-stove selection and the structure of the marketing model for businesses providing household energy services (Jagger and Das 2018). Here we focus on their experience with producing pellets. The firm's business model requires supplying enough biomass pellets to support the current customer base. Obtaining sustainably sourced feedstock of adequate type and quantity and maintaining a functional production facility are the major issues *Inyenyeri* has dealt with during its pilot phase from 2012 to 2018. The firm has experimented with a range of feedstock supply options, including a trade-in mechanism whereby rural biomass collectors exchange feedstock for pellets, and sourcing sawdust and other feedstocks from larger-scale operations in relatively close proximity to the pelletising plant in north-western Rwanda. The logistics of storing, transporting and drying feedstock have provided additional complexity to the operation. *Inyenyeri* is currently in negotiations with the Government of Rwanda to source feedstock from pine plantations in order to maintain a consistent supply of high-quality material for pelletising. The capital investment required for building large pelletising factories and the challenges of maintenance and repair in central Africa have been obstacles to scaling-up pellet production. Relying on a single pelletiser

Case Study 7.1 (cont.)

is a problem in an environment where capacity for equipment maintenance is low. The efficacy of a decentralised versus centralised system of pellet production (i.e. a few large-scale factories versus several small- to medium-scale enterprises) should be considered.

Inyenyeri's experience over the past five years illustrates the complexity of building a market for a clean cooking solution involving processed woodfuels. The potential of pellet and fan micro-gasification cooking should continue to be explored, particularly in settings where widespread distribution of affordable LPG and electric cooking systems will be realised in the distant future. *Inyenyeri* represents an important test case for understanding how to achieve a clean cooking system using a renewable biomass source in sub-Saharan Africa.

Case Study 7.2 Jatropha Biofuel

Jatropha curcas is a shrub promoted in several low- and middle-income countries as a source of biofuel, with co-benefits of improving rural employment opportunities, diversifying income, securing biodiversity and regenerating degraded lands (Brittaine and Lutaladio 2010, Reubens et al. 2011, Valdés-Rodríguez et al. 2014, von Maltitz and Setzkorn 2012). It survives well in harsh climatic and soil conditions, making it attractive in areas where agricultural production is marginal. Several governments have provided financial incentives to promote *J. curcas* cultivation by smallholders and larger-scale plantations with the aim of fostering a market for biofuels (Jull et al. 2007, Pradhan and Ruysenaar 2014, Soto et al. 2015). Several *J. curcas* cultivation projects were initiated in Central America, where the plant is indigenous, and throughout tropical Africa and Asia in the early 2000s. Evidence of the impact of *J. curcas* cultivation for livelihoods and sustainability is mixed.

In a comparison of smallholder and plantation-based *J. curcas* production, van Eijck et al. (2013) found that smallholder production is associated with more secure land rights, GHG balance, improved biodiversity and effectiveness in the number of people reached. Smallholder projects tend to be more resilient, likely because of lower start-up and production costs (Kgathi et al. 2017). However, because government subsidies for *J. curcas* cultivation tend to go to households with more resources and better risk-coping strategies (Soto et al. 2015), the poorest households are often excluded from government programmes. Low-income households are most vulnerable to negative social effects of *J. curcas* cultivation and are most likely to abandon the

Case Study 7.2 (cont.)

crop (Soto et al. 2018). Plantations, on the other hand, are associated with decreased food security, loss of land rights and decreased biodiversity (van Eijck et al. 2013). Plantation-based production creates more initial employment opportunities and higher incomes, but for a smaller number of people. The economic viability of plantations is limited in many settings because of high upfront costs, slow crop maturation and low yields, causing many projects to collapse before their yields can stabilise (Romijn et al. 2014, Gasparatos et al. 2015).

When cultivation involves clearing natural forest, impacts include deforestation, decreased biodiversity and threatened water sources (Creutzig et al. 2012, Fargione et al. 2008, Laurance et al. 2014, Wu et al. 2014). However, when *J. curcas* is planted on degraded lands, plantations have resulted in reduced soil erosion and renewed stimulation of biological activity, and thus improved soil quality, without competing with food production or depleting natural resources (Wani et al. 2012). Overall, small-scale *J. curcas* production on already degraded land not currently used for crop production has the best social and environmental impacts on forests and forest peoples (Skutsch et al. 2011).

A challenge for this study is the disaggregation of forest versus non-forest-based liquid biofuels. For example, while ethanol and biodiesel produced from agricultural residues are important in China, Brazil and Sweden, they are outside the scope of this study because they are not forest-based. Other liquid biofuels – for example, those derived from palm in Indonesia – play a major role in meeting liquid biofuel targets, but are considered a contributor to deforestation when primary forest is cleared to establish plantations (Case Study 7.3).

Renewable portfolio standards across countries and sectors influence the role of forest products in meeting renewable energy targets, with standards taking different shape depending on the sector, country and policy environment. To date, the EU Renewable Energy Directive (RED) may be the most impactful policy on forest-based bioenergy (Searchinger et al. 2018). RED is a binding target, though member states set their own (frequently non-binding or flexibly binding) domestic goals. Biofuels frequently play different roles in electricity and transportation sectors, ranging from wood pellets burned for electricity generation and household heating to liquid biofuels replacing fossil-fuel reserves in the transportation sector. For cooking, renewable energy

Case Study 7.3 Palm Biodiesel in Indonesia

The development of the biodiesel sector in Indonesia is driven by multiple factors, including (1) a national agenda to support energy security in response to heavy dependence on imported crude oil (Dermawan et al. 2012, Kharina 2016, Naylor and Higgins 2017), (2) expectations that developing the biodiesel sector contributes to efforts to mitigate climate change (da Silva Araujo 2014, McBride et. al. 2011, Sedjo 2011), (3) Indonesia's position as the world's largest producer of palm oil, and (4) a mechanism for mitigating risk associated with fluctuations in the global price of palm oil (Danny 2018, Nurfatriani et al. 2018). The National Energy Policy issued in 2014 mandates that new and renewable energy, including biodiesel, make up 24.5 per cent of the national energy mix by 2025 and 31 per cent by 2050. The main policy to develop the biodiesel sector has been the blending target of 30 per cent of biodiesel use by 2020. The blending target is applicable to the transportation, industrial and electricity sectors, with its main emphasis on the transport sector.

Estimating the impacts of palm oil production for biodiesel on deforestation in Indonesia is empirically challenging for several reasons (Obidzinski et al. 2012). First, the proportion of Indonesian palm oil that goes into biodiesel production is small. In 2017, 2.7 million tonnes – less than 10 per cent of total production – went into biodiesel (Wright and Rahmanulloh 2017). Second, palm oil is a product with multiple uses; biodiesel is only one of them. Large producers of palm oil derivatives can shift the palm oil from one purpose (e.g. food) to another (e.g. biodiesel) depending on economic conditions (Eynck et al. 2013).

Socio-economic analysis of the role of the palm oil sector with specific reference to biofuels is hindered by a lack of traceability of palm oil value chains. Biodiesel producers may receive palm oil from the company mills, from third-party corporate suppliers and from independent smallholders. Changes in biodiesel demand affects producers' allocation of palm oil; however, impacts on forests and smallholders depend more broadly on palm oil demand, which may or may not relate to demand for palm oil-derived biodiesel.

portfolios distinguish between sustainably produced pellet fuels that replace unsustainably produced charcoal and continued use of traditional biomass energy sources.

7.5.2 Sustainability

Bioenergy is controversial for its potential competition with crop production and because of potential links to deforestation (e.g. palm biodiesel, Case Study 7.3). A range of sustainability standards and monitoring frameworks have been developed since the USA and the EU each implemented bioenergy trade rules and regulations in 2007/2008 (Scarlat and Dallemand 2011). Bioenergy plays a particularly challenging role in renewable energy portfolios when sourced from forest products. Policies in place to prioritise waste, residues and specific crops help guide the monitoring and verification of liquid biofuel products, yet significant challenges remain to avoid unintended land-use changes resulting from renewable energy portfolio policies that incentivise bioenergy. Sustainability concerns include biodiversity impacts, landscape impacts, soil nutrients and protective functions, water impacts and GHG emissions. Renewable energy portfolio standards could increase forest product demand for bioenergy initiatives that produce electricity, transportation, heat and chemicals. Improved monitoring and verification of sources of processed woodfuels and liquid biofuels would provide a way to track the use of unsustainable forest products contributing to energy demand.

7.5.3 Livelihoods

Production of processed woodfuels and liquid biofuels is employment intensive, providing jobs at all stages in the value chain. REN21 (2010) estimates there were approximately 1.5 million direct jobs in 2010 for biomass production, operation, harvesting and transportation; biomass facility processing and upgrading; conversion plan construction, operation and maintenance; and distribution of final energy products. Due to the growing demand for bioenergy, the European Renewable Energy Council (EREC) and Greenpeace estimate the creation of 2.1 million new jobs in the sector by 2030 (EREC 2008). In many developed countries, regional policies support development of the bioenergy sector to enhance employment opportunities in rural economies (Halder et al. 2014). Similar potential for growth in the sector exists in low and middle-income countries; however, given the reliance of local populations on forests for a wide range of goods and services, energy and rural development policies should ensure that local populations are not harmed by development of the sector. Cultivation of some biofuel feedstocks is similar to other large-scale monoculture cropping schemes, having large impacts on the supply of goods and services provided by natural forests.

Buongiorno et al. (2011) modelled the aggregate effects of bioenergy on the forestry sector on both local and regional economies. The global forest products model (GFPM) projects the consequences of the global forest sector doubling the rate of growth of bioenergy demand relative to a base scenario, all else being equal. Doubling bioenergy demand leads to the convergence of the price of firewood and industrial roundwood, raising the projected price of industrial roundwood by nearly 30 per cent by 2030. The price of sawn wood and panels would be 15 per cent higher. The price of paper would be 3 per cent higher. Concurrently, the demand for all manufactured wood products would be lower in all countries, but production would rise in countries with competitive advantage. Global value added in wood-processing industries would be 1 per cent lower in 2030; forest stock would be 2 per cent lower for the world and 4 per cent lower for Asia. Estimated effects vary substantially by country. Overall, the analysis implies that development of the bioenergy sector will negatively affect forest product prices and forest sustainability in a number of countries.

7.5.4 Link to Energy Transition Theories and SDG 7

Processed woodfuels and liquid biofuels may play a major role in the realisation of SDG 7; however, pricing and market development for other renewables, along with the regulatory framework surrounding mandated portfolios and certification, will influence how their role evolves. In most low-income countries, there are few examples of processed woodfuels or forest-derived liquid biofuels utilised at a significant scale for cooking and heating, though the use of agriculture and forest-derived liquid biofuels in the transportation sector is common. Most likely, for residential and small-scale industrial use, processed woodfuels and liquid biofuels will be part of an energy stacking transition that also includes traditional woodfuels in low- and middle-income countries, and more diversified portfolios, including other renewables in higher-income countries. For the realisation of SDG 7, several challenges emerge. Modern woodfuels need companion heating and cooking technologies (e.g. improved stoves) to ensure that the energy is clean (i.e. achieving efficiency gains, emission reductions and associated health benefits). In many countries, the development of both processed woodfuels and forest-derived biofuels involves establishing entirely new supply chains or importing high volumes of biomass. The overall sustainability and economic feasibility of developing supply chains is complex.

7.6 Biopower Cogeneration for Electricity

7.6.1 Context

Biopower and combined heat and power systems (cogeneration) use biomass in the form of pellets or other wood products to generate electricity. In the USA, overall biomass electricity generation is increasing in total production but decreasing in share of the electricity mix, possibly due to the rapidly declining cost of natural gas and alternative renewable energy sources. Bioenergy is promoted for electricity generation as a way to decarbonise the electricity sector, reduce emissions and meet Intergovernmental Panel on Climate Change (IPCC) climate targets (Davis et al. 2018). In this context, many countries are exploring retrofitting coal plants to combust bioenergy for heat and power applications. In Brazil, biomass-derived charcoal could substitute for coal in the steel sector. To meet industry demands and phase out coal, millions of hectares of forest are necessary (Sonter et al. 2015). Despite infrastructure and pressure on forest resources, demand for biomass electricity continues to grow in Europe and Japan. Canada and the USA export a significant amount of wood pellets to supply UK and European markets. Dwivedi et al. (2014) estimate a 50–68 per cent decrease in GHG intensity for electricity from wood pellets used for electricity in the UK.

7.6.2 Sustainability

A major challenge for sustainable forest management and biopower production is to ensure the use of waste and residue biomass products before using virgin materials for electricity generation or district heating. Certification of sustainably sourced biomass for electricity generation is a challenge. The UK and the EU have introduced new requirements to sustainably source biomass for electricity, focusing on waste and residues rather than pure wood (European Commission 2016). Future targets that adhere to these priority measures can reduce pressure on forests. RED established non-binding criteria, including banning the use of biomass from land converted from high biodiversity forest areas and favouring national biofuel support schemes. Despite these reporting efforts, monitoring of the origin of biomass consumed in the EU remains a challenge to sustainably managing megawatt-scale biomass heat and power initiatives.

The IPCC Working Group Report includes biomass as a critical electricity generation technology along with carbon capture and storage (bioenergy carbon capture and storage or BECCS) in its models as one of the few ways to maintain two degrees of global warming without incurring significant costs to the electricity system. Future models of decarbonised electricity systems place the levelised cost of biomass electricity in a range similar to renewable electricity systems today – though it may require further integration for cost-effective, low-carbon biomass systems (Sanchez et al. 2015). The affordability and viability of such emissions reductions remain a point of debate and uncertainty, primarily due to the lack of alternative electricity supply options and the assumption that carbon sequestration remains cheaper than alternative generating sources, including solar and wind which do not have the same 'negative' emissions potential. Realising emission reduction strategies through BECCS technologies would require significant technological innovation and could impose higher costs than the IPCC estimates. This could significantly affect demand for bioenergy forest products and place pressure on forests in Africa and the Amazon region. Not all BECCS is produced and stored at the same location, which poses challenges to monitor and verify the emission reductions and avoid double counting.

Finally, an important consideration for the future of biopower in realising SDG 7 is the rapidly declining cost of solar, wind, geothermal and battery storage (Kittner et al. 2017). Renewable energy alternatives may affect demand for biopower in the future, but near-term generation indicates continued consumption of electricity from (mainly agriculture-sourced) biomass feedstocks in USA, China, Germany and Brazil. If expanded beyond agricultural capacity, there could be indirect effects on forests, such as the conversion of forestland to produce biopower crops or fast-growing wood pellet farms.

7.6.3 Livelihoods

Evidence of the livelihood impacts of the growth of biopower within the energy sector is limited. Government subsidies that support BECCS could induce conversion from natural forests to plantations to produce bioenergy, which may threaten forests or people with forest-based livelihoods. However, market stimulation of increased biopower energy demand may not have localised effects. For example, if bioenergy products for power generation in the EU are imported in pellet form from exporting nations such as the USA and Canada, employment generation may occur, but not in places where demand for biopower is realised.

7.6.4 Link to Energy Transition Theories and SDG 7

To the extent that biopower will replace other energy sources, particularly for district heating, it supports the stacking hypothesis. The RED set up legally binding mandates to target a certain percentage of energy consumption from renewable sources, and similar policies are in place in the UK. While electricity production using wood pellets will increase, it is unlikely to fully displace current modes of energy production.

Most notably, the RED has generated a large demand for wood pellets used in electricity generation and district heating for urban areas. The RED sets a binding target of 20 per cent final energy consumption from renewable sources by 2020, which includes biomass energy. All EU member states have created action plans, and a number of individual states with large heating demands and forest resources have turned to wood pellets as an energy technology to meet this target. Cogeneration of electricity and

Case Study 7.4 Biopower and Cogeneration in Southeast Europe

Most existing coal power plant infrastructure could transition at a relatively low cost to burning biomass pellets. Switching from coal to biomass pellets using existing infrastructure alleviates the financial burden of financing new infrastructure projects and has gained significant attention in the USA, Europe and China. Eastern European countries maintain large production and consumption shares of forest bioenergy for district heating and cogeneration. In particular, wood chips overtook natural gas in Lithuania as primary district heating fuels in 2017 (REN21 2017). Other countries in Southeast Europe – including Kosovo, Bosnia and Herzegovina, Serbia and Croatia – may continue this trend as they address rising air pollution and associated health burden concerns from burning lignite coal, and can switch fuels without significantly altering boiler technologies (Kittner et al. 2016). For emerging economies, biomass presents a dual challenge. The large area of forest cover in Kosovo provides a cost-effective alternative to lignite coal for household heating and electricity generation if managed domestically. However, sustainability issues remain, and a significant expansion of biomass reliance could increase demand for imported biomass, placing pressure on nations seeking extra revenue from wood product exports.

In Kosovo and the western Balkans, household heating remains a critical challenge to achieving SDG 7. It is expensive and difficult to provide affordable and reliable energy during the winter months, when temperatures can drop below freezing, and there is a high dependence on lignite coal for heating. The region has large areas of forest, allowing for the production and use of higher-quality woodfuels containing fewer toxic pollutants than lignite coal. However, lack of access to quality woodfuels has hindered availability for residents across the country. Switching from lignite towards cleaner bioenergy options could also reduce exposure to toxic trace metals, including chromium and arsenic (Kittner et al. 2018). Efforts to achieve the health benefits of burning cleaner heating fuels should pay special attention to the management and governance of land dedicated to growing fuel wood. Alternatively, if electric heat pumps are widely adopted, as they have been in other European nations, there could be better opportunities to use electricity for household heating and reduce the demand for woodfuels from forests. A significant expansion of the woodfuel market without domestic management could cause larger-scale woodchip operations and imports from as far away as the south-eastern United States, where fast-growing trees for wood pellets have surged in production over the past 10 years.

heat used for distribution throughout cities or buildings has emerged as a low-cost method to deliver critical renewable energy services to European households.

7.7 Large-Scale Energy Infrastructure Development and Impacts on Forests and Forest-Reliant People

Many emerging economies with low levels of electricity access view hydropower as a way to meet SDG 7 goals. There are an estimated 450 planned hydropower dams expected to generate dozens of gigawatts of electricity capacity across the Amazon, Democratic Republic of Congo and the Mekong River Basin (Myanmar, Laos and Vietnam) in the coming decade (Winemiller et al. 2016). If built to satisfy SDG 7 targets of clean and affordable energy without design precautions and consideration of environmental and social safeguards, these plants could drastically alter forest cover, biodiversity and local livelihoods. Widespread forest cover loss and concerns about displacing people from their homes are major concerns (Winemiller et al. 2016). For example, in Brazil, hydropower supplies more than two-thirds of electricity. Forest-dependent populations are displaced by dam construction, and new roads associated with dam development indirectly lead to agricultural expansion and increased forest cover losses (Barber et al. 2014, Zarfl et al. 2015). Plants are often justified as providing electricity to affected rural populations, even though they frequently fail to serve low-income or last-mile populations.

Hydropower often draws the attention of climate financiers that consider it a low-carbon electricity source. However, hydropower projects greater than 1 megawatt in size carry a substantial land footprint and require reservoirs spanning several hundred square kilometres, as is the case along the Amazon where reservoirs displace tropical forests to meet Brazil's demand for electricity (de Faria et al. 2015). Carbon emissions associated with these hydropower reservoirs include methane off-gassing, the carbon release from converted tropical forestland during dam construction and associated ecological changes in land use along the riparian zones (de Faria et al. 2015, Räsänen et al. 2018). In the Mekong, some hydropower reservoirs rival GHG emissions from fossil-fuel plants when considering the methane flux from reservoirs (Räsänen 2018).

Hydropower is also controversial due to uncertainty about whether plants can provide low-cost electricity access when alternative technologies are available (de Faria and Jaramillo 2017, Deshmukh et al. 2018). This includes options to use forest-based biomass for electrification or gasification and the adoption of solar or small hydropower-based mini-grids. At present, hydropower is appealing as basic solar home systems often fail to meet the demand required for rice milling or cooking that many populations without electricity access desire. Smaller, more ecologically friendly types of hydropower dams exist that can meet SDG 7 goals without destroying forests and displacing people. Mini hydropower projects with localised distribution are likely to have a far lower impact than large-scale efforts. Higher capacity mini-grids in Nepal, Myanmar and Laos provide new opportunities to utilise larger-scale solar photovoltaics or hydropower dams in complementary ways. A focus on the diversity of renewable energy options available, including those from solar, wind and biomass, can mitigate larger risks for land management, tropical forests and people who are seeking access to electricity.

7.8 SDG 7 and Its Relationship to Other SDGs

In order to understand the implications of fulfilling SDG 7 as it relates to other SDGs, we consider each of the four forest energy pathways reviewed and present the hypothesised impacts for both forests and forest-reliant peoples should SDG 7 be realised (see Table 7.3). Our assumption is that as progress towards SDG 7 increases, the role of traditional woodfuels will decline and the role of modern woodfuels and biofuels will increase.

7.9 Conclusion

This chapter reviews the role that forest-derived energy will play in the realisation of SDG 7, focusing on four pathways for forests to contribute to energy service provision: traditional woodfuels, modern woodfuels, liquid biofuels and biopower/cogeneration. Energy transitions in low- and middle-income countries will likely involve reductions in traditional woodfuel reliance for heating, cooking and small-scale industrial energy provision, whereas countries currently seeking to diversify renewable energy portfolios may see an increase in forest-based bioenergy as long as it remains competitive and cost-effective. The cost of other renewables will play a major role in determining how important forest-based energy sources are for electricity, heating, cooking and transportation. A recent and growing literature addresses various aspects of SDG 7 and the role of forests. Calzadilla and Mauger (2017) cite wind and solar as the most promising energy sources for developing countries while indicating concerns about the lack of attention to equity issues in case studies from Chile, India, Kenya and Mexico. In most settings, our expectation is a transition that involves the diversification of energy sources that households and businesses rely on rather than a complete transition away from current fuels and technologies. Baptista and Plananska (2017) cite problems of path dependence and inertia in the implementation of energy

Table 7.3 Trade-offs and synergies between fulfilling SDG 7 and other SDGs

SDG	Reduction in use of traditional woodfuels		Increase in processed woodfuels, liquid biofuels and biopower/cogeneration	
	Forests	People	Forests	People
1 No poverty	Reduced pressure on forests improves ecosystem services (+)	Reduced woodfuel reliance (+); Loss of employment (–)	Loss of ecosystems services (–)	Employment in renewables sector (+); Poor and last-mile populations left out of the transition (–)
2 Zero hunger	Reduced degradation allowing forest foods to flourish (+)	More efficient technologies requiring less fuel for cooking, more frequent/ diverse cooked meals (+)	Land degradation and loss of agricultural land from pressure to develop biofuels sector (–)	Potential decrease in food security in biofuel planation development areas (–)
3 Good health and well-being	Preservation of forests supporting human health and well-being (+)	Reduced exposure to household air pollution (+); Reduced risk of injury/ harm (+)	Loss of natural areas due to development of bioenergy (–)	Reductions in exposure to household air pollution (+)
4 Quality education		Reduced fuel collection and cooking time freeing people to go to school (+)		

Table 7.3 (cont.)

SDG	Reduction in use of traditional woodfuels		Increase in processed woodfuels, liquid biofuels and biopower/cogeneration	
	Forests	People	Forests	People
5 Gender equality	Improved access for society to women's forest management capabilities (+)	Reduced fuel-collection time freeing women of drudgery; Improved cooking conditions increasing safety (+)		
6 Clean water and sanitation	Reduced impact on forest ensuring high-quality water (+)		Water tables affected by emphasis on fast-growing species (–)	Reduced cost and time to treat water by boiling (+)
8 Decent work and economic growth		Reduced harvest time for woodfuels decreasing dangerous activity (+); Loss of connection with forests and social aspects of woodfuel collection (–)		New sector development, employment generation (+)

Table 7.3 (cont.)

SDG	Reduction in use of traditional woodfuels		Increase in processed woodfuels, liquid biofuels and biopower/cogeneration	
	Forests	People	Forests	People
9 Industry, innovation and infrastructure		Transition away from inefficient technologies (+)	New innovations in forest plantation use (+)	Emergence of biofuels sector as new in many countries – leading to diversified economies (+)
10 Reduced inequalities		Closing gap between those reliant on biomass and those with access to modern energy (+)	New opportunities for engagement in forest management (+)	
11 Sustainable cities and communities	Reduced pressure on urban trees and forests in rural areas supporting more sustainable environments (+)	Household adoption of modern fuels (+)		Commitment to renewable energy portfolios reduces household air pollution (+)
12 Responsible consumption and production	Reduces pressure on forests (+)			Increased use of more efficient technologies (+)

Table 7.3 (cont.)

SDG	Reduction in use of traditional woodfuels		Increase in processed woodfuels, liquid biofuels and biopower/cogeneration	
	Forests	People	Forests	People
13 Climate action	Reduced GHG emissions from deforestation/ forest degradation and from improved combustion processes (+)	Mitigation of ambient and household air pollution exposure (+)		
14 Life below water	Reduced land degradation leading to less run-off and water pollution (+)		Increased pressure on water resources to irrigate bioenergy crops (–)	
15 Life on land	Greater biodiversity results from reducing woodfuel pressure on forest (+)	Securing ecosystem services for human well-being (+)	Increased role of biofuels in energy portfolios (+/–)	
16 Peace, justice and strong institutions	Reduced corruption in traditional woodfuel sector leading to decreased deforestation and forest degradation (+)	Reduced rent-seeking behaviour with respect to traditional woodfuels to improve livelihoods (+)	Increased focus on forest plantations as energy source reinforcing property rights (+)	Support for renewable energy targets and links to global climate institutions can enhance economies (+)
17 Partnerships for the goals				

initiatives in sub-Saharan Africa, suggesting that transitions will be slow and likely support the use of multiple energy sources, making the energy stacking hypothesis most plausible. The case studies highlight the different trade-offs to consider when implementing SDG 7 targets and provide insights into the challenge of integrating forests into the transition to cleaner and more affordable energy systems.

Recognising the co-benefits associated with forest-based energy pathways generally supports the realisation of other SDGs. Partnerships with other SDGs that acknowledge the role of forests in energy service provision are particularly essential to improving livelihoods and conditions in forest regions (Gratzer and Keeton 2017). In contrast, if SDG 7 is realised through the promotion of large-scale energy infrastructure projects, including hydropower and land-intensive solar and wind farms, forest ecosystems and forest livelihoods could be at risk, compromising other SDG outcomes.

References

Angelsen, A., Jagger, P., Babigumira, R. et al. 2014. Environmental income and rural livelihoods: A global comparative analysis. *World Development* 64(Supplement 1):S12–28.

Arnold, J. E. M., Köhlin, G. and Persson, R. 2006. Woodfuels, livelihoods, and policy interventions: changing perspectives. *World Development* 34(3):596–611. doi:10.1016./j.worlddev.2005.08.008.

Bailis, R., Drigo, R., Ghilardi A. and Masera, O. 2015. The carbon footprint of traditional woodfuels. *Nature Climate Change* 5:266–72.

Baptista, I. and Plananska, J. 2017. The landscape of energy initiatives in sub-Saharan Africa: Going for systemic change or reinforcing the status quo? *Energy Policy* 110:1–8.

Barber, C. P., Cochrane, M. A., Souza, C. M. and Laurance, W. F. 2014. Roads, deforestation, and the mitigating effect of protected areas in the Amazon. *Biological Conservation* 177:203–9.

Bonjour, S., Adair-Rohani, H., Wolf, J. et al. 2013. Solid fuel use for household cooking: Country and regional estimates for 1980–2010. *Environmental Health Perspectives* 121(7):784–90.

Brittaine, R. and Lutaladio, N. 2010. *Jatropha: A smallholder bioenergy crop – the potential for pro-poor development. Integrated Crop Management*, vol. 8. Rome: Food and Agriculture Organization of the United Nations (FAO).

Bruce, N., Perez-Padilla, R. and Albalak, R. 2000. Indoor air pollution in developing countries: A major environmental and public health challenge. *World Health Organization Bulletin* 78:1078–92.

Buongiorno, J., Raunikar, R. and Zhu, S. 2011. Consequences of increasing bioenergy demand on wood and forests: An application of the global forest products model. *Journal of Forest Economics* 17:214–29.

Calzadilla, P. V. and Mauger, R. 2017. The UN's new sustainable development agenda and renewable energy: The challenge to reach SDG7 while achieving energy justice. *Journal of Energy & Natural Resources Law* 36(2):233–54.

Caniato, M., Cariliez, D. and Thulstrup, A. 2017. Challenges and opportunities of new energy schemes for food security in humanitarian contexts: A selective review. *Sustainable Energy Technologies and Assessments* 22:207–19.

Cooke, P. Köhlin, G. and Hyde, W. F. 2008. Fuelwood, forests and community management – evidence from household studies. *Environment and Development Economics* 13(1):103–35. doi:10.1017/S1355770X0700397X.

Creutzing, F., Popp, A., Plevin, R. et al. 2012. Reconciling top-down and bottom-up modelling on future bioenergy deployment. *Nature and Climate Change* 2:320–7.

Danny, W. 2018. *Efektifitas dana sawit dalam mendukung industry sawit berkenaljutan dan berkeadilan*. Presentation in the workshop on Optimizing the CPO Fund in Supporting Sustainable Palm Oil Industry and Avoiding Deforestation, Jakarta, 13 March 2018.

da Silva Araujo, F., Araujo, I. C., Costa, I. et al. 2014. Study of degumming process and evaluation of oxidative stability of methyl and ethyl biodiesel of *Jatropha curcas* L. oil from three different Brazilian states. *Renewable Energy* 71:495–501.

Davis, S. J., Lewis, N. S., Shaner, M. et al. 2018. Net-zero emissions energy systems. *Science* 360 (6396):eaas9793.

de Faria, F. A. and Jaramillo, P. 2017. The future of power generation in Brazil: An analysis of alternatives to Amazonian hydropower development. *Energy for Sustainable Development* 41:24–35.

de Faria, F. A., Jaramillo, P., Sawakuchi, H. O., Richey, J. E. and Barros, N. 2015. Estimating greenhouse gas emissions from future Amazonian hydroelectric reservoirs. *Environmental Research Letters* 10(12):124019.

de Montalembert, M. R. and Clement, J. 1983. *Fuelwood supplies in the developing countries*. FAO Forestry Paper. Rome: FAO.

Dermawan, A., Obidzinski, K. and Komarudin, H. 2012. *Withering before full bloom? Bioenergy development in Southeast Asia*. CIFOR Working Paper No. 94, Bogor, Indonesia: Center for International Forestry Research (CIFOR).

Deshmukh, R., Mileva, A. and Wu, G. C. 2018. Renewable energy alternatives to mega hydropower: A case study of Inga 3 for Southern Africa. *Environmental Research Letters* 13(6).

Drigo, R. Bailis, R., Ghilardi, A. and Masera, O. 2015. *Analysis of woodfuel supply, demand and sustainability in Honduras*. WISDOM Case Studies. Available at: www.cleancookingalliance.org/resources/425.html (Accessed 12 March 2019).

Dwivedi, P., Khanna, M., Bailis, R. and Ghilardi, A. 2014. Potential greenhouse gas benefits of transatlantic wood pellet trade. *Environmental Research Letters* 9(2):024007.

Eckholm, E. 1975. *The other energy crisis: Fuelwood*. Worldwatch Paper 1. Washington, DC: Worldwatch.

Eckholm, E., Foley, G., Barnard, G. and Timberlake, L. 1984. *Fuelwood: The energy crisis that won't go away*. Washington, DC: Earthscan.

Ellabban, O., Abu-Rub, H. and Blaabjerg, F. 2014. Renewable energy resources: Current status, future prospects and their enabling technology. *Renewable and Sustainable Energy Reviews* 39:748–64. doi:10.1016/j.rser.2014.07.113.

European Commission 2016. *Proposal for a directive of the European Parliament and of the Council on the promotion of the use of energy from renewable sources.* COM/2016/0767 final – 2016/0382 (COD). Brussels: European Commission.

European Renewable Energy Council (EREC) 2008. *2008 Renewable Energy Technology Roadmap 20% by 2020.* Brussels: EREC.

Eynck, C., Shrestha, D., Vollmann, J. et al. 2013. Sustainable oil crops production. In Singh, B. P. (ed.) *Biofuel crop sustainability*. West Sussex, UK: John Wiley & Sons, pp. 165–204.

FAOSTAT 2015. Forestry production and trade. Available at: http://faostat3.fao.org/faostat-gateway/go/to/download/F/*/E (Accessed 27 July 2019).

Fargione, J., Hill, J., Tilman, D., Polasky, S. and Hawthorne, P. 2008. Land clearing and the biofuel carbon debt. *Science* 319:1235–8.

Foley, J. A. 2005. Global consequences of land use. *Science* 309(5734):570–4. doi:10.1126/science.1111772.

Forest Resource Assessment (FRA) 2015. *FRA 2015 Terms and Definitions.* Forest Resource Assessment Working Paper Number 180. Rome: Food and Agriculture Organization of the United Nations.

Foster, V., Tre, J-P., and Wodon, Q. 2000. *Energy prices, energy efficiency, and fuel poverty*. World Bank Working Paper. Washington, DC: World Bank.

Fourouzanfar, M. H., Alexander, L, Anderson, H. R. et al. 2015. Global, regional, and national comparative risk assessment of 79 behavioural, environmental and occupational, and metabolic risks or clusters of risks in 188 countries, 1990–2013: A systematic analysis for the Global Burden of Disease Study 2013. *The Lancet* 386:2287–323.

Gasparatos, A., von Maltitz, G. P., Johnson, F. X. et al. 2015. Biofuels in sub-Sahara Africa: Drivers, impacts and priority policy areas. *Renewable and Sustainable Energy Reviews* 45:879–901.

Ghilardi, A., Bailis, R., Mas, J. F., et al. 2016. Spatiotemporal modeling of fuelwood environmental impacts: Towards an improved accounting of non-renewable biomass. *Environmental Modelling & Software* 82:241–54.

GEA (Global Energy Assessment) 2012. *Global energy assessment: Toward a sustainable future.* Cambridge UK and New York, USA: Cambridge University Press and the International Institute for Applied Systems Analysis, Laxenburg, Austria.

Gratzer, G. and Keeton, W. S. 2017. Mountain forests and sustainable development: The potential for achieving the United Nations' 2030 Agenda. *Mountain Research and Development* 37(3):246–53.

Gupta, G. and Köhlin, G. 2006. Preferences for domestic fuel: Analysis with socio-economic factors and rankings in Kolkata, India. *Ecological Economics* 57(1):107–21. doi:10.1016/j.ecolecon.2005.03.010.

Haile, F. 1991. *Women fuelwood carriers in Addis Ababa and the peri-urban forest: Report to International Development Research Centre (IDRC) and National Urban Planning Institute (NUPI).* Geneva: International Labour Organization.

Halder, P., Paladinic, E., Stevanov, M. et al. 2014. Energy wood production from private forests – non-industrial private forest owners' perceptions and attitudes in Croatia and Serbia. *Renewable and Sustainable Energy Reviews* 35:515–26.

Hansfort, S. and Mertz, O. 2011. Challenging the woodfuel crisis in West African woodlands. *Human Ecology* 39(5):583–95.

Hanssen, S., Duden, A. S., Junginger, M. et al. 2017. Wood pellets, what else? Greenhouse gas parity times of European electricity from wood pellets produced in the south-eastern United States using different softwood feedstocks. *Global Change Biology Bioenergy* 9(9):1406–1411. doi:10.1111/gcbb.12426.

Health Effects Institute 2018. *State of Global Air 2018. Special Report*. Boston: Health Effects Institute.

Heltberg, R. 2005. Factors determining household fuel choice in Guatemala. *Environment and Development Economics* 10(3):337–61. doi:10.1017/s1355770x04001858.

Hiemstra-van der Horst, G. and Hovorka, A. J. 2008. Reassessing the 'energy ladder': Household energy use in Maun, Botswana. *Energy Policy* 36(9):333–44. doi:10.1016/j.enpol.2008.05.006.

Hiemstra-van der Horst, G. and Hovorka, A. J. 2009. Fuelwood: The 'other' renewable energy source for Africa? *Biomass and Bioenergy* 33(11):1605–16. doi:10.1016/j.biombioe.2009.08.007.

Hosonuma, N., Herold, M., Veronique, D. S. et al. 2012. An assessment of deforestation and forest degradation drivers in developing countries. *Environmental Research Letters* 7(4):4009.

Jagger, P. and Das, I. 2018. Implementation and scale-up of a biomass pellet and improved cookstove enterprise in Rwanda. *Energy for Sustainable Development* 46:32–41. doi:10/1016/j.esd.2018.06.005.

Jagger, P. and Shively, G. 2014. Land use change, fuel use and respiratory health in Uganda. *Energy Policy* 67:713–26. doi:10.1016/j.enpol.2013.11.068.

Jull, C, Redondo, P. C., Mosoti, V. and Vapnek, J. 2007. Recent trends in the law and policy of bioenergy production, promotion and use. *FAO Legislative Study* No. 95. Rome: FAO.

Kgathi, D. L., Mmopelwa, G., Chanda, R., Kashe, K. and Murray-Hudson, M. 2017. A review of the sustainability of Jatropha cultivation projects for biodiesel production in southern Africa: Implications for energy policy in Botswana. *Agriculture, Ecosystems and Environment* 246:314–24.

Kharina, A., Malins, C. and Searle, S. 2016. *Biofuels policy in Indonesia: Overview and status report*. Washington, DC: International Council on Clean Transportation.

Kittner, N., Dimco, H., Azemi, V., Tairyan, E. and Kammen, D. M. 2016. An analytic framework to assess future electricity options in Kosovo. *Environmental Research Letters* 11(10):104013.

Kittner, N., Fadadu, R. P., Buckley, H. L., Schwarzman, M. R. and Kammen, D. M. 2018. Trace metal content of coal exacerbates air-pollution-related health risks: The case of lignite coal in Kosovo. *Environmental Science & Technology* 52(4):2359–67. doi:10.1021/acs.est.7b04254.

Kittner, N., Lill, F. and Kammen, D. M. 2017. Energy storage deployment and innovation for the clean energy transition. *Nature Energy* 2(9):17125.

Laurance W. F., Sayer, J. and Cassman, K. G. 2014. Agricultural expansion and its impacts on tropical nature. *Trends in Ecology & Evolution* 29:107–16.

Leach, G. and Mearns, R. 1988. *Beyond the woodfuel crisis: People, land, and trees in Africa*. London: Earthscan.

Lewis, J. J. and Pattanayak, S. K. 2012. Who adopts improved fuels and cookstoves? A systematic review. *Environmental Health Perspectives* 120(5):637–45. doi:10.1289/ehp.1104194.

Masera, O., Ghilardi, A., Drigo, R. and Trossero, M. A. 2006. WISDOM: A GIS-based supply demand mapping tool for woodfuel management. *Biomass and Bioenergy* 30(7):618–37. doi:10.1016/j.biombioe.2006.01.006.

Masera, O. R. and Navia, J. 1997. Fuel switching or multiple cooking fuels? Understanding inter-fuel substitution patterns in rural Mexican households. *Biomass and Bioenergy* 12(5):347–61. doi:10.1016/S0961-9534(96)00075-X.

Masera, O. R., Saatkamp, B. D. and Kammen, D. M. 2000. From linear fuel switching to multiple cooking strategies: A critique and alternative to the energy ladder model. *World Development* 28(12):2083–103. doi:10.1016/S0305-750X(00)00076-0.

Mayaux, P., Pekel, J. F., Desclee, B. et al. 2013. State and evolution of the African rainforests between 1990 and 2010. *Philosophical Transactions of the Royal Society B: Biological Sciences* 368(1625):1–10.

McBride, A. C., Dale, V. H. Baskaran, L. M. et al.2011. Indicators to support environmental sustainability of bioenergy systems. *Ecological Indicators* 11:1277–89.

Mentis, D., Howells, M., Rogner, H. et al. 2017. Lighting the world: The first application of an open source, spatial electrification tool (OnSSETT) on sub-Saharan Africa. *Environmental Research Letters* 12:085003.

Munro, P., van der Horst, G. and Healy, S. 2017. Energy justice for all? Rethinking Sustainable Development Goal 7 through struggles over traditional energy practices in Sierra Leone. *Energy Policy* 105:635–41.

Nansaior, A., Patanothai, A., Rambo, A. T. and Simaraks, S. 2011. Climbing the energy ladder or diversifying energy sources? The continuing importance of household use of biomass energy in urbanizing communities in Northeast Thailand. *Biomass and Bioenergy* 35(10):4180–8. doi:10.1016/j.biombioe.2011.06.046.

Naylor, R. and Higgins, M. 2017. The political economy of biodiesel in an era of low oil prices. *Renewable and Sustainable Energy Reviews* 77:695–705.

Nurfatriani, F., Ramawati, Sari, G. K. and Komarudin, H. 2018. *Optimalisasi Dana Sawit dan Pengaturan Instrumen Fiskal Penggunaan Lahan Hutan untuk Perkebunan dalam Upaya Mengurangi Deforestasi*. Working Paper No. 238. Bogor, Indonesia: CIFOR.

Obidzinski, K., Andriani, R., Komarudin, H. and Andrianto, A. 2012. Environmental and social impacts of oil palm plantations and their implications for biofuel production in Indonesia. *Ecology and Society* 17(1):25.

O'Brien, C. 2006. Introducing alcohol stoves to refugee communities; a case study from Kebrebeyah, Ethiopia. *Boiling Point* 52:16–18.

Oluwafemi, O., Oluwatofunmi, O. O., Godson, A. A. and Olopade, C. O. 2012. Indoor air pollution from biomass fuels: A major health hazard in developing countries. *Journal of Public Health* 20:565–75.

Pradhan, S. and Ruysenaar, S. 2014. Burning desires: Untangling and interpreting 'pro-poor' biofuel policy processes in India and South Africa. *Environmental Planning* 46:299–317.

Räsänen, T. A., Varis, O., Scherer, L. and Kummu, M. 2018. Greenhouse gas emissions of hydropower in the Mekong River Basin. *Environmental Research Letters* 13(3):034030.

Rehfuess, E. A., Briggs, D. J., Joffe, M. and Best, N. 2010. Bayesian modelling of household solid fuel use: Insights towards designing effective interventions to promote fuel switching in Africa. *Environmental Research Letters* 110(7):725–32. doi:10.1016/j.envres.2010.07.006.

REN21 (Renewable Energy Policy Network for the 21st Century) 2010. *Renewables 2010 Global Status Report* Paris: REN21 Secretariat. Available at: www.ren21.net/ (Accessed 13 March 2019).

REN21 2017. *Renewables 2017 Global Status Report.* Paris: REN21 Secretariat. Available at: www .ren21.net/ (Accessed 13 March 2019).

REN21 2018. *Renewables 2018 Global Status Report.* Paris: REN21 Secretariat. Available at: www .ren21.net/ (Accessed 13 March 2019).

Reubens, B., Achten, W. M. J., Maes, W. H. et al. 2011. More than biofuel? *Jatropha curcas* root system symmetry and potential for soil erosion control. *Journal of Arid Environments* 75:201–5.

Riahi, K., Dentener, F., Gielen, D., et al. 2012. Energy pathways for sustainable development. In *Global energy assessment – Toward a sustainable future.* Vienna: International Institute for Applied Systems Analysis, and Cambridge: Cambridge University Press, pp. 1203–6.

Romijn, H., Heijnen, S., Colthoff, J. R., Jong, B. and Van Eijck, J. 2014. Economic and social sustainability performance of jatropha projects: Results from field surveys in Mozambique, Tanzania and Mali. *Sustainability* 6(9):6203–35.

Sanchez, D. L., Nelson, J. H., Johnston, J., Mileva, A. and Kammen, D. M. 2015. Biomass enables the transition to a carbon-negative power system across western North America. *Nature Climate Change* 5(3):230.

Scarlat, N. and Dallemand, J. F. 2011. Recent developments of biofuels/bioenergy sustainability certification: A global overview. *Energy Policy* 39(3):1630–46.

Searchinger, T. D., Beringer, T., Bjart Holtsmark, D. M. et al. 2018. Europe's renewable energy directive poised to harm global forests. *Nature Communications* 9(1). doi:10.1038/s41467-018-06175-4.

Sedjo, R. A. 2011. *Carbon neutrality and bioenergy: A zero-sum game?* Resources for the Future Discussion Paper No. 11–15. Washington, DC: Resources for the Future.

Singh, G., Rawat, G. S. and Verma, D. 2010. Comparative study of fuelwood consumption by villagers and seasonal 'Dhaba owners' in the tourist affected regions of Garhwal Himalaya, India. *Energy Policy* 38(4):1895–9.

Skutsch, M., de los Rios, E., Solis, S. et al. 2011.Jatropha in Mexico: Environmental and social impacts of an incipient biofuel program. *Ecology and Society* 16(4):11–38.

Smeets, E. M. W. and Faaij, A. P. C. 2007. Bioenergy potentials from forestry in 2050. *Climatic Change* 81(3):353–90. doi:10.1007/s10584-006-9163-x.

Smith, K. R., Bruce, N., Balakrishnan, K. et al. 2014a. HAP CRA Risk Expert Group. Millions dead: How do we know and what does it mean? Methods used in the comparative risk assessment of household air pollution. *Annual Review of Public Health* 35:185–206.

Smith, P., Bustamante, M., Ahammad, H. et al. 2014b. Agriculture, forestry and other land use (AFOLU). In Edenhofer, O., Pichs-Madruga, R. and Sokona, Y. et al. (eds.) *Climate Change 2014: Mitigation of Climate Change. Contribution of Working Group III to the Fifth Assessment Report of the Intergovernmental Panel on Climate Change*. Cambridge: Cambridge University Press, pp. 811–922.

Sonter, L. J., Barrett, D. J., Moran, C. J. and Soares-Filho, B. S. 2015. Carbon emissions due to deforestation for the production of charcoal used in Brazil's steel industry. *Nature Climate Change* 5(4):359.

Soto, I., Achten, W. M. J., Muys, B. and Mathijs, E. 2015. Who benefits from energy policy incentives? The case of jatropha adoption by smallholders in Mexico. *Energy Policy* 79:37–47.

Soto, I., Ellison, C., Kenis, M. et al. 2018. Why do farmers abandon jatropha cultivation? The case of Chiapas, Mexico. *Energy for Sustainable Development* 42:77–86.

Thulstruo, A. W., Habimana, D., Joshi, I. and Oduori, S. M. 2018. Uncovering the challenges of domestic energy access in the context of weather and climate extremes in Somalia. *Weather and Climate Extremes* XX:1000185.

United Nations (UN) 2015. *Sustainable Development Goal 7*. Available at: http://sustainabledevelopment.un.org/sdg7 (Accessed 27 July 2019).

UN DESA (United Nations, Department of Social Affairs, Population Division) 2015. *World population prospects: The 2015 revision, key findings and advanced tables*. Working Paper No. ESA/P/WP.241. New York: United Nations.

UNESCAP (United Nations Economic and Social Commission for Asia and the Pacific) 2017. *Asia-Pacific progress in sustainable energy: A global tracking framework 2017 regional assessment report*. Bangkok: UNESCAP.

Valdés-Rodríguez, O., Pérez-Vázquez, A. and Muñoz Gamboa, C. 2014. Drivers and consequences of the first *Jatropha curcas* plantations in Mexico. *Sustainability* 6:3732–46.

van Eijck, J., Romijn, H., Smeets, E. et al. 2013. Comparative analysis of key socio-economic and environmental impacts of smallholder and plantation based production based jatropha biofuel production systems in Tanzania. *Biomass and Bioenergy* 61:25–45.

von Maltitz, G. and Setzkorn, K. 2012. Potential impacts of biofuels on deforestation in Southern Africa. *Journal of Sustainable Forestry* 31(1–2):80–97. doi:10.1080/10549811.2011.566114.

Wani S. P., Chander, G., Sahrawat, K. L. et al. 2012. Carbon sequestration and land rehabilitation through *Jatropha curcas* plantation in degraded lands. *Agriculture, Ecosystems and Environment* 161:112–20.

WHO (World Health Organization) 2016. *Burning opportunity: Clean household energy for health, sustainable development, and wellbeing of women and children*. Geneva: WHO.

Winemiller, K. O., McIntyre, P. B., Castello, L. et al. 2016. Balancing hydropower and biodiversity in the Amazon, Congo, and Mekong. *Science* 351(6269):128–9. doi:10.1126/science.aac7082.

Wright, T. and Rahmanulloh, A. 2017. *Indonesia biofuel annual report 2017*. Foreign Agricultural Service Global Agricultural Information Network No ID1619. Gain Report. Jakarta: USDA Foreign Agricultural Service.

Wu, M., Zhang, Z. and Chiu, Y. W. 2014. Life-cycle water quantity and water quality implications of biofuels. *Current Sustainable/Renewable Energy Reports* 1:3–10.

Yang, F. and Yang, M. 2017. Rural electrification in sub-Saharan Africa with innovative energy policy and new financing models. *Mitigation and Adaption Strategies for Global Change* 23(6):933–52.

Zarfl, C., Lumsdon, A. E., Berlekamp, J., Tydecks, L. and Tockner, K. 2015. A global boom in hydropower dam construction. *Aquatic Sciences* 77(1):161–70.

Chapter 8 SDG 8: Decent Work and Economic Growth – Potential Impacts on Forests and Forest-Dependent Livelihoods

Dietmar Stoian*, Iliana Monterroso and Dean Current

Key Points

- Diverse combinations of predominant development paradigms lead to differentiated SDG 8 target prioritisations, with mixed impacts on forests and people.
- Significant trade-offs are expected for countries focused on the growth of agriculture, energy and mining: the decoupling of economic growth from forest-related environmental degradation will be a major challenge. Global trade-offs are anticipated on climate change.
- Synergy potentials exist where growth strategies and associated policies explicitly target the forest sector with a view on tree plantations, timber and non-timber forest products (NTFPs) from natural forests, ecotourism and payments for environmental services.
- Improvements in the enabling environment can help minimise trade-offs and maximise synergies by reconciling government policies and private sustainability standards, formalising community stewardship of tropical forests, addressing the informality prevalent in forest product value chains and providing specific incentives for youth to become involved in forest-based economic activities.
- Advancing decent work in the forest sector requires addressing gender, ethnicity and other social differentiation factors, as well as mobility aspects and labour safety.
- Alternative, more integrated measurements of economic growth and decent work are needed, accounting for broader environmental and social impacts.

* Lead author.

8.1 Introduction

8.1.1 Conceptual Foundations of Decent Work and Economic Growth

Since the early twentieth century, the core of prevalent socio-economic and political systems has been economic growth. This has fundamentally and irreversibly reshaped societies and the entire planet (Schmelzer 2017). In the aftermath of World War II, emerging theories and paradigms for developing non-industrialised countries in the Global South were largely based on the premise of economic growth (e.g. modernisation theory), if not overtly labelled as such (economic growth theory).

The concept of decent work, however, only appeared at the end of the twentieth century. Tensions between economic relationships and their social context arose in the late nineteenth century (Rodgers 2007), addressed by the emerging trade unions. Yet it was not until 1999 that the International Labour Organization (ILO) launched the concept of decent work. Its emergence reflected new development theories and paradigms that sought to overcome the limitations of the modernisation and economic growth theories and associated policies. Nurtured by Maslow's theory of human motivation, the 1970s laid the foundation for a development theory of human needs (Max-Neef et al. 1992). The resulting basic-needs paradigm reformed development programmes in the 1980s by integrating social indicators into the measurement of economic growth. Environmental indicators were added in the 1990s as the sustainable development paradigm emerged in response to global environmental degradation (Lélé 1991).

Policies and strategies based on diverse development theories and paradigms have varied effects on forests and forest-dependent livelihoods. This chapter introduces an analytical framework illustrating the evolution of such theories and paradigms, the contextual conditions shaped by them and the principal drivers determining the impact of SDG 8 implementation on forests and people dependent on them (Section 8.2). We then present key actors and stakeholders in the forest and other natural resource sectors and the latter's contributions to national economies. The subsequent assessment of anticipated impacts (Section 8.3) addresses forest-based economic growth and decent work in forest product value chains (FPVC) from livelihoods, enterprise and conservation perspectives. Synergies and trade-offs are then discussed, within SDG 8 implementation, and with other SDGs (Section 8.4). We conclude with an outlook on how overarching development paradigms lead to varying prioritisations of SDG 8 targets, and how decoupling economic growth from forest-related degradation will continue to be a challenge for countries seeking economic growth in natural resource sectors other than the forest sector (Section 8.5).

8.1.2 Indicators for Decent Work and Economic Growth

The SDG 8 targets and indicators reflect a firm grounding in economic growth paradigms and, to some extent, the paradigmatic evolution to a more integrated set of indicators, including social and environmental aspects (Table 8.1).

Table 8.1 SDG 8 targets and indicators

Targets	Indicators
8.1 Sustained per capita economic growth – min. 7 % GDP growth per annum in least developed countries	8.1.1 Annual growth rate of real GDP/capita
8.2 Higher levels of economic productivity	8.2.1 Annual growth rate of real GDP/employed person
8.3 Development-oriented policies for formalized micro-, small- and medium-sized enterprises	8.3.1 Proportion of informal employment in non-agriculture employment (by sex)
8.4 Global resource efficiency in consumption and production and decoupling economic growth from environmental degradation	8.4.1 Material footprint (per capita and GDP) 8.4.2 Domestic material consumption (per capita and GDP)
8.5 Full and productive employment, decent work for all women and men, and equal pay for work of equal value	8.5.1 Average hourly earnings of female and male employees (by occupation, age and persons with disabilities) 8.5.2 Unemployment rate (by sex, age and persons with disabilities)
8.6 By 2020, proportion of youth not in employment, education or training substantially reduced	8.6.1 Proportion of youth not in education, employment or training
8.7 Forced labour, modern slavery, human trafficking and worst forms of child labour eradicated (by 2025 child labour in all its forms)	8.7.1 Proportion and number of children engaged in child labour (by sex and age)
8.8 Labour rights protected and safe and secure working environments for all workers	8.8.1 Frequency rates of occupational injuries (by sex and migrant status) 8.8.2 National compliance of labour rights (freedom of association and collective bargaining, by sex and migrant status)

Table 8.1 (cont.)

Targets	Indicators
8.9 Policies for sustainable tourism (job creation, promotion of local culture and products)	8.9.1 Tourism direct GDP (proportion of total GDP and growth rate) 8.9.2 Number of jobs in tourism industries (proportion of total jobs and growth rate, by sex)
8.10 Domestic financial institutions strengthened (enhanced access to banking, insurance and financial services)	8.10.1 Number of commercial bank branches and ATMs (per 100 000 adults) 8.10.2 Proportion of adults with an account at a bank or other financial institution or with a mobile-money-service provider
8.A Aid for Trade support for developing countries increased, in particular for least developed countries	8.A.1 Aid for Trade commitments and disbursements
8.B By 2020, a global strategy for youth employment and the Global Jobs Pact of ILO implemented	8.B.1 Total government spending in social protection and employment programmes (proportion of the national budgets and GDP)

Note: Targets are for 2030 unless stated otherwise.
Source: Adapted from UN (2019)

The SDG 8 targets and indicators align with the ILO (2018a) framework on the measurement of decent work, which is closely linked to four strategic pillars: (1) international labour standards and fundamental principles and rights at work, (2) employment creation, (3) social protection and (4) social dialogue and tripartism.

8.2 Framework for Analyses

8.2.1 Development Paradigms Driving Policies, Institutions and Markets

How 'less developed' countries can follow the pathway of 'developed' countries, or how they can be 'modernised', has been disputed by social scientists for more than a century. This debate has influenced development thinking and policymaking around the globe. In response to challenges and opportunities facing the developing world in the postcolonial era, the modernisation paradigm was developed in the 1950s. Advocates such as Rostow (1959) assumed a universal evolutionary path from traditional societies to the age of high mass consumption. The underlying assumption of relatively uniform, linear modernisation pathways was soon criticised. It was argued that the societal boundaries, political and economic institutions, and the social distribution of power underlying the absence of 'modern' societies had to be identified and solutions be developed accordingly (Tipps 1973).

Despite such criticism, the modernisation paradigm has strongly influenced development policies and strategies worldwide, with economic growth at its core and a simplistic equation: industrial transformation equals economic growth, which, in turn, allows poorer countries to catch up with industrialised countries. Eventually, economic growth would generate broader societal wealth through trickle-down effects (Thornton et al. 1978).

The modernisation and growth paradigms have had a strong imprint on tropical forests and, to some extent, temperate ones too. Starting in the 1950s, governments have increasingly treated forests as a natural capital reserve to be exploited for industrial transformation – initially through log sales, and progressively through value-added products. Processes of forest-based industrialisation occurred in several waves across forest-rich regions along the tropical belt: West Africa (1950s–70s), Southeast Asia (1960s–80s) and, more recently, the Congo and Amazon basins. However, the prevalent view of development as a purely economic phenomenon securing jobs and economic opportunities for the masses soon turned out to be a myth (Arndt 1983). When the waves started to ebb in the late 1970s it became evident that industrialisation based on natural resource processing is likely to perpetuate the pattern of dualism and inequality present in typical resource-rich countries, rather than leading to efficient growth, employment creation, greater equity and economic independence (Roemer 1979). The model of a 'dual economy' (Lewis 1954), which assumes that the agricultural sector (including forestry) generates the capital necessary for take-off towards industrialisation before becoming redundant, was shown to ignore the positive link between growth in industry and agricultural growth (Verner and Blunch 1999).

Over the past decades, new theories and models have emerged that provide a more nuanced picture of development processes: 'basic needs' (1970s), 'sustainable development' (early 1990s), 'sustainable livelihoods' and 'green growth' (both late 1990s). These are reflected in the Millennium Development Goals (2000), followed by the Sustainable Development Goals (2015). Lately, it has been suggested that alternative measures to GDP are needed, such as the Genuine Progress Indicator (GPI),[1] World Values Survey (WVS),[2] Happy Planet Index (HPI)[3] and Better Life Index[4] (Costanza et al. 2014).

Despite these conceptual advances, many factors that have long been driving economic policy in relation to natural resources are still at play. While protected areas in tropical countries have been expanded and forest ecosystem services figure more prominently on political agendas, development policies, institutions and markets continue to be largely driven by modernisation and economic growth paradigms. This prompts the question of what effects these paradigms have on the natural resource base of tropical countries and economic activities based on them – a question this chapter addresses with a view on forests and FPVC.

8.2.2 Analytical Framework

Our analytical framework for assessing the potential impacts of SDG 8 on forests and forest-dependent livelihoods has been derived from our own reflection, insight and conceptualisation. It accounts for the development theories and paradigms outlined in the previous section as well as contextual conditions that determine the impact of principal drivers:

- political–legal framework: principal policies geared towards forest, agriculture, energy, mining, tourism sectors;
- institutional support environment: technical and financial assistance by government agencies, non-governmental organisations (NGOs), the private sector;
- macroeconomic conditions: composition and evolution of GDP, formal and informal employment, standard of living;
- market forces: supply–demand patterns in the forest, agriculture, energy, mining, tourism sectors;
- status of the forest resource base: forest cover, degradation and deforestation; institutional arrangements for protecting forests;
- cultural contexts: importance of forests at societal, communal levels.

---

[1] In addition to GDP measurement, the GPI considers the cost of the negative effects related to economic activity (e.g. resource depletion).

[2] Based on nationally representative surveys in almost 100 countries, the WVS provides cross-national time series on human beliefs and values.

[3] Drawing on existing metrics, the HPI accounts for well-being, life expectancy, inequality of outcomes and ecological footprint.

[4] OECD's Better Life Index measures 11 parameters (income, jobs, housing, health, access to services, environment, education, safety, civic engagement and governance, community and life satisfaction).

The analytical framework also accounts for interactions with policies and trends in relation to other natural resource sectors (agriculture, energy, mining) and associated services (tourism, provision of ecosystem services).

Within the forest sector, our assessment focuses on the following stakeholders:

- national and local governments
- international and local NGOs
- multinational and national companies
- corporate associations
- small and medium enterprises
- Indigenous and non-Indigenous groups dependent on forests

We assess the impact of SDG 8 on the forest sector with a view on the forest industry, forest-dependent people and the forest resource base (Figure 8.1).

8.2.3 Key Actors and Stakeholders

Our analytical framework distinguishes between key actors influencing the design and implementation of policies and strategies for achieving SDG 8 on the one hand, and forest-sector stakeholders affected by these policies

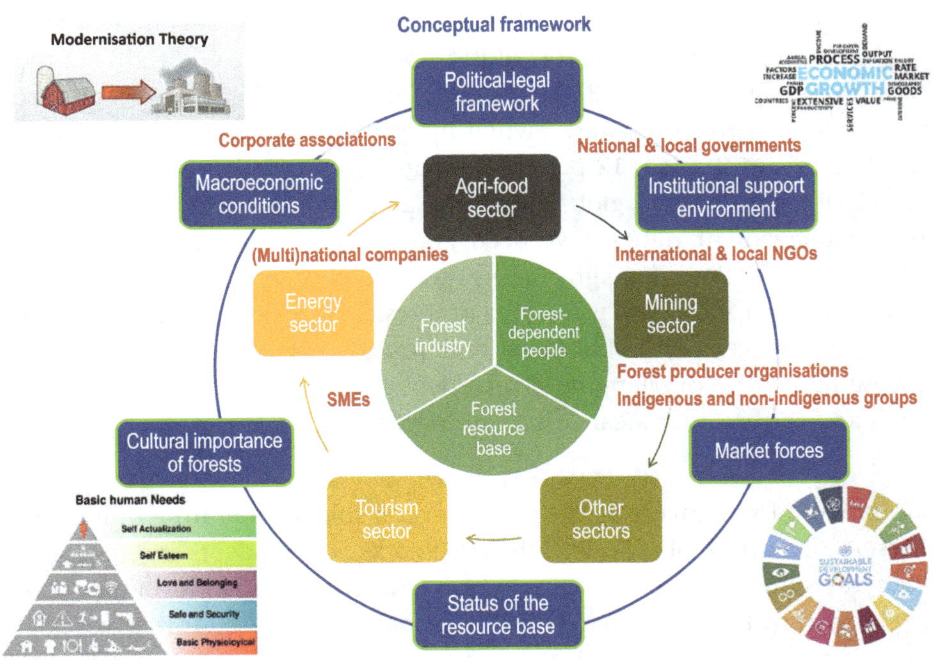

Figure 8.1 Analytical framework for assessing the impact of SDG 8 on forests and forest-dependent people.

and strategies on the other. Key actors in the political arena are national and local governments, with their legislative, regulatory and executive powers varying according to governance structures and the degree of decentralisation. Principal stakeholders are forest users, including the forest industry and forest-dependent people. Further stakeholders are civil society organisations, ranging from local and international NGOs to corporate associations, and society as a whole. Depending on their role inside or outside of FPVC and their influence on political–legal and regulatory decisions, given groups may be both key actors and stakeholders. Their interactions are complex, as are the forest-tenure arrangements underlying interactions. In our assessment of statutory forest tenure (Table 8.2) and subsequent analyses, we focus on the top 10 countries with largest forest area (Top10-LFA) which account for 50 per cent of the global forest area and represent both industrialised (Australia, Canada, Russian Federation, United States) and emerging or developing economies (Brazil, China, DR Congo, India, Indonesia, Peru).

Table 8.2 reflects that, across the 58 countries studied, 72 per cent of the forest area in 2017 was publicly administered, while an increasing portion is owned by Indigenous peoples and local communities (13 per cent) or has been designated to them for their use (2.3 per cent). Forest tenure varies significantly across the Top10-LFA, ranging from countries where most of the forests are publicly owned and controlled (Russian Federation, DR Congo, India, Indonesia) to countries with incipient (Canada), moderate (Australia, Peru) or strong (Brazil, China) devolution of forest tenure and rights to local communities. In Australia, Brazil, Canada, India and the United States, significant portions of forests are owned by individuals and firms and, across the 58 countries studied, 12 per cent of the forest area was privately owned in 2017 (RRI 2018). On a global scale, the area of publicly owned forests for which management rights have been granted to private companies under concessionary or licensing agreements has increased from 6 per cent in 1990 to 14 per cent in 2010, as has the area of forests within protected areas which reached 16 per cent of global forest area in 2015 (FAO 2016). Not reflected in these figures are overlapping claims between publicly or privately owned forests and those under local customary rights.

NATIONAL AND LOCAL GOVERNMENTS

National and local governments play a central role in the future of forests as representatives of publicly owned forests and as policymakers and regulators. At the international level, coordinated government efforts contribute to meeting global (e.g. Bonn Challenge) and regional (e.g. 20 x 20 Initiative) goals. At the country level, the political–legal and regulatory frameworks reflect the role of forests in national development strategies, both as regards the forest sector and vis-à-vis other sectors. The principal management objective of the

Table 8.2 Statutory forest tenure (millions of ha) in the top 10 countries with largest forest area and the 58 countries included in the study of RRI (2018) which account for 50% and 86% of the global forest area, respectively, 2002 and 2017

	Government administered		Designated for indigenous peoples and local communities		Owned by indigenous peoples and local communities		Privately owned by individuals and firms	
	2002	2017	2002	2017	2002	2017	2002	2017
Australia	94.0	83.3	0.0	9.1	20.9	12.1	14.0	20.2
Brazil	341.0	238.4	10.7	40.4	75.3	118.1	94.3	99.9
Canada	319.3	318.3	0.2	0.3	6.6	6.8	21.7	21.6
China	76.1	75.2	–	–	103.1	124.3	–	–
DR Congo	157.3	152.4	–	0.2	–	–	–	–
India	56.0	59.3	–	–	–	1.1	9.4	9.8
Indonesia	97.7	85.4	0.3	0.8	–	0.01	1.5	4.9
Peru	58.8	54.4	1.6	5.0	10.5	12.8	5.3	0.1
Russian Federation	809.3	814.9	0.0	0.0	0.0	0.0	–	–
United States	129.2	130.0	–	–	7.3	7.5	166.6	172.6
World (58 countries)	**2 748**	**2 482**	**18.2**	**80.5**	**357.8**	**447.4**	**403.4**	**418.5**

Notes: Dashes (–) denote situations in which the tenure category is not legally possible under national law. For forest area, 'RRI largely relies on … data submitted by national governments to the FAO as input to the Global Forest Resources Assessment, which is published every five years. However, RRI may instead utilise alternate data concerning countries' total forest area where more recent or accurate information is available through other sources' (RRI 2018: 26).

Source: Own elaboration based on data for 58 countries by RRI (2018); share of global forest area based on comparison between 2017 data from RRI (2018) and 2015 data from FAO (2016)

world's forests in 2015 as defined by governments was the supply of forest products (31 per cent), protection of soil and water (31 per cent), multiple use (28 per cent) and conservation of biodiversity (13 per cent) (FAO 2016).[5] Important differences exist between countries with centralised governance and those with strong decentralisation (Larson and Soto 2008). Countries also differ regarding interactions between government agencies in charge of forests and those looking after other sectors, with implications for the role of forests in development strategies and the potential for private investments (Agrawal et al. 2013).

INTERNATIONAL AND LOCAL NGOS

NGOs often play an important role in technical assistance, capacity building and advocacy in support of liaisons between local communities, value chain stakeholders and governments. They have been instrumental in developing voluntary standards for sustainable forest management (SFM) and tree crop sectors (e.g. oil palm, cocoa). In countries where financial and human resources of government agencies have been downscaled due to structural adjustments, NGOs are critical for promoting development alternatives. In remote forest areas, they may be the only providers of technical and other services. Given their capacity to mobilise financial resources, international NGOs tap into funding streams that local NGOs find difficult to access. The latter, in turn, provide the expertise and local embeddedness needed for international NGOs to run impactful projects.

MULTINATIONAL AND NATIONAL COMPANIES

The corporate sector is another key actor, from multinational companies to those operating in national domains. Given the intrinsic nature of global FPVC, companies may make important contributions to SDG 8 in terms of employment generation, decent work, and sustainability of the natural resource base. Multinational companies are well placed to contribute to international agendas, such as the New York Forest Declaration where they represent 57 out of a total of 191 endorsing organisations (Climate Focus 2017). Private companies can also access emerging finance opportunities aligned with goals such as zero deforestation and programmes aimed at generating employment in the forest sector (FAO 2018). Transitioning towards SFM by private companies requires a combination of regulatory policies and financial instruments to restructure operations and spur technological innovation. Through public–private partnerships, multinational and national companies can shape the design and implementation of sustainability standards and practices along different nodes of FPVC.

---

[5] In some cases, multiple objectives have been set for forests – shares therefore do not add up to 100 per cent.

CORPORATE ASSOCIATIONS

Individual key actors and stakeholders may organise into corporate associations for better representation of their interests. In the natural resource sectors, such associations include the chambers of commerce, industry associations (wood, oil, gas, coal, tourism) and commodity associations (metals and minerals, crops, livestock). These associations may advance corporate social responsibility and, in the forest sector, may represent community forestry groups (e.g. Federation of Community Forestry Users, Nepal and Association of Forest Communities of Petén, Guatemala). Community-based associations can play important advocacy roles for SFM through community stewardship (Stoian et al. 2019). However, corporate associations may also promote agendas that effectively lead to deforestation – for example, through expansion of tree crop production (Benami et al. 2018) or hydroelectric dams (Fearnside 2016).

SMALL AND MEDIUM ENTERPRISES

Small and medium forest enterprises (SMFEs) often make up 80–90 per cent of the forest enterprises in tropical countries, and account for more than 50 per cent of forest-related jobs (Macqueen 2008). Their operations span timber, NTFP and ecotourism activities. Exact figures are scarce due to the high degree of informality in the SMFE sub-sector. Quantifying and addressing informal sectors in forestry can increase the availability of decent work among disenfranchised populations (FAO 2018). For SMFEs to develop into viable businesses, enabling environments are required that promote legal access to forest resources; incentives for sound forest management and value-adding; and the building of human, social, physical and financial capitals for sustainable production of timber and NTFPs (Donovan et al. 2006). Community forest enterprises (CFEs) are a subset of SMFEs with specific challenges: (1) legal entities that rarely address their realities and needs; (2) low levels of productivity and quality due to limited processing and management skills; (3) undercapitalisation; (4) long phases to reach maturity (often 20–40 years); and (5) limited participation of women in leadership positions and decision making (Stoian et al. 2009).

INDIGENOUS AND NON-INDIGENOUS GROUPS DEPENDENT ON FORESTS

The number of people relying on forests for some part of their livelihood and income is estimated at 1–1.6 billion, but reliable quantitative data for global estimates are not available (Agrawal et al. 2013). Many forest-dependent people are not organised or, if they are, are informally organised. In addition to legally constituted CFEs, there is a considerable number of unregistered forest producer organisations that can make important contributions to SFM and the generation of employment and income. This potential is yet to be fully exploited as local communities own or have been assigned use rights for more than 500 million ha of forests (Table 8.2), along with significant portions of forest they manage under customary rights that are not yet formally recognised. In many tropical countries, statutory and customary tenure regimes overlap, leaving local communities in a weak legal status (Wily 2011).

8.2.4 Natural Resource Sectors and Their Contributions to National Economies

SECTORS DEPENDENT ON NATURAL RESOURCES

The forest sector is but one of the natural resource sectors contributing to national economies and it often competes with these for land, investments and human resources. Such sectors include agriculture, energy (large-scale fossil fuel exploitation and hydro dams), minerals and metals, and ecotourism. When anticipating the impact of SDG 8 on the forest sector it is important to account for the contributions of these sectors to GDP, as governments may pursue growth policies focusing on natural resource sectors other than forestry given their weight in national economies.

Natural capital is the most or second-most important asset in low-income and lower-middle-income countries, constituting 47 and 27 per cent, respectively, of wealth in 2014 (Lange et al. 2018). It comprises both renewable (agricultural land, forests, protected areas) and non-renewable resources (fossil fuels and minerals/metals). Contributions of the non-renewables sectors to GDP generally outweigh those from the forest sector (Table 8.3), often coupled with significant environmental pressure (Schandl et al. 2016).

Table 8.3 shows that in the Top10-LFA, contributions of forest rents to GDP are low relative to those of other natural resource sectors – typically below 1 per cent.[6] Most of these countries rely on natural resources other than forests, with contributions to GDP of up to 11 per cent. Excluding DR Congo, forest rents contribute less than one-tenth of total natural resource rents in these countries. At the same time, household income in forest-rich regions often relies to a much higher extent on forest resources (Angelsen et al. 2014). Accounting for these additional contributions requires addressing informality in the forest sector and developing national-level statistics beyond GDP measurement. Indicators 8.1.1 and 8.2.1 alone will not capture the full value of forest-sector contributions to economic growth and productivity.

FOREST SECTOR

On a global scale, the formal forest sector is a relevant source of employment and gross value added, totalling 13.2 million employees and USD 606 billion, respectively, in 2011 (Table 8.4).

[6] The case of DR Congo is ambiguous, given the significant difference between World Bank and FAO data.

Table 8.3 Contributions of forest and other natural resource sectors to GDP in top 10 countries with largest forest area and the world in 2016, by forest area as per cent of land area

	Forest area (million ha)	% of land area	Rents (% of GDP)						Forest as % of total nat. res. rents
			Total natural resources	Oil	Gas	Coal	Mineral	Forest	
DR Congo	152.6	67	32.7	0.4	0	0	13.2	19.1 *	58
Brazil	493.6	59	3.1	1	0	0	1.4	0.7	23
Peru	74.0	58	7.7	0.5	0.2	0	6.7	0.2	3
Indonesia	91.0	53	2.5	0.6	0.2	0.5	0.7	0.5	20
Russian Federation	815.0	50	11.5	7	2.7	0.3	1	0.4	3
Canada	347.1	38	1.0	0.3	0	0.1	0.6	0.1	10
United States	310.1	34	0.3	0.1	0	0.1	0.1	0	0
India	70.7	24	1.9	0.4	0.1	0.8	0.4	0.3	16
China	208.3	22	1.1	0.3	0	0.3	0.4	0.1	9
Australia	124.8	16	5.5	0.2	0.2	0.5	4.5	0.1	2
World	**3999**	**27**	**1.9**	**1**	**0.2**	**0.1**	**0.4**	**0.2**	**11**

Notes: Natural resource rents are World Bank staff estimates based on sources and methods described in Lange et al. (2018). Forest rents are roundwood harvest times the product of regional prices and a regional rental rate.

*This figure is inconsistent with the 18.6% contributions to GDP reported by World Bank (2018a) for the combined agricultural, fishery and forest sectors in DR Congo for 2016; it is also in stark contrast to the 0.6% reported by FAO (2014) as contributions of the forest sector to GDP in DR Congo in 2011 – World Bank (2017) reports 18.5% for 2011.

Source: Own elaboration based on data for forest area in 2015 (FAO 2016) and World Development Indicators for contributions of natural resources to GDP in 2016 (World Bank 2017)

Table 8.4 Contribution of the formal forest sector to employment and GDP in top 10 countries richest in forest area and the world, 2011

	Employment					Gross value added				
	Roundw. prod.	Wood proces.	Pulp and paper	Total for the forest sector		Roundw. prod.	Wood proces.	Pulp and paper	Total for the forest sector	
	('000)	('000)	('000)	('000)	% of total labour force	USD million	USD million	USD million	USD million	% of contr. to GDP
Australia	11	40	15	67	0.6	1119	3975	2587	7682	0.9
Brazil	133	434	205	772	0.7	7036	5802	9676	22513	1.1
Canada	47	112	75	234	1.2	5759	6679	7351	19789	1.2
China	1021	1304	1516	3841	0.5	32386	41120	53013	126519	1.6
DR Congo	15	1	–	16	0.1	29	56	–	85	0.6 *
India	246	246	215	707	0.1	28097	352	2509	30958	1.7
Indonesia	103	211	131	445	0.4	5904	1805	6860	1470	1.7
Peru	37	14	10	61	0.4	212	192	912	1,316	0.8
Russian Federation	228	261	111	600	0.8	2767	5108	5200	13075	0.8
United States	122	327	378	827	0.5	20264	22100	53300	95664	0.6
Total World	**3456**	**5439**	**4339**	**13233**	**0.4**	**169488**	**170131**	**266334**	**605953**	**0.9**

*see footnote to Table 8.3
Source: FAO (2014)

Table 8.4 illustrates that, across the Top10-LFA, 0.1–1.2 per cent of the total labour force are employed in the forest sector – about a quarter above the global average of 0.4 per cent. However, these figures refer to formal employment in the wood-based industry only. Including informal employment and NTFP-based activities would result in considerably higher figures. The number of employees in formal and informal forest enterprises (including family businesses) is estimated at 45 million (Shackleton et al. 2011). Similarly, forest-sector contributions to GDP would probably be at least twice the official figures when allowing for the informal or non-monetary economy (Lebedys and Yanshu 2014). According to the official data in Table 8.4, the combined contributions of roundwood production, wood processing, and pulp and paper production to GDP vary by 0.6–1.7 per cent across the ten countries – roughly one-fifth above the global average of 0.9 per cent. Forest-sector contributions are relatively highest in the developing Asia-Pacific region (1.5% of GDP), followed by sub-Saharan Africa (1.2% of GDP) (Lebedys and Yanshu 2014).

AGRICULTURAL SECTOR

The agricultural sector is the principle source of employment in many countries of the Global South, with widely varying contributions to GDP (Table 8.5).

Table 8.5 reflects that, despite its decreased importance relative to other sectors over the past decades, the agricultural sector is the principal source of employment in the countries with the largest forest area in the Global South, with important contributions to GDP. Reductions in agricultural employment were more pronounced in emerging economies (Brazil, China, Indonesia) than in less-developed countries (DR Congo, India, Peru), but the agricultural sector still employs 10–31 per cent of the total labour force in the former and 28–82 per cent in the latter. From 2000 to 2016, sector contributions to GDP in the Top10-LFA decreased from 11 to 8 per cent. Still, they are roughly double those at the global scale, which decreased from 5 to 4 per cent. In most of these countries, the decrease of GDP contributions was well below that of employment, pointing at increased efficiencies and higher value added in the agricultural sector vis-à-vis other sectors. DR Congo diverts from this trend as contributions to GDP reflect a marked decline despite the sector's fairly stable importance in terms of employment relative to other sectors. At the same time, Brazil's agricultural sector slightly increased GDP contributions, although employment as part of the total labour force decreased by about a half.

The combined contributions of agriculture, forestry and fishing to GDP (Table 8.5) can be compared with the specific forest-sector contributions (Table 8.4). With the caveat that datasets and reference years vary, it can be

Table 8.5 Employment in the agricultural sector between 2000 and 2017 and its contribution to GDP from 2000 to 2016 in the top 10 countries with largest forest area and the world

	Employment (% of total employment)				Contribution to GDP (%)			
	2000	2010	2017	Trend 2000–2017	2000	2010	2016	Trend 2000–2016
Australia	5	3	3	–47%	3	2	2	–22
Brazil	21	16	10	–52%	5	4	5	+3
Canada	3	2	2	–41%	n.d.	1	n.d.	n.d.
China	44	26	18	–60%	15	10	9	–42
DR Congo	85	80	82	–3%	32	21	19	–42
India	60	52	43	–28%	22	18	16	–25
Indonesia	45	39	31	–31%	16	14	13	–14
Peru	35	28	28	–20%	8	7	7	–13
Russian Federation	14	8	7	–54%	6	3	4	–27
United States	2	2	2	–9%	1	1	1	–12
Total World	**39**	**31**	**26**	**–32%**	**5**	**4**	**4**	**–29**

Notes: contribution of agricultural sector to GDP as value added of agriculture, forestry and fishing; n.d. = no data available
Source: Own elaboration based on World Development Indicators (World Bank 2018a, 2018b)

deducted that contributions of agriculture (and fishing) to GDP in the Top10-LFA are, on average, more than ten times those of the forest sector. At the same time, agricultural commodities such as soy, palm oil and beef are closely linked with deforestation. The associated loss of ecosystem services (e.g. in Brazil, Indonesia and Peru) points towards unsustainable development pathways (Carrasco et al. 2017a). On a global scale, agriculture contributes about one-quarter of greenhouse gas emissions, and decoupling these from agricultural production will remain a major challenge (Bennetzen et al. 2016).

TOURISM SECTOR

Tourism is another important source of employment, contributions to GDP and economic growth. In 2017, direct and total contributions of Tourism and Travel to global employment were 3.8 and 9.9 per cent; and to GDP, 3.2 and

10.4 per cent, respectively (WTTC 2018). The tourism sector significantly encourages economic growth, but it also degrades the quality of the environment (Danish and Wang 2019). This general picture needs to be modified with a view on nature-based tourism and, specifically, ecotourism. However, the exact delimitation and size of these sub-sectors are difficult to establish, and there is no universally accepted definition of ecotourism. In connection with the effects of SDG 8 on forests and Target 8.9 (policies to promote sustainable tourism), different types of ecotourism need to be considered. In the Global South, the focus should shift from modernist forms of ecotourism, aimed at mere economic development, to more locally controlled, participatory forms of community-based ecotourism (Regmi and Walter 2017).

In terms of environmental performance, there is evidence for both positive and negative impacts of ecotourism (Buckley 2018). A recent global systematic review of ecotourism impacts on forests in biodiversity hotspots found that ecotourism, as typically practised, leads to local deforestation due to increased demand for fuelwood, food and timber; but when accompanied by conservation mechanisms (e.g. protected area, ecosystem service payments, monitoring/enforcement), it can protect forests (Brandt and Buckley 2018). As ecotourism often implies visiting relatively remote areas, and significant numbers of ecotourists use long-haul air travel to reach to tropical destinations, its carbon footprint can be considerable (see Gale 2016).

MINING SECTOR

On a global scale, the mining of metals and minerals contributes to regional and national economies by generating budget and export revenues, employment and infrastructure development, while it is also responsible for causing a wide range of adverse environmental and social impacts (e.g. disruption of river flows, degradation of land and forest resources, impacts on livelihoods in local communities and disturbance of indigenous people's traditional lifestyles) (Yakovleva 2017). Compared with other industrial sectors, the mining industry faces some of the most difficult sustainability challenges (Azapagic 2004). Mining of minerals and metals is an important driver of deforestation in tropical countries, often far beyond operational lease boundaries and paired with contamination of soils and aquifers (Mwitwa et al. 2012). Different types of mining impacts can be distinguished (Megevand et al. 2013): the *direct* impact on forest cover may be fairly limited, but *indirect* impacts tied to larger infrastructure developments (e.g. power plants, dams, roads) can be considerable, as can be *induced* impacts associated with a large influx of workers (e.g. subsistence agriculture, logging, poaching) and widespread aquatic contamination through the use and discharge of toxic substances, as well as *cumulative* impacts related to artisanal mining, where many small individual mining sites add up to significant impacts.

ENERGY SECTOR

Given the diversity of the energy sector, a differentiated analysis is required with a view on SDG 8. Many countries rely heavily on fossil fuels and, to some extent, nuclear energy. At the same time, the renewable energy sector is growing around the globe, based on increasing use of hydropower, wind, solar energy, wood, residues from agriculture and forestry, biogas and various types of biofuels. Energy production is in direct competition with forests for land, and is potentially linked with environmental contamination, when involving open-pit coal mining, oil exploitation, the construction of hydro dams in forested river basins such as the Amazon, Congo and Mekong (Winemiller et al. 2016) and the production of certain biofuels. In the Brazilian Amazon, both hydro dams and mining threaten protected areas, boosted by a political–legal framework supportive of both sectors (Ferreira et al. 2014).

While economic growth and decent work will progressively be sought in the renewable energy sector, the constant push back of peak oil, gas and coal as new reserves become exploitable slows down progress in the energy transition. In 2015, only 17.5 per cent of global final energy consumption was produced through renewable energies, and only 55 per cent of this share was derived from modern forms of renewable energy, i.e. other than fuelwood and charcoal (UN 2018). The fossil fuel sector continues to be a major source of employment and economic growth, with a high environmental footprint (see O'Rourke and Connolly 2003). Similar trade-offs are reported for agricultural crops producing biofuels, such as sugarcane (Jusys 2017) and oil palm (Vijay et al. 2016).

Fossil fuel production and consumption are also the primary source of greenhouse gas emissions, with significant, mainly adverse effects on forests through climate change (Allen et al. 2010). The impacts of biofuel production and consumption on forests are more ambiguous. Biofuels produced from crops using conventional agricultural practices will likely not mitigate the impacts of climate change caused by the use of petroleum fuels, and will exacerbate stresses on water supply and quality as well as land use (Delucchi 2010).

8.3 Assessment of Anticipated Impacts

Before addressing the anticipated impacts of SDG 8 action on forests and forest-dependent people, it is insightful to look into progress towards SFM as recently reported with relation to SDG 15 (Life on Land). Although the forest area in the world continues to decline, the rate of loss has dropped by 25 per cent from 2000–5 to 2010–15, accompanied by a general trend towards protection of forest and terrestrial ecosystems (UN 2018). But the same report also stresses that escalating biodiversity loss requires urgent actions to protect and restore ecosystems and the biodiversity they support. While net forest area gains are reported for Central/Southern Asia and Eastern/South-Eastern Asia from 2005–10 to

2010–15, this does not necessarily imply that deforestation has halted altogether. The ongoing reduction in above-ground biomass stock in forests reported for these two regions points to continued loss or degradation of natural forests. In fact, part of the progress in net forest area change is due to the advance of plantations, which provide a very different level of ecosystem services compared to natural forests. Similarly, the largely positive trend in the proportion of forest area under legally established protected areas or long-term forest management plans does not ensure effective change on the ground unless accompanied by viable enforcement mechanisms and other enabling conditions.

8.3.1 Forest-Based Economic Growth

FAO (2018) holds that forests and FPVC are of critical importance for sustained economic growth, full and productive employment and decent work for all, especially in remote areas. While at the national level manufacturing, services and other natural resource sectors are principal sources of economic growth and employment, remote forest areas rely to a much higher extent on the forest sector (Angelsen et al. 2014). Yet, many tropical countries struggle to achieve sustainable work opportunities and economic growth based on their forest riches (Swamy et al. 2018). Only a few countries (e.g. Thailand and Malaysia) have successfully used their forest resources to trigger broad-based economic growth spilling over to other natural resource sectors, and from there to manufacturing and services. The nominal GDP contributions of the forest sector do not motivate policies to develop its untapped potential. At the same time, the absence of policies promoting forest-sector development underlies its widespread stagnation in tropical regions. Political disinterest can also be attributed to a general dearth of reliable data on overall contributions of forests to national economies, spanning formal and informal employment, and timber and NTFP value chains.

Fairly robust data are available for the formal forest sector, particularly as regards the wood-based industry. Global employment in the formal forest sector decreased by about 6 per cent over the past decade, from 14.0 million people in full-time equivalents in 2000 to 13.2 million in 2011 (Lebedys and Yanshu 2014). The decline was most pronounced in the forestry sub-sector[7] (21 per cent) and in the developed regions. Losses were partly offset by increased formal forestry employment in developing regions, where most of the estimated 41 million people in the informal forest-sector work (FAO 2014).

In the Global South, roundwood production is relatively more important than wood processing and the production of pulp and paper. Global

[7] FAO (2014) distinguishes three sub-sectors: forestry (roundwood production), wood processing, and pulp and paper.

value-adding across the three sub-sectors has only slightly increased (5 per cent), from USD 583 billion in 2000 (at 2011 prices and exchange rates) to USD 606 billion in 2011. The pulp and paper industry contributes most to the global gross value-added (44 per cent), followed by the wood industry and the forestry sub-sector (28 per cent each). Overall distribution of value-added across the sub-sectors remained stable in the 2000s, but the share of the pulp and paper industry has recently declined (FAO 2014).

Refined metrics are needed to fully document actual and potential contributions of the forest sector to economic growth and decent work, as are sub-national policies and strategies to promote forest-sector development in regions where there are few alternatives. These, in turn, need to promote formal employment opportunities, especially for young people, reduce labour market inequality (gender pay gap), promote safe and secure working environments, and improve access to financial services to ensure sustained and inclusive economic growth (UN 2018). Given the intrinsic differences between timber and NTFP value chains, gender-differentiated approaches are required that increase employment and income opportunities for women, particularly in NTFP value chains (see Sunderland et al. 2014).

The positive impact of SDG 8 on the forest sector may be supported by voluntary sustainability standards. Forest certification has had positive effects on indicators related to decent work, particularly regarding social security and forest worker safety (see Cashore et al. 2006). There is little evidence, however, that forest certification leads to significant economic growth in tropical countries (Romero et al. 2017). Similarly, the advance of 'zero deforestation' and similar eco-labels for agricultural commodities associated with deforestation (e.g. soy, palm oil, cocoa) have yet to show significant impacts on forest loss (van der Ven et al. 2018).

Looking forward, the potential of the forest sector to contribute to SDG 8 largely relates to developing markets and value chains for sustainable timber and NTFPs that stimulate economic growth, provide gender- and age-differentiated employment opportunities, and ensure forest conservation. In tropical and other forest regions, sustainable tourism has the potential for promoting both economic growth and decent work. Where robust mechanisms can be established, such endeavours may be complemented with payments for environmental services.

8.3.2 Decent Work along Forest Product Value Chains

ILO's guiding framework for international labour standards comprises 8 fundamental, 4 governance and 177 technical conventions. One of the fundamental conventions and 6 technical ones were crafted after ILO's proclamation of 'decent work' in 1999. The fundamental one relates to the Worst Forms of Child Labour Convention (1999), and the pertinent technical ones are Maternity Protection (2000), Safety and Health in Agriculture (2001) and Promotional Framework for Occupational Safety and Health (2006) (ILO 2018b). While the other fundamental conventions predate the proclamation of decent work, they are closely linked to the concept. Table 8.6 illustrates the status of these conventions for the Top10-LFA.

Table 8.6 shows that five of the Top10-LFA have ratified all eight of ILO's fundamental conventions: Canada, DR Congo, Indonesia, Peru and the Russian Federation. Two conventions crucial for SMFE development have yet to be ratified by several countries: Freedom of Association & Protection of the Right to Organise, and Right to Organise & Collective Bargaining. In addition, effective enforcement mechanisms need to be in place to ensure these rights, as shown in countries where conventions have been ratified but implementation is lacking (e.g. DR Congo).

Several ILO conventions have direct links with SDG 8 targets and indicators, particularly Targets 8.3 (SMEs), 8.5 (productive employment and decent work), 8.6 (youth employment), 8.7 (forced labour) and 8.8 (labour rights and safety). While most Top10-LFA have ratified the conventions pertinent to Target 8.5 (except China and the United States), they lag behind in the ratification of those relating to Targets 8.3 (rights to organise), 8.6 (night work of young persons) and 8.8 (night work of women, migrant workers). The latter is particularly problematic as women and migrant workers play prominent roles in the informal forest sector. This underlines the importance of enabling environments comprising the ratification of international conventions, their translation into national policies and the formalisation of all nodes of FPVC.

Decent work may also be offered in tree crop value chains by enterprises engaged in larger schemes that replace natural forests: for example, oil palm, rubber, timber and cocoa plantations. Employment effects are often significant but, in addition to environmental externalities, labour conditions may be precarious (e.g. labour safety in sawmills, exposure to agrochemicals). Labour conditions in the first nodes of forest product and tree crop value chains may become less attractive to the younger generation which, through increased availability of educational services in rural areas, are prompted to search out livelihood options other than forestry or agriculture. At the same time, better formal education fosters skills required for developing SMFE that provide local opportunities for educated youth. Along with better access to modern information and communication technologies (ICT), there will be increased opportunities for them as managers of such enterprises.

Table 8.6 Status of ILO's 8 fundamental conventions in top 10 countries with largest forest area (R=ratified, NR not ratified)

Convention	29	87	98	100	105	111	138	182
	Forced labour	Freedom of association & protection of the right to organise	Right to organise & collective bargaining	Equal remu-neration	Abolition of forced labour	Discrimi-nation	Minimum age	Worst forms of child labour
Australia	R	R	R	R	R	R	NR	R
Brazil	R	NR	R	R	R	R	R	R
Canada	R	R	R	R	R	R	R	R
China	NR	NR	NR	R	NR	R	R	R
DR Congo	R	R	R	R	R	R	R	R
India	R	NR	NR	R	R	R	R	R
Indonesia	R	R	R	R	R	R	R	R
Peru	R	R	R	R	R	R	R	R
Russian Federation	R	R	R	R	R	R	R	R
United States	NR	NR	NR	NR	R	NR	NR	R

Source: Own elaboration based on data from ILO (2017)

8.3.3 Forest-Dependent Livelihoods

Forest-dependent people vary widely, as do their relationships with forests (e.g. formal or customary rights), their livelihoods (e.g. forest and non-forest activities) and their demands on forests and the broader resource system (e.g. products and ecosystem services). Official accounts rarely capture the intrinsic nature of such relationships and tend to underestimate the monetary and non-monetary values of forests to these people and society. As forests continue to be degraded or converted to other land uses, forest-dependent people may be forced into other livelihood activities in their respective locations, move further into the forest or out-migrate altogether. At the same time, the rights of forest-dependent communities are progressively being recognised (RRI 2017), providing an institutional environment enabling them to determine livelihood shifts more autonomously rather than responding to externally driven pressures.

8.3.4 Gender, Intergenerational and Ethnic Equity

Empowerment of forest-dependent communities, including participation in decision making and strengthening of livelihoods, is closely linked to gender, age and ethnicity. Access to land and natural resources is a critical entry point for empowering women and marginalised groups as it defines social status and political power and structures relationships within and outside the household (Agarwal 1994). Official statistics on differentiated access to forest resources are scarce, as is information on women, youth and other vulnerable groups in relation to their participation as labour force in the forest sector. Given the often seasonal nature of forest-based activities, their participation tends to be intermittent, informal and underpaid (FAO 2018).

This lack of recognition influences policymaking. In many cases, women, indigenous peoples and other marginalised groups are not considered beneficiaries unless programmes target them explicitly (Larson et al. 2018). Despite important gender differentiation in the collection of forest products, with distinctive 'male' and 'female' roles (Sunderland et al. 2014), forests continue to be perceived as a 'male domain' and development interventions are often designed accordingly. Combined, these factors tend to perpetuate the position of women and other marginalised groups in the informal part of the forest sector, with the associated underestimation of their contributions to sector development. In countries such as Guatemala and Cameroon, where community rights to forests are formalised and where internal governance structures do not unduly favour male dominance or that of certain economic strata, women, youth and indigenous peoples can play important roles in the management of CFEs (see Belibi et al. 2015, Stoian et al. 2019).

8.3.5 Forest Conservation

Despite a 15 per cent reduction in the global rate of net forest loss from 1990 to 2015 (FAO 2016), forests and associated biodiversity continue to be threatened. While SDG 8 seeks to decouple economic growth from environmental degradation, there is also an opportunity to *couple economic growth with forest conservation and sustainable resource management*. For example, community-based forest management can link forest conservation with economic growth and livelihoods improvement through SMFE development and tourism (Macqueen et al. 2018). The relationship between forest concessions managed by private companies, forest conservation and local economic development is less clear (see FAO and EFI 2018). While inappropriate logging can lead to forest degradation, the primary drivers of deforestation lie outside the forest sector: commercial and subsistence/local agriculture, followed by infrastructure development, mining and urban sprawl (Hosonuma et al. 2012).

Forest conservation needs to be achieved from both outside, by checking extra-sectoral drivers of deforestation, and from within through SFM or preservation with limited human intervention. A comparative analysis of 40 protected areas and 33 community-managed forests showed that annual deforestation rates in the latter were lower and less variable than those in protected forests; forest conservation strategies should therefore encompass a regional differentiation of land use types, tenure rights, social and economic needs of local inhabitants, and local capacities (Porter-Bolland et al. 2012).

In regions where deforestation has been reduced or halted, opportunities exist for initiating landscape restoration. Diverse combinations of agroforestry systems and timber plantations can stimulate economic growth and recover ecosystem services, with the bottom line that forest, agroforestry and plantation forestry options generate income comparable to alternative land uses (Appanah 2016).

8.3.6 Anticipated Impacts within the Framework of SDG 8

Principal synergies between SDG 8 and forests can be expected for areas where national policies for economic growth and decent work explicitly target the forest sector, and where these are paired with adequate legislation for sustainable management and conservation of forests and effective enforcement mechanisms. Several tropical countries provide monetary incentives for carbon sequestration through reforestation (Baker et al. 2019), with payments typically varying between USD 800 and USD 1500 per ha. The establishment of timber plantations has important employment effects, as shown for Brazil, China and Indonesia (Tomberlin et al. 2001). Some countries also provide incentives for managing natural forests for timber and NTFPs (Agrawal et al.

2018), generating local employment and value added. Community-based forest management and processing of forest products by CFEs can be combined with ecotourism to generate additional employment and income, as documented for Petén, Guatemala (Stoian et al. 2019).

While such synergies have important employment effects at local and regional levels, their impact at national and global scales will continue to be modest in light of the limited contributions of the forest sector to GDP – 0.2 per cent worldwide and 0.7 per cent in lower-middle-income countries (World Bank 2017). Accounting for the informal sector could double forest-sector contributions to GDP (World Bank 2016), but contributions of other natural resource sectors will continue to outperform those of the forest sector in many tropical countries. This holds particularly for the agricultural and mining sectors and, depending on the energy mix in a given country, the energy sector. Most governments have long pursued growth strategies based on modernisation and economic growth theories. Without a major paradigm shift, national policies will continue to prioritise the natural resource sectors that contribute most to GDP.

Principle trade-offs within SDG 8 implementation, particularly as regards deforestation, relate to policies for competing natural resource sectors. But some trade-offs are also expected for areas of potential synergy. For example, a systematic review of the socio-economic impacts of large-scale tree plantations found ambiguous impacts: slightly positive for employment, mixed regarding livelihoods and negative for land and intertwined social factors (Malkamäki et al. 2018). Strong global evidence on long-term socio-economic impacts of large-scale tree plantations remains limited (Malkamäki et al. 2018). Similarly, the evidence base for inferring positive effects between ecotourism and forests is insufficient (Brandt and Buckley 2018). A recent review on development policies in relation to the SDGs found that many commonly applied development interventions do not explicitly consider natural resources, let alone forests, leading to suboptimal, unsustainable outcomes; even if interventions tackle both development and conservation goals, they often lack coordination and sufficient levels of natural capital to ensure long-term sustainability (Miteva 2019).

8.4 Synergies and Trade-Offs between SDG 8 and Other SDGs

With the aim to maximise synergies and minimise trade-offs with other SDGs, partnerships for working towards SDG 8 have emerged at global, regional and national levels. As of February 2019, the SDG Knowledge Platform lists 770 partnerships in relation to SDG 8 (UN 2019). The World Bank Group alone reports more than 570 active projects with a jobs angle, representing investments of close to USD 75 billion, reaching nearly 2 million new beneficiaries and leveraging additional investments through global partnerships (World Bank 2018c). The United Nations Conference on Trade and Development (UNCTAD) and its partners seek to mainstream the promotion of investment in SDG sectors and to build capacity for SDG-related projects (UNCTAD 2018).

For assessing potential synergies and trade-offs, we developed a matrix that juxtaposes SDG 8 with other SDGs. We first reviewed connecting targets and assessed interaction intensity (high, medium, low) and then, from the perspective of the forest resource base, forest-dependent people and the forest industry, we considered the nature of the interactions (synergies, neutral, trade-offs) in dependence on political-strategic priorities (Table 8.7).

Table 8.7 illustrates that, in relation to the forest sector, SDG 8 has strong interactions with SDGs 1, 2, 5, 7, 9, 10, 12, 13, 15, and 17. At the same time, interactions between SDG 8 and SDGs 3, 6, 11, 14 and 16 are less strong, and those with SDG 4 are important but relatively indirect. Our assessment is in line with the literature showing that forests are important to the success of many of the sectors and functions represented within the context of the SDGs. In an assessment depicting the SDGs as a network of linked targets, SDG 8 ranks third as regards the number of SDGs it is connected with (10), including strong links with SDGs 9 and 10 (3 linked targets each), followed by SDGs 12 and 14 (2 targets each) and SDGs 1, 2, 4, 13, 15 and 16 (1 target each) (Le Blanc 2015). Diversions from our assessment owe to our focus on the forest sector.

Synergies between SDG 8 and other SDGs regarding forest-dependent people, the forest industry and the forest resource base are likely in countries where policies and strategies explicitly focus on the forest sector and are accompanied by safeguards for SFM and forest conservation. Such synergies can be expected as regards poverty reduction (SDG 1), clean water (SDG 6), 'modern' renewable energies (SDG 7), forest industry development (SDG 9), reduced (gender) inequalities in FPVC (SDGs 5 and 10), safe and affordable housing based on materials derived from forests (SDG 11), responsible consumption of forest products (SDG 12), economic growth through forests managed and protected as carbon sinks (SDG 13) and sustainable forest products and ecotourism (SDG 15).

Trade-offs for forests are anticipated in countries where policies and strategies focus on other natural resource sectors, particularly agriculture, energy and mining. A principal challenge is the significant increase of global food production required to feed the world's growing population (FAO 2018). Major trade-offs are anticipated between SDG 8 on the one hand, and SDGs 2, 13 and 15 on the other. Other trade-offs exist between the forest and mining

Table 8.7 Intensity and nature of interactions between SDG 8 and other SDGs in relation to the forest sector and depending on political-strategic priorities

SDG 8 SDGs	Intensity of interactions High (dark grey) Medium (mid-grey) Low (light grey)	Nature of interactions depending on political-strategic priorities Synergies (yellow) Neutral (orange) Trade-offs (red)
SDG 1 – No Poverty		*Nature of interactions depends on sector focus of policies and strategies to foster employment and income for the poor:*
		Growth of the forest sector and ecotourism can generate additional employment and income for poor forest-dependent people.
		Growth of manufacturing and service sectors with low demand for natural resources may generate limited additional employment and income for poor forest-dependent people and be largely neutral to the forest resource base.
		Growth of other natural resource sectors may jeopardise livelihoods of poor forest-dependent people due to deforestation.
SDG 2 – Zero Hunger		*Nature of interactions depends on the type of agriculture promoted:*
		Sustainable intensification of agriculture and lower demand for animal-based food can contribute to zero deforestation over time, retaining options for forest-sector growth.
		Some forms of commercial and subsistence agriculture are primary drivers of deforestation, limiting forest-sector growth.

Table 8.7 (cont.)

SDG 8 SDGs	Intensity of interactions High (dark grey) Medium (mid-grey) Low (light grey)	Nature of interactions depending on political-strategic priorities Synergies (yellow) Neutral (orange) Trade-offs (red)
SDG 3 – Good Health and Well-being	(dark grey)	*Nature of interactions depends on the sector focus of policies and strategies to foster economic growth and decent work:*
		(yellow) Advancing decent work in the forest sector reduces occupational injuries; investments by CFEs in health facilities and services can improve the health of their members.
		(orange) Growth of manufacturing and service sectors with low demand for natural resources may be largely neutral to the health and well-being of forest-dependent people.
		(red) Growth of other natural resource sectors may compromise health of forest-dependent people: malaria (hydro dams); contamination with heavy metals (mining) and agrochemicals (agriculture).
SDG 4 – Quality of Education	(light grey)	*Nature of interactions depends on the type, quality and location of educational facilities and services available to forest-dependent people:*
		(yellow) Forest sector and ecotourism growth may curb outmigration from forest areas and spur reinvestment in education and expansion of educational services; these, in turn, can help upgrade capacities and skills for managing forests and SMFE.
		(orange) Upgrading general educational facilities and services is largely neutral to the forest resource base and, hence, does not affect forest-sector growth.
		(red) Improved access to educational facilities may lead youth to search out livelihood options other than forest-based activities.

SDG 5 – Gender Equality		*Nature of interactions depends on the sector focus of policies and strategies to foster economic growth and decent work:*
		Growth of forest-based ecotourism and NTFP value chains may foster gender equality; equal representation and participation in decision making of women and men in the management of forest enterprises may boost their economic and social performance.
		Growth of certain segments of the energy and agricultural sectors may be largely neutral to forest-dependent people and, hence, not affect gender equality among them.
		Growth of the mining sector and timber and fuelwood value chains may perpetuate gender inequalities.
SDG 6 – Clean Water and Sanitation		*Nature of interactions depends on watershed management regulations and the sector focus of policies and strategies to foster economic growth and decent work:*
		Forest sector growth based on SFM helps to maintain or restore forests as water-related ecosystems.
		Growth of forest-based sustainable ecotourism may be largely neutral in terms of water availability and quality and, hence, not affect forest-sector growth.
		Growth of the agricultural, mining and energy sectors and unsustainable tourism may induce deforestation and, thus, compromise water availability and quality; watershed management regulations may impose restrictions limiting forest-sector growth.

Table 8.7 (cont.)

SDG 8 SDGs	Intensity of interactions High (dark grey) Medium (mid-grey) Low (light grey)	Nature of interactions depending on political-strategic priorities Synergies (yellow) Neutral (orange) Trade-offs (red)
SDG 7 – Affordable and Clean Energy		*Nature of interactions depends on the type of energy promoted:*
		Growth of 'modern' renewable energies may reduce pressure on forests exploited for firewood and charcoal and, thus, provide opportunities for alternative forest-sector growth.
		Promotion of solar and wind energy may be largely neutral to forestry industry, forest-dependent livelihoods and the forest resource base in areas where firewood extraction is insignificant.
		Construction of large-scale hydro dams may increase deforestation and, thus, limit forest-sector growth.
SDG 9 – Industry, Innovation and Infrastructure		*Nature of interactions depends on the type of infrastructure promoted and the sector focus of policies and strategies to foster economic growth and decent work:*
		Promoting growth of the forest and ecotourism industries can create additional employment and income; road construction in forest areas can improve market access.
		Developing infrastructure for economic growth in urban areas may be largely neutral to the forest resource base and, hence, not affect forest-sector growth.
		Promoting growth of the agri-food, mining and energy industries and construction of roads and hydro dams may increase deforestation and, thus, limit forest-sector growth.

SDG 10 – Reduced Inequalities		*Nature of interactions depends on the sector focus of policies and strategies to foster economic inclusion:*
		There is high potential for economic inclusion in the forest sector where about 75% of employment is informal.
		Economic inclusion in the tourism sector may be largely neutral to the forest resource base and, hence, not affect forest-sector growth.
		Economic inclusion in the agricultural, mining and energy sectors may increase deforestation and, thus, limit forest-sector growth.
SDG 11 – Sustainable Cities and Communities		*Nature of interactions depends on the sectors affected by urban sustainability policies and strategies:*
		Increased demand for safe and affordable housing based on wood and other materials from the forest can spur forest-sector growth; demand for peri-urban forests as green public spaces can foster growth of ecotourism.
		Urban transport and settlement policies may be largely neutral to the forest resource base and, hence, not affect forest-sector growth.
		Urban water demand may impose restrictions on management of forest resources in nearby watersheds; demand for non-polluting energies may increase deforestation through expansion of hydro dams and, thus, limit forest-sector growth.

Table 8.7 (cont.)

SDG 8 SDGs	Intensity of interactions High (dark grey) Medium (mid-grey) Low (light grey)	Nature of interactions depending on political-strategic priorities Synergies (yellow) Neutral (orange) Trade-offs (red)
SDG 12 – Responsible Consumption and Production		*Nature of interactions depends on the sectors targeted by responsible consumption and production policies and strategies:*
		Promotion of building materials derived from wood and other forest resources, along with standards attesting their sustainability, provides incentives for forest sector and ecotourism growth based on SFM.
		Advance of sustainability standards attesting zero deforestation in agricultural commodity chains ensures neutrality to the forest resource base and, thus, does not affect forest-sector growth.
SDG 13 – Climate Action		*Nature of interactions depends on climate change regulations and the sector focus of policies and strategies to foster economic growth and decent work:*
		Forest-sector growth based on SFM helps to maintain or restore forests as carbon sinks and, thus, reduces greenhouse gas emissions; successful climate change mitigation measures support forest health and, thus, forest industry and forest-dependent people.
		Growth of the agricultural, mining, energy and tourism sectors may increase deforestation and/or greenhouse gas emissions; climate change regulations may impose restrictions limiting forest-sector growth.

SDG 14 – Life Below Water		*Nature of interactions depends on ocean management regulations and the sector focus of policies and strategies to foster economic growth and decent work:*
		Growth of inland aquaculture may be largely neutral to the forest resource base and, hence, not affect forest-sector growth.
		Forest-sector growth may be limited by: (1) expansion of fish and shrimp farming in coastal areas in response to regulations restricting fishing in open waters, leading to deforestation of mangroves and other coastal forest types; (2) regulations restricting the use of forest resources in coastal regions; (3) discharge of materials from areas deforested through agriculture and mining may affect growth of mangrove forests and associated forest-sector development.
SDG 15 – Life on land		*Nature of interactions depends on protected area regulations and the sector focus of policies and strategies to foster economic growth and decent work:*
		SFM sustains forest-sector growth; expansion and management of protected areas can foster growth of sustainable ecotourism.
		Growth of the agricultural, mining and energy sectors may increase deforestation and protected area regulations may restrict forest management and, thus, limit forest-sector growth.
SDG 16 – Peace, Justice and Strong Institutions		*Nature of interactions depends on the focus of policies and strategies to foster peace, justice and institutional strengthening:*
		Strong local institutions promote SFM and SMFE; reduction of illicit forest-related activities (illegal logging, wildlife trade, drug trafficking) linked with deforestation supports forest-sector growth.
		Non-forest-related regulations are essentially neutral to the forest resource base and, hence, do not affect forest-sector growth.

Table 8.7 (cont.)

SDG 8 SDGs	Intensity of interactions High (dark grey) Medium (mid-grey) Low (light grey)	Nature of interactions depending on political-strategic priorities Synergies (yellow) Neutral (orange) Trade-offs (red)
SDG 17 – Partnerships for the Goals		*Nature of interactions depends on the sector focus of policies and strategies to foster economic growth, decent work and partnerships:*
		Private, public and civil society partnerships can foster FPVC development; enhanced finance, technology, capacity development, trade and systemic support for the forest sector will spur its growth and decent work.
		Enhanced finance, technology, capacity development, trade and systemic support for the ecotourism sector may be largely neutral to the forest resource base and, hence, not affect forest-sector growth.
		Enhanced finance, technology, capacity development, trade and systemic support for the agricultural, mining and energy sectors may increase deforestation and, thus, limit forest-sector growth.

sectors. World demand for mineral resources has increased significantly since 2000 and continued growth in the technology, transportation and construction sectors will further fuel demand (Megevand et al. 2013). Untapped mineral resources are concentrated in the Amazon, Congo and Mekong basins. Growth of the mining sector implies important trade-offs between SDG 8 and SDGs 3, 5, 6, 12, 13, and 15 and, in view of armed conflicts around mineral resources in forested regions (Switzer 2001), with regard to SDG 16.

Countries rich in natural resources have long been known for strong links between resource extraction and environmental degradation. This 'resource curse' often leads to forest degradation, undermining Target 8.4, which seeks to decouple economic growth from environmental degradation (Swamy et al. 2018). In Ethiopia, for example, the government has embarked on a rapid economic growth trajectory emphasising agriculture and energy development, requiring vast tracts of land and competing with other ecosystems including forests (Weitz et al. 2014).

Progress towards all SDGs depends on multiple interactions among diverse stakeholder groups. Beyond potential synergies among SDGs, there is a need for identifying common interests among key stakeholders and developing mechanisms for coordinated, collaborative efforts. Cross-sector, multi-stakeholder dialogue is also needed to address conflicting interests and minimise trade-offs. Conflict resolution and stakeholder-informed prioritisation are required as there is considerable risk that SDG actions undermine one another and compromise the sustainable use of natural resources (Swamy et al. 2018).

8.5 Conclusions

As most governments in the Global South are pursuing 'growth and trickle-down' strategies to reduce poverty, while largely relying on natural capital, both renewable and non-renewable resources will be in high demand to achieve rapid economic growth (FAO 2010). As a result, significant trade-offs can be expected between achieving SDG 8 and SDGs focused on the conservation of natural resources, particularly SDGs 6, 13 and 15. At the country level, the effects of SDG 8 on forests will depend on the prioritisation of overarching development paradigms (modernisation, economic growth, basic needs, sustainable development) which, in turn, will lead to a differentiated prioritisation of SDG 8 targets by governments, the private sector, investors and civil society. Even in forest-rich countries such as Russia, Canada, Brazil, Indonesia and DR Congo, the contributions of the forest sector to GDP and overall economic growth are modest at best. Growth strategies are likely to continue to rely on natural resource sectors other than forestry (agriculture, energy, and mining), which, in many cases, directly compete with the forest sector for land, investments and human resources.

Efforts to factor forest ecosystem services into national accounts of natural capital are laudable and will play an increasing role in shaping the debate about the value of forests vis-à-vis that of alternative land uses. Still, for the foreseeable future such economic valuation of forests will hardly become a game-changer for halting deforestation and forest degradation at the global scale, given that competing sectors hold higher potential for generating economic growth, employment and, arguably, decent jobs. Decoupling economic growth from forest-related environmental degradation will continue to be a principal challenge in countries where such growth is sought by developing sub-sectors of agricultural (e.g. cereals, sugarcane, soy, beef) and tree crops (e.g. rubber, oil palm), the energy sector through expansion of hydro dams and the mining sector through granting licenses in extensive forest areas.

Conflicting growth policies and those neglecting environmental trade-offs reflect, to some extent, the architecture of the SDGs. It is argued that the global negotiation process for SDG formulation resulted from 'political' mapping rather than biophysical and socio-economic considerations; consequently, the political framework does not explicitly reflect the multiplicity of links that matter for policy purposes, and in practice the SDGs will be of limited use in providing guidance to address those various links (Le Blanc 2015). Furthermore, despite the notion of equivalence among SDGs, interventions and investments will be guided by societal goal prioritisation. Such a process, in turn, will reflect overarching development paradigms and, depending on the access to information and decision making among stakeholder groups, varying degrees of inclusiveness as regards prioritisation.

The nature of such processes will determine the extent to which the synergetic dimension of SDG 8 will figure more prominently. This would be reflected, for example, in significant numbers of poorer people involved in the establishment and management of timber plantations, the management of natural forests for timber and NTFPs, their processing into value-added products, and services related to ecotourism and the provision of environmental services. The impact of such endeavours can be expected to grow over time as the 'green' economy, public and private sustainability standards, and impact investments advance. Increasing the community stewardship of tropical forests will further contribute to synergies between forest conservation and livelihood development, with equitable benefit sharing between men, women and youth. Community-based approaches will allow for gender-differentiated opportunities in timber and NTFP value chains. Involvement of the youth and decent work in the forest sector can increase with better access to skills development programmes and modern ICT for running forest-based enterprises, ensuring long-term engagement and better positioning in FPVC.

However, in many countries and for many years to come, the societal prioritisation of SDGs will continue to follow established patterns: putting economic goals over environmental and social goals. For example, donor agencies such as the Department for International Development (DFID) are returning to an economic growth agenda. Such a strategy may achieve growth in partner countries but, without sufficient conceptual rigour, regulatory oversight or attention to the 'connective fabric' between growth and development, may yield to state–corporate interests and not achieve progressive, just development outcomes (Mawdsley 2015). In general, business-as-usual approaches to economic growth will perpetuate trade-offs with regard to the conservation of forests and the livelihoods dependent on them. However, the prioritisation of goals and actions that negatively affect forests may be unavoidable in given situations. In these cases, the overall SDG outcomes need to be acceptable from a broader societal development perspective ('justified deforestation') – for example, when the benefits from agricultural conversion outweigh the environmental costs (see Carrasco et al. 2017b). For this to happen, access to information and education needs to improve, societal debate needs to become more inclusive and paradigm shifts need to happen ('decent work and *decent* growth'). These shifts would reflect, and can be nurtured through, alternative approaches for measuring forest-sector impact – beyond GDP contributions, employment generation and the hectarage of forests set aside as protected areas. Such approaches allow for a more integrated measurement of economic growth, assessing its environmental impact by accounting for the conservation or depletion of overall natural capital stocks and its social impact by using metrics that account for gender, age and other factors of social differentiation. Results of integrated measuring can contribute to individual and institutional learning, foster innovative cross-sector partnerships and, based on these, more informed prioritisation of goals and better targeted interventions and investments by public and private sectors and civil society.

References

Agarwal, B. 1994. Gender and command over property: A critical gap in economic analysis and policy in South Asia. *World Development* 22(10):1455–78.

Agrawal, A., Cashore, B., Hardin, R., Shepherd, G., Benson, C. and Miller, D. 2013. *Economic contributions of forests*. Background paper prepared for the 10th session of the United Nations Forum on Forests held in Istanbul, 8–19 April 2013.

Agrawal, A., Hajjar, R., Liao, C., Rasmussen, L. V. and Watkins, C. 2018. Editorial overview: Forest governance interventions for sustainability through information, incentives, and institutions. *Current Opinion in Environmental Sustainability* 32:1–7.

Allen, C. D., Macalady, A. K., Chenchouni, H. et al. 2010. A global overview of drought and heat-induced tree mortality reveals emerging climate change risks for forests. *Forest Ecology and Management* 259(4):660–84.

Angelsen, A., Jagger, P., Babigumira, R. et al. 2014. Environmental income and rural livelihoods: a global-comparative analysis. *World Development* 64:S12–28.

Appanah, S. 2016. *Forest landscape restoration for Asia-Pacific forests*. Bangkok: FAO/RECOFTC.

Arndt, H. W. 1983. The 'trickle-down' myth. *Economic Development and Cultural Change* 32(1):1–10.

Azapagic, A. 2004. Developing a framework for sustainable development indicators for the mining and minerals industry. *Journal of Cleaner Production* 12(6):639–62.

Baker, J. S., Wade, C. M., Sohngen, B. L., Ohrel, S. and Fawcett, A. A. 2019. Potential complementarity between forest carbon sequestration incentives and biomass energy expansion. *Energy Policy* 126:391–401.

Belibi, M. B., van Eijnatten, J., Mala, W. A. and Ingram, V. 2015. Empowering women and ethnic minority groups to collectively market non timber forest products from community forests in Cameroon. *Journal of Life Sciences* 9(8):381–90.

Benami, E., Curran, L. M., Cochrane, M. et al. 2018. Oil palm land conversion in Pará, Brazil, from 2006–2014: Evaluating the 2010 Brazilian Sustainable Palm Oil Production Program. *Environmental Research Letters* 13(3):034037.

Bennetzen, E. H., Smith, P. and Porter, J. R. 2016. Decoupling of greenhouse gas emissions from global agricultural production: 1970–2050. *Global Change Biology* 22(2):763–81.

Brandt, J. S. and Buckley, R. C. 2018. A global systematic review of empirical evidence of ecotourism impacts on forests in biodiversity hotspots. *Current Opinion in Environmental Sustainability* 32:112–18.

Buckley, R. 2018. Impacts positive and negative: links between ecotourism and environment. In Buckley, R. (ed.) *Environmental impacts of ecotourism*. Wallingford: CABI Publishing, pp. 5–14.

Carrasco, L. R., Le Nghiem, T. P., Chen, Z. and Barbier, E. B. 2017a. Unsustainable development pathways caused by tropical deforestation. *Science Advances* 3(7):e1602602.

Carrasco, L. R., Webb, E. L., Symes, W. S., Koh, L. P. and Sodhi, N. S. 2017b. Global economic trade-offs between wild nature and tropical agriculture. *PLoS Biology* 15(7): e2001657.

Cashore, B., Gale, F., Meidinger, E. and Newsom, D. (eds.) 2006. *Confronting sustainability: Forest certification in developing and transitioning countries*. New Haven: Yale School of Forestry and Environmental Studies.

Climate Focus 2017. *Progress on the New York Declaration on Forests: Finance for forests–Goals 8 and 9*. Assessment report prepared by Climate Focus in cooperation with the New York Declaration on Forest Assessment Partners with support from the Climate and Land Use Alliance.

Costanza, R., Kubiszewski, I., Giovannini, E. et al. 2014. Time to leave GDP behind. *Nature* 505:283–5.

Danish and Wang, Z. 2019. Dynamic relationship between tourism, economic growth, and environmental quality. *Journal of Sustainable Tourism* 26(11):1928–43. doi:10.1080/09669582.2018.1526293.

Delucchi, M. A. 2010. Impacts of biofuels on climate change, water use, and land use. *Annals of the New York Academy of Sciences* 1195(1):28–45.

Donovan, J., Stoian, D., Grouwels, S. and Macqueen, D. 2006. *The business side of sustainable forest management: Small and medium forest enterprise development for poverty reduction*. Natural Resource Perspectives 104. London: ODI.

FAO 2010. *Asia-Pacific forests and forestry to 2020*. Report of the second Asia-Pacific Forestry Sector Outlook Study. Rome: FAO.

FAO 2014. *The state of the world's forests – Enhancing the socio-economic benefits from forests*. Rome: FAO.

FAO 2016. *Global Forest Resources Assessment 2015 – How are the world's forests changing? 2nd edition*. Rome: FAO.

FAO 2018. *The State of the World's Forests 2018 – Forest pathways to sustainable development*. Rome: FAO.

FAO and EFI (European Forest Institute) 2018. *Making forest concessions in the tropics work to achieve the 2030 Agenda: Voluntary guidelines*. FAO Forestry Paper 180. Rome: FAO.

Fearnside, P. M. 2016. Environmental and social impacts of hydroelectric dams in Brazilian Amazonia: Implications for the aluminum industry. *World Development* 77:48–65.

Ferreira, J., Aragão, L. E. O. C., Barlow, J. et al. 2014. Brazil's environmental leadership at risk. *Science* 346(6210):706–7.

Gale, T. 2016. Thinking globally about ecotourism impact: The contribution of ecological footprint analysis. In Hill, J. and Gale, T. (eds.) *Ecotourism and environmental sustainability*. London: Routledge, pp. 49–66.

Hosonuma, N., Herold, M., De Sy, V. et al. 2012. An assessment of deforestation and forest degradation drivers in developing countries. *Environmental Research Letters* 7(4):044009.

ILO 2017. *Information System on International Labour Standards – Ratification of ILO Conventions*. International Labour Organization. Available at: www.ilo.org/dyn/normlex/en/f?p=1000:11001:::NO::: (Accessed 29 December 2018).

ILO 2018a. *ILOSTAT – the world's leading source of labour statistics*. International Labour Organization. Available at: www.ilo.org/ilostat/faces/ilostat-home/metadata?_adf.ctrl-state=vwps62w1y_57&_afrLoop=796086079109955#! (Accessed 29 December 2018).

ILO 2018b. *Decent work and the Sustainable Development Goals: A guidebook on SDG labour market indicators*. Geneva: ILO, Department of Statistics (STATISTICS).

Jusys, T. 2017. A confirmation of the indirect impact of sugarcane on deforestation in the Amazon. *Journal of Land Use Science* 12(2–3):125–37.

Lange, G. M., Wodon, Q. and Carey, K. (eds.) 2018. *The changing wealth of nations 2018: Building a sustainable future*. Washington, DC: World Bank.

Larson A. M., Monterroso, I. and Canturias, P. 2018. *Gender and formalization of native communities in the Peruvian Amazon*. CIFOR InfoBrief 238. Bogor, Indonesia: CIFOR.

Larson, A. M. and Soto, F. 2008. Decentralization of natural resource governance regimes. *Annual Review of Environment and Resources* 33:213–39.

Lebedys, A. and Yanshu, L. 2014. *Contribution of the forestry sector to national economies, 1990–2011*. Forest Finance Working Paper 09. Rome: FAO.

Le Blanc, D. 2015. Towards integration at last? The Sustainable Development Goals as a network of targets. *Sustainable Development* 23(3):176–87.

Lélé, S. M. 1991. Sustainable development: a critical review. *World Development* 19(6):607–21.

Lewis, W. A. 1954. Economic Development with Limited Supplies of Labour. *Manchester School of Social Science* 22:139–91.

Macqueen, D. 2008. *Supporting small forest enterprises – A cross-sectoral review of best practice*. IIED Small and Medium Forestry Enterprise Series No. 23. London: IIED.

Macqueen, D., Bolin, A., Greijmans, M., Grouwels, S. and Humphries, S. 2018. Innovations towards prosperity emerging in locally controlled forest business models and prospects for scaling up. *World Development* https://doi.org/10.1016/j.worlddev.2018.08.004

Malkamäki, A., D'Amato, D., Hogarth, N. J. et al. 2018. A systematic review of the socio-economic impacts of large-scale tree plantations, worldwide. *Global Environmental Change* 53:90–103.

Mawdsley, E. 2015. DFID, the private sector and the re-centring of an economic growth agenda in international development. *Global Society* 29(3):339–58.

Max-Neef, M., Elizalde, A. and Hopenhayn, M. 1992. Development and human needs. In Elkins, P. and Max-Neef, M. (eds.) *Real-Life economics: Understanding wealth creation*. London: Routledge, pp. 197–213.

Megevand, C., Mosnier, A., Hourticq, J. et al. C. 2013. *Deforestation trends in the Congo Basin: Reconciling economic growth and forest protection*. Washington, DC: World Bank.

Miteva, D. A. 2019. The integration of natural capital into development policies. *Oxford Review of Economic Policy* 35(1):162–81.

Mwitwa, J., German, L., Muimba-Kankolongo, A. and Puntodewo, A. 2012. Governance and sustainability challenges in landscapes shaped by mining: Mining–forestry linkages and impacts in the Copper Belt of Zambia and the DR Congo. *Forest Policy and Economics* 25:19–30.

O'Rourke, D. and Connolly, S. 2003. Just oil? The distribution of environmental and social impacts of oil production and consumption. *Annual Review of Environment and Resources* 28(1):587–617.

Porter-Bolland, L., Ellis, E. A., Guariguata, M. R. et al. 2012. Community managed forests and forest protected areas: An assessment of their conservation effectiveness across the tropics. *Forest Ecology and Management* 268:6–17.

Regmi, K. D. and Walter, P. 2017. Modernisation theory, ecotourism policy, and sustainable development for poor countries of the Global South: perspectives from Nepal. *International Journal of Sustainable Development & World Ecology* 24(1):1–14.

Rodgers, G. 2007. *Labour market flexibility and decent work*. DESA Working Paper 47. New York: United Nations.

Roemer, M. 1979. Resource-based industrialization in the developing countries: a survey. *Journal of Development Economics* 6(2):163–202.

Romero, C., Sills, E. O., Guariguata, M. et al. 2017. Evaluation of the impacts of Forest Stewardship Council (FSC) certification of natural forest management in the tropics: A rigorous approach to assessment of a complex conservation intervention. *International Forestry Review* 19(S2):36–49.

Rostow, W. W. 1959. *The stages of economic growth and the problems of peaceful co-existence*. Center for International Studies. Cambridge, MA: Massachusetts Institute of Technology.

RRI 2017. *Securing community land rights: Priorities and opportunities to advance climate and Sustainable Development Goals*. Washington, DC: RRI.

RRI 2018. *At a crossroads: Consequential trends in recognition of community-based forest tenure from 2002 to 2017*. Washington, DC: RRI.

Schandl, H., Hatfield-Dodds, S., Wiedmann, T. et al. 2016. Decoupling global environmental pressure and economic growth: scenarios for energy use, materials use and carbon emissions. *Journal of Cleaner Production* 132:45–56.

Schmelzer, M. 2017. History, hegemony, and the contested making of economic growthmanship. In Caradonna, J. L. (ed.) *Routledge handbook of the history of sustainability*. London: Routledge, pp. 164–86.

Shackleton, C., Shackleton S. and Shanley, P. (eds.) 2011. *Non-timber forest products in the global context*. Berlin: Springer.

Stoian, D., Donovan, J. and Poole, N. 2009. *Unlocking the development potential of community forest enterprises: Findings from a comparative study in Asia, Africa, Latin America, and the United States*. Paper presented at the XIII World Forestry Congress held in Buenos Aires on 18–23 October 2009.

Stoian, D., Rodas, A., Butler, M. Monterroso, I. and Hodgdon, B. 2019. *The forest concessions in Petén, Guatemala: A systematic analysis of the socio-economic performance of the community enterprises in the Maya Biosphere Reserve*. Nairobi: Bioversity International, CIFOR, Rainforest Alliance, World Agroforestry.

Sunderland, T., Achdiawan, R., Angelsen, A. et al. 2014. Challenging perceptions about men, women, and forest product use: A global comparative study. *World Development* 64:S56–66.

Swamy, L., Drazen, E., Johnson, W. R. and Bukoski, J. J. 2018. The future of tropical forests under the United Nations Sustainable Development Goals. *Journal of Sustainable Forestry* 37(2):221–56.

Switzer, J. 2001. *Armed conflict and natural resources: The case of the minerals sector*. London: IIED.

Thornton, J. R., Agnello, R. J. and Link, C. R. 1978. Poverty and economic growth: Trickle down peters out. *Economic Inquiry* 16(3):385–94.

Tipps, D. C. 1973. Modernization theory and the comparative study of national societies: A critical perspective. *Comparative Studies in Society and History* 15(2):199–226.

Tomberlin, D., Buongiorno, J., Alegría, J. A., Korhonen, K. and Palo, M. 2001. Timber plantations, timber supply and forest conservation. In Palo, M., Uusivuori, J. and Mery, G. (eds.) *World forests, markets and policies*. Dordrecht: Springer, pp. 85–96.

UN 2018. *The Sustainable Development Goals Report 2018*. New York: United Nations.

UN 2019. *Sustainable Development Goals Knowledge Platform*. Available at: https://sustainabledevelopment.un.org/ (Accessed 15 February 2019).

UNCTAD 2018. *World Investment Report 2018 – Investment and new industrial policies*. New York: UNCTAD.

van der Ven, H., Rothacker, C. and Cashore, B. 2018. Do eco-labels prevent deforestation? Lessons from non-state market driven governance in the soy, palm oil, and cocoa sectors. *Global Environmental Change* 52:141–51.

Verner, D. and Blunch, N. H. 1999. *Sector growth and the dual economy model: Evidence from Cote d'Ivoire, Ghana, and Zimbabwe*. Washington, DC: The World Bank.

Vijay, V., Pimm, S. L., Jenkins, C. N. and Smith, S. J. 2016. The impacts of oil palm on recent deforestation and biodiversity loss. *PLoS ONE* 11(7):e0159668. https://doi.org/10.1371/journal.pone.0159668.

Weitz, N., Nilsson, M. and Davis, M. 2014. A nexus approach to the post-2015 agenda: Formulating integrated water, energy, and food SDGs. *SAIS Review of International Affairs* 34(2):37–50.

Wily, L. A. 2011. 'The law is to blame': The vulnerable status of common property rights in sub-Saharan Africa. *Development and Change* 42(3):733–57.

Winemiller, K. O., McIntyre, P. B., Castello, L. et al. 2016. Balancing hydropower and biodiversity in the Amazon, Congo, and Mekong. *Science* 351(6269):128–9.

World Bank 2016. *World Bank Group Forest Action Plan FY16–20*. Washington, DC: World Bank.

World Bank 2017. *World Development Indicators: Contribution of natural resources to gross domestic product*. Available at: http://wdi.worldbank.org/table/3.14 (Accessed 29 December 2018).

World Bank 2018a. *World Development Indicators: Agriculture, forestry, and fishing, value added (% of GDP)*. Available at: https://data.worldbank.org/indicator/NV.AGR.TOTL.ZS (Accessed 29 December 2018).

World Bank 2018b. *World Development Indicators: Employment in agriculture (% of total employment) (modeled ILO estimate)*. Available at: https://data.worldbank.org/indicator/SL.AGR.EMPL.ZS (Accessed 29 December 2018).

World Bank 2018c. *Implementing the 2020 Agenda – 2018 Update*. Washington, DC: World Bank.

WTTC (World Travel & Tourism Council) 2018. *Travel & Tourism Economic Impact 2017: World*. London: WTTC.

Yakovleva, N. 2017. *Corporate social responsibility in the mining industries*. London: Routledge.

Chapter 9 SDG 9: Industry, Innovation and Infrastructure – Anticipating the Potential Impacts on Forests and Forest-Based Livelihoods

Maria Fernanda Tomaselli*, Joleen Timko*, Robert Kozak*, Justin Bull, Sean Kearney, Jack Saddler, Susan van Dyk, Guangyu Wang and Xinxin Zhu

Key Points

- Target 9.1 and its corresponding indicators risk irreversible and widespread forest degradation and deforestation; the short- and long-term environmental and social costs of this goal need to be better assessed.
- The impacts of other indicators on forests (e.g. Target 9.3, Target 9.C) will largely depend on how they are implemented.
- Major trade-offs exist between SDG 9 and SDG 15 (Life on Land), especially if economic expansion and increasing planetary impacts remain coupled.
- Target 9.4 and its corresponding indicator should go beyond greenhouse gas emissions and intensity-based measures to ensure absolute reductions in ecological or material impact, as higher global material use will mean more pressure and competing demands on forests, likely impacting these ecosystems in negative ways.
- SDG 9 should be reformulated to promote and support alternative socio-economic models that are not based on indefinite economic growth or reliant on the ongoing expansion of infrastructure. In this light, the maintenance of ecosystem services and forests could be seen as essential building blocks of a green and sustainable economy.

9.1 Introduction

Sustainable Development Goal (SDG) 9 is centred on three main pillars: industry, infrastructure and innovation. With 8 targets and 12 indicators (broadly summarised in Table 9.1), SDG 9 will certainly have multiple impacts on forests, forest-based livelihoods and forest-based economies. This chapter explores some of the potential implications of this goal as currently proposed – within the context of forested landscapes – and examines possible interactions, synergies and trade-offs for implementation. In addition, it

* Lead authors.

Table 9.1 Summary of targets and the main focus of the indicators for SDG 9
9.1. Infrastructure development (road and transportation expansion)
9.2. Industry and manufacturing (increase of manufacturing value added and employment)
9.3. Small-scale industry integration to markets and finance (proportion of small-scale enterprises in total value added and greater access to credit)
9.4. Clean and environmentally sound industry and resource efficiency (carbon intensity)
9.5. Research and development (R&D expenditure as fraction of GDP and number of researchers)
9.A. Financial, technological and technical support to LDCs and others (ODA and other financial flows to infrastructure)
9.B. Technology, research and innovation support to developing nations (proportion of medium- and high-tech industry value added)
9.C. Access to information and telecommunications in LDCs (proportion of population covered by a mobile network)
Source: https://sustainabledevelopment.un.org/sdg9

explores the potential implications of alternative socio-economic pathways for forests and forest-dependent peoples.

SDG 9 is seen as essential to achieving economic growth, making it inextricably linked to the aims of SDG 8 (Decent Work and Economic Growth). It acknowledges that industrialisation must be inclusive, environmentally sound and sustainable; that infrastructure must be resilient; and that technology must play a central role in achieving these aims through resource- and energy-efficiency and access to digital technologies.

As currently proposed, SDG 9 is embedded in an 'ecological modernisation' narrative, which places a greater emphasis on the role of science and technology in ensuring the compatibility between economic growth and environmental sustainability (Tracy et al. 2017). These assumptions can be viewed as contentious, especially as the human population – now exceeding 7.5 billion – grows at an annual rate of 1.1 per cent (UNEP 2016) and our global ecological footprint continues to increase, while global biocapacity is in decline (Wackernagel and Rees 1996). Moreover, the world is experiencing amplified income and wealth inequality: in 2015, the wealth of the richest 1 per cent surpassed that of the remaining 99 per cent (OXFAM 2016). These are important considerations when evaluating the potential impacts of SDG 9 on forests, forest-dependent peoples and forest-based economies, especially since four out of nine planetary boundaries are estimated to have been crossed: climate change, biosphere integrity (e.g. loss of biodiversity), land system change and alterations to biochemical flows (e.g. nitrogen and phosphorus cycles) (Steffen et al. 2015).

The UN (2017a) and the World Bank (2017) recognise some signs of global progress towards achieving SDG 9, including increases in manufacturing value added as a share of gross domestic product (GDP), growth in air transit, moderate gains in research and development investments, increases in development assistance for infrastructure projects (mainly transport and energy) and declines in CO_2 emissions per unit of manufacturing value added. Ninety-five per cent of the world's population lives within the range of a mobile-cellular signal and 50 per cent have access to the Internet, although only 11 per cent of the population in least-developed countries (LDCs) has access to the Internet. Also, basic infrastructure needs – sanitation, electrical power and water – remain unmet in many LDCs, especially in remote areas where many forests are found (Mead 2017). In this context, the UN (2017a) is calling for a renewed investment in infrastructure and a doubling of industry's share of GDP contributions in LDCs by 2030.

Some countries with significant forest cover have documented their progress towards SDG 9 in their Voluntary National Reviews (VNRs).[1] For instance, in Brazil's VNR, investment in energy is seen as central to development efforts, especially the generation of renewable energy. In Indonesia's VNR, infrastructure improvement and expansion, especially transportation (e.g. roads, railways, ports), is seen as central to reducing the remoteness of rural areas and to the nation's development plan. In India's VNR, it is reported that all forms of transportation (including non-motorised transport) are being rapidly expanded. India is also engaged in expanding manufacturing, promoting small and medium-sized enterprises (SMEs), improving rural access to energy, encouraging foreign direct investment (FDI) and expanding internet penetration. In China's executive summary, development – specifically in the form of innovative, low-carbon options – is seen as the main priority, with the major goals being to lift 50 million people out of poverty and double GDP and per capita income. For China, South to South cooperation is seen as fundamental, with investment in infrastructure playing an important role. Although these reports briefly mention environmental quality and protection, in most there is no mention of forests – neither of how these may contribute to the new economy, nor how they may be impacted or shielded from the impacts of industrialisation. Notably, in terms of environmental sustainability, Indonesia's VNR expresses a commitment to replace the linear economy with a circular one.

[1] Voluntary National Reviews can be accessed at https://sustainabledevelopment.un.org/vnrs/

While many nations are prioritising and promoting industrialisation and the expansion of infrastructure, manufacturing and trade, other contextual conditions occurring at the macro level influence the implementation and uptake of SDG 9 – most notably, factors related to the state of the economy, investment and governance. Between 1970 and 2010, the global economy tripled in size, from USD 15.4 trillion to USD 51.7 trillion (at 2005 constant prices), growing at an average annual rate of 3.1 per cent (UNEP 2016). This is due, in no small part, to a rise in economic openness globally, which has been shown to have a positive impact on economic growth (Costantini and Monni 2008). However, uniform liberalisation can also lead to deindustrialisation, impacting sectors in their early stages. The rate of industrialisation itself is dependent on a number of contextual factors related to competitive advantage. For instance, industrialisation is faster in countries with strong export performance and large domestic markets and in countries with undervalued exchange rates (Guadagno 2016). The expansion of industrial capacity – and the concomitant increases in output and employment – depend on levels of domestic and foreign direct investment (Agosin and Machado 2005, Szkorupová 2015).

The impacts of industrialisation on forests and the environment are complex. Greater income and affluence increase energy use and domestic material consumption (UNEP 2016), oftentimes affecting the environment in negative ways. For example, China's rapid industrialisation has led to rising energy use, particularly the use of coal, increasing the country's greenhouse gas (GHG) emissions. Industrialisation also tends to increase the use of minerals in relation to the use of biomass (UNEP 2016). Impacts on forests can be diverse. In some countries, increased economic development has led to increased forest areas as rural inhabitants emigrate to urban and semi-urban hubs to pursue non-farm jobs. While domestic impacts may be reduced in these cases, the global impact may grow if countries increase their imports of wood, food and other products.

Governance is an important factor in determining how SDG 9 plays out (Costantini and Monni 2008, Guadagno 2016). For instance, in contexts with weak governance (as in rural regions of many tropical nations), local people may be particularly affected by the expansion of large-scale infrastructure projects, especially in areas with unclear tenure regimes and weak property rights. Given the current power dynamics and corrupt practices in many countries, benefits from such projects may not be equitably distributed, possibly even leading to the fulfilment of a resource-curse hypothesis. Meanwhile, weak law enforcement may enable the spread of illegal activities and the unsustainable exploitation of forest resources.

Notably, all of these contextual factors come to bear when viewed through the lens of forests and forest-based livelihoods. Clearly, the implementation of SDG 9 in forest-dependent regions offers economic and employment opportunities. At the same time, the uptake of SDG 9 could lead to increasing pressures on the life-supporting systems – such as forest ecosystems and biodiversity – on which our societies and economies depend. This chapter aims to examine many of the complexities involved and address some of these nuanced synergies and trade-offs by exploring the potential impacts of implementing some of the targets and indicators proposed in SDG 9.

9.2 Potential Impacts of SDG 9 on Forests and Forest-Based Livelihoods

Enacting SDG 9 as currently proposed will have numerous and varied impacts on forest and forest-dependent peoples as a consequence of expanding infrastructure (Target 9.1), increasing manufacturing (Target 9.2), growing the SME sector (Target 9.3), developing cleaner and more efficient industries (Target 9.4) and increasing access to digital technology and telecommunications (Target 9.C). This section examines some of the potential outcomes, trade-offs and synergies of implementing these targets as currently proposed. Particular emphasis is given to Target 9.1, which may have impacts on forests that are not only considerable but potentially irreversible. The chapter also includes a brief discussion surrounding the possibilities of decarbonising air transit (Box 9.1). Table 9.2 broadly summarises the main conclusions of this analysis regarding the potential impacts of implementing some SDG 9 targets and indicators on forest ecosystems and forest-based livelihoods.

9.2.1 Expanding Infrastructure

IMPACTS OF ROAD EXPANSION

Indicator 9.1.1 focuses on increasing the proportion of rural people who live within 2 km of an all-season road. Roads have been shown to improve transportation (e.g. reduce costs, shorten travel times), facilitate access to markets and expand trade, encourage entrepreneurship and diversification of livelihoods, improve social integration and increase income and economic growth (Alamgir et al. 2017, Bucheli et al. 2017, Campbell et al. 2017). They have also been linked to better education and health, as they facilitate access to these services (Alamgir et al. 2017, Bucheli et al. 2017, Hettige 2006). For rural farmers, roads can link them to urban markets, enable access to agricultural inputs (e.g. fertilisers), raise crop prices and improve agricultural technology (Laurance and Burgues 2017). A recent study from Ghana found that improved roads led to more agricultural productivity while decreasing farm size (Acheampong et al. 2018).

The relationship between people and infrastructure is complex, as benefits and costs are often context-dependent, diverse and moderated by multiple factors.

Table 9.2 Summary of analysis reflecting the potential impacts on forest ecosystems and forest-based livelihoods of implementing some SDG 9 targets and indicators

Target / Indicator	Potential Impact	
	Forest Ecosystems	Forest-Based Livelihoods
9.1. Infrastructure development (road and transportation expansion) [Section 9.2.1]	Largely negative.	Mixed, depending on the location and characteristics of specific group affected. Likely positive for forest industry.
9.2. Industry and manufacturing (increase of manufacturing value added and employment) [Section 9.2.2]	Mixed. Negative if overall environmental impact of economies increases (thus impacting forests directly or indirectly). Positive if greater value added is obtained from the same or lesser amount of resources.	Positive if greater value is added to forest products, possibly increasing forest-based employment in rural and urban areas.
9.3. Small-scale industry integration to markets and finance (proportion of small-scale enterprises in total value added and greater access to credit) [Section 9.2.3]	Mixed, depending on which types of SMEs are supported and their corresponding ecological footprints.	Positive, as greater employment and other social benefits could be generated through forest SMEs (including community-forest enterprises).
9.4. Clean and environmentally sound industry and resource efficiency (carbon intensity) [Section 9.2.4]	Mixed. Negative if environmental gains due to greater efficiency are offset by economic growth (i.e. rebound effect). Positive if absolute impact of industries and products is reduced.	N/A
9.C. Access to information and telecommunications in LDCs (proportion of population covered by a mobile network) [Section 9.2.5]	Mixed, depending on how mobile networks are employed.	Mixed, depending on how mobile networks are employed.

While roads can improve food access and diversity, they can also lead to lower nutrition as more processed foods become available (Bucheli et al. 2017). Roads do not de facto alleviate poverty, as effects are moderated by access to different modes of transport, which in turn could be moderated by income. Bryceson et al. (2008: 3) caution that, 'applied uncritically to rural areas', the assumption that roads automatically alleviate poverty 'could easily slide into naivety about the power of road investment to catalyse development and a reductionism that casually assumes poverty reduction will necessarily follow'. Other studies show that impacts vary across socio-demographic groups (Bucheli et al. 2017).

In the context of forests, roads can be viewed as beneficial or detrimental, depending upon whether their impacts are viewed from a business, social or ecological perspective. Roads may also be viewed differently by different local groups – whether they are colonist populations, traditional communities with a long-term history in a place or Indigenous peoples. In the forest sector, poor infrastructure and road conditions are frequently cited as an important challenge facing small and medium forest enterprises (SMFEs), hindering the timely delivery of products and their competitive pricing (Macqueen 2008). Thus, infrastructure development is an important aspect of the enabling environment required for SMFEs to flourish (Macqueen 2008). Through improved access to markets, the expansion of all-season roads could ease the operations of many SMFEs in addition to facilitating agricultural activities. Moreover, large-scale forest operations could probably benefit from road expansion, facilitating access to new forest frontiers with valuable timber. Yet, this may increase the risk of future encroachment and deforestation in contexts of weak governance.

In terms of social impacts, roads can greatly affect rural incomes. Empirical evidence from Ethiopia shows that access to all-season roads reduced poverty by 6.9 per cent and increased consumption growth by 16.3 per cent (Dercon et al. 2009). In addition, some studies have also reported positive perceptions about roads and road expansion in rural communities, although rural dwellers recognise some of the downsides of road expansion (Clements 2013, Fyumagwa et al. 2013).

The deforestation and colonisation that often follow road building have irreversibly affected many forest-dependent Indigenous groups in the Amazon (Finer et al. 2008). Contact often translates into high mortality and other health implications, especially for people living in voluntary isolation (Finer et al. 2008, Napolitano and Ryan 2007), as roads facilitate the transmission of diseases (Alamgir et al. 2017). Road-building projects can increase social costs such as corruption and vulnerability to social exploitation, eroding traditional social structures (Alamgir et al. 2017, Hettige 2006). Other negative externalities include pollution, road hazards, threat to cultural sites and the perpetuation of car-centric development approaches (Bucheli et al. 2017).

Road expansion is associated with large ecological costs (Barber et al. 2014). A leading driver of habitat loss and ecosystem fragmentation and degradation (Ibisch et al. 2016), roads threaten much of the world's remaining wilderness. They are directly or indirectly linked to increased fire risk, proliferation of extractive – sometimes illegal – activities, over-exploitation of resources, increased wildlife mortality and biodiversity loss (Alamgir et al. 2017, Barber et al. 2014, Benítez-López et al. 2010, Ibisch et al. 2016, Laurance et al. 2014).

Roads frequently lead to agricultural expansion – the leading global driver of deforestation – as they are often built to promote agricultural production and food security (Laurance et al. 2014). The economic returns from agriculture motivates the clearing of forests (Busch and Ferretti-Gallon 2017). In Amazonia, 95 per cent of all deforestation occurs in close proximity to transportation networks: within 5.5 km of a road or 1 km of a river (Barber et al. 2014). Similar patterns have been found elsewhere (Alamgir et al. 2017).

The current expansion of road networks is unprecedented in human history (Campbell et al. 2017, Ibisch et al. 2016). Roads have already fragmented the Earth into more than 600 000 pieces of areas without roads, with only 7 per cent of these being larger than 100 km^2 (Ibisch et al. 2016). They have been described as highly contagious, in that they spread into secondary and tertiary roads. For every kilometre of legal road in the Amazon, there are about 3 km of illegal, unmapped ones (Barber et al. 2014), illustrating the lack the proper governance or the means to plan, monitor and control road networks in many countries (Ibisch et al. 2016). Their total length is expected to increase 60 per cent in the next 30 years (Alamgir et al. 2017), with 90 per cent of this expansion occurring in the Global South (in highly biodiverse tropical and subtropical regions, where a large share of forest-dependent communities live) (Laurance and Burgues 2017).

In response to these staggering numbers, some researchers are calling for a comprehensive global strategy for planned and strategic road expansion. They suggest constructing or improving roads in areas where these can generate higher social or human development returns (e.g. settled areas with higher agricultural potential, urban or peri-urban lands) and avoiding areas with high environmental values and lower agricultural potential (Campbell et al. 2017, Laurance 2018, Laurance et al. 2014). Other authors have made a call to leave remote areas roadless (or at least leave roads unpaved) and to strengthen governance (i.e. enforcement, monitoring) in areas that have long-established roads (Ibisch et al. 2016).

If faithfully implemented, Indicator 9.1.1 would continue fuelling the current road-building spree and risk irreversible and widespread forest degradation. As written, it ignores the environmental and social costs and trade-offs associated with road development. Target 9.1 should be rewritten to emphasise the need for roads to be well-planned and strategic (i.e. where to locate them to maximise benefits and minimise costs, as proposed by Laurance et al. 2014 and Campbell et al. 2017). Road-expansion costs need to be carefully assessed, especially since road-building proponents tend to overemphasise the benefits (Alamgir et al. 2017) and traditional environmental impact assessments (EIAs) tend to underestimate project costs and challenges (Laurance and Burgues 2017).

ANTICIPATED IMPACTS OF OTHER PROPOSED INFRASTRUCTURE PROJECTS

Multiple development projects are being planned, implemented or upgraded in Africa, Asia and Latin America, aimed at improving agricultural output and food security, mining exports and economic integration, among others. In South America, about 600 infrastructure projects are being planned, are underway or are already implemented in the energy, transportation (e.g. ports, railways) and telecommunication sectors, among others (COSIPLAN 2017). For example, oil projects now cover more than two-thirds of the Ecuadorian and Peruvian Amazon, many overlapping Indigenous territories and areas where people live in voluntary isolation (Finer et al. 2008, Napolitano and Ryan 2007). Indigenous groups in the region that oppose oil development on their lands have, in many cases, successfully ended projects (Finer et al. 2008).

Hydropower expansion is also underway across South America. Currently, there are plans to expand the number of hydro dams in the Andean foothills from 48 to 152 in the next 20 years, causing major disruptions in connectivity between 5 of the 6 major Andean tributaries and the Amazon River (Finer and Jenkins 2012, Gibson et al. 2017). In the Amazon basin, there are currently 191 dams, while another 246 are planned or are under construction (Gibson et al. 2017). The accumulated effects of current and proposed dams mean massive disturbances to the Amazon floodplain, South America's northeast coast and the regional climate (Latrubesse et al. 2017). Although the long-term impacts on biodiversity of mega-dams have been overlooked, Benchimol and Peres (2015) expose recent major local extinction threats to vertebrate species. Similarly, the impacts on forests should not be underestimated. Analysing the ecological impacts of current and potential dams, Finer and Jenkins (2012: 1) conclude that more than 80 per cent of the proposed projects in the Amazon 'would drive deforestation due to new roads, transmission lines or inundation'. In their review of green energy, Gibson et al. (2017: 928) conclude that 'the substantial greenhouse gas emissions and pronounced disruption of terrestrial and aquatic ecosystems from hydropower dams raise serious questions as to whether they should be considered "green energy" at all'.

Furthermore, China plans to expand infrastructure in Eurasia and around the Global South. It is currently investing about USD 100 billion annually for transport, energy and mining infrastructure in Africa (Alamgir et al. 2017, Laurance 2017), and its FDI increased tenfold between 2005 and 2015, largely for infrastructure development projects and resource extraction (Tracy et al. 2017). One of these major projects is the Belt and Road Initiative (BRI, also known as One Belt, One Road).

The BRI, announced in 2013, refers to the Silk Road Economic Belt and the 21st Century Maritime Silk Road, a significant development strategy intended to promote economic cooperation among countries along the proposed Belt and Road routes. The initiative aims to connect Asia, Europe and Africa along five routes, including international transport routes, core cities and key ports, and six international economic cooperation corridors. The BRI is open to all countries, as well as international and regional organisations; however, official maps and documents emphasise the importance of 71 countries in Asia, the Middle East, Eastern Africa and Eastern Europe (Figure 9.1). Unprecedented in scale (Tracy et al. 2017), the initiative has been identified as one of the 17 emerging issues that could affect global biodiversity, ecosystem services and conservation (Sutherland et al. 2018).

Most investments generated from the BRI have thus far been in infrastructure, energy and mining, ranging from a standard-gauge railway in Kenya to hydropower projects in Cambodia, and from the Prairie Road between China and Mongolia to lignite coal deposits in Pakistan. The BRI will increase investment and foster economic collaboration in the ancient Silk Road area; however, little attention has been paid to the ecological impacts generated from the massive construction of infrastructure and natural resources investments. Although China has been pursuing green investment opportunities (e.g. solar, hydropower), the country has not released any overarching guidelines for the sustainability requirements of BRI projects beyond individual institutions (Pike 2017). Moreover, the official document outlining the BRI's vision and actions (NDRC 2015) references environmental protection only in passing, with no mention of EIAs or strategic environmental assessments (SEAs) (Tracy et al. 2017: 74), which is particularly concerning since the 21st Century Maritime Silk Road passes through many South and Southeast Asian countries holding a high concentration of global biodiversity hotspots and forest-dependent communities. Likewise, many of the BRI's proposed routes cross protected areas (Sutherland et al. 2018) and will 'open for exploitation unique old-growth forests' (Tracy et al. 2017: 76).

While environmental protection has not yet been emphasised in the BRI (Sutherland et al. 2018), Chinese and foreign NGOs have committed to helping China develop guidelines under the umbrella of the China Green Leadership: Belt and Road Green Development project, which has resulted in the BRI Ecological Protection Cooperation Plan, issued in May 2017. In addition,

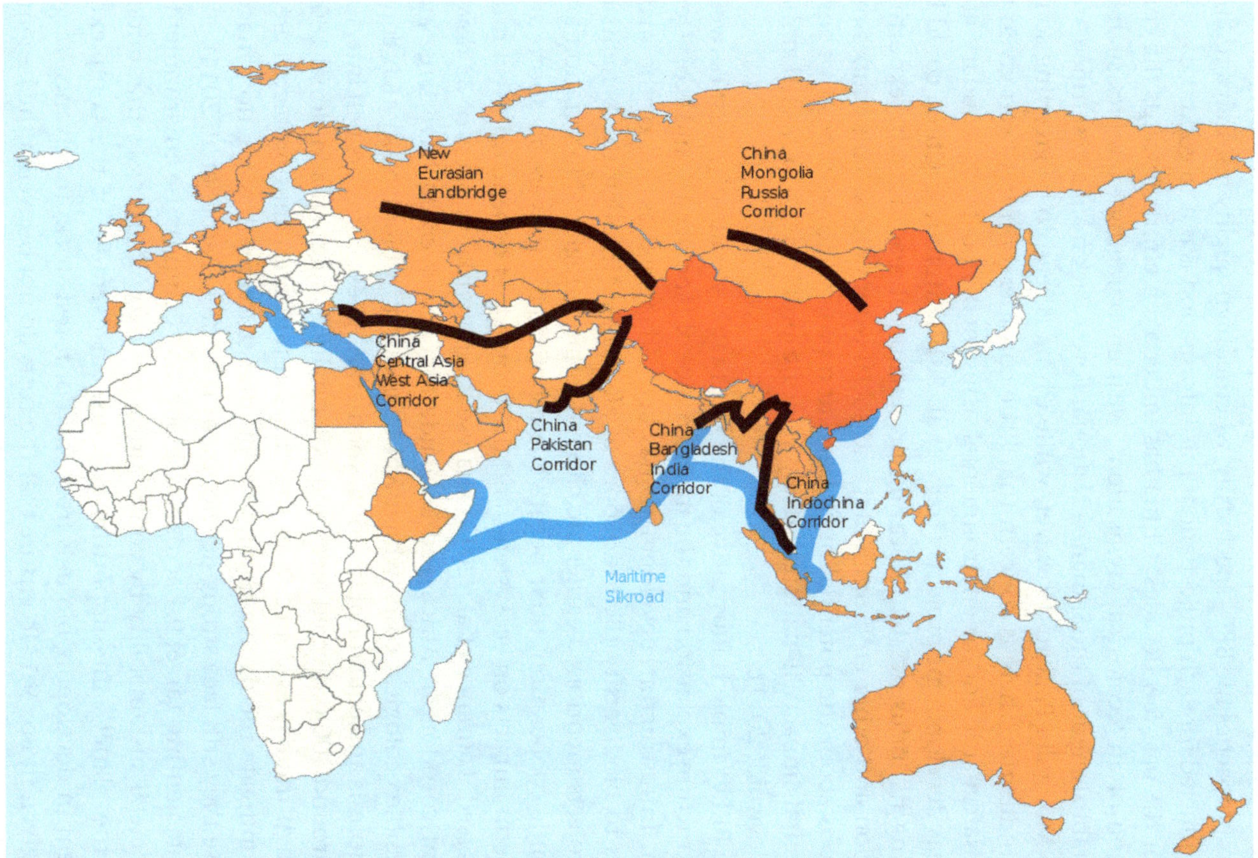

Figure 9.1 China's Belt and Road Initiative. Source: Creative Commons 'One Belt One Road' by Lommes, licensed under CC BY-SA 4.0.

President Xi has also called for creating a 'big data' service platform on environmental protection promising support for countries adapting to climate change (Normile 2017). For the new Silk Road to catalyse a new era of Chinese global resource stewardship and sustainable development depends largely on how China approaches the BRI – specifically, whether high-quality research and EIAs are conducted for each project and if this information is put to good use.

Over the past few decades, China has undertaken efforts towards the construction of an 'ecological civilisation', with encouraging examples such as the establishment of the Saihanba National Forest Park from a desertified area in the Mongolia Plateau. However, while China is seen to be greening some of its industries, there is concern that little consideration has been given to social safeguards and/or environmental assessments on transboundary and overseas development projects (Tracy et al. 2017). Moreover, China could be greening its industries by relocating production abroad, thereby exporting pollution and other environmental and social externalities. This echoes reservations about the potential of conservation projects to yield positive environmental impacts if nations merely relocate problems to others (Lambin and Meyfroidt 2011).

The BRI example is illustrative of the fact that key trade-offs exist between infrastructure expansion and the maintenance of biodiversity and ecosystem health. Infrastructure development has been identified as one of the main threats to biodiversity (Benítez-López et al. 2010). Although the benefits of the projects mentioned earlier are clear in terms of regional integration and economic cooperation, their negative and potentially irreversible short- and long-term impacts on ecosystems and the people that depend on them must be assessed. Laurance et al. (2015) analyse the potential impacts of 33 development corridors in Africa and conclude that many could have large and irreversible ecological costs, which will be greatest in biodiversity-rich equatorial forests and equatorial savanna woodlands. These corridors will intersect with around 400 protected areas and potentially damage an additional 1800. Although there is evidence from the Amazon rainforest that protected areas could mitigate the damaging impacts of infrastructure, they are no panacea because they still face strong development pressures (Barber et al. 2014).

Implementing Target 9.1 across the globe may compromise environmental and societal sustainability by contributing to ongoing processes that undermine the planet's life-supporting systems. An example of the complexity inherent in developing biofuels from food or forest stocks to advance Target 9.1 is given in Box 9.1. To ensure that the costs do not outweigh the benefits, more effective planning is necessary (Laurance and Burgues 2017). If infrastructure is to be sustainable and resilient, it must not harm the ecological services on which the economy and society depend.

Box 9.1 What Role Could the Forest Sector Play in Decarbonising Air Transit?

Indicator 9.1.2 focuses on passenger and freight volumes for different modes of transport. In 2017, the transportation sector accounted for 23 per cent of global energy-related GHG emissions (IEA 2017). To decarbonise, transport must either use green electricity or switch to biofuels. The expanded use of bioethanol and biodiesel will likely continue in nations where substantial production already exists, such as Brazil and the USA. Although there will be an ongoing food-versus-fuel debate as biofuels are increasingly used, groups such as the FAO and the International Energy Agency (IEA) have advocated for a food-and-fuels approach, with diversification of farmers' markets being one of several advantages to this approach (Michalopoulus 2017, Scott-Thomas 2015).

Ongoing research on using forest and agricultural residues to make advanced cellulosic-derived biofuels is likely to increase the volume of available biofuels over the mid- to long-term (IEA 2017). Biojet fuels for aviation illustrate the importance of the dynamics at play. In 2017, 4.1 billion passengers were carried by airlines (ATAG 2018). This is the fastest growing transportation sector globally and its GHG emissions are predicted to increase incrementally. Many airlines, aircraft manufacturers and industry associations have committed to voluntary, aspirational goals to collectively achieve carbon-neutral growth by 2020 and a 50 per cent reduction in GHG emissions by 2050 (relative to 2005 levels) (IRENA 2017). Such significant, longer-term emission reductions will only be achieved if airlines increasingly use renewable and sustainable aviation fuels (IRENA 2017). Unlike ground transportation, where there are alternatives such as electric-powered vehicles, aviation has no other ways to reduce its GHG emissions in the near term (IATA 2018).

Currently, the vast majority of global biojet fuels are derived from lipid feedstocks, such as vegetable oil, animal fats and used cooking oil (IATA 2015); these face a number of supply-side constraints. Advanced technologies using lignocellulosic biomass, such as forest or agricultural residues, have the potential to provide biojet fuel at the scale needed to meet long-term goals (IATA 2015). Theoretically, saw/pulp mill and forest residues could be supplied in a cost-effective and sustainable manner, piggybacking on the supply chains established by the wood-pellet companies and existing forest certification processes to provide a major source of the feedstock biomass to make drop-in biofuels/biojet fuels. To ensure sustainability, current forest certification mechanisms must be updated to incorporate the sustainable removal and use of residues (Larock 2017).

9.2.2 Promoting Industrialisation: Increasing Manufacturing Value Added and Related Employment

Target 9.2 promotes inclusive and sustainable industrialisation, with key indicators related to increasing manufacturing value added as a proportion of GDP and per capita (Indicator 9.2.1) and as a proportion of total employment (Indicator 9.2.2). The goal for LDCs is to double industry's share of GDP by 2030.

Manufacturing has a higher material intensity than the service industry (UNEP 2016). Between 1970 and 2010, global material use tripled, initially growing on average 2.7 per cent annually and accelerating to 3.7 per cent between 2000 and 2010. Per capita material use grew from 6.4 tonnes annually in 1970 to 7.9 tonnes in 2000 and to 10.1 tonnes in 2010. The increase in material intensity experienced in the 2000–2010 period is explained by a shift in manufacturing from more materially efficient economies (e.g. Europe, USA, Japan) towards less efficient ones (e.g. China, India, Brazil) (UNEP 2016). Greater overall material and energy use translates into greater environmental pressures (UNEP 2016), which likely means more pressures on natural forests and already stressed natural ecosystems.

Achieving Target 9.2 sustainably will require businesses, both large and small, to adopt efficient and environmentally benign process all along the value chain, from procuring raw materials to manufacturing goods to transporting finished products. The measures of success must extend well beyond our current preoccupation with measuring CO_2 emissions as a sole indicator of environmental impact (Gaussin et al. 2013). The uptake of a wide range of sustainability indicators for manufacturing, including how socio-economic benefits are distributed along global supply chains, will be essential in achieving this target.

Target 9.2 recognises that value-added manufacturing is one means of potentially achieving these goals. When applied to the context of forests and forest products, the term value added refers to a variety of solid wood products that extend beyond the traditional commodity products – logs, lumber, panel products and pulp and paper – typically manufactured by large, multinational corporations. These include engineered building products, finished building products, joinery, mouldings, millwork, cabinetry, furniture and other appearance products (Gaston and Pahkasalo 2017). The general premise underlying the promotion of value-added products within Target 9.2 is that more value can be derived and more jobs created per volume of wood harvested. Consequently, stakeholders – Indigenous peoples, governments, industry, organised labour, communities, environmental groups – embrace it as a sound conservation-based strategy and a viable alternative to commodity production (Grace et al. 2018, Kozak 2007).

Critics argue that value-added products represent a fairly inconsequential economic sector, perhaps a reflection of value-added producers generally being smaller in scale than lumber, panel and pulp and paper companies (Grace et al. 2018). The value of the global furniture sector alone is approximately USD 420 billion (wood furniture accounts for about one-third), and the growth trajectories for markets are more robust compared to upstream commodity goods (Gaston and Pahkasalo 2017). Interestingly, a sizeable share of value-added production occurs in urban settings and is sold to local markets (Gilani et al. 2018). This is an important result within the SDG 9 context. Increased urbanisation – especially in developing regions – may come with opportunities for small-scale value-added wood producers vis-à-vis increasingly accessible markets, decreasingly complex supply chains, less of a reliance on capital and the use of locally sourced materials.

Value-added products can also refer to the growing basket of bio-economy products, ranging from renewable energy to wood-based chemicals, which are derived from forest fibre and residues. The promise of the bio-economy presents an interesting opportunity for the future of forest producers (Roos and Stendahl 2016, Stern et al. 2018), especially since differentiation and innovation have clearly been shown to lead to higher levels of firm competitiveness within the forest sector (Hansen 2016, Korhonen et al. 2018). Several challenges surrounding this burgeoning sector remain, including questions of economics and long-term viability, requirements for robust policies that promote the substitution of fossil fuels with bio-based alternatives and increased collaboration needs across sectors to achieve success (Guerrero and Hansen 2018, Roos and Stendahl 2016).

9.2.3 Access of Small-Scale Industry to Finance and Market Integration

Target 9.3 focuses on increasing small enterprises' access to markets and financial services. Indicator 9.3.1 centres on increasing the proportion of small industry relative to total industry value, while Indicator 9.3.2 focuses on their access to credit or loans. SMEs are often labelled as the backbone of economies. Globally, they occur in large numbers and employ a significant share of the population, but this is especially the case in emerging economies (Creech et al. 2014). In these countries, most SMEs engage in the trade and manufacture of goods (Scott 2000).

The impacts of non-forestry-based SMEs on the environment and forests is an understudied topic (Nulkar 2014, Scott 2000); one of the few published studies finds mixed results (Scott 2000). In Zimbabwe's brick-making industry, small producers using wood-based fuels contributed to deforestation while large-scale producers using coal as an energy source contributed more CO_2 and SO_2 emissions per unit of output (i.e. number of bricks). Similarly, in Bangladesh's textile industry, small-scale dyers generated more water pollution per unit of output, although large-scale dyers generated greater overall pollution. The study shows that the environmental impacts of SMEs depend on the technologies employed, the types of impacts measured, the specific sector, and national regulations and enforcement capacities.

SMEs are widespread in the forest sector; estimates suggest that they provide about 50 per cent of employment and make up between 80 per cent and 90 per cent of forest-based businesses in the Global South (Macqueen 2008). It has been argued that SMFEs are beneficial to forest-dependent people because they generate local income and promote the sustainable use of forests. Although not always successful, community-based forest businesses have been shown to generate benefits for local communities, such as providing supplementary income, creating local employment, providing greater access to training and capacity-building, improving community infrastructure (e.g. schools, roads) and enhancing community-level governance and empowerment (Schreckenberg and Luttrell 2009, Tomaselli et al. 2014). Small-scale community forestry has had marginally better environmental outcomes in forest cover than other management options or open-access areas, although research is needed to establish more definitive conclusions (Burivalova et al. 2017).

One of the greatest challenges facing SMFEs relates to insufficient access to finance, due partly to high transaction costs and difficulties providing collateral (Kozak 2007, Spantigati and Springfors 2005). Thus, access to financial services, as proposed by Target 9.3, could prove beneficial for some SMFEs and forest-dependent people, especially if those funds are directed towards businesses dedicated to sustainable or regenerative activities creating positive societal externalities. Microfinance can fund more ecologically sensitive activities, such as renewable energy, organic agriculture and climate resilient projects (Allet and Hudon 2013), with green microfinance gaining increasing attention (Huybrechs et al. 2015). It is difficult to predict what impacts the broad promotion of SMFEs and microfinance may have on forests, as it will largely depend on the types of activities that are prioritised by governments and/or financial institutions and their respective ecological footprints. Notably, if microcredit is invested in agricultural expansion, it could have detrimental effects on forests.

9.2.4 Clean and Environmentally Sound Industry

Target 9.4 focuses on increased resource-use efficiency and greater adoption of clean and environmentally sound technologies and industry, with CO_2 emissions per unit of value added as the only indicator. Trends related to greening industries, businesses and the economy have gained traction in recent decades. At the company level, environmental corporate social responsibility (ECSR) has grown as a response to greater environmental awareness and increased expectations from the public (Chuang and Huang 2018). ECSR incorporates social, environmental and financial goals into the company's strategy and often involves practices covering a broad range of activities, including energy efficiency, recycling, certifications and greater stakeholder engagement (Chuang and Huang 2018). An increasing number of multinational firms are generating sustainability reports, although greater disclosure is not necessarily related to better environmental performance (Aragon-Correa et al. 2016).

Prominent, broad trends include notions of a circular economy, the bio-economy and the green economy (D'Amato et al. 2017). The circular economy refers to reducing the material inputs and waste outputs generated in product life cycles, while the bio-economy places more emphasis on the use of renewable biological resources as industrial inputs, with a central role for research and innovation (D'Amato et al. 2017). In comparison, the green economy is a broader, more global narrative that includes social equity as well as environmental sustainability goals and centres on 10 sectors (forestry being one of them) seen as key in the transition to sustainability (UNEP 2011). Despite the differences, they all have in common a trust in technological solutions as the means of change and a belief in the possibilities of green growth (D'Amato et al. 2017).

A central SDG 9 indicator of green industry is carbon intensity (i.e. CO_2 emissions per unit of value added). Many advances have occurred since the 1990s, with most countries reducing their carbon intensities. For instance, 0.47 kg of CO_2 were emitted per unit of GDP in 1990, while carbon intensity fell to 0.35 kg of CO_2 per unit in 2013 (at 2011 constant prices) (Ritchie and Roser 2018). Although carbon efficiency has improved greatly, critics caution that efficiency measures may not reduce emissions in absolute terms due to the rebound effect[2] (Korhonen et al. 2018). Although the global economy's carbon intensity has dropped, total emissions have not; they reached a plateau in 2014, increasing again in 2017. Hence, intensity-based indicators as proposed by SDG 9 may not be effective for tackling climate change or reducing environmental impact if the rebound effect is not taken into account. Efficiency gains should more than offset economic growth, and should ideally be accompanied by adequate policies to reduce consumption.

[2] The rebound effect occurs when increased efficiency lowers the cost of producing a good or service, which in turn increases consumption of this good or service, partially offsetting the beneficial effects of the new technology (Lambin and Meyfroidt 2011).

An underlying and fundamental assumption of SDGs 8 and 9 is that economic growth and environmental sustainability can be made compatible by decoupling environmental impacts from GDP growth. Economic decoupling refers to de-linking environmental degradation and resource consumption from economic growth.[3] The Environmental Kuznets Curve (EKC) is often used as evidence to show that as GDP rises in a country, pollution decreases. However, for indicators other than local air and water pollution (e.g. GHG emissions, biodiversity loss, soil degradation), the evidence for the EKC is not very strong (Raworth 2017). A recent meta-analysis concludes that 'early influential studies favoring EKCs are counterbalanced by recent estimates that do not corroborate the EKCs for deforestation' (Choumert et al. 2013: 26).

As for global material use, data shows that in the past century (1900s–2000), relative decoupling has occurred as material intensity decreased from 3.5 kg/USD in 1900 to 1.2 kg/USD in 2000.[4] However, since the 2000s, material intensity has increased, working 'against the hypothesis of decoupling' (UNEP 2016: 16). For example, while 1.2 kg of materials were needed per USD of GDP in the year 2000, by 2010 intensity had increased to 1.4 kg of materials per USD of GDP (UNEP 2016). Similarly, the World Bank (2017: 48) concludes that for the period 1990–2015, not only have very few countries achieved strong decoupling,[5] 'most countries show weak decoupling or intensified coupling'.[6] Current evidence for absolute decoupling is weak at best: 'there is little indication that any fundamental decoupling of raw economic growth from material use has occurred' (UNEP 2016: 89). Moreover, if the current trajectory of resource use continues (even stabilising resource use in high-income countries), global resource extraction will triple again by 2050 (Fischer-Kowalski and Steinberger 2017).

Material flows tend to increase with industrialisation (UNEP 2016), reflecting some of the ecological costs that achieving SDG 9 may bring to already stressed natural ecosystems. Higher global material use likely means more pressure and competing demands on forests and biodiversity due to extractive activities, such as mining and oil exploration, as well as a greater demand for agricultural products. Moreover, if GHG emissions are not curbed or reduced,

[3] Relative decoupling often relates to declining ecological impact per unit of GDP, while absolute decoupling refers to an absolute decline of ecological impact (Jackson 2011).

[4] Although material intensity decreased, absolute material flows increased 7.3-fold globally, while global GDP (in real terms) increased 19-fold.

[5] Indicators of environmental impact in the World Bank (2017) report include GHGs emissions, the unsustainable harvesting of forests and premature death due to environmental problems.

[6] Intensified coupling means that environmental impact increases even faster than economic growth.

the varied and multifaceted impacts on forests and forest-based communities will worsen (Kirilenko and Sedjo 2007, Nobre et al. 2016). Overall, due to the high global resource use, Fischer-Kowalski and Steinberger (2017: 386) suggest that 'decoupling well-being from biophysical resource use is more achievable than decoupling biophysical resource use from economic activity'. The challenge for high-income nations is even greater since they need to substantially reduce their use of material resources (Fischer-Kowalski and Steinberger 2017). The green industry needs to go beyond GHG emissions and resource efficiency to consider the absolute impact of industries and products within the global economy, possibly using more comprehensive indicators of sustainability, such as the ecological footprint (Wackernagel and Rees 1996) or the material footprint of consumption (UNEP 2016).

9.2.5 Expansion of Information and Communication Technologies

Target 9.C seeks to 'significantly increase access to information and communications technology and strive to provide universal and affordable access to the Internet in least developed countries by 2020' (UN 2017a). The intention here is to increase information availability, economic opportunity and connectivity to the global information society (UN 2017b). To measure progress against this goal, the proportion of a population covered by a mobile network is measured using data provided by the International Telecommunication Union (ITU 2017).

The core function of mobile networks is to transmit data. Data connectivity implies that textual or numerical information can be shared as well as visual information (like pictures or videos) or audio information (such as live voice calls or recordings). The impacts of this data-sharing on forest landscapes, biodiversity and communities can be both positive and negative. Data itself is neutral; how data is used determines impacts. For example, the rapid advancement of digital technologies in the forest sector is profoundly impacting forests and forest-dependent people, potentially improving livelihoods and empowering sustainable management. Mobile networks and information and communication technology (ICT) can work in conjunction to allow forest managers, forest-dependent communities and civil society to more effectively measure and report on forest health and activities in forested areas (Fry 2011). They can also be used to improve the livelihoods of communities by improving their access to information about markets, prices and other economic indicators that enable more equitable economic arrangements (Stienen 2007). Underserved forest-dependent peoples in low-income brackets can also use emerging financial technologies to join the formal economy and benefit from financial credit, easy and secure financial transactions and other banking services (Mbogo 2010).

On the other hand, the same technologies that allow for the monitoring and protection of forest landscapes could enable their exploitation and degradation. Higher quality maps and instant communication enable illegal logging operators and others operating outside forest governance regimes to better coordinate their activities, avoid monitoring and evade law enforcement measures. There are also material impacts from establishing and operating mobile networks. Physical infrastructure is required, typically towers with transmitting receivers at their peak, as well as connectivity to the electrical grid and transportation networks to conduct maintenance and upgrades. Mobile networks are also a significant and growing source of energy consumption globally (Fehske et al. 2011), meaning that further establishment and expansion of mobile networks will lead to increased carbon emissions and climate change adversely impacting forest health (Trumbore et al. 2015). The growth of mobile networks and the environmental impacts are well understood, and efforts are underway to ameliorate these impacts by designing more efficient networks that transmit more data using less energy per unit transmitted (Hilty et al. 2009, Wang et al. 2012).

Two of the most influential and quickly changing digital technologies that have clear applications in a forestry context are distributed web-connected devices (e.g. smart-phones) and remote sensing data. Distributed devices have become exponentially more powerful, interconnected and affordable, opening up opportunities for field data collection by trained professionals and the public alike. Mobile technologies can enable and empower Indigenous communities, citizens and other civil society actors interested in protecting forest landscapes to monitor illegal forest activities or map tenure rights (Swamy et al. 2018). In the Amazon rainforest in Brazil, local communities and civil society have collaborated with Google to develop tools that leverage machine learning technology on mobile devices to detect evidence of illegal logging by monitoring for the sounds emitted by chainsaws (White 2018).

Remote sensing data include passive reflectance data (i.e. imagery) collected from satellites, aircraft, drones or ground cameras as well as active data such as laser scanning (LiDAR) and radar, which can be collected from the sky or the ground. The temporal and spatial resolution of remotely sensed data have improved rapidly and, combined with the proliferation of cost-free imagery, have substantially increased the capacity for forest monitoring over the past decade, especially in less-industrialised countries (Romijn et al. 2015). Deforestation can now be monitored in near real time, and open cloud-based platforms can mitigate the storage and analysis challenges of the massive datasets required for such monitoring (Reiche et al. 2016). For example, in

2018 the FAO announced that it is collaborating with Google to provide free access to satellite data repositories and cloud computing for the 2020 Global Forest Resources Assessment at the national level. Open-source and cloud-based processing can improve the monitoring and management capacity of local governance organisations as well; anyone with a computer and an internet connection can undertake detailed and complex spatial analyses using remotely sensed data, provided they have the necessary competencies.

Given the complexity of mobile networks and ICT, it is no surprise that the pursuit of SDG 9.C is not uniformly positive or negative for forest landscapes, biodiversity or forest-dependent communities. Inequitable access to digital technology can increase the risk of forest degradation, conflict and over-exploitation of the resources upon which forest-dependent people rely (Fisher et al. 2018, Fox et al. 2008, Swamy et al. 2018). However, mobile data collection tools, open-source software and free or low-cost remote sensing data can lead to more equitable control and access to digital technologies. Moreover, recent developments in crowdsourcing – the creation of citizen-generated datasets – can not only increase the quantity of data collected (e.g. for remote sensing applications) at very low cost, but can also provide diverse stakeholder perspectives that may not be well-captured in traditional scientific field campaigns (Schepaschenko et al. 2015). In order to ensure successful uptake, these efforts must be coupled with decentralised training and capacity-building that is accessible to a diverse range of user groups (Fisher et al. 2018).

9.3 Synergies and Trade-offs Between SDG 9 and Other SDGS

Table 9.3 outlines the most prominent synergies and trade-offs, both current and potential, between SDG 9 (mainly Target 9.1, infrastructure expansion) and other SDGs. Some of the most salient synergies occur with SDG 8 (Decent Work and Economic Growth), as infrastructure (especially for transportation) tends to increase trade and thus consumption, which increases economic growth. Indicators 9.1.1 and 9.1.2 have a strong reinforcing effect with SDG 8. Similarly, Target 9.3 (promoting SMEs) could have a positive impact on SDG 8, especially regarding the creation of decent jobs. Another important synergy occurs with SDG 1 (No Poverty), as roads (indicator 9.1.1) tend to increase consumption and reduce income poverty. Likewise, SMEs could play an important role in reducing poverty and supporting the creation of sustainable cities and communities (SDG 11). Another clear synergy occurs between Indicator 9.4.1 (carbon intensity) and SDG 13 (Climate Action).

Table 9.3 Current and potential synergies and trade-offs between SDG 9 (mainly Target 9.1, infrastructure expansion) and other SDGs (based on a framework developed by Nilsson et al. 2016).

	Relationship	Interaction with Other SDGs	Explanation and Evidence
SYNERGIES	INDIVISIBLE	8 – Economic growth	Roads and infrastructure can expand trade, consumption and economic growth (Campbell et al. 2017).
	REINFORCING	1 – Poverty	Roads can increase income of rural populations, thus contributing to poverty-reduction efforts (Dercon et al. 2009).
	ENABLING	2 – Food security	Roads could improve the capacity to feed people as they have a positive relationship with agricultural production (Acheampong et al. 2018, Laurance 2016).
		3 – Good health/well-being	Roads could enable forest-dependent people to more easily access health services (Alamgir et al. 2017).
		4 – Quality education	More or better roads could mean easier access to quality education (Alamgir et al. 2017).
		5 – Gender equality	Access to education could increase with better roads, which could positively affect gender equity, as women might be able to gain better education, resulting in better capacity to defend/define their own rights.
		8 – Economic growth	If adequately supported, SMEs could generate decent jobs for forest-dependent communities and rural inhabitants.
		10 – Reduced inequalities	Inequality could be reduced by generating economic opportunities for rural inhabitants and forest-dependent communities.

TRADE-OFFS	CONSTRAINING	1 – Poverty	Roads could trigger conflict and uncontrolled 'frontier expansion and associated poverty' in areas inhabited by traditional people (Ibisch et al. 2016, supplementary material).
		2 – Food security	Roads could indirectly contribute to climate change (via forest degradation and deforestation), compromising food security over the long term. In remote regions, roads can lead to unsustainable exploitation of wildlife, making bush meat scarce for local residents. Roads may bring access to more food, but not necessarily more nutritious foods (Bucheli et al. 2017).
		3 – Good health/well-being	Ecosystem services that are central to people's health and well-being could be put at risk with roads (e.g. medicinal plants could become scarce with forest degradation/deforestation). Roads may constrain the achievement of Indicator 3.6.1 related to halving deaths in road accidents. Roads facilitate the incursions of human and animal pathogens and disease vectors (Alamgir et al. 2017) and could be at odds with some indicators of Target 3.3 (e.g. reducing HIV, malaria).
		5 – Gender equality	The ability to grow SMEs is important to women, but the benefits depend on the kind of control they can have over their own involvement and its implications for forest sustainability (e.g. are men making the decisions on pricing and location, thus disempowering women producers?).
		6 – Clean water and sanitation	Road expansion could impact water quality via soil erosion and sediments (Laurance and Burgues 2017).

Table 9.3 (cont.)

	Relationship	Interaction with Other SDGs	Explanation and Evidence
		10 – Reduced inequalities	Inequalities could increase for forest-dependent communities and other rural people if the resources upon which they depend are exploited, in the context of unclear tenure rights and disempowerment.
		14 – Life below water	Road building on flood lands or steep terrain could impact water quality and fish breeding sites, causing negative externalities on fisheries (Laurance and Burgues, 2017).
	COUNTERACTING	13 – Climate action	Roads are a 'major proximate driver of habitat loss and fragmentation, wildfires, overhunting and other environmental degradation, often with irreversible impacts on ecosystems' (Laurance et al. 2014: 229). Deforestation contributes a significant proportion of GHGs.
	CANCELLING	15 – Life on land	Roads penetrating into wilderness often have irreversible impacts on ecosystems and are a major proximate cause of fragmentation and habitat and biodiversity loss (Benítez-López et al. 2010, Laurance et al. 2014). Avoiding roads is one of 'the most cost-effective of all conservation strategies' (Alamgir et al. 2017: 1131).

As mentioned earlier, these interactions are highly contextual and are moderated by multiple factors. The interactions among goals can be complex and could play out in conflicting ways. For example, the impact of Target 9.1 on SDG 3 (Good Health and Well-Being) could be mixed. Roads are believed to facilitate 'incursions of human and animal pathogens and disease vectors' (Alamgir et al. 2017: 1135). At the same time, more roads could enable better access to health services for rural populations; however, more roads could simultaneously constrain progress on Indicator 3.6.1 (reducing road injuries) and Target 3.3 (on ending epidemics such as HIV and malaria). Similar potential conflicting pathways in the short and long term have been identified between Indicator 9.1.1 and SDG 2 (Zero Hunger), and even between Indicator 9.4.1 and SDG 13 if the rebound effect is not taken into account.

Important trade-offs include that road and transportation expansion could cancel out the achievements of SDG 15 (Life on Land), especially Indicator 15.1.1 (expanding forest area), Targets 15.2 (halting deforestation), 15.5 (reducing habitat degradation and loss of biodiversity), 15.7 (reducing poaching) and 15.8 (reducing the impact of invasive alien species). As discussed in Section 9.2.1, in the context of tropical and subtropical landscapes, roads are usually inconsistent with the conservation of remaining natural forests (Figure 9.2). With the potentially negative impacts of Target 9.1 on tropical forests (Swamy et al. 2018), keeping wilderness areas road-free is seen by some as the best strategy for their preservation (Barber et al. 2014, Laurance et al. 2014) because 'limiting forest access is the primary deterrent of land clearing' (Barber et al. 2014: 208). SDG 9 (Target 9.1) may also counteract SDG 13 as

Figure 9.2 Impacts of roads on biodiversity. Adapted from: Ibisch et al. (2016) in supplementary material.

tropical deforestation accounts for 25 per cent of GHGs emissions (more than all cars and trucks combined) (Barber et al. 2014). This will likely increase if the impacts of roads on reducing forest cover continues unabated.

9.4 An Alternative to Business as Usual: Exploring Different Socio-Economic Pathways

Given some of the serious and potentially irreversible impacts on forests from some SDG 9 targets and indicators, alternative socio-economic models and new development paradigms could be considered to mitigate some of these effects. Economic growth is increasingly recognised as a major driver of environmental impact, motivating the reassessment of growth's central role in our economies (Pacheco et al. 2018, Ripple et al. 2018). For instance, increased material wealth in industrialised countries is failing to deliver larger gains in well-being and life satisfaction (Jackson 2011); distinguishing and valuing qualitative aspects of well-being, quality of life and prosperity from the quantity of goods and services produced in the economy is imperative. This becomes increasingly relevant and urgent since we may be in a period of uneconomic growth, where the costs of economic expansion may well exceed the benefits (Daly 2013). A new paradigm is needed – one that delivers well-being and basic social standards while respecting the limits of our planet (Raworth 2017).

Various proposals are gaining traction in their attempts to redefine the primary goals of our economic systems and societies. Some of these include sustainable degrowth, the steady-state economy and other post-growth discourses (Raworth 2017, Schneider et al. 2010, Van den Bergh 2017); *buen vivir, sumak kawsay* or *suma qamaña*[7] (Ramirez 2012); the conservation economy (Ripple et al. 2018); and indicators such as the Genuine Progress Indicator, Gross National Happiness and the Happy Planet Index (De Graaf and Batker 2011, Kubiszewski et al. 2013).

Rethinking development and prosperity entails changing the way we measure progress towards forestry goals, which could put less emphasis on increasing production and GDP and greater focus on other indicators, such as decent employment, well-being, sustainability and other forms of wealth (e.g. cultural, social, spiritual, natural) (Tomaselli et al. 2017). Some community-forest operations have successfully incorporated goals and values into their *raison d'etre* that go beyond the profit motive, including the preservation of cultural practices, ecosystem restoration and political empowerment, among others (Hajjar et al. 2013, Trosper 2009). To this end, much could be learned from Indigenous

[7] Indigenous philosophies focused on the good living.

peoples around the world, who have long-held views as stewards of natural forests and opponents to forest conversion (Pokorny and De Jong 2015).

Forestry, as part of the conservation economy, could play an important role in encouraging lowered consumerism and reducing the global ecological footprint. As discussed in Section 9.2.4, more inclusive measures of sustainability are needed, beyond focusing solely on carbon emissions and reducing impacts per unit of value added.

> 'Forest-based businesses could contribute to the goal of one-planet living by refusing planned obsolescence as a built-in characteristic of products, instead prioritizing and guaranteeing the commercialization of long-lasting forest goods ... In addition, focusing on forest-based ecosystem services such as nature-based tourism and recreation, carbon sequestration, and watershed preservation, can, if done mindfully, enhance local economies while limiting material expansion' (Tomaselli et al. 2017: 146–7).

Locally controlled forestry could play an essential role in this transition (Tomaselli et al. 2017). Small- and medium-sized forest operations (including community-based businesses) tend to possess a stronger sense of place and deeper local ecological knowledge, especially if they have inhabited the same place for generations (Rockwell and Kainer 2015). By encouraging local economic activities, wealth could be distributed more locally and regionally (Pokorny and de Jong 2015), generating high-quality employment opportunities and improving rural livelihoods (Macqueen 2008).

Undoubtedly, natural forests will fare better in an economic and political system that more accurately recognises and internalises the value of nature and the innumerable direct and indirect services forests provide to society. Valuing nature should not necessarily be seen as a synonym for 'setting a price' or monetisation, but rather an attempt to better capture, protect and sustainably use the wealth afforded by nature. In this context, the maintenance of ecosystem services, forests and roadless areas could be seen as essential building blocks of a green and sustainable economy (based on strong sustainability concepts) that ensures well-being and healthy lives (see Chapter 3), rather than as stumbling blocks to development.

Capturing the costs or negative externalities of economic activities would also be central to a new economy. For instance, pricing fossil fuels (i.e. carbon pricing) closer to their true societal cost has been proposed as an important step for moving towards a conservation economy (Ripple et al. 2018). Curbing carbon and other GHG emissions will not only reduce the rate of climate change, but will also mitigate the negative impacts that climate change is having on the health of forest ecosystems – for example, the increased severity of forest fires in temperate and boreal forests (Hansen et al. 2013) and the increased risks of destabilising the Amazon rainforest if certain temperature thresholds are surpassed (Nobre et al. 2016). Having said that, if the growth economy and energy demands continue unabated, even green and renewable energy may have severe impacts on forest ecosystems and the people that directly depend on them (as discussed for hydropower in Section 9.2.1).

Other SDG 9 targets may also be well suited within an alternative socio-economic pathway. As mentioned in Section 9.2.5, information and communication technologies are not inherently positive or negative; their impacts depend on how they are employed and implemented. If our economic and political goals are to increase consumption and growth, then technology will most likely be used to achieve this. Indeed, technology has been incredibly effective at facilitating market access and spreading consumer culture worldwide. However, if our goals were to shift from material consumption towards sustainable well-being, then technology would likely play a central role facilitating this transition.

9.5 Conclusions

Our analysis shows that some SDG 9 targets will clearly impact forests in negative, and possibly irreversible, ways (especially Target 9.1), while for others it will depend on how they are employed (e.g. Target 9.C) or implemented (e.g. Target 9.3). SDG 9 does not seriously consider the overall environmental costs of industrialisation and how forests are thereby impacted, with the possible exception of accounting for a reduction of CO_2 emissions per unit of value added. Moreover, the premise of economic decoupling on which SDGs 8 (Recent Work and Economic Growth) and 9 are based is not strongly supported by current empirical evidence. This points to a potentially inherent contradiction between SDGs 8 and 9, on one hand, and SDG 15 (principal focus on the maintenance of forests and biodiversity) and possibly SDG 13 (Climate Action), on the other hand.

If SDG 9 were to seek and support alternative socio-economic models (possibly not based on indefinite economic growth or on ones that rely so heavily on the expansion of infrastructure), the maintenance of forests and ecosystem services would be seen as essential for a green and sustainable economy. Humanity is already exceeding the Earth's sustainable capacity (e.g. ecological footprint, loss of biodiversity, deforestation, climate change), so it is imperative to question what it would mean to continue expanding the consumer culture across the globe. If material consumption is to increase in LDCs and other less-industrialised nations, then should it not be reduced elsewhere to bring the human economy into a sustainable scale? While many nations currently do not satisfy the basic needs of their citizens and many could be seen to under-consume, many other countries over-consume the planet's limited resources and have even been called 'overdeveloped'. A great challenge lies in changing the current economic logic of these latter countries, where greater marginal consumption does not translate into significantly better quality of life. The SDGs do not seem to put any serious focus on this other side of the equation.

In this sense, SDG 9 does not seriously consider limits to the biophysical scale of the economy. This is a key question of sustainability (Daly and Farley 2011) and could be central to the long-term maintenance of natural forests and biodiversity. Moreover, issues of 'sufficiency' as a path to sustainability for industrialised economies are not really addressed by any of the SDGs (not even SDG 12, Responsible Consumption and Production).

Indicators are important because they influence and guide governmental policies, organisational norms and, ultimately, societal actions. 'Indicators arise from values (we measure what we care about), and they create values (we care about what we measure)' (Meadows 1998: 2). Although SDG 9 incorporates concepts such as *resilient, sustainable* and *equitable*, the indicators do not reflect any radical departure from 'business as usual' industrialisation, nor do they fundamentally challenge the economic *status quo*. This is problematic for the sustainability of forests, their biodiversity and the people who depend on them.

References

Acheampong, E. O., Sayer, J. and Macgregor, C. J. 2018. Road improvement enhances smallholder productivity and reduces forest encroachment in Ghana. *Environmental Science and Policy* 85(April):64–71.

Agosin, M. R. and Machado, R. 2005. Foreign investment in developing countries: Does it crowd in domestic investment? *Oxford Development Studies* 33(2):149–62.

Alamgir, M., Campbell, M. J., Soan, S. et al. 2017. Economic, socio-political and environmental risks of road development in the tropics. *Current Biology* 27(20):R1130–40. https://doi.org/10.1016/j.cub.2017.08.067.

Allet, M. and Hudon, M. 2013. Green microfinance: Characteristics of microfinance institutions involved in environmental management. *Journal of Business Ethics* 126(3):395–414.

Aragon-Correa, A., Marcus, A. and Hurtado-Torres, N. 2016. The natural environmental strategies of international firms: Old controversies and new evidence on performance and disclosure. *Academy of Management Perspectives* 30(1):1–16. http://dx.doi.org/10.5465/amp.2014.0043.

ATAG 2018. *Facts and Figures*. Air Transportation Action Group. Available at: www.atag.org/component/factfigures/?Itemid= (Accessed 30 April 2018).

Barber, C. P., Cochrane, M. A., Souza, C. M. and Laurance, W. F. 2014. Roads, deforestation, and the mitigating effect of protected areas in the Amazon. *Biological Conservation* 177:203–9. http://dx.doi.org/10.1016/j.biocon.2014.07.004.

Benchimol, M. and Peres, C. A. 2015. Widespread forest vertebrate extinctions induced by a mega hydroelectric dam in lowland Amazonia. *PLoS ONE* 10(7):1–15.

Benítez-López, A., Alkemade, R. and Verweij, P. A. 2010. The Impacts of roads and other infrastructure on mammal and bird populations: A meta-analysis. *Biological Conservation* 143(6):1307–16. http://dx.doi.org/10.1016/j.biocon.2010.02.009.

Bryceson, D. F., Bradbury, A. and Bradbury, T. 2008. Roads to poverty reduction? Dissecting rural roads' impact on mobility in Africa and Asia. *Development Policy Review* 26(4):1–38.

Bucheli, J. R., Bohara, A. K. and Villa, K. 2017. Paths to development? Rural roads and multidimensional poverty in the hills and plains of Nepal. *Journal of International Development* 30(3):430–56.

Burivalova, Z., Hua, F., Koh, L. P., Garcia, C. and Putz. F. 2017. A critical comparison of conventional, certified, and community management of tropical forests for timber in terms of environmental, economic, and social variables. *Conservation Letters* 10(1):4–14.

Busch, J. and Ferretti-Gallon, K. 2017. What drives deforestation and what stops it? A meta-analysis. *Review of Environmental Economics and Policy* 11(1):3–23.

Campbell, M., Alamgir, M. and Laurance, W. 2017. Roads to ruin. *Australasian Science* 38(2): 40–41.

Choumert, J., Combes Motel, P. and Dakpo, H. K. 2013. Is the environmental Kuznets Curve for deforestation a threatened theory? A meta-analysis of the literature. *Ecological Economics* 90:19–28. http://dx.doi.org/10.1016/j.ecolecon.2013.02.016.

Chuang, S.-P. and Huang, J.-H. 2018. The effects of environmental corporate social responsibility on environmental performance and business competitiveness: The mediation of green information technology capital. *Journal of Business Ethics* 150:991–1009.

Clements, G. R. 2013. *The environmental and social impacts of roads in Southeast Asia* (PhD thesis). James Cook University. Available at: https://researchonline.jcu.edu.au/31888/ (Accessed 30 April 2018).

COSIPLAN 2017. *Project portfolio 2017*. Available at: www.iirsa.org/admin_iirsa_web/Uploads/Documents/CARTERA_DIGITAL_INGLES.pdf (Accessed 17 February 2019).

Costantini, V. and Monni, S. 2008. Environment, human development and economic growth. *Ecological Economics* 64(4):867–80.

Creech, H., Paas, L., Gabriel, H. G., Voora, V., Hybsier, C. and Marquard, H. 2014. Small-scale social-environmental enterprises in the green economy: Supporting grassroots innovation. *Development in Practice* 24(3):366–78. http://dx.doi.org/10.1080/09614524.2014.899561.

D'Amato, D., Droste, N., Allen, B. et al. 2017. Green, circular, bio economy: A comparative analysis of sustainability avenues. *Journal of Cleaner Production* 168:716–34.

Daly, H. 2013. A further critique of growth economics. *Ecological Economics* 88:20–24. Available at: www.sfu.ca/~poitras/Daly_Economic.pdf (Accessed 29 July 2019).

Daly, H. and Farley, J. 2011. *Ecological economics: Principles and applications*. 2nd ed. Washington DC: Island Press.

De Graaf, J. and Batker, D. 2011. *What's the economy for anyway? Why it's time to stop chasing growth and start pursuing happiness*. New York: Bloomsbury Press.

Dercon, S., Gilligan, D., Hoddinott, J. and Woldehanna, T. 2009. The impact of agricultural extension and roads on poverty and consumption growth in fifteen Ethiopia villages. *American Journal of Agricultural Economics* 91(4):1007–21.

Easterlin, R. 2001. Income and happiness: Towards a unified theory. *The Economic Journal* 111(473):465–84.

Fehske, A., Malmodin, J., Biczók, G. and Fettweis, G. 2011. The global footprint of mobile communications – The ecological and economic perspective. *IEEE Communications Magazine* 49(8):55–62.

Finer, M. and Jenkins, C. N. 2012. Proliferation of hydroelectric dams in the Andean Amazon and implications for Andes-Amazon connectivity. *PLoS ONE* 7(4):1–9.

Finer, M., Jenkins, C. N., Pimm, S. L., Keane, B. and Ross, C. 2008. Oil and gas projects in the western Amazon: Threats to wilderness, biodiversity, and Indigenous peoples. *PLoS ONE* 3(8): e2932.

Fischer-Kowalski, M. and Steinberger, J. K. 2017. Growth and sustainability in a material world: The self-reinforcing cycle of population, GDP and resource use. In Victor, P. and Dolter, B. (eds.) *Handbook on Growth and Sustainability*. Cheltenham: Edward Elgar Publishing Limited, pp. 372–93.

Fisher, R. P., Hobgen, S. E., Haleberek, K., Sula, N. and Mandaya, I. 2018. Free satellite imagery and digital elevation model analyses enabling natural resource management in the developing world: Case studies from eastern Indonesia. *Singapore Journal of Tropical Geography* 39(1):45–61.

Fox, J., Suryanata, K., Hershock, P. and Pramono, A. H. 2008. Mapping boundaries, shifting power: The socio-ethical dimensions of participatory mapping. In Goodman, M. K., Boykoff, M. T. and Evere, K. T. (eds.) *Contentious geographies: Environmental knowledge, meaning, scale*. New York: Ashgate Publishing, pp. 203–19.

Fry, B. P. 2011. Community forest monitoring in REDD+: The 'M' in MRV? *Environmental Science and Policy* 14(2):181–7. http://dx.doi.org/10.1016/j.envsci.2010.12.004.

Fyumagwa, R., Gereta, E., Hassan, S. et al. 2013. Roads as a threat to the Serengeti ecosystem. *Conservation Biology* 27(5):1122–5.

Gaston, C. and Pahkasalo, T. 2017. Value-added wood products. In FAO/UNECE. *Forest Products Annual Review 2016–2017*, Geneva, Switzerland: United Nations, pp. 114–25.

Gaussin, M., Hu, G., Abolghasem, S. et al. 2013. Assessing the environmental footprint of manufactured products: A survey of current literature. *International Journal of Production Economics* 146(2):515–23.

Gibson, L., Wilman E. N. and Laurance, W. F. 2017. How green is 'green' energy? *Trends in Ecology and Evolution* 32(12):922–35. http://dx.doi.org/10.1016/j.tree.2017.09.007.

Gilani, H., Kozak, R. and Innes, J. 2018. A change management model for the adoption of chain of custody certification in the British Columbia value-added wood products sector. *Journal of Change Management* 18(3):240–56.

Grace, P., Nelson, H. and Kozak, R. 2018. Understanding SME success in the value-added forest products sector: Insights from British Columbia. *BioProducts Business* 3(9):107–17.

Guadagno, F. 2016. *The determinants of industrialisation in developing countries, 1960–2005*. UNU-MERIT Working Papers. http://doi.wiley.com/10.1111/j.1467-629X.1980.tb00220.x.

Guerrero, J. and Hansen, E. 2018. Cross-sector collaboration in the forest products industry: A review of the literature. *Canadian Journal of Forest Research* 48:1269–78.

Hajjar, R., Kozak, R., El-Lakany, H. and Innes, J. 2013. Community forests for forest communities: integrating community-defined goals and practices in the design of forestry initiatives. *Land Use Policy* 34:158–67.

Hansen, E. 2016. Responding to the bioeconomy: business model innovation in the forest sector. In Kutnar, A. and Muthu, S. (eds.) *Environmental impacts of traditional and innovative forest-based bioproducts*. Singapore: Springer Science + Business Media, pp. 227–48.

Hansen, M. C., Potapov, P. V., Moore, R. et al. 2013. High-resolution global maps of 21st-century forest cover change. *Science* 342(November):850–53.

Hettige, H. 2006. *When do rural roads benefit the poor and how? An In-depth analysis based on case studies*. Manila: Asian Development Bank.

Hilty, L. M., Coroama, V., Osses de Eicker, M. et al. 2009. *The Role of ICT in Energy Consumption and Energy Efficiency*. Technology and Society Lab, EMPA, Swiss Federal Laboratories for Materials Testing and Research, Switzerland.

Huybrechs, F., Bastiaensen, J. and Forcella, D. 2015. Guest editorial: An introduction to the special issue on green microfinance. *Enterprise Development and Microfinance* 26(3):211–14.

IATA 2015. *IATA sustainable aviation fuel roadmap*. International Air Transport Association. Available at: www.iata.org/whatwedo/environment/Documents/safr-1-2015.pdf (Accessed 17 February 2019).

IATA 2018. *Sustainable aviation fuels factsheet*. International Air Transport Association. Available at: www.iata.org/pressroom/facts_figures/fact_sheets/Documents/fact-sheet-alternative-fuels .pdf (Accessed 17 February 2019).

Ibisch, P. L., Hoffmann M. T., Kreft, S. et al. 2016. A global map of roadless areas and their conservation status. *Science* 354(6318):1423–27.

IEA 2017. *Tracking Clean Energy Progress 2017*. International Energy Agency.

IRENA (International Renewable Energy Agency) 2017. *Biofuels for Aviation: Technology Brief*. Abu Dhabi.

ITU 2017. Goal 9. Infrastructure, industrialization, innovation. Available at: www.itu.int/en/sustainable-world/Pages/goal9.aspx (Accessed 17 February 2019).

Jackson, T. 2011. *Prosperity without growth: economics for a finite planet*. London: Earthscan.

Kahneman, D. 2011. *Thinking, fast and slow*. New York: Farrar, Straus and Giroux.

Kirilenko, A. P. and Sedjo, R. A. 2007. Climate change impacts on forestry. *PNAS* 104(50):19697–702. www.pnas.org/content/104/50/19697.abstract.

Korhonen, J., Honkasalo, A. and Seppälä, J. 2018. Circular economy: The concept and its limitations. *Ecological Economics* 143:37–46. https://doi.org/10.1016/j .ecolecon.2017.06.041.

Korhonen, J., Humrmekoski, E., Hansen, E. and Toppinen, A. 2018. Firm-Level Competitiveness in the forest industries: review and research implications in the context of bioeconomy strategies. *Canadian Journal of Forest Research* 48:141–52.

Kozak, R. 2007. *Small and medium forest enterprises: Instruments of change in the developing world.* Washington, DC: Rights and Resources Initiative.

Kozak, R. and Maness, T. 2001. Quality assurance for value-added wood producers in British Columbia. *Forest Products Journal* 51(6):47–55.

Kubiszewski, I., Costanza, R., Franco, C. et al. 2013. Beyond GDP: Measuring and achieving global genuine progress. *Ecological Economics* 93:57–68.

Lambin, E. F. and Meyfroidt, P. 2011. Global land use change, economic globalization, and the looming land scarcity. *Proceedings of the National Academy of Sciences* 108(9):3465–72.

Larock, F. 2017. *The potential of increasing the use of BC forest residues for bioenergy and biofuels* (MSc thesis) University of British Columbia, Vancouver.

Latrubesse, E. M., Arima, E., Dunne, T. et al. 2017. Damming the rivers of the Amazon basin. *Nature* 546(7658):363–69. http://dx.doi.org/10.1038/nature22333.

Laurance, W. F. 2018. Conservation and the global infrastructure tsunami: Disclose, debate, delay! *Trends in Ecology and Evolution* 33(8):568–71.

Laurance, W. F. and Burgues I. 2017. Roads to riches or ruin? *Science* 358(6362):442–4.

Laurance, W. F., Clements, G. R., Sloan, S. et al. 2014. A global strategy for road building. *Nature* 513(7517):229–32. http://dx.doi.org/10.1038/nature13717.

Laurance, W. F. Sloan, S., Weng, L. and Sayer, J. A. 2015. Estimating the environmental costs of Africa's massive 'development corridors'. *Current Biology* 25(24):3202–8. http://dx.doi.org/10.1016/j.cub.2015.10.046.

Macqueen, D. 2008. *Supporting small forest enterprises: A cross-sectoral review of best practice.* London: International Institute for Environment and Development (IIED).

Mbogo, M. 2010. The impact of mobile payments on the success and growth of micro-business: The case of M-Pesa in Kenya. *Journal of Language, Technology & Entrepreneurship in Africa* 2(1). Available at: www.ajol.info/index.php/jolte/article/view/51998 (Accessed 28 July 2019).

Mead, L. 2017. How can progress on infrastructure, industry and innovation contribute to achieving the SDGs? *IISD SDG Knowledge Hub.* Available at: http://sdg.iisd.org/commentary/policy-briefs/how-can-progress-on-infrastructure-industry-and-innovation-contribute-to-achieving-the-sdgs/ (Accessed 20 March 2018).

Meadows, D. 1998. *Indicators and information systems for sustainable development.* Hartland Four Corners: The Sustainability Institute.

Michalopoulus, S. 2017. FAO official: Food-based biofuels not necessarily bad. *EURACTIV.* Available at: www.euractiv.com/section/biofuels/news/fao-official-food-based-biofuels-not-necessarily-bad/ (Accessed 19 September 2018).

Napolitano, D. A. and Ryan, A. 2007. The dilemma of contact: voluntary isolation and the impacts of gas exploitation on health and rights in the Kugapakori Nahua Reserve, Peruvian Amazon. *Environmental Research Letters* 2(4). Available at: https://iopscience.iop.org/article/10.1088/1748-9326/2/4/045005 (Accessed 29 July 2019).

NDRC (National Development and Reform Commission, People's Republic of China) 2015. *Vision and actions on jointly building Silk Road Economic Belt and 21st-Century Maritime Silk Road.* Available at: web.archive.org/web/20170130085453/http://en.ndrc.gov.cn/newsrelease/201503/t20150330_669367.html (Accessed 19 September 2018).

Nilsson, M., Griggs D. and Visback, M. 2016. Map the interactions between Sustainable Development Goals. *Nature* 534(15):320–2.

Nobre, C. A., Sampaio, G., Borma, L. S. et al. M. 2016. Land-use and climate change risks in the Amazon and the need of a novel sustainable development paradigm. *PNAS* 113(39):10759–68.

Normile, D. 2017. China's Belt and Road infrastructure plan also includes science. *Science*. www.sciencemag.org/news/2017/05/china-s-belt-and-road-infrastructure-plan-also-includes-science (Accessed 30 April 2018).

Nulkar, G. 2014. SMEs and environmental performance – A framework for green business strategies. *Procedia – Social and Behavioral Sciences* 133:130–40.

OXFAM 2016. *An Economy for the 1% (Summary)*. 210 OXFAM Briefing Papers. Oxford: Oxfam.

Pacheco, L. F., Altrichter, M., Beck, H., Buchori, D. and Owasu, E. H. 2018. Economic growth as a major cause of environmental crisis: comment to Ripple. *BioScience* 68(4):238.

Pike, L. 2017. Will China's new Silk Road be green? *China dialogue (blog)*. Available at: www.chinadialogue.net/blog/9775-Explainer-Will-China-s-new-Silk-Road-be-green-/en (Accessed 17 February 2019).

Pokorny, B. and de Jong, W. 2015. Special Issue: Smallholders and forest landscape transitions: Locally devised development strategies of the tropical Americas. *International Forestry Review* 17(1):1–19.

Ramirez, R. 2012. *La vida (buena) como riqueza de los pueblos: Hacia una socioecología política del tiempo*. Quito, Ecuador: INEC.

Raworth, K. 2017. *Doughnut economics: Seven ways to think like a 21st century economist*. Vermont: Chelsea Green Publishing.

Reiche, J., Lucas, R., Mitchell, A. L. et al. 2016. Combining satellite data for better tropical forest monitoring. *Nature Climate Change* 6(2):120–2.

Ripple, W. J., Wolf, C., Galetti, M. et al. 2018. The role of scientists' warning in shifting policy from growth to conservation economy. *BioScience* 68(4):239–40.

Ritchie, H. and Roser, M. 2018. CO_2 and other greenhouse gas emissions. *Our world in data*. Available at: https://ourworldindata.org/co2-and-other-greenhouse-gas-emissions (Accessed 15 April 2018).

Rockwell, C. and Kainer, K. 2015. Local and scientific perspectives on the bamboo-dominated forest in Acre, Brazil: A complementary knowledge base for multiple-use forest management. *International Forestry Review* 17:51–64.

Romijn, E., Lantican, C. B., Herlod, M. et al. 2015. Assessing change in national forest monitoring capacities of 99 tropical countries. *Forest Ecology and Management* 352: 109–23.

Roos, A. and Matti, S. 2016. The emerging bio-economy and the forest sector. In Panwar, R., Kozak, R. and Hansen, E. (eds.) *Forests, business and sustainability*. London: Earthscan/Routledge, pp. 179–201.

Schepaschenko, D., See, L., Lesiv, M. et al. 2015. Development of a global hybrid forest mask through the synergy of remote sensing, crowdsourcing and FAO statistics. *Remote Sensing of Environment* 162:208–20.

Schmink, M. 2004. Communities, forests, markets and conservation. In Zarin, D., Putz, F. E., Schmink, M. and Alavalapati, J. (eds.) *Working forests in the tropics: Conservation through sustainable management?* New York: Columbia University Press, pp. 119–29.

Schneider, F., Kallis, G. and Martinez-Alier, J. 2010. Crisis or opportunity? Economic degrowth for social equity and ecological sustainability. Introduction to this special issue. *Journal of Cleaner Production* 18(6):511–8.

Schreckenberg, K. and Luttrel, C. 2009. Participatory forest management: A route to poverty reduction? *International Forestry Review* 11(2):221–38.

Scott, A. 2000. Small-scale enterprises and the environment in developing countries. In Hillary, R. (ed.) *Small and medium sized enterprises and the environment: business imperatives*. Sheffield: Greenleaf Publishing, chapter 22.

Scott-Thomas, C. 2015. *FAO chief urges 'paradigm shift' toward sustainable food*. Available at: www.foodnavigator.com/Article/2015/01/20/FAO-chief-urges-paradigm-shift-toward-sustainable-food (Accessed 18 September 2018).

Spantigati, P. and Springfors, J. 2005. *Microfinance and small-scale forest-based enterprises*. Rome: FAO.

Steffen, W., Richardson, K, Rockström, J. et al. 2015. Planetary boundaries: Guiding human development on a changing planet. *Science* 347(6223):736–47.

Stern, T., Ploll, U., Spies, R. et al. L. 2018. Understanding perceptions of the bioeconomy in Austria – An explorative case study. *Sustainability* 10(4142):1–17.

Stienen, J. 2007. *How ICT Can Make a Difference in Agricultural Livelihoods*. The Hague: International Institute for Communication and Development (IICD).

Sutherland, W. J., Barnard, P., Broad, S. et al. 2018. A 2018 horizon scan of emerging issues for global conservation and biological diversity. *Trends in Ecology and Evolution* 33(1):47–58. http://dx.doi.org/10.1016/j.tree.2017.11.006.

Swamy, L., Drazen, E., Johnson, W. R. and Bukoski, J. J. 2018. The future of tropical forests under the United Nations Sustainable Development Goals. *Journal of Sustainable Forestry* 37(2):221–56.

Szkorupová, Z. 2015. Relationship between foreign direct investment and domestic investment in selected countries of Central and Eastern Europe. *Procedia Economics and Finance* 23:1017–22. http://linkinghub.elsevier.com/retrieve/pii/S2212567115003500.

Tomaselli, M. F., Hajjar, R., Ramon Hidalgo, A. E. and Vasquez Fernandez, A. M. 2017. The problematic old roots of the new green economy narrative: How far can it take us in re-imagining sustainability in forestry? *International Forestry Review* 19(S1):1–13.

Tomaselli, M. F., Kozak, R. A., Hajjar, R. et al. K. 2014. Small forest-based enterprises in The Gambia: Opportunities and challenges. In Katila, P., Galloway, G., de Jong, W. and Pacheco, P. (eds.) *Forests under pressure: Local responses to global issues*. IUFRO World Series no. 32, pp. 315–28.

Tracy, E. F., Shvarts, E., Simonov, E. and Babenko, M. 2017. China's new Eurasian ambitions: The environmental risks of the Silk Road Economic Belt. *Eurasian Geography and Economics* 58(1):56–88. http://dx.doi.org/10.1080/15387216.2017.1295876.

Trosper, R. 2009. *Resilience, reciprocity and ecological economics: Northwest coast sustainability.* London: Routledge.

Trumbore, S., Brando, P. and Hartmann, H. 2015. Forest health and global change. *Science* 349(6250): 814–18.

UN 2017a. Sustainable Development Goal 9. *Sustainable Development Knowledge Platform.* Available at: https://sustainabledevelopment.un.org/sdg9 (Accessed 30 April 2018).

UN 2017b. *The Sustainable Development Goals Report.* Available at: https://unstats.un.org/sdgs/report/2017/overview/ (Accessed 30 April 2018)

UNEP 2011. *Towards a green economy: Pathways to sustainable development and poverty eradication – a synthesis for policymakers.* Available at: http://archive.ipu.org/splz-e/rio+20/rpt-unep.pdf (Accessed 27 July 2019).

UNEP 2016. *Global material flows and resource productivity.* Assessment Report for the UNEP International Resource Panel. Available at: https://wedocs.unep.org/handle/20.500.11822/21557 (Accessed 27 July 2019).

Van den Bergh, J. 2017. Don't worry, be happy. *Alternatives Journal* 43(1):22–24.

Wackernagel, M. and Rees, W. 1996. *Our ecological rootprint: Reducing human impact on the Earth.* Gabriola Island, BC: New Society Publishers.

Wang, X., Vasilakos, A. V., Chen, M., Liu, Y. and Taekyoung Kwon, T. 2012. A survey of green mobile networks: Opportunities and challenges. *Mobile Networks and Applications* 17(1):4–20.

White, T. 2018. *The fight against illegal deforestation with TensorFlow* (blog). Available at: https://blog.google/topics/machine-learning/fight-against-illegal-deforestation-tensorflow/ (Accessed 3 May 2018).

World Bank 2017. *Atlas of Sustainable Development Goals 2017 from World Development Indicators.* doi:10.1596/978-1-4648-1080-0.

Chapter 10 SDG 10: Reduced Inequalities – An Environmental Justice Perspective on Implications for Forests and People

Bimbika Sijapati Basnett*, Rodd Myers and Marlène Elias

Key Points

- SDG10 has the potential to reflect a core commitment within the SDGs – 'leave no one behind' – and to recognise that the dominant economic system exacerbates inequalities within countries through rules that reinforce the current global distribution of wealth.
- In principle, considerable synergies and complementarities exist between SDG 10 targets and principles of environmental justice (distributive, representative and recognition). However, there is a disjuncture between SDG 10 and SDGs on environmental sustainability, which may undermine efforts to promote environmental justice.
- A key gap in SDG 10 is the failure to include trade in spite of the heightened intensification of markets for forest products driving unsustainable forest resource extraction while exacerbating distributive principles of justice between upstream and downstream actors in global production networks.
- For SDG 10 to properly address inequality structures, it must improve distributive, representative and recognition justice for marginalised populations. This would have a positive impact on forest-dependent populations.
- Addressing migration-related indicators in SDG 10 sheds light on the importance of these issues in forestry policy and research and challenges simplistic assumptions informing existing research. Whether this amounts to significant reduction in environmental injustices would, however, depend on what informs the framing of SDG 10 – concerns for making migration work for development or narrow nationalist fears of looming migrant crisis.

* Lead author.

10.1 Introduction

As we embark on the great collective journey, we pledge that no one will be left behind. Recognizing that dignity of the human person is fundamental, we wish to see the goals and targets met for all nations and peoples and all segments of society. (UN General Assembly 2015*: 1)*

SDG 10 calls for reducing inequalities in income as well as those based on age, sex, disability, race, ethnicity, origin, religion or economic or other status within a country. The Goal also addresses inequalities among countries, including those related to representation, migration and development assistance (UN 2018). This is an ambitious goal with many overlapping and distinct targets, as is reflected in Table 10.1. Inequality is understood in this chapter as 'the state of not being equal, especially in status, rights, and opportunities' (UN DESA 2015: 1). The goal recognises the importance of combating economic, social and cultural dimensions of inequalities at the individual, group and societal levels (Kabeer and Santos 2017). Proponents argue that SDG 10 closely reflects one of the core agendas of Agenda 2030 on Sustainable Development – 'leave no one behind' – and shows clear signs of lessons learned from fundamental criticisms levelled against the Millennium Development Goals (MDGs), predecessor to the SDGs (Kabeer 2015, Stuart and Woodroofe 2016, Willis 2016). The MDGs were criticised for focusing narrowly on halving extreme poverty rather than addressing the underlying inequalities that generate poverty in the first place. Hence, SDG 10 is a reaction to growing disparities in income and socio-economic well-being despite overall increases in gross domestic product (GDP), with roots in inclusive-growth approaches (UNDP 2013).

Likewise, the targets and associated indicators pertaining to inequalities among countries recognise that the dominant economic system exacerbates inequalities within countries through rules that reinforce the current global distribution of wealth (IIED 2016). As demonstrated in Table 10.1, SDG 10 focuses on enhancing representation of developing countries in global markets, managing migration and increasing the flow of funds to poor countries through foreign direct investment and official development flows. In this way, addressing inequalities among countries is both an end in itself as well as a means to reducing inequalities within countries. SDG 10 acknowledges the role that migration and remittances can play in furthering the global development agenda by contributing to the reduction of inequalities at the country and cross-country levels. However, critics point out that SDG 10's emphasis on managing migration reflects growing anxieties over the migrant crisis in Europe, North America and Australia rather than leveraging migration

Table 10.1 SDG 10 targets

Inequality within countries	Inequalities among countries
10.1 Progressively achieve and sustain income growth of the bottom 40 per cent at a higher rate than the national average	10.5 Improve the regulation and monitoring of global financial markets and institutions and strengthen the implementation of such regulations
10.2 Empower and promote the social, economic and political inclusion of all irrespective of age, disability, race, ethnicity, origin, religion or economic or other status	10.6 Ensure enhanced representation and voice for developing countries in decision-making in global economic and financial institutions in order to deliver effective, credible, accountable and legitimate institutions
10.3 Ensure equal opportunity and reduce inequalities of outcome, including by eliminating discriminatory laws, policies and practices and promoting appropriate legislation, policies and action in this regard	I0.A Implement the principle of special and differential treatment for developing countries, in particular least-developed countries, in accordance with World Trade Organization agreements
	10.B Encourage official development assistance and financial flows, including foreign direct investment, to states where there is greatest need, in particular least-developed countries, African countries, small island developing countries, and landlocked developing countries
	10.7 Facilitate orderly, safe, regular and responsible migration and mobility of people, including through implementation of planned and well-managed migration policies
10.4 Adopt policies, especially fiscal, wage and social protection policies, and progressively achieve greater equality	10.C Reduce to less than 3 per cent the transaction costs of migrant remittances and eliminate remittance corridors with costs higher than 5 per cent

Adapted from: https://sustainabledevelopment.un.org/sdg10

and development for socially just, environmentally responsive development (Nijenhuis and Leung 2017).

This chapter evaluates the potential effects of addressing SDG 10 on forest outcomes using an 'environmental justice lens' (Forsyth 2014, Forsyth and Sikor 2013, Fraser 1995, 2009; Martin 2017, Sikor and Newell 2014). We focus on three principles of environmental justice, which relate to 'distribution', 'representation' and 'recognition'. This perspective shows considerable synergies and complementarities between SDG 10 and environmental justice goals; it also exposes pitfalls, contradictions and trade-offs.

First, we show that because SDG 10 narrowly translates its aspirational goals into easily measurable indicators, addressing SDG 10 may risk undermining representative, distributive and recognition notions of justice. Second, the disjuncture between SDG 10 and the other SDGs may mean that resource distribution, recognition and representation could be sidelined. Third, a key gap in SDG 10 is the absence of trade despite an intensification of markets for forest products driving forest resource extraction while exacerbating distributive justice between upstream and downstream actors in production networks (Myers and Muhajir 2015, Myers et al. 2017, Myers et al. forthcoming).

We combine an overall evaluation of SDG 10 with a more detailed focus on two clusters of targets and indicators related to inequalities within countries (SDG Targets 10.1–10.4) and migration and remittances (10.7 and 10C), and we draw on existing literature on these topics in forestry. The scholarship concerned with inequalities in forestry has long recognised that there is no direct causal relationship between inequalities and forest outcomes (Agrawal and Benson 2011, Agrawal and Gibson 1999, Agrawal and Ostrom 2001, Johnson 2004, Poteete and Ostrom 2004). Institutions, or formal and informal rules and norms that govern how forests are managed, mediate the effects of inequalities on forests and people. We will show that this literature remains divided between those concerned with inequalities to the extent that they influence environmental outcomes and those who view pre-existing social relations as fundamentally constituting institutions. The latter suggests that for institutions to properly promote environmental justice requires addressing the socio-economic, cultural and symbolic inequalities (re)produced in institutions. From this perspective, if SDG 10 successfully reduces broad-based inequalities, as outlined in Targets 10.1–10.4, this will in turn reduce environmental injustices on marginalised individuals and groups in forests. But this hinges on whether, to what extent and how SDG 10 will amount to such substantive reductions in inequalities.

We will point out that addressing the migration-related targets and indicators in SDG 10 may contribute to elevating the importance of focusing on these issues in the forestry literature too. Most forestry research either ignores the role that migration, mobility and remittances play in current forest transitions or makes far-fetched conclusions about its effects on forests (Hecht et al. 2015). However, it is unclear whether SDG 10, in the way it is framed, opens room for promoting environmental justice for poor and marginalised groups through well-managed migration and remittances.

In what follows, we outline the conceptual framework that informs our analyses (Section 10.2). We then provide an overall evaluation of the gaps and openings in SDG 10 and dive deeper into two clusters of issues – inequalities within countries (Section 10.3) and migration and remittances (Section 10.4). Section 10.5 offers a summary and concluding remarks about potential trade-offs, shortcomings and new openings.

10.2 Principles of Environmental Justice: An Approach to Evaluating SDG 10

Environmental outcomes include forest productivity, availability of natural resources, biological diversity and carbon sequestration (World Bank 2009). Human well-being encompasses economic, social and political dimensions. We focus on the human well-being outcomes related to inequality, which encompass economic, social and political dimensions. An environmental justice perspective is particularly well-suited as it acknowledges the inherent synergies, tensions and trade-offs of equality and environmental goals and the need to find a balance between the two, rather than assume win–win outcomes.

Justice is broadly defined as fairness (Rawls 1999), yet what fairness means is contested (Sen 2009). Environmental justice, as a theory and a practice, has a long, rich history in the Global North, particularly in the USA (Agyeman 2005, Bullard 2005, Cole and Foster 2001). More recently, social movements, international organisations and businesses in the Global South widely use the language of justice to lend credibility to their struggles. Examples include local communities and environmental activists resisting dispossession from customary land, opposing polluting industries and struggling for fair distribution of natural-resource revenues. These also include international donors and governments seeking to promote a rights-based approach, and/or rectify past injustices (Newell 2006, Sikor 2010, Sikor and Stahl 2011, Walker 2009).

Different actors bring different notions of (in)justice in environmental struggles depending on the historical circumstances they inherit and the contemporary political economy they must navigate (Forsyth 2014, Forsyth and

Sikor 2013, Martin 2013). Certain assertions about justice find public support; others are rejected outright as illegitimate concerns (Kumar 2014, Mulvaney 2014, Sikor and Newell 2014). There is often a gulf between justice principles and implementation efforts, with national and local political and economic factors often subverting original intentions (Mehta et al. 2014). The challenge, as Sikor and Newell point out (2014: 153), is to ask 'what kinds of justice are being asserted, by whom and for whom'.

Notwithstanding the plurality of conceptions and practices of justice, as Schlosberg (2004, 2007) points out, everyday struggles and mobilisations around justice are about distribution of environmental goods and bads, recognition of particular group identities and histories, and participation in decision-making. Within this, three notions of justice are apparent: distribution, recognition and representation. These principles build on Fraser's (1995, 2009) understanding of justice.

Distribution notions of justice emerge from socio-economic injustice in the form of exploitation (having the fruits of one's labour appropriated by others); economic marginalisation (being confined to poorly paid work or dearth of income-generating opportunities); and/or deprivation (being denied basic living conditions).The remedies to distributive injustices call for political-economic restructuring, which might involve redistributing income, altering division of labour and/or transforming other economic structures.

Recognition justice aims to remedy symbolic-cultural injustices: revaluing identities against cultural domination, valorising socio-cultural diversity against non-recognition or transforming society against disrespect. Such injustices take many forms, but some are ascribed from birth and hence harder to shed (Fraser 1995, 2009). The remedies may involve cultural and symbolic change such as upwardly revaluing certain identities, recognising and publicly valorising socio-cultural diversity, and/or wholesale transformation of societal representation in ways that would change everyone's sense of self.

Representation justice is about parity of participation in processes: society is fair to the extent that it makes participation possible for all members in institutionalised values and norms, in deliberation processes about distribution rules and in social interactions. These parity principles apply to all spheres of life: from family, market, and informal and formal politics to various civil society associations (Fraser 2009).

All three justice notions are intertwined. Cultural norms biased against marginalised social groups are underpinned by material support from the state or market. For instance, the caste system in Nepal, introduced in the eighteenth century by the ruling elite, resulted in the subjugation of a diverse population, with systematic political favours to some and marginalisation to others (Guneratne 2002, Hoffer 1979, Levine 1987). Meanwhile, economic disadvantage impedes parity of participation in cultural and social life. Continuing with the example from Nepal, a major reason why low-caste members and minorities are under-represented in natural-resource management derives from high opportunity costs to participating meaningfully in such processes (Agarwal 2016, Bennett et al. 2013). While the three justice notions reinforce one another, some distinctions are important as one notion is not a precondition for another. Accepting some notions of justice by powerful actors can mean jeopardising others. For instance, Myers and Mujhair's (2015) research in Indonesia found that while Indigenous peoples living in or adjacent to the national park decry the lack of material benefit from the park, they have resisted the state's offers for material benefits on the grounds that consenting to such offers would constitute acceptance of state authority over their customary land and its continued non-recognition of their rights. A justice understanding of inequality highlights the importance of how (in) equalities are framed and by whom.

We use these three notions of environmental justice to evaluate the effects of SDG 10 on marginalised individuals and social groups in forests. We provide an overall evaluation of SDG 10 and then focus in-depth on two clusters of targets and indicators dealing with inequalities and migration.

10.3 An Overview of Complementarities and Gaps between SDG 10 and Environmental Justice

SDG 10 Targets 10.1–4 are articulated in ways that recognise how a wide range of economic, social and political variables are distributed among individuals, between social groups, and across multiple and intersecting groups (Kabeer and Santos 2017). Target 10.1 recognises income disparities within a country and Target 10.3 recognises opportunity distribution; both share considerable complementarities with distributive notions of justice. Target 10.2 calls for empowering and promoting the social, economic and political inclusion of all (irrespective of age, disability, race, ethnicity, origin, region or economic or other social status), which is compatible with recognition and representation notions of justice. Target 10.4 acknowledges policies to address diverse social, economic and political inequalities in line with all three notions of environmental justice.

Despite this, there is a disjunction between the way SDG 10 targets are articulated and the indicators selected for measuring target progress. While

Target 10.2 aims for social, economic and political inclusion for all 'irrespective of age, sex, disability, race, ethnicity, origin, religion or economic or other status', Indicator 10.2.1 measures progress in economic terms only and in regards to 'age, sex and persons of disabilities'. The indicator is arguably easier to measure because countries likely collect statistics on income, which in turn can be disaggregated by age, gender and perhaps by disability. Social and political inclusion of all is difficult to translate into a measurable indicator. Many countries do not collect adequate data on race, ethnicity and/or religion, especially if these relations are highly politicised and/or a narrow subset is officially recognised (Sijapati Basnett 2018, UN Women 2018).

The extent to which representative and recognition notions of justice can be addressed through SDG 10 will be restricted if countries are only held accountable for progress against the very narrow Indicator 10.2.1. An environmental justice perspective would prompt attention to a broader array of potential reforms related to recognition of who the marginalised are and what sustains their marginalisation; distribution of broader social, economic and political resources beyond income; and parity of participation of marginalised social groups in decision-making.

SDG 10 is disconnected from SDGs pertaining to environmental sustainability (such as SDGs 12–15). The potential tensions, trade-offs and synergies between SDG 10 and these environmental SDGs remains unacknowledged. Such a disjuncture is problematic because environmental justice scholarship is increasingly concerned with the effects environmental solutions spurred by global environmental challenges have on localised struggles (Sikor and Newell 2014). For instance, SDG 15 (Life on Land) – the SDG that deals directly with forests – only alludes to distributive justice once, in Target 15.6, and specifically in the context of access and utilisation of genetic resources. Meanwhile, Target 15.A calls to 'mobilise and significantly increase financial resources from all sources to conserve and sustainably use biodiversity and ecosystems'. In this regard, the environmental justice and political ecology literature (Li 2017, Peluso and Lund 2011, Ribot and Peluso 2003, Schoenberger et al. 2017, White et al. 2012) points to the very unequal distributions of access to, and control over, natural resources that such pursuits generate, and the ways in which they create and exacerbate place-based conflicts related to cultural recognition and political self-determination (Newell 2005, 2007). The ensuing conflicts are then as much about whose notions of justice and framings of environmental problems are privileged as about competition over access to and control over material resources (Fraser 2009, Jasanoff and Martello 2004, Sikor 2013).

10.3.1 'Trade' as a Major Lacunae in SDG 10

Trade is not explicitly mentioned in SDG 10. It is arguably a precursor to SDG 8 (Decent Work and Economic Growth) which only explicitly mentions increasing aid for trade (Target 8.A). SDG 17 (Partnerships for the Goals) includes a set of targets related to trade but only mentions strengthening the World Trade Organization (WTO) trading system (Target 17.10), improving the market share of developing countries (Target 17.11) and reducing trade barriers (Target 17.12). Forest resource extraction is largely about markets and trade. Trade has also been a major driver of inequality among countries since colonial times when the dominance of colonial powers was imposed over resource-rich (natural and human) (Hickel 2017). SDG 10 has a lacuna: the neoliberal global trading system produces inequalities both among and within countries (Hickel 2017). While this was certainly an argument that Marx (1867) made, contemporary scholars continue to make the point of the incompatibility between the global system of trade and equality (see Harvey 2004, Moore 2010, Thurow 1975, Williamson 2000).

Trade of forest products includes timber, non-timber plant species, animal products and carbon (which reinforces the place of conservation in markets). Heightened intensification of markets for forests and forest products in the contemporary era means that a wider range and combination of actors are now involved in regulating global forest trading regimes, each with their own implications for others (Maryudi et al. 2015, Maryudi and Myers 2018, Myers 2015). These include different layers of the state, various private sector interests, global NGOs and organised and spontaneous civil mobilisations. Such processes add pressure on some actors and bestow more control on others. Hence, the globalisation of these markets is fertile ground for understanding inequalities and resulting environmental injustices at global and local levels (Bair 2005).

Global markets also affect access to non-timber forest products, and thereby forest-dependent livelihoods (Belcher and Schreckenberg 2007). Studies show a direct correlation between the value of a product and the extent to which powerful actors exercise control over harvesting, production and trading (Dove 1994, Wollenberg 2001). Markets are a significant driver for local peoples to engage with forests (Ruiz-Pérez et al. 2004). As global market prices increase, so does the involvement of powerful, connected and risk-tolerant actors, which then pushes out less-powerful actors. Examples include gaharu (swiftlet nests) in Southeast Asia (Marcone 2005, Soehartono and Newton 2002), and shea in West Africa (Elias and Arora-Jonsson 2017, Elias and Carney 2007). These processes create and exacerbate distributive injustices especially, and, as we show shortly, can also have consequences for representative and recognition justices.

Anti-deforestation strategies of the 1970s–1990s relied mainly on state- or bilaterally enforced log-export bans and protection of vulnerable species. Now, supranational policies and actions address illegal logging by cutting off the markets for illegally harvested timber. The 2003 EU Forest Law Enforcement, Governance and Trade (FLEGT) Action Plan outlines a system in which (almost) all wood and timber products entering the EU must be proven legal. Efforts to reduce import of illegal timber have also resulted in laws in the USA (US Lacey Act 2008 amendment), Australia (Australian Illegal Logging Prohibition Act 2012) and Japan (Japanese Clean Wood Act 2017). These initiatives are based on a notion of legality that is often rooted in hegemonic notions of forests as a global good – frequently at odds with the interests of local communities (Myers et al. forthcoming).

The EU FLEGT Action Plan and the EU Timber Regulations dictate that traders must demonstrate timber legality through document trails and verification systems. This can have significant implications for small-scale producers (Cerutti and Tacconi 2008, Hajjar 2014, Maryudi and Myers 2018). Maryudi and Myers (2018) show that the increased administration and costs of verifying legality in Indonesia have exacerbated the concentration of power among the bigger and wealthier manufacturers and exporters. Distributive injustices arise when processing is increasingly focused around larger actors, based on the island of Java, with smaller actors, especially on peripheral islands, unable to comply with or make arrangements to otherwise adapt to increasing requirements.

Private certification of sustainable forest products by groups such as the Forest Stewardship Council (FSC) and Programme for the Endorsement of Forest Certification (PEFC) serves as yet another emerging form of forest governance that bypasses the state to implement a non-state solution to the problem of unsustainable forestry practices (Schepers 2009). The FSC, for instance, was a response to the lack of progress after Rio in 1992. The World Wide Fund for Nature (WWF) and a coalition of environmental and social NGOs formed the FSC (Bartley 2007, Cashore et al. 2007, Espach and Ralph 2006). Like FLEGT, FSC represents a relatively new way of governing forests by cutting across local, national and international levels in novel ways to connect the Global North and South. While such private certification enables more direct linkages among actors, it is only available to those who can afford it and who have the required knowledge to navigate the complicated process to comply with standards. In this way, private certification reinforces or creates new forms of distributive injustices.

While the state's role in creating and exacerbating injustices has been well documented (Byrne et al. 2016, Lund 2016, Peluso 1994, Scott 1999), the unprecedented influence of non-state actors is also problematic because concerns of rights and equality are on the periphery of the current policy focus. Furthermore, these recent developments disrupt normative systems of accountability between the state and its citizens without offering alternatives to advance democratic citizenship and parity of participation for marginalised forest-dependent peoples.

The inequalities embedded in globalised trade regimes have led to counterclaims of social and environmental injustices – human rights violations, ecological integrity threats and future capabilities harm (Forsyth and Sikor 2013, Martin 2017, Schlosberg 2013, Sikor 2013, Sikor and Stahl 2011, Walker 2012). These claims of injustice provide the impetus for initiatives to facilitate collective action beyond the nation state (Bernstein and Cashore 2007), such as the SDGs, a number of which touch on forest issues. In response, new markets have emerged for forest resources that aim to pay local communities for the provision of ecosystem services – protecting forests to store carbon – e.g. Reducing Emissions from Forest Degradation and Deforestation (REDD+). Such conservation efforts use distributive notions of justice to compensate local communities for the opportunity costs or the loss of access to customary forestland from which they derive food, medicines and materials.

However, critics point out that such compensatory mechanisms (financial payments and livelihoods training) insufficiently recognise customary rights and community identities tied to forestlands. Furthermore, engaging local and Indigenous peoples to participate in predetermined notions of sustainable forest management is insufficient to satisfy representation claims of injustice when local perspectives have no place in decision-making or defining forest outcomes (Schroeder et al. 2014). While the claims of recognition injustices are far from new, the legitimisation of these claims is historically unprecedented. These calls challenge the dominant conceptions of justice enshrined in tropical forest governance and call into question whose version of forest governance, shaped by trade, is adopted at global and local levels. They push the boundaries of what equality means for different actors. The question then is whether SDG 10's principles of equality, as outlined in the targets and corresponding indicators, are sufficient.

10.4 Inequalities within Countries and Forests – Targets 10.1–10.4

Unlike many topics covered in SDG 10 that have received less attention in forestry/environment scholarship (such as migration and remittances, global financial markets, global institutions), the relationship between inequality within countries (Targets 10.1 and 10.4) and environmental outcomes have long been a matter of debate (Martin 2017). Environmental degradation, including forest loss and degradation, can reinforce socio-economic and cultural inequalities, and inequalities influence environmental and forest management (Islam 2015). In what follows, we draw from two schools of thought – 'commons' and 'entitlement' scholarship (Johnson 2004) – concerned with this relationship. They differ with regard to their normative commitments, interpretation of institutions and conceptual and methodological approaches. We will demonstrate that the 'entitlement' school of thought offers more insights on whether and under what conditions reducing inequalities, as outlined in SDG 10, contributes to environmental justice.

A first school of thought, which Johnson (2004) refers to as 'commons scholars', is concerned with the influence of inequality on the efficiency and environmental health of the commons. It largely consists of modelling individual behaviour based on rational choice theories to explore how institutions can keep users from overexploiting or degrading the commons. Within this literature, authors offer different views of the relationship between socio-economic inequality or heterogeneity, collective action or local governance institutions, and environmental outcomes. Their research is largely informed by theory and case studies, and, in many of the works cited, positive ecological outcomes from collective action and/or well-functioning forest-governance institutions are assumed rather than empirically demonstrated (Andersson and Agrawal 2011).

Within 'commons' scholarship, a large body of research demonstrates a positive relationship between socio-economic equality, largely assessed on the basis of wealth holdings and forest condition (Budhathoki 2004, Trawick 2001). Its authors argue that socio-economic inequality leads to exclusionary decision-making (representative injustice) (Neupane 2003), distrust (Seabright 1993) and an unequal distribution of benefits from commonly managed resources (distributive injustice) (Moore 1993). Lack of social cohesion and disincentives for the rich – who can benefit from private resources – to contribute to collective action can hinder collective action and lead to inefficient resource governance (Chatterton and Chatterton 2001, Corbera et al. 2007, Mukhopadhyay 2004, Smith 2004), perverse resource-use strategies and less ecologically sustainable governance outcomes. One of the few multi-locality, cross-country comparative studies of forest-user groups (n=228) empirically demonstrates that economic inequalities within and across these groups consistently lead to negative forest outcomes (Andersson and Agarwal 2011).[1]

[1] The study draws from data from the International Forestry Resources and Institutions (IFRI) network. The IFRI network (http://ifri.forgov.org/) brings together 14 Collaborating Research Centers (CRCs) worldwide that examine how governance arrangements affect forests and forest-dependent peoples. Using a common methodology, these CRCs have collected data on biodiversity, livelihoods, institutions and forest carbon for more than 250 sites in 15 countries since 1992.

Other commons scholars posit the contrary: that socio-economic inequality (linked to notions of distributive justice) can be positively related to environmental outcomes. This literature builds on Olson's (1965) influential work on privileged groups, which sees group heterogeneity as supporting collective action assuming that rich individuals bring benefits to the rest of the group (Sandler 2015). As the argument goes, there can be high costs to initiating and maintaining collective action, which may more readily be borne by better-resourced individuals than in a decentralised manner among a homogenous group (Hardin 1982). Wealthier individuals may bear these costs either for the sake of a well-functioning institution or for a greater share of benefits from common pool resources (Baland and Platteau 1999). In this view, complementarities amid socio-economically differentiated groups can promote cooperative resource management (Quiggin 1982), and social exclusions can encourage collective action among the disenfranchised (Jodha 1985), leading to more favourable resource management outcomes.

Finally, some commons scholars demonstrate a non-linear, U-shaped relationship between equality, local governance and ecological outcomes (Baker 1998, Dayton-Johnson and Bardhan 2002, Molinas 1998, Zapata et al. 2014). In a comparative study of 104 local peasant cooperative institutions in Paraguay, Molinas (1998) identifies lower levels of cooperation among both the least and the most unequal (based on endowments) organisations. In the low inequality cases, the capacity to bear the costs of collective action was low, whereas resentment and out-migration occurred in highly unequal contexts and impeded cooperation. Similarly, Baker (1998) notes that some inequality can allow certain individuals to carry a bigger share of the organising costs of collective action, but too much inequality can diminish shared interests in the collective good. Zapata et al. (2014) underscore that any such analysis must consider how power is distributed within communities, as well as the interests of the more powerful actors. Using panel data from the Bolivian Amazon, they find a negative correlation between income inequality at the village level and deforestation at the household level, which they claim supports the idea that 'unilateral conservation' can occur when wealthier actors perceive more benefits than costs from environmental conservation, and vice versa.

In one of the few studies explicitly and empirically examining the three-way relationship between heterogeneity, collective action and forest outcomes, Varughese and Ostrom (2001) find that spatial, caste and ethnic, and wealth inequalities – linked to recognition and distributive notions of justice, respectively – among forest-user-group members in Nepal pose challenges but do not in themselves determine the success of collective action. When groups have autonomy to make their own rules, they can overcome stressful heterogeneities through institutional arrangements adapted to their local circumstances. In Nepal, Adhikari and Lovett (2006) echo that the effects of heterogeneity on collective action can vary widely and recommend flexible systems of governance that can allow management regimes to be adapted to local conditions. Poteete and Ostrom (2004), in turn, based on a multi-country review of IFRI studies, note that heterogeneity and the size of user groups affect collective action in contextually specific ways. The authors argue that the importance of group characteristics may depend on other attributes of the resource and resource users, and that interaction effects and contingent relations play a role in explaining divergent findings. Inequality thus interacts with forest market conditions, management rights and rules, for example, to shape forest outcomes.

Influenced by the literature on entitlements (Sen 1981, Leach et al. 1999), 'entitlements scholars' (Johnson 2004) bring a different focus to these analyses. They elucidate that positive forest outcomes will rarely equally benefit all users, including poor and marginalised social groups that rely most on the forest and have the least voice and influence in decision-making processes. Hence, while recognising that environmental outcomes are critical, they argue for local institutions that enable poor and marginalised groups to access and use the commons to their benefit, thereby furthering distributive, representative and recognition notions of justice in their own right (Agarwal 2001, 2002, 2010, Bandiaky 2008, Cleaver 2002, Moore 1993, Peluso 1994, Sikor and Lund 2010).

Taking a historical perspective, entitlement scholars focus on the structures operating across scales (micro to macro) that shape social relations and reinforce (in)equalities in the commons. They underscore the complex, uncertain and dynamic processes and relations underpinning access to and control over resources, environmental management and social and ecological change (Scoones 1999). In this perspective, policies and multi-scalar socio-political processes, as well as historical and path-dependent patterns of resource use, non-place-based relations (e.g. markets) and the presence and power of different state and non-state actors and authorities, contribute to shaping forest outcomes (Coomes et al. 2016, Mearns 1996, Rangan and Kull 2009, Robbins 2004, Zimmerer and Bassett 2003).

This scholarship begins with the recognition that power relations, and thus inequalities, are embedded in formal or bureaucratic institutions (introduced through organisations and legal structures) as well as informal or socially embedded institutions (based on culture, social organisation and everyday practices) that govern common property resources (Cleaver 2002). Because institutional processes of formation, maintenance and change are inherently power laden and asymmetrical, institutional arrangements do not affect nor are they affected by inequalities; rather, they are constituted by them, as inequalities are at the very heart of institutional building and change (Agrawal and Gibson 1999). Ideologically embedded inequalities linked to gender and/or ethnicity or caste (recognition notions of justice), as well as those linked to political participation (representative justice) and access to and distribution of resources (distributive justice), are thus reproduced and reinforced in forest-governance institutions (Agarwal 2002, 2010, Cleaver 2002). These constituent inequalities problematise assumptions that local institutions are rooted in moral economies based on equity, social welfare and security (Li 1996, Mosse 1997).

Studies across regional contexts show that collective resource management institutions and initiatives often reinforce representative injustices, excluding poorer households and groups marginalised on the basis of gender, caste or landlessness (recognition justice), thereby reinforcing existing power disparities and hierarchies (Agarwal 2001, 2010, Astuti and McGregor 2016, Beck and Nesmith 2001, Hébert and Rosen 2007, Taylor 2000). Even when such groups are formally represented in committees of forest-user groups, skewed power relations and their lack of recognition in these spaces typically result in limited participation and influence over decision-making processes and related outcomes (Agarwal 2001). For example, Sunam and McCarthy's (2010) work on community-forest-user groups in Nepal shows that procedures for electing committee members favour better off, male, upper caste residents, and the interests of powerful elites rather than marginalised forest-dependent groups. Still in Nepal, Chaudhary et al. (2018) demonstrate that such exclusions in participation and representation result in an inequitable distribution of benefits from the forest.

Likewise, in Tanzania the framing of the community-based forest-management policy in technical and procedural terms reproduced intra-village inequalities, as it called for local expertise and required literacy and administrative skills to respond to the system's bureaucratic obligations. As only wealthy and powerful villagers were considered 'experts', this village elite dominated local decision-making processes and conservation narratives as well as distributive benefits from forest-related income (Green and Lund 2014; see also Khatun et al. 2015 on elite control in REDD+ Tanzania, and Kopnina 2017 on elite capture of commodified natural resources and forest ecosystem services). In other contexts, women and less-powerful men have been systematically excluded from collective management institutions (Mosse 1997, Taylor 2000), or have participated only nominally, without real voice or influence (Baynes et al. 2015).

Of concern from an environmental justice perspective is not only how inequality affects access to and influence in governance processes (representative justice), but also whether or how processes of institutional formation and change create and sustain resource access for vulnerable groups (distributive justice). For instance, Agarwal (2001, 2010) points out that in India and Nepal, rules of entry in forest-user committees, participation in decision-making and principles of distribution of resources reproduce inequalities on the basis of gender and caste in seemingly participatory community-forestry user groups. In Kenya, the distribution of REDD+ benefits accrued first and disproportionately to larger landowners, with patterns of land ownership reflecting land dispossession processes rooted in colonial times (Chomba et al. 2015). A review of the social impacts of neoliberal conservation underscores that the 'commodification and marketisation of nature creates new rents and incomes for formal or informal appropriation by elites and patron-client networks' (Holmes and Cavanaugh 2016: 205). Li (2018, personal communication) argues that the current forest-tenure reform championed in Indonesia (Siscawati 2017) only favours social groups that can conform to state-sanctioned definitions of indigeneity (recognition justice), which may not easily correspond to complex social realities, thereby creating new forms of contestations and exclusions. Similarly, a comparative assessment of forest-tenure reforms in 30 countries across Asia and Africa shows that new statutory rights do not automatically translate into rights in practice, and that institutional weaknesses and policy distortions thwart environmental justice (Larson et al. 2010). Additionally, the justice that is to be translated through international and national programmes such as REDD+ is typically framed in global narratives that focus on participation (representative justice) or material concerns (e.g. benefit-sharing from REDD+) (distributive justice), overlooking local understandings of justice and the experiences of dispossession of identity (recognition justice) tied to nature and land (Suiseeya 2017).

10.5 Migration and Remittances – Targets 10.7, 10.C and 17.3

Many observers hail the SDGs for explicitly recognising migration for the first time in the global development agenda. By devoting two indicators to migration and remittance out of a total of nine indicators in SDG 10, it is clear that migration and remittance are being framed in the SDGs as a way of addressing inequalities within and between countries. Target 17.3 further views remittances as 'mobilising additional financing resources for developing countries from multiple sources'. The SDG framework, therefore, recognises that migration and remittances potentially contribute to the GDP of poor countries and help reduce income inequalities within and between countries (Appave and Sinha 2017).Target 10.7 emphasises 'planned and well-managed migration policies' (orderly, safe, regulated and responsible migration and mobility), even as the corresponding Indicator (10.7.1) only mentions number of countries that have implemented well-managed migration policies but does not define what this means.

Forestry policies and academic scholarship have been slow to catch up despite the prominence of migration in the global development agenda and the fact that enhanced mobility and the remittances generated by mobility are key elements of current transitions in forested landscapes of developed countries. As Hecht et al. (2015) found through a review of existing literature, much forest-based research assumes that rural households and communities are largely static and spatially bounded, while issues of migration, mobility and remittances lie outside the attention of most natural-resource-governance policymakers. Where forestry and environmental policy do acknowledge migration, it is often seen as disruptive or a sign of livelihood failure. The scholarship that does exist remains premised on simplistic and dichotomous assumptions that view in-migration into forested areas as a cause of tropical deforestation and out-migration as leading to forest regeneration and growth (Carr 2009a, 2009b, Hecht et al. 2012, 2014, Wunder et al. 2014a, 2014b). Each of these potential impacts of migration on forests is possible, but there has been limited research on the many mediating factors that influence these outcomes. These may include a wide variety of social, economic, political and environmental factors playing a role in determining outcomes for forests and peoples, including from an environmental justice perspective. These include time and type of incorporation, feedback loops of economic development and technological change, changing tenure systems, differences in cultural norms and different forms of migration. Instead of these simplistic views, Black et al. (2011) rightly point to the importance of examining the multiple drivers and net effects of migration/environmental change.

The literature on forests and migration in Indonesia is a case in point. Based on an extensive review of literature, Thung and Juniwaty (2018) find that in the 1980s and 1990s, deforestation and forest degradation were often blamed on environmentally destructive migrants, with Secrett's 1986 article in The Ecologist often credited with initiating such a discursive trend. Since the decline of the state-sponsored transmigration programme, through which poor Javanese were resettled in the forest frontiers of Indonesia's remoter islands, attention has mostly shifted towards spontaneous migration (see, for instance, Potter 2012), with many researchers drawing conclusions similar to previous studies and perpetuating negative images about migrants. A number of scholars have also sought to quantify the effects of in-migration on forest cover change through various GIS and statistical analyses. For instance, coupling satellite imagery on forest cover change between 2000 and 2007 with results of the housing and population censuses of 2000 and 2010, Darmawan et al. (2016) found a strong correlation between deforestation and migration in Indonesia. Similar conclusions were drawn from another longitudinal case study in Sulawesi, which concludes that a 1 per cent increase in population due to in-migration leads to a 0.93 per cent increase in forest encroachment for agricultural purposes (Maertens et al. 2002).

In comparison, others who have looked more carefully at the relationship between in-migration and forest change in the Indonesian context point out that correlations do not necessarily imply causation. Even where there is evidence that in-migration has coincided with a corresponding rise in deforestation and forest degradation, the role of a wide range of mediating factors is more important than migration per se. Examples include factors that drive people to migrate to forest frontiers (such as commodity booms, infrastructure projects, government policies) and customary/local institutions that are in place to govern forests (see Thung and Juniwaty 2018 for an extensive review of this literature). Abe's (2006) research on land-use practices among migrant communities in Sumatra's peat swamp forests found that migrant communities for generations have prioritised short-term cultivation of coconut without much regard for long-term sustainability. This is because migrants remain frontier communities, or strangers, because of the difficult biophysical environment of the forest frontier, the cultural and social features of migrants and of receiving communities, and the wider social-economic-political conditions under which migrants operate. For Abe (2006) the wider policy environment is a key determining factor – government policies incentivise migrants to move spontaneously to peat swamp forests but do not address their needs for education or healthcare that would enable them to settle and cultivate over a longer time. Eghenter's (2006) ethnography of new movements of people into the interior of Borneo for large-scale exploitation of particular forest products (gaharu, or aloeswood) also serves as an illustration of the interplay between local institutions and new migration trends in determining environmental outcomes. Eghenter found that the growing movement of outsiders and return migrants in search of quick and lucrative returns from gaharu stretched the ability of the Apo Kayan's customary institutions to physically monitor the arrival of more collectors, and jeopardised internal abilities to develop equitable solutions in the management of gaharu and other forest resources.

Similar studies focusing on effects of other types of migration – out-migration, circular migration and multi-local livelihoods – point to potential synergies and trade-offs between migration and forest outcomes and the role of interacting factors in mediating the relationship between the two (Hecht et al. 2014). Robson and Berkes (2011) found that rural out-migration in Oaxaca,

Mexico, had contributed to extensive forest resurgence, but also a gradual loss of the forest-agriculture mosaic, resulting in a decline in biodiversity. In this instance, forest resurgence did not automatically translate into biodiversity gain. Parry et al.'s (2010) study in the state of Amazonas in Brazil found that the effect of depopulation of remote areas improved forest cover, but this was largely offset by new threats from logging, gold mining and resource extraction. Padoch et al. (2008) showed that in Amazonia most rural communities have established homes in urban centres, and the movement between rural and urban areas is frequent and commonplace. These households maintain their consumption patterns of forest products even when they move to cities, and they continue to play a role in rural forest-use decisions. In this sense, just as the definition of households is getting fuzzy as they stretch between rural and urban areas, so are patterns of land- and forest-use change.

A subset of literature on the links between remittances (financial and social) and forests mirrors the findings outlined above in the wider scholarship on migration and forests. While some attribute remittances to positive change in forests, others argue that remittances in fact contribute to deforestation and/or forest degradation (Hecht et al. 2014). In Latin America, migration from rural areas and receipt of remittances has led, in some cases, to an increase in cash incomes and agricultural retraction that has produced forest resurgence. In comparison, Montefrio et al.'s (2014) research among Filipino oil palm workers in Malaysia found that the flow of ideas on land-use decisions from migrants to their family members back in the Philippines influenced farmers' decisions to engage with the oil palm industry in the migrants' home countries. This resulted in a switch from smallholder agricultural practices to large-scale, monocultural plantations. Still others, such as Gray and Bilsborrow (2014), point out that remittances and out-migration had mixed, countervailing and weak effects on agricultural and forestry activities.

From an environmental justice perspective, however, even the more careful and nuanced studies do not provide much insights on implications of migration and remittances for distributive, representative and recognition justice. One recent exception is Peluso and Purwanto's (2018) research in Java, Indonesia, which explored the effects of remittances sent by poor and landless women (who were previously presiding illegally in government monopolised forestlands in Java) on state–society power dynamics in forest governance. The study finds that an increasing number of these women are migrating to Hong Kong and other prosperous Asian cities to work as maids and domestic labourers; they send remittances back to their husbands, who remain as formal forest labourers. The remittances pay for everyday household expenses and are eventually invested in agrarian resources that generate income. This has led to an unprecedented increase in investment in cows, with a subsequent surge in planting of elephant grass as fodder for the cows in light gaps in the forest understory. Slowly and unintentionally, these women are redressing historical injustices in access to forest resources by making inroads into the control of state monopolised forestlands.

Meanwhile, research on the effects of male out-migration on those left behind in forested landscapes in Nepal is less optimistic and more ambiguous. There are comparatively more studies emanating from Nepal because of the recent surge in large-scale male out-migration for employment purposes in Malaysia and the Gulf countries since the early 2000s (Shrestha 2017), Nepal's reputation as a global leader in innovative participatory environmental governance (Baynes et al. 2015, Fox 2018) and well-established movements for greater social, economic and political justice following the Maoist movement and ensuing civil conflict in the country (Bennett et al. 2013, Sijapati Basnett 2011).

In Nepal, scholars have inquired whether and to what extent male out-migration contributes to environmental justice for marginalised women who are left behind. While some point out that women and marginalised social groups are now able to exercise unprecedented voice and influence in community-wide decision-making processes related to forests (Adhikari and Hobley 2011), others find that such voice is largely offset by a surge in paid and unpaid work burdens that these social groups must now assume. Others rightly recognise that the effects of migration on distributive and representative justice for individuals and social groups are influenced by pre-existing gender and social relations (Sijapati Basnett 2011), a generational divide between migrants and the elderly male population that is left behind (Lama et al. 2017), and the structure and composition of the household (Giri and Darnhofer 2010).

Likewise, in a noteworthy study, Sunam and McCarthy (2015) find that while migration has the potential to lift households and families out of poverty, its effects on distributive justice are mediated by modes of incorporation into migration processes. Those who are better off, with greater social networks, are able to access lucrative migration opportunities, whereas others take up highly risky jobs with limited prospects for upward mobility. Poor and marginalised households incur significant debt in order to pay for migration. Agriculture and forest-based livelihoods subsidise the direct and indirect costs of migration for them. There is limited evidence that migrants invest any surplus income on agriculture and forestry or diversify their livelihoods in a way that lowers their growing dependence on remittance. In this context, the prospects of poor and marginalised migrant workers and their families are limited and highly contingent on the vagaries of the highly unequal market for migrant labour (Fox 2018, Sijapati Basnett and Manandhar 2018).

Other scholars also point out that not all remittances are invested back into forests or into maintaining or enhancing agrarian capital. This is because of 'opportunity costs of other potential applications of funds and diverse interests of households' (Hecht et al. 2015: 16). For instance, in peri-urban landscapes, significant agricultural and forest retraction has occurred, and migrant investment in real estate and housing speculation is widespread. These landscapes of immigration have been documented in Central America and Mexico (Kandel and Cuellar 2012), in the Andes (Bebbington and Batterbury 2001, Rudel 2006) and in Nepal (Sunam and McCarthy 2015). The effects have not only been a reduction in forest cover and/or decline in incentives to manage commonly held forests, but also heightened environmental injustices. Sunam and McCarthy (2015), for instance, point to aggravated inequalities in access to private and public lands, with potential consequences for food security and nutrition for poor and marginalised social groups.

10.6 Summary and Concluding Remarks

This chapter has discussed the potential effects of addressing SDG 10 on forest outcomes from an environmental justice perspective, focusing on two major clusters of issues in SDG 10: inequalities within countries, and migration and remittances. We have defined environmental justice as constituting three notions of justice: distribution, representation and recognition. While distribution notions of justice emerge from socio-economic injustice, recognition relates to symbolic and/or cultural injustice. Representation justice is concerned with parity of participation in institutionalised values and norms, deliberation about rules in distribution and social interactions more broadly. All three notions of justice are intertwined, and yet one is not a precondition for another. Authoritative definitions of environmental justice may clash with and, therefore, have trade-offs with poor and marginalised local peoples' visions of what is just.

We find considerable overlaps among the three notions of justice and SDG 10 targets pertaining to inequalities within countries. However, target aspirations are not fully translated into the corresponding indicators, limiting SDG 10's contribution to environmental justice. Likewise, SDG 10 remains disconnected from SDGs pertaining to environmental sustainability, even when these have bearing on the achievement of SDG 10 and on the promotion of environmental justice.

One of the major gaps within SDG 10 is that trade is not explicitly mentioned, despite mounting concerns about the incompatibility between global systems of trade and goals of equality. A rise in global prices for forest products is often accompanied by unsustainable resource extraction alongside consolidation of power and wealth by a few. Rules governing global trade of forest products are increasingly influenced by international and supranational actors, thereby bypassing the state. Such rules are rooted in hegemonic notions of good governance for the 'global good', which are often at odds with the interests of local communities. Alternatively, complying with such rules is only possible for those who can afford to navigate increasingly complicated and costly systems. The consequence has been that smaller actors are being pushed out while normative systems of accountability between the state and marginalised citizens are being disrupted. Such inequalities in the global trading regime have given rise to unprecedented movements for environmental justice. However, current efforts to address these calls remain preoccupied by the need to compensate forest-dependent communities for loss and/or incentivise their action (distributive). Such terms and ideas foreclose the application of representative and recognition notions of justice for poor and marginalised people in forested landscapes.

Therefore, by omitting trade, SDG 10 excludes a major source of inequality within and between countries. It also misses options for redressing historical injustices levelled against poor marginalised social groups and supporting environmental justice movements that truly represent them.

10.6.1 Targets and Indicators Related to Inequalities within Countries

The literature on forestry, and on common pool resources more broadly, includes a vast and growing body of scholarship examining the linkages between inequalities and forest outcomes, though epistemological and normative commitments within this scholarship vary considerably. While some argue that there is a clear trade-off or synergy between levels of inequalities and forest outcomes, others view a U-shaped relationship between these two variables. Still others point out that pre-existing levels of inequalities can diminish as long as local people have autonomy over formation of rules to govern their resources and that the institutional arrangements established reflect and/or are adaptable to local conditions. Meanwhile, critical scholarship in the field of environmental justice and political ecology, among others, begins with a recognition that power relations, and thus inequalities, are embedded in formal or bureaucratic institutions as well as informal or socially embedded institutions that govern common property resources. Institutional arrangements do not merely affect nor are they merely affected by inequalities; rather, inequalities are at the very heart of institutional building and change. Deeply rooted relations of caste, class, gender and/or ethnicity (recognition justice), as well as those linked to political participation (representative justice) and access to and distribution of resources (distributive justice), are thus reproduced and reinforced in forest-governance institutions.

If SDG 10 successfully reduces broad-based structural inequalities, it is likely to have positive spill-over effects for furthering distributive, representative and recognition justice for forest-dependent populations. However, whether, to what extent and how this will happen remains unclear. Addressing the rising divide between the rich and the poor would require a broad-based political and social agenda, which encompasses reforms in health and education, land tenure, tax and transfer systems, and jobs, to name just a few (see Picketty and Goldhammer 2014, World Bank 2016), while transforming state–society–market relations (see Hall et al. 2011). All these reforms are in fact covered across the SDG framework, but despite pledges to integrate actions across goals (UN General Assembly 2015), each goal operates in silo. And even within SDG 10, there is a disconnect between the aspirations behind the targets and the choice of corresponding indicators to measure progress against their achievement. Such disjunctions within SDG 10 and between SDG 10 and other SDGs limit the extent to which SDG 10 will amount to transformative changes within and between countries.

10.6.2 Targets and Indicators Related to Migration and Remittances

Forestry literature either largely ignores migration and remittances or narrowly frames them as a problem or solution for forests. Even the more nuanced and careful analyses offer limited insights from an environmental justice perspective. The small body of studies that do address these gaps, however, point to an array of mediating variables, such as intra-household dynamics, modes of incorporation into migration, opportunity costs of remittance investment, etc.

In such a context, addressing the migration- and remittance-related targets and indicators in SDG 10 may contribute to greater focus on these issues in the forestry scholarship too. But for SDG 10 to contribute meaningfully towards promoting environmental justice for marginalised individuals and social groups in forested landscapes, policies aimed at 'managing migration' (as outlined in SDG 10) need to consider measures to redress pre-existing injustices in gender and generational lines, offer increased opportunities for poor and socially marginalised groups to migrate and invest their remittances productively, and safeguard access rights of those who are left behind, among other considerations. However, it is unclear whether Target 10.7's focus on planned and well-managed migration policies reflects developmentalist views or narrow nationalist ones. There is a growing consensus among migration and development scholars that 'migration alone cannot independently set in motion broader processes of human and economic development' (de Haas 2012: 14). This literature (see de Haan and Roglay 2002, de Haas 2012, Rigg 2006) implies that migration can produce and exacerbate distributive, representative and recognition notions of justice in developing countries between those who are able to migrate and those who cannot afford to; those who migrate to further enhance their livelihood portfolio versus those for whom migration is low return and survivalist (Razavi 2009); and between failed and successful migrants (Sunam and McCarthy 2015). Critics such as Nijenhuis and Leung (2017: 11) caution that such emphasis on managed migration is a reflection of the growing anxiety in Europe and America about the migration 'crisis' and the inflow of refugees in developed countries, rather than a concern for reducing inequalities and promoting justice within migrant-sending developing countries and between migrant-receiving (developed) and migrant-sending countries. In this regard, whether or not addressing the migration-related targets and indicators will amount to significant reduction in environmental injustices would depend on what informs the framing of SDG 10: concerns for making migration work for development, or narrow nationalist fears of looming migrant crisis.

References

Abe, K. 2006. We come to grow coconuts, but not to stay: Temporary migrations into the peat swamp forests of Sumatra. In de Jong, W., Ken-ichi, A. and Tuck-Po, L. (eds.) *The social ecology of tropical forests: Migration, populations and frontiers*. Melbourne, Australia: Trans Pacific Press, pp. 247–61.

Adhikari, B. and Lovett, J. C. 2006. Transaction costs and community-based natural resource management in Nepal. *Journal of Environmental Management* 78(1):5–15.

Adhikari, J. and Hobley, M. 2011. *Everyone is leaving – Who will sow our fields? The effects of migration from Khotang District to the Gulf and Malaysia*. Kathmandu: Nepal Institute of Development Studies (NIDS).

Agarwal, B. 2001. Participatory exclusions, community forestry and gender: An analysis for South Asia and a conceptual framework. *World Development* 29(10):1623–48.

Agarwal, B. 2002. The hidden side of group behaviour: A gender analysis of community forestry groups. In Heyer, J., Stewart, F. and Thorp, E. (eds.) *Group behaviour and development: Is the market destroying cooperation?* Oxford: Oxford University Press, pp. 185–208.

Agarwal, B. 2010. *Gender and green governance: The political economy of women's presence within and beyond community forestry*. Oxford: Oxford University Press.

Agarwal, B. 2016. *Environmental change and collective action*. Vol. 3 of Gender challenges. Oxford: Oxford University Press.

Agrawal, A. and Benson, C. S. 2011. Common property theory and resource governance institutions: Strengthening explanations of multiple outcomes. *Environmental Conservation* 38(2):199–210.

Agrawal, A. and Gibson, C. G. 1999 enchantment and disenchantment: The role of community in natural resource conservation. *World Development* 27(4):629–49.

Agrawal, A. and Ostrom, E. 2001. Collective action, property rights and decentralisation in resource use in India and Nepal. *Politics & Society* 29(4):485–514.

Agyeman, J. 2005. *Sustainable communities and the challenge of environmental justice*. New York: New York University Press.

Andersson, K. and Agrawal, A. 2011. Inequalities, institutions and forest commons. *Global Environmental Change* 21:866–75.

Appave, G. and Sinha, N. (eds.) 2017. *Migration in the 2030 Agenda*. Geneva: International Organization for Migration. Available at: https://publications.iom.int/system/files/pdf/migration_in_the_2030_agenda.pdf (Accessed 8 May 2018).

Astuti, R. and McGregor, A. 2016. Indigenous land claims or green grabs? Inclusions and exclusions within forest carbon politics in Indonesia. *The Journal of Peasant Studies* 44(2):445–66.

Bair, J. 2005. Global capitalism and commodity chains: Looking back, going forward. *Competition & Change* 9:153–80.

Baker, J. M. 1998. The effect of community structure on social forestry outcomes: Insights from Chota Nagpur, India. *Mountain Research and Development* 18:51–62.

Baland, J. M. and Platteau, J. P. 1999. The ambiguous impact of inequality on local resource management. *World Development* 27(5):773–88.

Bandiaky, S. 2008. Gender inequality in Malidino Biodiversity Community-based Reserve, Senegal: Political parties and the 'village approach'. *Conservation and Society* 6(1):62–73.

Bartley, T. 2007. How foundations shape social movements: The construction of an organizational field and the rise of forest certification. *Social Problems* 54:229–55.

Baynes, J., Herbohn, J., Smith, C., Fisher, R. and Bray, D. 2015. Key factors which influence the success of community forestry in development countries. *Global Environmental Change* 35:226–38.

Bebbington, A. J. and Batterbury, S. P. J. 2001. Transnational livelihoods and landscapes: Political ecologies of globalization. *Cultural Geographies* 8(4):369–80.

Beck, T. and Nesmith, C. 2001. Building on poor people's capacities: The case of common property resources in India and West Africa. *World Development* 29(1):119–33.

Belcher, B. and Schreckenberg, K. 2007. Commercialisation of non-timber forest products: A reality check. *Development Policy Review* 25(3):355–77.

Bennett, L., Sijapati, B. and Thapa, D. 2013. *Gender and social exclusion in Nepal: Update*. Kathmandu, Nepal: Himal Books.

Bernstein, S. and Cashore, B. 2007. Can non-state global governance be legitimate? An analytical framework. *Regulation & Governance* 1:347–71.

Black, R., Adger, W. N., Arnell, N. W. et al. 2011. The effect of environmental change on human migration. *Global Environmental Change* 21:S3–S11.

Brockington, D., Duffy, R. and Igoe, J. 2008. *Nature unbound: Conservation, capitalism and the future of protected areas*. London: Earthscan/James & James.

Budhathoki, P. 2004. Linking communities with conservation in developing countries: Buffer zone management initiatives in Nepal. *Oryx* 38:334–41.

Bullard, R. D. 2005. *The quest for environmental justice: Human rights and the politics of pollution*. Berkeley: University of California Press.

Byrne, S., Nightingale, A. J. and Korf, B. 2016. Making territory: War, post-war and the entangled scales of contested forest governance in mid-western Nepal. *Development and Change* 47(6):1269–93.

Carr, D. 2009a. Rural migration: The driving force behind tropical deforestation on the settlement frontier. *Progress in Human Geography* 33(3):355–78.

Carr, D. 2009b. Population and deforestation: Why rural migration matters. *Progress in Human Geography* 33(3):355–78.

Carr, E. R. 2005. Placing the environment in migration: Environment, economy and power in Ghana's central region. *Environment and Planning A* 37:925–46.

Cashore, B. 2002. Legitimacy and the privatization of environmental governance: How non-state market-driven (NSMD) governance systems gain rule-making authority. *Governance* 15(4):503–29. Available at: https://onlinelibrary.wiley.com/doi/abs/10.1111/1468-0491.00199 (Accessed 27 July 2019)

Cashore, B., Auld, G., Bernstein, S. and McDermott C. 2007. Can non-state governance 'ratchet up' global environmental standards? Lessons from the forest sector. *Review of European Comparative & International Environmental Law* 16(2):158–72.

Cerutti, P. O. and Tacconi, L. 2008. Forests, illegality, and livelihoods: The case of Cameroon. *Society & Natural Resources* 21(9):845–53.

Chatterton, B. and Chatterton, L. 2001. The Australian water market experiment. *Water International* 26(1):62–7.

Chaudhary, S., McGregor, A., Houston, D. and Chettri, N. 2018. Environmental justice and ecosystem services: A disaggregated analysis of community access to forest benefits. *Ecosystem Services* 29:99–115.

Chomba, S., Kariuki, J., Lund, J. F. and Sinclair, F. 2015. Roots of inequity: How the implementation of REDD+ reinforces past injustices. *Land Use Policy* 50:202–13.

Cleaver, F. 2002. Reinventing institutions: Bricolage and the social embeddedness of natural resource management. *European Journal of Development Research* 14(2):11–30.

Cole, L. and Foster, S. 2001. *From the ground up: Environmental racism and the rise of the environmental justice movement*. New York: New York University Press.

Coomes, O. T., Takasaki, Y. and Rhemtulla, J. M. 2016. Forests as landscapes of social inequality: Tropical forest cover and land distribution among shifting cultivators. *Ecology and Society* 21(3):20.

Corbera, E., Kosoy, N. and Martinez Tuna, M. 2007. Equity implications of marketing ecosystem services in protected areas and rural communities: Case studies from Meso-America. *Global Environmental Change* 17:365–80.

Darmawan, R., Klasen, S. and Nuryartono, N. 2016. Migration and deforestation in Indonesia. *EFForTS Discussion Paper Series* No. 19. Goettingen: University of Goettingen.

Dayton-Johnson, J. and Bardhan, P. 2002. Inequality and conservation on the local commons: A theoretical exercise. *The Economic Journal* 112(481):577–602.

De Haan, A. and Rogaly, B. 2002. Introduction: Migrant workers and their role in rural change. *Journal of Development Studies* 38:1–14.

De Haas, H. 2012. The migration and development pendulum: A critical view on research and policy. *International Migration* 50(3):8–25.

Djoudi, H. and Brockhaus, M. 2011. Is adaptation to climate change gender neutral? Lessons from communities dependent on livestock and forests in northern Mali. *International Forestry Review* 13:123–35.

Dove, M. 1994. Marketing the rainforest: 'Green' panacea or red herring? *Asia Pacific Issues: Analysis from the East-West Center* 13.

Eghenter, C. 2006. Movements, social entitlements and economic fortunes in the forests of the interior of Borneo. In de Jong, W., Tuck-Po, L. and Ken-Ichi, A. (eds.) *The social ecology of tropical forests: Migration, population and frontiers*. Kyoto: Kyoto University Press, pp. 228–46.

Elias, M. and Arora-Jonsson, S. 2017. Negotiating across difference: Gendered exclusions and cooperation in the shea value chain. Environment Planning D: *Society and Space* 35(1):107–25.

Elias, M. and Carney, J. 2007. African shea butter: A feminized subsidy from nature. *Africa* 77(1):37–62.

Espach, R. and Ralph, E. 2006. When is sustainable forestry sustainable? The Forest Stewardship Council in Argentina and Brazil. *Global Environmental Politics* 6:55–84.

Forsyth, T. 2014. Climate justice is not just ice. *Geoforum* 54:230–2.

Forsyth, T. J. and Sikor, T. 2013. Forests, development and the globalisation of justice. *Geographical Journal* 179:114–21.

Fox, J. 2018. Community forestry, labor migration and agrarian change in a Nepali village: 1980 to 2010. *Journal of Peasant Studies* 45(3):610–29.

Fraser, N. 1995. From redistribution to recognition? Dilemmas of justice in a 'post-socialist' age. *New Left Review* I/212, July–August.

Fraser, N. 2009. *Scales of justice: Reimagining political space in a globalizing world*. New York: Columbia University Press.

Giri, K. and Darnhofer, I. 2010. Outmigrating men: A window of opportunity for women's participation in community forestry? *Scandinavian Journal of Forest Research* 25(9):55–61.

Gray, C. L. and Bilsborrow, R. E. 2014. Consequences of out-migration for land use in rural Ecuador. *Land Use Policy* 36(1):182–91.

Green, K. E. and Lund, J. F. 2014. The politics of expertise in participatory forestry: A case from Tanzania. *Forest Policy and Economics* 60:27–34.

Guneratne, A. 2002. *Many tongues, one people: The making of Tharu identity in Nepal*. Ithaca: Cornell University Press.

Hajjar, R. 2014. Advancing small-scale forestry under FLEGT and REDD in Ghana. *Forest Policy and Economics* 58:12–20.

Hall, D., Hirsch, P. and Murray Li, T. 2011. *Powers of exclusion: Land dilemmas in Southeast Asia*. Honolulu: University of Hawaii Press.

Hardin, R. 1982. *Collective action*. Baltimore: Johns Hopkins Press.

Harvey, D. 2004. The 'New' Imperialism: Accumulation by Dispossession. *Socialist Register* 40. Available at: www.socialistregister.com/index.php/srv/article/view/5811 (Accessed 24 February 2019).

Hébert, M. and Rosen, M. G. 2007. Community forestry and the paradoxes of citizenship in Mexico: The cases of Oaxaca and Guerrero. *Canadian Journal of Latin American and Caribbean Studies* 32:9–44.

Hecht, S. B., Kandel, S. and Morales, A. 2012. *Migration, livelihoods and natural resources*. San Salvador: IDRC PRISMA.

Hecht, S. B., Morrison, K. D. and Padoch, C. (eds.) 2014. *The social lives of forests: Past, present, and future of woodland resurgence*. Chicago: University of Chicago Press.

Hecht S., Yang A. L., Sijapati Basnett B., Padoch C. and Peluso N. L. 2015. *People in motion, forests in transition: Trends in migration, urbanization and remittances and their effects on tropical forests*. CIFOR Occasional Paper 142. Bogor, Indonesia: CIFOR.

Hickel, J. 2017. *The Divide: A Brief Guide to Global Inequality and its Solutions*. William London: Heinemann.

Hoffer, A. 1979. *The caste hierarchy and the state in Nepal: A study of the Mulki Ain in 1854*. Innsbruck: Universitätsverlag Wagner (Khumbu Himal: Ergebnisse des Forschungsunternehmens Nepal Himalaya).

Holmes, G. and Cavanagh, C. J. 2016. A review of the social impacts of neoliberal conservation: Formations, inequalities, contestations. *Geoforum* 75:199–209.

IIED 2016. *Five considerations for national evaluation agendas informed by SDGs*. Briefing. 2016. Available at: http://pubs.iied.org/17374IIED (Accessed 8 May 2018).

Islam, S. N. 2015. *Inequality and environmental sustainability*. DESA Working Paper No. 145. New York: United Nations Department of Economic and Social Affairs (DESA).

Jasanoff, S. and Martello, M. L. (eds.) 2004. *Earthy Politics: Local and Global in Environmental Governance*. Cambridge, MA: MIT Press.

Jodha, N. S. 1985. Population growth and the decline of common property resources in Rajasthan, India. *Population and Development Review* 11:247–64.

Johnson, C. 2004. Uncommon ground: The 'poverty of history' in common property discourse. *Development and Change* 35(3):407–34.

Kabeer, N. 2015. *Gender equality, the MDGs and the SDGs: Achievements, lessons and concerns*. International Growth Center. Available at: www.theigc.org/blog/gender-equality-the-mdgs-and-the-sdgs-achievements-lessons-and-concerns/ (Accessed 1 May 2018).

Kabeer, N. and Santos, R. 2017. *Intersecting inequalities and the Sustainable Development Goals: Insights from Brazil*. Working Paper 14. International Inequalities Institute, London School of Economics and Political Science. Available at: www.wider.unu.edu/sites/default/files/Publications/Working-paper/PDF/wp2017-167.pdf (Accessed 15 September 2018).

Kandel, S. and Cuellar, N. 2012. Migration Dynamics, rural Livelihoods and Challenges for Territorial Management: Lessons from El Salvador. In Hecht, S. B., Kandel, S. and Morales, A. (eds.) *Migration, Livelihoods and Natural Resources San Salvador*. IDRC, pp. 125–146.

Khatun, K., Gross-Camp, N., Corbera, E., Martin, A. and Glory Massao, S. B. 2015. When participatory forest management makes money: Insights from Tanzania on governance, benefit sharing and implications for REDD+. *Environment and Planning* 47:2097–2112.

Kopnina, H. 2017. Commodification of natural resources and forest ecosystem services: Examining implications for forest protection. *Environmental Conservation* 44(1):24–33.

Kumar, K. 2014. The sacred mountain: Confronting global capital at Niyamgiri. *Geoforum* 54:196–206.

Lama, A. S., Kharel, S. and Ghale, T. 2017. When the men are away: Migration and women's participation in Nepal's Community Forestry. *Mountain Research Development* 37(3):263–70.

Larson, A. M., Barry, D., Dahal, G. R. and Colfer, C. J. P. (eds.) 2010. *Forests for people: Community rights and forest tenure reform*. London: Earthscan.

Leach, M., Mearns, R. and Scoones, I. 1999. Environmental entitlements: Dynamics and institutions in community-based natural resource management. *World Development* 27(2):225–47.

Levine, N. E. 1987. Caste, state and ethnic boundaries in Nepal. *Journal of Asian Studies* 46(1):81–8.

Li, T. M. 1996. Images of community: Discourse and strategy in property relations. *Development and Change* 27(3):501–27.

Li, T. M. 2017. Rendering land investible: Five notes on time. *Geoforum* 82:276–8.

Lund, C. 2016. Rule and rupture: State formation through the production of property and citizenship. *Development and Change* 47(6):1199–228.

Maertens, M., Zeller, M. and Birner, R. 2002. *Explaining agricultural land use in villages surrounding the Lore Lindu National Park in Central Sulawesi, Indonesia*. STORMA Discussion Paper Series No. 4. Bogor, Indonesia: Stability of Rain Forest Margins (STORMA).

Marcone, M. F. 2005. Characterization of the edible bird's nest the 'Caviar of the East'. Food Research International 38:1125–34.

Martin, A. (ed.) 2013. Global environmental in/justice, in practice: Introduction. *Geographical Journal* 179(2):98–104.

Martin, A. 2017. *Just conservation: Biodiversity, wellbeing and sustainability*. London: Routledge.

Marx, K. 1867. *Capital: A Contribution to the Critique of Political Economy*, (ed.) Frederick Engels. Moscow: Progress Publishers.

Maryudi, A. and Myers, R. 2018. Renting legality: How FLEGT is reinforcing power relations in Indonesian furniture production networks. *Geoforum* 97:46–53.

Maryudi, A., Nawir, A. A., Permadi, D. B. et al. 2015. Complex regulatory frameworks governing private smallholder tree plantations in Gunungkidul District, Indonesia. *Foreign Policy and Economics* 59:1–6.

Mearns, R. 1996. Environmental entitlements: Pastoral natural resource management in Mongolia. *Cahiers des Sciences Humaines* 32(1):105–31.

Metha, L., Allouche, J., Nicol, A. and Walnycki, A. 2014. Global environmental justice and the right to water: The case of peri-urban Cochabamba and Delhi. *Geoforum* 54:158–66.

Milanovic, B. 2012. Global inequality: From class to location, from proletarians to migrants. *Global Policy* 3(2):125–34.

Molinas, J. R. 1998. The impact of inequality, gender, external assistance and social capital on local-level cooperation. *World Development* 26:413–31.

Montefrio, M. J. F., Ortiga, Y. Y. and Josol, M. 2014. Inducing development: Social remittances and the expansion of oil palm. *International Migration Review* 48(1):216–42.

Moore, D. S. 1993. Contesting terrain in Zimbabwe's Eastern Highlands: Political ecology, ethnography, and peasant resource struggles. *Economic Geography* 69(4):380–401.

Moore, J. W. 2010. The End of the Road? Agricultural Revolutions in the Capitalist World-ecology, 1450–2010. *Journal of Agrarian Change* 10(3):389–413.

Mosse, D. 1997. The symbolic making of a common property resource: History, ecology and locality in a tank-irrigated landscape in South India. *Development and Change* 28(3):467–504.

Mukhopadhyay, L. 2004. Inequality, differential technology for resource extraction and voluntary collective action in commons. *Ecological Economics* 49(2):215–30.

Mulvaney, D. 2014. Are green jobs just jobs? Cadmium narratives in the life cycle of photovoltaics. *Geoforum* 54:178–86.

Myers, R. 2015. *Access in a Global Rattan Production Network*. Norwich: University of East Anglia.

Myers, R., Intarini, D., Sirait, M. T. and Maryudi, A. 2017. Claiming the forest: Inclusions and exclusions under Indonesia's 'new' forest policies on customary forests. *Land Use Policy* 66:205–13.

Myers, R., McDermott, C. L., Rutt, R. et al. Forthcoming. *Imposing legality: The hegemonic notions of forest governance in the EU Forest Law Enforcement, Governance and Trade (FLEGT).*

Myers, R. and Muhajir, M. 2015. Searching for justice: rights vs 'benefits' in Bukit Baka Bukit Raya National Park, Indonesia. *Conservation & Society* 13:370–81.

Neupane, H. 2003. Contested Impact of community forestry on equity: Some evidences from Nepal. *Journal of Forest and Livelihoods* 2(2):55–62.

Newell, P. 2005. Race, class and the global politics of environmental inequality. *Global Environmental Politics* 5(3):70–93.

Newell, P. 2006. Environmental justice movements: Taking stock, moving forward. *Environmental Politics* 15(4):656–60.

Newell, P. 2007. Trade and environmental justice in Latin America. *New Political Economy* 12(2):237–59.

Nijenhuis, G. and Leung, M. 2017. Rethinking migration in the 2030 Agenda: Towards a de-territorialized conceptualization of development. *Forum for Development Studies* 44(1):51–68.

Okereke, C. and Dooley, K. 2010. Principles of justice in proposals and policy approaches to avoided deforestation: Towards a post-Kyoto climate agreement. *Global Environmental Change* 20:82–95.

Olson, M. 1965. *The logic of collective action: Public goods and the theory of groups*. Cambridge, MA: Harvard University Press.

Padoch, C., Brondizio, E., Costa, S., Pinedo-Vasquez, M., Sears, R. R. and Siqueira, A. 2008. Urban forest and rural cities: Multi-sited households, consumption patterns, and forest resources in Amazonia. *Ecology and Society* 13(2):2.

Parry, L., Day, B., Amaral, S. and Peres, C. A. 2010. Drivers of rural exodus from Amazonian headwaters. *Population and Environment* 32:137–76.

Peluso, N. L. 1994. *Rich forests, poor people: Resource control and resistance in Java*. Berkeley: University of California Press.

Peluso, N. L. and Lund, C. 2011. New frontiers of land control: Introduction. *Journal of Peasant Studies* 38(4):667–81.

Peluso, N. L. and Purwanto, A. B. 2018. The remittance forest: Turning mobile labour into agrarian capital. *Singapore Journal of Tropical Geography* 39(1):6–36.

Piketty, T. and Goldhammer, A. 2014. *Capital in the twenty-first century*. Cambridge, MA: The Belknap Press of Harvard University Press.

Poteete, A. R. and Ostrom, E. 2004. Heterogeneity, group size and collective action: The role of institutions in forest management. *Development and Change* 35(3):435–61.

Potter, L. 2012. New transmigration 'paradigm' in Indonesia: Examples from Kalimantan. *Asia Pacific Viewpoint* 53(3):272–87.

Quiggin, J. 1982. A theory of anticipated utility. *Journal of Economic Behavior & Organization* 3(4):323–43.

Rangan, H. and Kull, C. A. 2009. What makes ecology 'political'? Rethinking 'scale' in political ecology. *Progress in Human Geography* 33(1):28–45.

Ratha, D. 2005. Remittances: A lifeline for development. *Finance and Development* 42(4).

Rawls, J. 1999. *A theory of justice*, rev. ed. Cambridge, MA: Harvard University Press.

Razavi, S. 2009. Engendering the political economy of agrarian change. *Journal of Peasant Studies* 36(1):197–226.

Ribot, J. C. and Peluso, N. L. 2003. A theory of access. *Rural Sociology* 68(2):153–81.

Rigg, J. 2006. Land, farming, livelihoods, and poverty: Rethinking the links in the rural South. *World Development* 34(1):180–202.

Robbins, P. 2004. *Political ecology: A critical introduction*. Oxford: John Wiley & Sons Ltd.

Robson, J. P. and Berkes, F. 2011. Exploring some of the myths of land use change: Can rural to urban migration drive declines in biodiversity? *Global Environmental Change* 21(3):844–54.

Roser, M. 2018. *Global economic inequality*. Available at: https://ourworldindata.org/global-economic-inequality (Accessed 15 November 2018).

Rudel, T. K. 2006. After the labor migrants leave: The search for sustainable development in a sending region of the Ecuadorian Amazon. *World Development* 34(5):838–51.

Ruiz-Pérez, M., Belcher, B., Achdiawan, R. et al. 2004. Markets drive the specialization strategies of forest peoples. *Ecology and Society* 9(2):4.

Sandler, T. 2015. *Collective action: Fifty years later. Public Choice* 164:195–216

Schepers, D. H. 2009. Challenges to legitimacy at the Forest Stewardship Council. *Journal of Business Ethics* 92:279–90.

Schlosberg, D. 2004. Reconceiving environmental justice: Global movements and political theories. *Environmental Politics* 13(3):517–40.

Schlosberg, D. 2007. *Defining environmental justice: Theories, movements, and nature*. Oxford: Oxford University Press.

Schlosberg, D. 2013. Theorising environmental justice: The expanding sphere of a discourse. *Environmental Politics* 22:37–55.

Schoenberger, L., Hall, D. and Vandergeest, P. 2017. What happened when the land grab came to Southeast Asia? *Journal of Peasant Studies* 44(4):697–725.

Schroeder, H. and McDermott, C. 2014. Beyond carbon: Enabling justice and equity in REDD+ across levels of governance. *Ecology and Society* 19:31.

Scoones, I. 1999. New ecology and the social sciences: What prospects for fruitful engagement? *Annual Review of Anthropology* 28:479–507.

Scott, J. C. 1999. *Seeing like a state: How certain schemes to improve the human condition have failed*. New Haven: Yale University Press.

Seabright, P. 1993. Managing local commons: Theoretical issues in incentive design. *Journal of Economic Perspectives* 7(4):113–34.

Secrett. C. 1986. The environmental impact of transmigration. *The Ecologist* 16(2/3):77–88.

Sen, A. 1981. *Poverty and famines*. Oxford: Oxford University Press.

Sen, A. 2009. *The idea of justice*. Cambridge, MA: Harvard University Press.

Shrestha, M. 2017. *The Impact of Large-Scale Migration on Poverty, Expenditures, and Labor Market Outcomes in Nepal*. Policy Research Working Paper No. 8232. Washington, DC: World Bank.

Sijapati Basnett, B. 2011. Linkages between gender, migration and forest governance: Re-thinking community forestry policies in Nepal. *European Bulletin of Himalayan Research* 38:9–34.

Sijapati Basnett, B. 2018. *UN Women's evaluation of gender in the SDGs: What's the role for the CGIAR?* CIFOR Infobrief No.229. Bogor, Indonesia: CIFOR.

Sijapati Basnett, B. and Manandhar, S. 2018. *Effects of large-scale male out-migration on the left behind populations and land: Insights from Nepal*. Paper presented at the CGIAR Annual Gender Conference and Capacity Development Workshop, Addis Ababa, Ethiopia, September 26, 2018.

Sikor, T. 2010. Forest justice: Towards a new agenda for research and practice? *Journal of Integrative Environmental Sciences* 7(4):245–50.

Sikor, T. (ed.) 2013. *The justices and injustices of ecosystem services*. London: Earthscan.

Sikor, T. and Lund, C. 2010. *The politics of possession: Property, authority, and access to natural resources*. Hoboken: Wiley-Blackwell,

Sikor, T. and Newell, P. 2014. Globalizing environmental justice? *Geoforum* 54:151–7.

Sikor, T. and Stahl, J. 2011. Introduction: The rights-based agenda in international forestry. In Sikor, T. and Stahl, J. (eds.) *Forests and people*. London: Earthscan, pp. 1–13.

Siscawati, M., Banjade, M. R., Liswanti, N. et al. 2017. *Overview of forest tenure reforms in Indonesia*. Working Paper No. 223. Bogor, Indonesia: CIFOR.

Smith, L. 2004. The murky waters of the second wave of neoliberalism: Corporatization as a service delivery model in Cape Town. *Geoforum* 35(3):375–93.

Soehartono, T. and Newton, A. C. 2002. The gaharu trade in Indonesia: Is it sustainable? *Economic Botany* 56:271–84.

Stuart, E. and Woodroffe, J. 2016. Leaving no-one behind: Can the Sustainable Development Goals succeed where the Millennium Development Goals lacked? *Gender & Development* 24(1):69–81.

Suiseeya, K. R. M. 2017. Contesting Justice in Global Forest Governance: The Premises and Pitfalls of REDDD+. *Conservation and Society* 15(2):189–200.

Sunam, R. K. and McCarthy, J. F. 2010. Advancing equity in community forestry: Recognition of the poor matters. *International Forestry Review* 12:370–82

Sunam, R. K. and McCarthy, J. F. 2015. Reconsidering the links between poverty, international labor migration, and agrarian change: Critical insights from Nepal. *Journal of Peasant Studies* 43(1):39–63.

Taylor, P. 2000. Producing more with less: Community forestry in Durango, Mexico, in an era of trade liberalisation. *Rural Sociology* 65:253–74.

Thung, P. H. and Juniwaty, K. S. 2018. *Missing links in the forest – migration nexus*. CIFOR Occasional Paper no. 186. Bogor, Indonesia: CIFOR.

Thurow, L. C. 1975. *Generating Inequality: Mechanisms of distribution in the US Economy*. London: Macmillan Press.

Trawick. P. B. 2001. Successfully governing the commons: Principles of social organization in an Andean irrigation system. *Human Ecology* 29(1):1–25

UN DESA (United Nations Department of Economic and Social Affairs) 2015. *Concepts of Inequality*. Development Issues No.1. Department of Economic and Social Affairs. Available at: www.un.org/en/development/desa/policy/wess/wess_dev_issues/dsp_policy_01.pdf (Accessed 13 January 2019).

UNDP (United Nations Development Programme) 2013. *Humanity divided: Confronting inequality in developing countries*. New York: UNDP. Available at: www.undp.org/content/undp/en/home/librarypage/poverty-reduction/humanity-divided–confronting-inequality-in-developing-countries.html (Accessed 27 July 2019).

United Nations General Assembly 2015. *Transforming our world: The 2030 Agenda for Sustainable Development*. Resolution adopted by the General Assembly on 25 September 2015, Available at: www.un.org/ga/search/view_doc.asp?symbol=A/RES/70/1&Lang=E (Accessed 15 November 2018).

UN Women 2018. *Turning promises into action: Gender equality in the 2030 Agenda for Sustainable Development*. New York: UN Women.

Varughese, G. and Ostrom, E. 2001. The contested role of heterogeneity in collective action: Some evidence from community forestry in Nepal. *World Development* 29(5):747–65.

Walker, G. 2009. Globalising environmental justice: The geography and politics of frame contextualization and evolution. *Global Social Policy* 9(3):355–82.

Walker, G. 2012. *Environmental justice: Concepts, evidence and politics*. London: Routledge.

White, B., Borras Jr., S. M., Hall, R., Scoones, I. and Wolford, W. 2012. The new enclosures. Critical perspectives on corporate land deals. *Journal of Peasant Studies* 39(3–4):619–47.

Williamson, J. A. 2000. Globalization and Inequality, Past and Present. In Frieden J. A. and Lake, D. A. (eds.) *International Political Economy*. Boston: Bedford/St. Martin's, pp. 405–16.

Willis, K. 2016. Viewpoint: International development planning and the Sustainable Development Goals (SDGs). *International Development Planning Review (IDPR)* 38(2):105–11.

Wollenberg, E. K. 2001. Incentives for collecting gaharu (fungal-infected wood of Aquilaria SPP., Thymelaeaceae) in East Kalimantan. *Economic Botany* 55:444–56.

World Bank 2009. *Roots for good forest outcomes: An analytical framework for governance reforms*. Washington, DC: World Bank Agriculture and Rural Development Department.

World Bank 2016. *Poverty and shared prosperity 2016: Taking on inequality*. Washington, DC: World Bank. Available at: www.worldbank.org/en/publication/poverty-and-shared-prosperity (Accessed 15 January 2019).

World Bank 2016. *Indonesia's Rising Divide*. World Bank, Jakarta. Available at: https://openknowledge.worldbank.org/handle/10986/24765 (Accessed on 23 February 2019).

Wunder, S., Angelsen, A. and Belcher, B. 2014a. Forests, livelihoods, and conservation: Broadening the empirical base. *World Development* 64:S1–S11.

Wunder, S., Borner, J., Shively, G. and Wyman, M. 2014b. Safety nets, gap filling and forests: A global comparative perspective. *World Development* 64:S29–S42.

Zapata-Ríos, O., Vadez, V., Godoy, R. et al. 2014. *Income inequality and deforestation: Evidence from a small-scale, pre-industrial society in the Bolivian Amazon*. TAPS Working Paper Series, Tsimane' Amazonian Panel Study.

Zimmerer, K. S. and Basset, T. (eds.) 2003. *Political ecology: An integrative approach to geography and environment-development studies*. New York: Guildford Press.

Chapter 11 SDG 11: Sustainable Cities and Communities – Impacts on Forests and Forest-Based Livelihoods

Tahia Devisscher*, Cecil Konijnendijk*, Lorien Nesbitt, Jennifer Lenhart, Fabio Salbitano, Zhaohua Cindy Cheng, Shuaib Lwasa and Matilda van den Bosch

Key Points

- Addressing global urban challenges through the implementation of SDG 11 depends on how cities prioritise resources and strategies over the next decade. This prioritisation is context-specific, relating to socio-economic development trajectories and spatio-temporal urbanisation patterns.
- Implementing SDG 11 will affect forests and forest livelihoods near and far from urban centres. The strategic inclusion of urban and peri-urban forests in city agendas and planning may help manage adverse effects, emphasising the role forests play in fostering productive rural–urban relationships.
- SDG 11 implementation needs to foster people–nature connections in cities to avoid the possible negative consequences for forests and forest-based livelihoods caused by the urbanisation of minds and attitudes.
- Many cities currently prioritise SDG 11 targets focused on basic services such as housing and transport, giving less attention to inclusive access to urban forests, protecting cultural and natural heritage or improving urban–rural linkages.
- SDG 11 shows synergies with all other SDGs, creating opportunities in and around cities. Synergies delivered through sound urban forestry approaches could benefit not only urban dwellers but also forest communities.
- The potential role of urban forests in achieving SDG 11 may be enhanced through the New Urban Agenda and global networks for collective stewardship. Benefits for forests and forest-dependent livelihoods largely depend on multi-scale governance and integrated territorial planning.

* Lead author.

11.1 Introduction

As the world continues to rapidly urbanise, the greatest sustainability challenges of the twenty-first century will likely be urban. With three-quarters of the global population projected to be living in cities by 2050 (UN 2014), we can expect an unprecedented pressure on the living environment, including freshwater resources, soils and vegetation cover, with direct and indirect consequences for social relations, security, energy and public health.

Cities' ecological footprints reach far beyond their physical boundaries. Globally, cities cover circa 3 per cent of the land surface, but account for 60–80 per cent of energy consumption, 70 per cent of carbon emissions and 75 per cent of resource consumption (UNDP 2018). Locally, climates are changed in cities through the urban heat island effect or by altered precipitation patterns (Arnfield 2003). As cities expand across land surfaces, they also impact biodiversity. Many cities are growing in areas adjacent to biodiversity hotspots, including biodiversity-rich coastal zones or forests, constraining conservation efforts (Seto et al. 2013).

Furthermore, the contemporary city often deals with several problems: poverty, social segregation and inequality, vulnerability to extreme weather events, energy inefficiency, poor performance of services and infrastructure, non-optimal waste management, misuse of land and non-renewable resources, air and water pollution and low safety (Moraci et al. 2018). Ambient air pollution is a major environmental hazard for urban residents, accounting for more than four million premature deaths annually (Cohen et al. 2017). According to modelled estimates based on measurements for about 3000 cities and towns worldwide, in 2014 only about 1 in 10 people were breathing clean air, as defined by the World Health Organization safety standards (WHO 2016).

Furthermore, urban expansion in areas of limited economic development and institutional capacity can expose local populations and economies to expanded natural and human-made hazards. In many instances these hazards are exacerbated by climate change, resulting in extreme events such as wildfires, flooding and heat waves affecting cities (Dickson et al. 2012, UN 2015). Key drivers of urban vulnerability to these hazards include: competition for land; environmental degradation; unplanned expansion of urban infrastructure and services; and unequal distribution of wealth, as well as access to urban space, services and security (UNISDR 2015). Communities constrained by lower access to these services and resources show inevitably higher levels of vulnerability.

While cities may be drivers of environmental degradation, they can also offer solutions to humanity's problems (Bettencourt and West 2010).

Addressing this century's urban challenges and many of humanity's problems greatly depends on how cities prioritise resources and urban planning strategies over the next decade. A recent global effort to foster more sustainable and resilient cities was endorsed by 193 countries in 2015 as part of the United Nations 2030 Agenda for Sustainable Development and its Sustainable Development Goal 11 (SDG 11) to 'make cities and human settlements inclusive, safe, resilient and sustainable'. By endorsing a stand-alone goal on cities ('the urban SDG'), the international community recognised urbanisation as a transformative force for development (UN 2017). The targets under SDG 11 (see Table 11.1) provide an opportunity to harness cities' transformational force for innovation and sustainable development, making them protagonists of the 'Future We Want'.[1] This chapter provides an analysis of SDG 11 implementation and explores potential effects on forests and forest-based livelihoods around the world, considering different contexts, synergies and trade-offs from local to global levels.

Table 11.1 SDG 11 targets
11.1: By 2030, ensure access for all to adequate, safe and affordable housing and basic services and upgrade slums
11.2: By 2030, provide access to safe, affordable, accessible and sustainable transport systems for all
11.3: By 2030, enhance inclusive and sustainable urbanisation and capacity for participatory, integrated and sustainable human settlement planning and management
11.4: Strengthen efforts to protect and safeguard the world's cultural and natural heritage
11.5: By 2030, significantly reduce the number of deaths and the number of people affected and substantially decrease the direct economic losses relative to global gross domestic product caused by disasters
11.6: By 2030, reduce the adverse per capita environmental impact of cities, including by paying special attention to air quality and municipal and other waste management
11.7: By 2030, provide universal access to safe, inclusive and accessible, green and public spaces, in particular for women and children, older persons and persons with disabilities

[1] See https://sustainabledevelopment.un.org/futurewewant.html

Table 11.1 (cont.)
11.A: Support positive economic, social and environmental links between urban, peri-urban and rural areas by strengthening national and regional development planning
11.B: By 2020, substantially increase the number of cities and human settlements adopting and implementing integrated policies and plans towards inclusion, resource efficiency, mitigation and adaptation to climate change, resilience to disasters
11.C: Support least developed countries, including through financial and technical assistance, in building sustainable and resilient buildings utilising local materials
Source: UNDP 2018

11.2 SDG 11 Synergies and Trade-offs for Forests and Forest Livelihoods

11.2.1 Critical Assessment of Target Prioritisation: Bias within the Goal

Despite global efforts to monitor SDG 11 implementation, the lack of consistent data collection and common methodological procedures generate inconsistencies, gaps and biases towards certain targets. The bias is linked to the UN's tracking capacity, but also reflects country needs when prioritising SDG 11 targets (Table 11.1). World data collected by the SDG tracker[2] only shows performance for SDG 11 targets on housing and basic services (11.1), disaster-related loss (11.5), air quality and waste (11.6) and regulation to manage disaster risk (11.B). For progress on SDG 11, the UN Statistics Division tracks only the proportion of urban population living in slums, the ratio of urban sprawl, air pollution levels and the proportion of countries implementing national urban policies. If priority continues to be given to SDG 11 targets linked to housing, waste management and transport, SDG 11 may represent a challenge for forests as cities keep expanding and densifying.

Prioritising grey infrastructure and basic services may miss opportunities to benefit from services provided by nature (i.e. natural or modified ecosystems), including urban forests. Urban forests are understood here as networks or systems comprising all woodlands, groups of trees and individual trees located in urban and peri-urban areas (Salbitano et al. 2016). Urban forests are part of green spaces. Green space, public or private, consists of predominantly

[2] See www.sdg-tracker.org

unsealed, permeable, 'soft' surfaces such as soil, grass, shrubs and trees (Swanwick et al. 2003). Ignoring or postponing consideration of green space, urban forests and nature-based solutions (NBS)[3] in urban planning could result in further environmental degradation, with cascading negative effects on human health and well-being.

Bias towards grey infrastructure and basic services and lack of consistency was also observed among Voluntary National Reviews (VNRs) prepared by countries to report on their SDG implementation efforts. Of the 112 VNRs submitted by 2018, there was no uniform reporting method. Some countries included all SDGs, others prioritised specific SDGs based on national interests. The High-Level Political Forum (HLPF) reviews the VNRs to assess progress on Agenda 2030. The HLPF assessed SDG 11 for the first time in July 2018 under the theme 'Transformation towards Sustainable and Resilient Societies'.[4]

In general, countries that included SDG 11 in their VNRs recognised cities as a socio-economic force, but highlighted challenges associated with rapid and/or unplanned urbanisation, segregation of urban dwellers and increasing air pollution. As part of SDG 11 implementation, countries prioritised access to adequate housing with interlinkages to health, education and employment (VNR 2017). Sustainable transport and mobility were frequently mentioned, including access to low-carbon public transport. Few countries highlighted policies to ensure safe, inclusive and accessible, green and public spaces (Target 11.7) or the protection of cultural and natural heritage in and around cities (Target 11.4), despite the manifold benefits urban forests and heritage can bring in terms of ecosystem services, enhancing sense of place, fostering recreational and physical activities, increasing aesthetic appreciation, inspiring artistic and spiritual expression, and generating additional income (FAO 2018). Worldwide, only 13 per cent of the 384 UNESCO World Heritage Sites mention trees, forests, gardens, parks or man–nature relationships in their description or management plans (FAO 2018).

Probably the only SDG 11 targets that promote clear synergies between forests and the social and economic considerations many countries prioritise are those aimed at reducing deaths and economic losses caused by disasters (Target 11.5) and increasing urban resilience to climate-change impacts and disasters (Target 11.B). Countries have achieved most progress in the formulation of policies for climate adaptation and mitigation, disaster-risk reduction and national-level urban policies. For example, 142 countries confirmed

[3] The Commission on Ecosystem Management of the International Union for Conservation of Nature (IUCN: see www.iucn.org/commissions/commission-ecosystem-management/our-work/nature-based-solutions) defines NBS as 'actions to protect, sustainably manage, and restore natural or modified ecosystems, that address societal challenges effectively and adaptively, simultaneously providing human well-being and biodiversity benefits'.

[4] See https://sustainabledevelopment.un.org/hlpf/2018

policy development and/or implementation in one or more of these areas (UN 2016). Synergies between these targets/policies and forests are particularly evident when NBS are considered to manage risk and build resilience of cities to disasters and climate change in a more integrated manner.

Recent global developments may help balance potential trade-offs among SDG 11 targets. In 2012, UN-Habitat created a monitoring tool to measure sustainability at the urban level: the City Prosperity Index (CPI). Tested in more than 400 cities, this index has the potential to become a global framework to monitor SDG 11 indicators and targets beyond national reporting. The CPI includes environmental sustainability as one of the six dimensions it tracks, thus providing the architecture needed to increase the weight of environment and NBS in urban planning.

Another promising initiative for fostering environmental sustainability in and around cities is the New Urban Agenda (NUA) adopted by 167 countries in 2016. The NUA's key principles provide a normative direction for the transformation of cities and their achievement of Agenda 2030. These principles recognise the need to integrate green growth considerations, decouple urban growth from resource use and its environmental impacts, and include environmental strategies in long-term urban planning. Several commitments in the NUA relate to promoting green spaces as drivers of social and economic development: leveraging natural and cultural heritage; emphasising multifunctional areas for social interaction and inclusion to positively affect human health, well-being and cultural expression; and supporting territorial systems that integrate urban and rural functions in more efficient ways.

11.2.2 Building on Synergies to Enhance Forest Opportunities

The CPI and NUA provide new tools and principles to catalyse synergies, not only among SDG 11 targets but also between SDG 11 and other SDGs. The SDG framework is explicit that the goals depend on each other.[5] Nonetheless, many complex interactions are not yet well-understood. Trade-offs exist that need to be minimised, while synergies should be fostered for the 2030 Agenda to deliver on its full potential (Nilsson et al. 2016). Some identified problems include: policymakers and planners often operate in silos with limited budgets; different public and private entities with competing priorities manage the sectors; and evidence is lacking on interventions that may help or hinder SDG integration (Nilsson et al. 2016, Weitz et al. 2017). This section briefly presents how some strategic interactions between SDG 11 and other SDGs could be fostered, focusing on SDG 11 targets with potential impacts on forests (Figure 11.1).

[5] See declaration of the Agenda 2030: https://sustainabledevelopment.un.org/post2015/transformingourworld

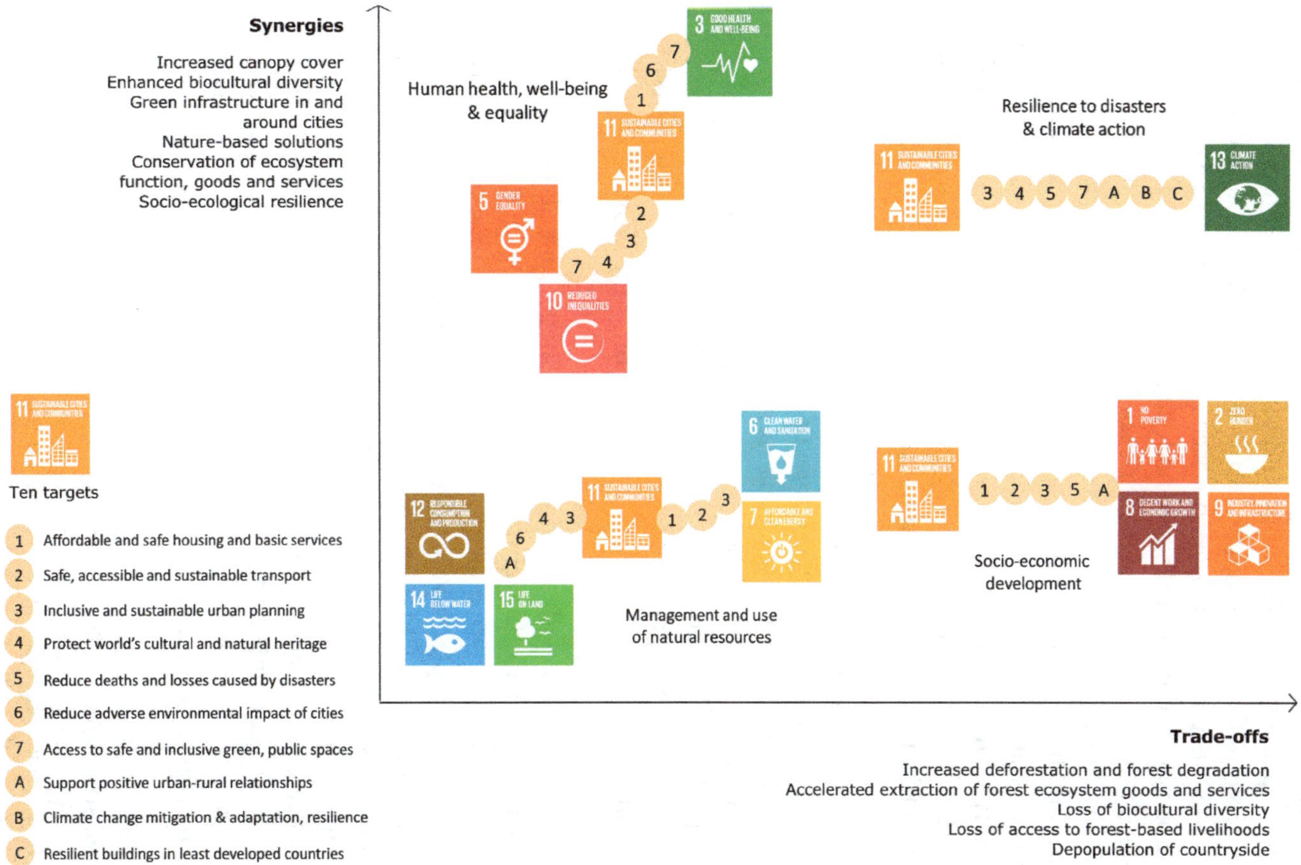

Figure 11.1 Synergies and trade-offs between SDG 11 targets and the other SDGs in relation to potential effects on forests and forest-based livelihoods. Some interactions may present more synergies (high in the 'synergies' y-axis). Others have more trade-offs (high in the 'trade-offs' x-axis). SDGs 4 (Quality Education), 16 (Peace, Justice and Strong Institutions) and 17 (Partnerships for the Goals) are cornerstones of potential synergies presented in the graph.

A clear area of synergy relates to human health and well-being. SDG 3 (Health and Well-Being) envisions better health for all human beings. This inclusive approach is closely linked to SDG 10 (Reduced Inequalities) and SDG 5 (Gender Equality). Achieving SDG 3 targets is still hampered by lack of safe water, sanitation and hygiene in urban areas. While most countries reporting VNRs have made progress on access to safe drinking water, challenges in urban areas remain linked to inadequate basic services in slums, as well as water shortfalls and management issues in the context of disasters (UN DESA 2015). Addressing Target 11.1 (access for all to adequate, safe and affordable housing and basic services and upgrading slums) in vulnerable slum areas would constitute an intervention with clear co-benefits for SDG 3, given that populations living in these conditions are the most vulnerable to urban health risks, such as respiratory problems, waterborne diseases and premature deaths.

Focusing on human health synergies can benefit forests in and around cities (Figure 11.1). This is possible through the equitable deployment of Target 11.7 (access to safe, inclusive and accessible, green and public spaces) and Target 11.4 (efforts to protect and safeguard the world's cultural and natural heritage). According to an increasing number of studies, access to green spaces and other aspects of the urban forest may play a pivotal role in maintaining and positively influencing human health through various ecosystem services,[6] including regulating (e.g. heat reduction), cultural (e.g. physical activity/recreation, stress relief and social cohesion) and provisioning (e.g. food for survival) (van den Bosch and Sang 2017).

Unfortunately, there is evidence of widespread inequitable access to urban green spaces around the world (Schwarz et al. 2015). This inequity is a missed opportunity to address the health and well-being of the most marginalised urban dwellers. For example, in Vancouver, Canada, urban tree canopy cover is lowest in lower-income neighbourhoods, depriving those communities of the ecosystems services provided by urban forests. This bias calls for implementing Target 11.7 in conjunction with SDG 10. Empowering communities to steward their local urban green spaces (including gardens and farms) with interventions supporting social, economic and political inclusion (SDG 10) holds potential synergies with increasing participatory urban planning, increasing equitable access to urban green space and improving overall community health.

[6] The ecosystem services framework became more prominent with the Millennium Ecosystem Assessment (2005), which defines ecosystem services as the benefits people obtain from ecosystems. They include provisioning services such as food and water, regulating services such as processes that affect climate and the water cycle, cultural services that provide recreational, aesthetic and spiritual benefits, and supporting services such as soil formation and photosynthesis. For more details, see www.millenniumassessment.org/documents/document.356.aspx.pdf

Other potential synergies pertain to building resilience and adaptive capacity to climate-related hazards (SDG 13 Climate Action). This can have positive impacts on urban and peri-urban forests if NBS are taken into consideration as a strategy to manage disaster risk (Raymond et al. 2017). NBS include strategies for city greening, from green roofs or green walls to wetland conservation, park establishments and street tree planting. As temperatures continue to rise due to climate change, the cooling effects of urban green spaces are likely to become increasingly important. By mitigating the urban heat island effect, urban forests can also improve energy efficiency at neighbourhood and city levels by reducing energy consumption for cooling. In the USA, the establishment of 100 million mature trees around residences is estimated to save about USD 2 billion annually in reduced energy costs (Akbari et al. 1988, Donovan and Butry 2009).

In addition to city greening, NBS and low-impact development create other synergies among SDGs that depend on rural–urban connections, such as SDG 6 (Clean Water and Sanitation), SDG 7 (Affordable and Clean Energy), SDG 12 (Responsible Consumption and Production) and SDG 15 (Life on Land). Low-impact development includes connecting water bodies and green spaces in and around cities to provide additional functions. These functions may involve water storage and recycling for irrigation and industrial use, as well as flood risk control, water purification and protection of wildlife habitat (Ghofrani et al. 2017). Focusing on water management (SDG 6) as an entry point can have positive impacts on forest conservation, where many urban water sources are located. A report by the World Bank and World Wildlife Federation's (WWF) Alliance for Forest Conservation and Sustainable Use (2003) found that about one-third of the world's largest cities obtain a significant proportion of their drinking water directly from protected forests. Interestingly, in many countries there has been a significant increase over the last 15 years in using urban protected areas as water sources (FAO 2018).

Finally, another area for synergies that can benefit forests is related to clean (i.e. low pollution) cooking fuels. This topic connects SDG 11 with SDG 7 and SDG 13. Cities are primary sources of greenhouse gas (GHG) emissions, requiring mitigation efforts via policies, communication and investment in technology and infrastructure (Hoornweg et al. 2011, Laukkonen et al. 2009, Lenhart 2015). While access to clean fuels and technologies for cooking climbed to 57.4 per cent in 2014, more than 3 billion people (particularly in Asia and sub-Saharan Africa) still cook without clean fuels and efficient technologies. In Bangui (Central African Republic) and the slums of Nairobi (Kenya), about 95 per cent of residents still use wood for cooking and lighting (Drigo and Salbitano 2008, FAO 2009). Traditional fuelwood has clear impacts on urban and peri-urban forests, and it exposes users to high levels of indoor air pollution. Transitions to clean fuels in cities can thus have positive impacts on urban forest systems and public health. For instance, raising awareness of cleaner cooking technologies in N'djamena (Chad) was paired with improved capacity for sustainable urban forest management (FAO 2012).

These clear synergies provide guidance for policymakers and urban planners as they try to integrate multiple SDGs in urban areas. Many cities are aiming to develop local policies and plans that work with and enhance one another and are explicitly considering synergies among SDG targets. Finding interventions that build on SDG interactions and can have positive impacts on forest and forest-based livelihoods calls for education to promote sustainable development (SDG 4), as well as strong institutions (SDG 16) and partnerships (SDG 17), as discussed in Section 11.4. Strengthening access to education in urban areas may increase pressures on urban resources if rural-to-urban migration increases. Appropriate attention needs to be placed on enhancing education and work opportunities in rural areas where there can be growth opportunities. A more integrated approach that strengthens rural–urban development planning (Target 11.A) is therefore key and is increasingly being recognised by national agendas (see Section 11.4).

11.3. Contextual Factors Guiding SDG 11 Target Prioritisation

11.3.1 Recognising Different Urbanisation Patterns

Urban areas are expanding across the globe, but urbanisation patterns vary and each situation comes with its own challenges, as shown in Box 11.1. Urban sprawl is a common phenomenon, with the expansion of urban land outpacing the growth of urban populations (Seto et al. 2013). This has placed increasing pressure on agricultural, forested and other land use. While some cities densify, others show trends of suburbanisation and peri-urbanisation.

As of 2018, the most urbanised regions of the world include North America (82 per cent urban population), Latin America and the Caribbean (80 per cent) and Europe (74 per cent).[7] Asia (50 per cent) and Africa (42 per cent) are still mostly rural, although urban settlements are expanding rapidly. By 2050, 64 per cent of Asia's and 56 per cent of Africa's population are projected to be urban. This means that 95 per cent of urban expansion in the coming decades will take place in the developing world (UNDP 2018). Intra-regional differences apply: some areas grow with many small to medium settlements (i.e. less than 500 000 inhabitants); others grow megacities of more than 10

[7] See UN Population Division World Population Prospects 2018 https://population.un.org/wup/Download/

million inhabitants, such as China's Beijing (17 million), Japan's Tokyo (32 million) and Indonesia's Jakarta (23 million) (Fensom 2015, UN 2014). Some urban areas are shrinking or transforming in population and economic activity (e.g. gentrification and urban rot), which raises new challenges and opportunities associated with vacant land, displacement and infrastructure (Seto et al. 2013).

Urbanisation also needs to be addressed from the perspective of urban–rural linkages (Box 11.1). Most development theory and practice are implicitly based on the dichotomy between rural and urban areas, populations and activities. This results in a division of policies along spatial and sector lines, with urban planners usually focusing on urban issues and paying little attention to rural-led development, while rural planners tend to ignore urban centres and define rural areas as consisting only of farms, villages and their agricultural land (Tacoli 1998). This dichotomy becomes blurry as urbanisation trajectories change rapidly and different patterns emerge, such as ex-urbanisation, multi-nucleation and rurbanisation, with various implications for urban–rural relationships (Box 11.1).

Box 11.1 Patterns in Urbanisation and Urban–Rural Relationships

- **Urban sprawl:** the expansion of urban land outpacing the growth of urban populations, placing increasing pressure on the countryside and natural landscapes.
- **Densification and compact city:** the opposite of urban sprawl. The compact (densified) city presents an urban form having a high density of settlements, a clear boundary from surrounding areas, mixed land use and a relative independence of government.
- **Conurbation and agglomeration:** a region comprising cities, large towns and other urban fragments that, through population growth and physical expansion, have merged to form one continuous urban or industrially developed area. In most cases, a conurbation is a polycentric urbanised zone in which transportation has developed to link areas to create a single urban context.
- **Suburbanisation:** the population shift from central urban areas into suburbs, resulting in the formation of (sub)urban sprawl. Suburbanisation is inversely related to urbanisation, which denotes a population shift from rural areas into urban centres.
- **Ex-urbanisation:** a mostly permanent transfer of activities (e.g. malls and shopping centres) from the city centre to the periphery and agglomeration of a city. This phenomenon is particularly widespread in major cities of

Box 11.1 (cont.)

industrialised countries, but also accompanies the growth of cities in emerging countries.

- **Peri-urbanisation:** the shift of urban populations from denser to less dense areas. It is spatially explicit as an extension process of urban agglomerations, in their periphery, resulting in a permanent transformation of rural areas.
- **Multi-nucleation:** the clustering of populations around several centres, rather than just one, in the same region.
- **Counter-urbanisation (or de-urbanisation):** a demographic and social process by which people migrate from urban to rural communities. Observed in developing and hyper-urbanising countries linked to unbearable stress of urban life or need of changing lifestyle.
- **Rurbanisation:** a process of rural transformation. Predominantly rural agriculture economies, forms of settlements, lifestyles and social attitudes are changing towards urban behaviours and a new *rurban* form is emerging. Clustering rural settlements brings benefits similar to those of urban areas, empowers rural people with urban facilities (e.g. electrification, Wi-Fi). Rurbanisation is catching the attention of urban planners as a prominent development process commonly witnessed in developing countries.
- **Gentrification and urban rot:** the social process of renewal of degraded urban areas by wealthier residents. It can improve the physical and material quality of a neighbourhood, while becoming a discriminatory process when it forces the move of current and established residents and businesses from a gentrified area, looking for low-cost housing and shops. Conversely, urban rot is the social process where part of a city or town becomes old or dirty or ruined because businesses and wealthy families have moved away from it.

As early as 1970, Johnson (1970: 28) noted that 'It is incorrect to assume that urban entrepreneurial decisions are wholly discrete and separable from rural decisions and choices.' This statement is even more relevant in modern times when urban–rural interactions have become more complex, diverse and multidimensional. The rural context can range from isolated housing or small settlements on the fringe of cities to remote villages or green-belt agriculture, to family farms or large extensive farming, forestry and grazing lands. In recent decades, 'ruralities' have closer economic and social relationships with urban cultures (Hiner 2016, Scott et al. 2007). Rural villages, small towns, exurbs, peri-urban areas within urban commuter sheds and the rural–urban fringe (Hiner 2016) are places where rural and urban identities are more entangled than in large urban centres (Taylor and Hurley 2016).

Recognising that cities' socio-economic development conditions and urbanisation patterns are of contextual importance to assess SDG 11 impacts on forests and people, we apply a three-dimensional framework to classify cities (Figure 11.2). First, we classify cities according to their pace of growth in the past two decades, with some cities showing rapid growth (several per cent per year) and others showing much slower growth (less than 1 per cent per year), or even experiencing population decline ('shrinking cities'). Second, we distinguish between cities that are affluent, with sufficient resources to manage and plan their growth and deal with urbanisation's negative consequences, and those that are less affluent. We base this dimension partly on the World Bank's country income groups, with countries that are at least in the upper-middle-income group being described as affluent. Third, we consider the spatial pattern of urban growth, identifying cities whose growth is primarily concentrated in the urban core (densifying) versus cities expanding outward (sprawling).

We recognise that classifications are simplifications and acknowledge, for example, that rapidly growing cities will often grow both in their core and in their perimeters, and that levels of affluence can be debated because of large

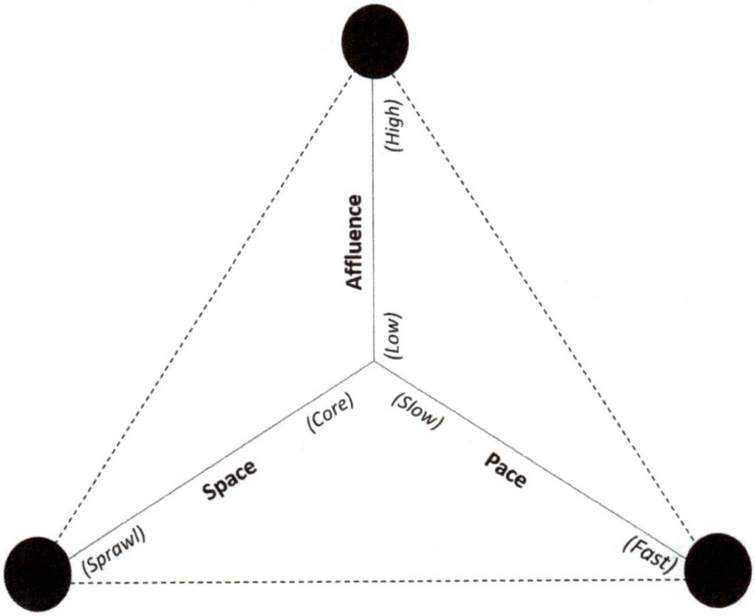

Figure 11.2 Framework used to categorise cities according to interrelated city growth dimensions and development characteristics.

discrepancies within the country and the city. Urban densification can be a planning strategy, but in other cases it can be the result of geographical and other limitations (e.g. a city being confined by neighbouring water bodies or steep slopes). We also acknowledge that this classification can change over time. For example, a city that used to grow at a very rapid pace can slow its growth over time. Nevertheless, the proposed framework to categorise the world's cities can be used to analyse similarities and differences in how SDG 11 targets are prioritised according to contextual factors and the associated impacts on forests and forest-based livelihoods within and outside city boundaries.

11.3.2 Anticipated Impacts of SDG 11 on Forests and Forest-Based Livelihoods

Applying the framework presented in Figure 11.2, and recognising the vast diversity in the conditions and development patterns of the world's cities, we analyse how different city types (in terms of growth and affluence) have a different way of prioritising SDG 11, thus affecting forests and forest-based livelihoods. First, we look within urban areas themselves and review the (potential) role of forests and forestry, for example in promoting more sustainable cities. Next, we shift the focus to the implications of urban development on rural forests and forest-based economies.

To conduct the analysis, we use case-study cities as proxies of city types characterised by distinct urbanisation patterns and urban–rural connections (see Table 11.2). These case studies were selected based on information from literature, and socio-economic data obtained from online databases such as the World Population Review.[8] For affluence level, we made adjustments based on the specific city's stand in relation to its national mean. For example, Medellín (Colombia) ranks in the lower affluence group and Curitiba (Brazil) in the higher affluence group, even though Colombia and Brazil are both ranked in the upper-middle-income group by the World Bank.

We also focus on cities' pace of growth over the past two decades, acknowledging it may differ from previous decades. Some cities, such as Milan (Italy), experienced heavy urbanisation early on (150 000 people in the fourteenth century, 200 000 in the seventeenth century, more than half a million in 1901), resulting in forest fragmentation over time and suburbanisation patterns combined with multi-nucleation in more recent decades. Other cities expanded differently: for example, Vancouver (Canada) had nearly 20 000 inhabitants in 1901 and then grew rapidly to 630 000 inhabitants in 2016, while its metro area had 40 000 residents in 1901 versus 2.5 million residents

[8] http://worldpopulationreview.com/

Table 11.2 Overview of case-study cities as examples of different combinations of population growth, affluence and spatial development

City type	Region where type is common	City case study	Urban growth and development	Recognition of forests and trees	Urbanisation pattern (Box 11.1)	Effects on urban–rural relationship	Social and landscape changes
(1) 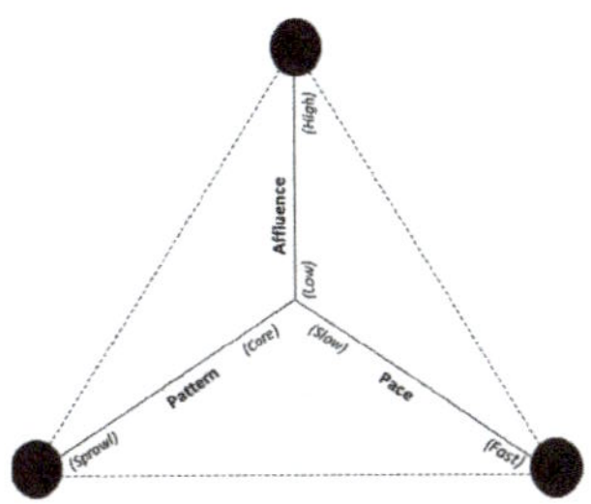 Fast growth High affluence Sprawl	Asia	Hangzhou, China	Fast growth, rapidly expanding Stronger planning control	Recognition of importance of forests and trees; Forest City and urban forestry programmes with large-scale afforestation	• Urban sprawl • Ex-urbanisation • Suburbanisation	• Large-scale land-use change • Loss of suburban forests and green space and forest fragmentation, resulting in a series of environmental and social issues, e.g. habitat loss, air pollution and health concerns • Soil sealing, flooding problems • Peri-urban forest fragmentation and habitat loss	• Urbanisation of minds and attitudes • Urban poverty and income disparities • Habitat fragmentation and loss

Table 11.2 (cont.)

City type	Region where type is common	City case study	Urban growth and development	Recognition of forests and trees	Urbanisation pattern (Box 11.1)	Effects on urban–rural relationship	Social and landscape changes
(2) 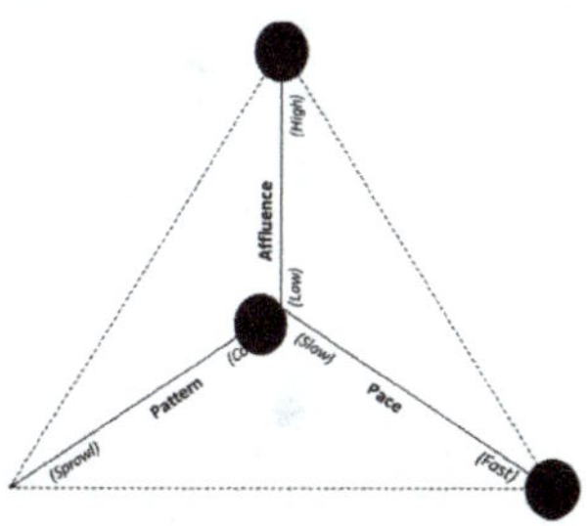 Fast growth High affluence Densifying core	North America, Australia	Vancouver, Canada	Fast growth, mostly through densification Stricter planning and regulation	Greenest City Vision and urban forestry programme; urban afforestation and efforts to prevent canopy loss; important role of forests in watersheds for freshwater provision	• Conurbanation • Counter-urbanisation • Ex-urbanisation	• Regional food system maintained • Green network development in the region, for example for recreation • Solid and liquid waste management • Affordable housing, providing clean drinking water • Air quality maintenance	• Urbanisation of minds and attitudes • Intensification of land use • Habitat fragmentation

(3) 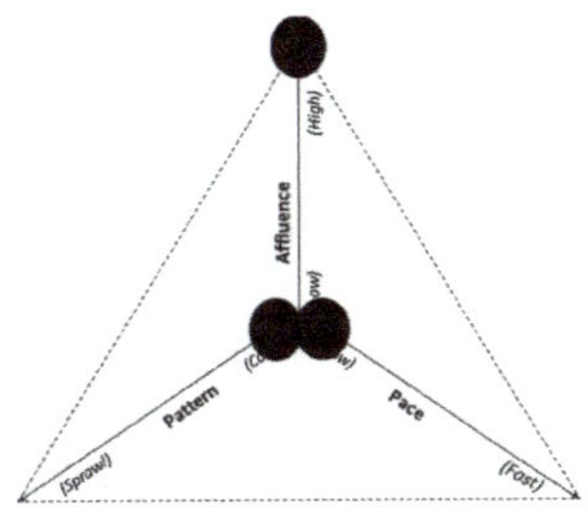 Slow growth High affluence Densifying core	Europe	Milan, Italy	Industrial change and densification Sub urbanisation slowed down, but experiencing urban renewal	Protection and enhancement of green areas; focus on afforestation in former industrial land; introduction of a system of parks, forests and wetlands	• Suburbanisation • Counter-urbanisation • Multi-nucleation	• Extensive green network, opportunities through transformation of former industrial land, agricultural land and forest parks in the metro region • Progressive transformation of the flat land area around the metropolitan area in cropland since the 17th century • Regional planning and policy recovering forest cover at the urban fringe	• Urbanisation of minds and attitudes • Habitat fragmentation and loss • Abandonment of land • Intensification of land use

Table 11.2 (cont.)

City type	Region where type is common	City case study	Urban growth and development	Recognition of forests and trees	Urbanisation pattern (Box 11.1)	Effects on urban–rural relationship	Social and landscape changes
(4) 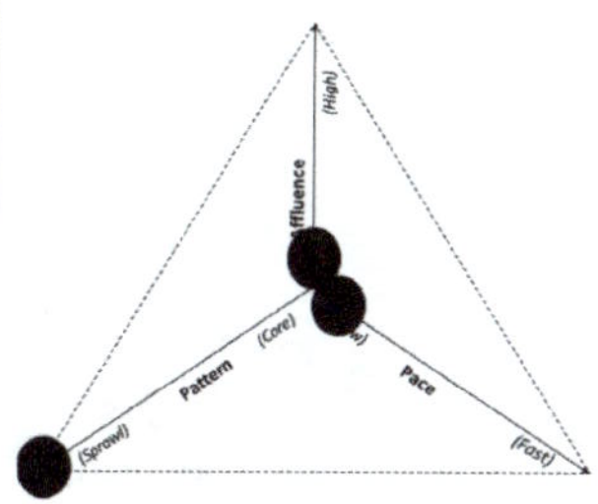 Slow growth Low/medium affluence Sprawl	Latin America	Medellín, Colombia	Slowing growth after rapid growth due to urban renaissance	Greening efforts as part of new urban planning; development of a Green Belt	• Urban sprawl • Peri-urbanisation	• Urban sprawl and landslides on the steep hillsides that surround the valley city • Green Belt around the city declared a nature reserve • Low soil sealing, no active forest management due to legal bans on logging and because of the strong resistance of urban dwellers to logging • Proactive management of peri-urban slums	• Urbanisation of minds and attitudes • Urban poverty • Intensification of land use • Habitat fragmentation and loss • Abandonment of land

(5) 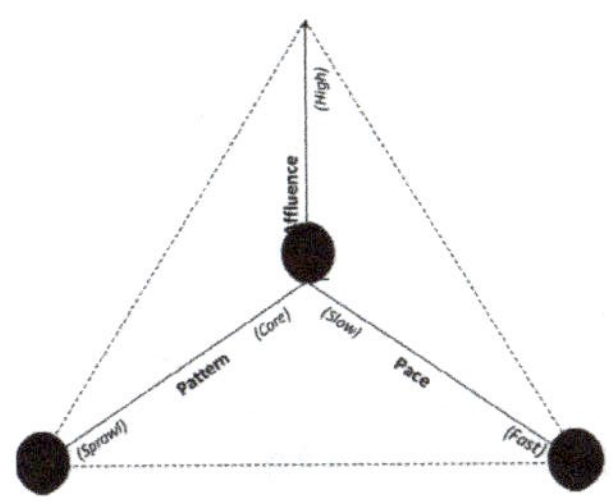 Fast growth Low/medium affluence Sprawl	Africa Southeast Asia	Kinshasa, Democratic Republic of Congo	Major, unplanned urban growth and sprawl	Loss and fragmentation of forests both inside and outside of the city; forests provide important products and services, including regulatory and provisioning (e.g. fuelwood, food) ecosystem services	• Urban sprawl • Conurbation • Multi-nucleation	• Loss and fragmentation of forests both inside and outside the city • Forests provide important products and services • Advancing agriculture frontier to fulfil need for fuelwood and food	• Urban poverty • Modification • Habitat fragmentation and loss

Table 11.2 (cont.)

City type	Region where type is common	City case study	Urban growth and development	Recognition of forests and trees	Urbanisation pattern (Box 11.1)	Effects on urban–rural relationship	Social and landscape changes
(6) 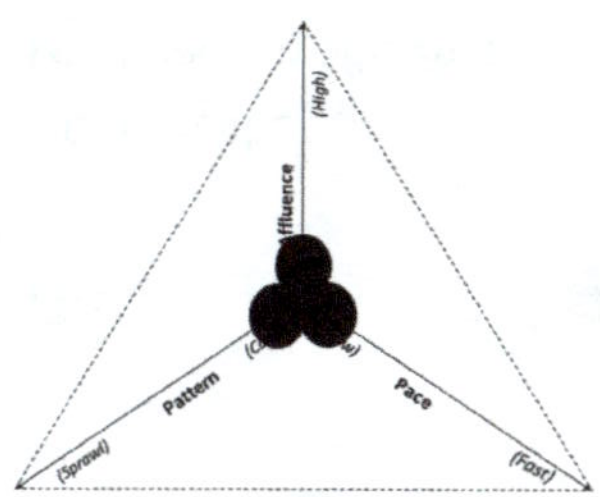 Slow growth Low/medium affluence Densifying core	Middle East	Tehran, Iran	Very dense cities with major challenges, e.g. in terms of pollution and mobility Growth is slowing down	Urban forests and green spaces are recognised for their important ecosystem services, but a more comprehensive approach to urban forestry and green infrastructure planning is still needed	• Urban densification • Suburbanisation	• Abrupt change from rural organisation to urban society • Soil sealing, arid fragile ecosystems • Loss of habitat, desertification risk	• Urbanisation of minds and attitudes • Urban poverty • Modification • Habitat loss, habitat degradation • Abandonment of land

(7) 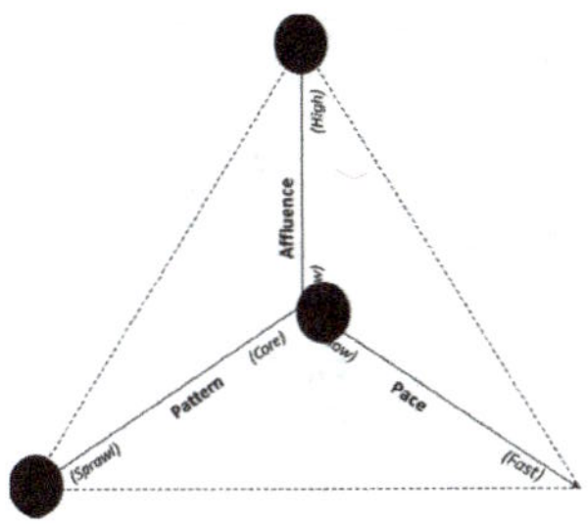 Slow growth High affluence Sprawl	Latin America (some affluent cities) North America (East mainly)	Curitiba, Brazil	Commitment to urban sustainability and strict planning, e.g. focusing on public transport Some sprawl and slum issues	Urban forest is seen as central to sustainable urban planning, for example for stormwater management	• Urban sprawl • Conurbation • Peri-urbanisation	• Advancing agricultural frontier • Landscape conservation through blue–green infrastructure networks • Decreasing active forest management for provision of multiple goods and services	• Urbanisation of minds and attitudes • Urban poverty • Modification • Habitat fragmentation
(8) 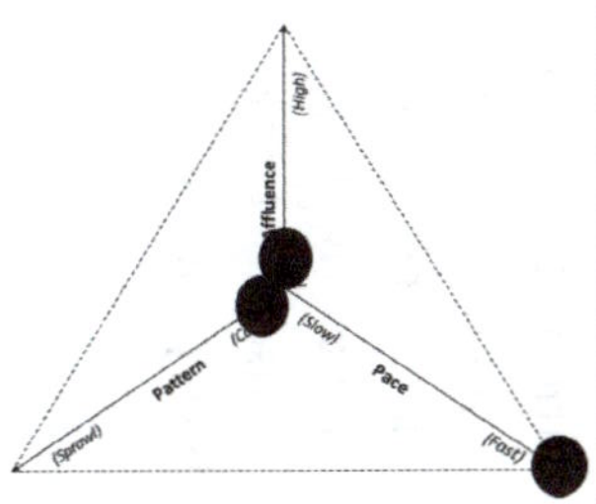 Fast growth Low/middle affluence Densifying core	Asia	Ho Chi Minh City, Vietnam	Mostly unplanned urban densification Major liveability and flooding issues	New climate-change adaptation policies recognise the importance of blue–green networks; major tree loss due to densification	• Urban densification • Conurbation • Peri-urbanisation	• Forest fragmentation • Entanglement of rural villages and smaller settlements • Soil sealing	• Urban poverty • Habitat fragmentation and loss, habitat degradation • Modification • Intensification

in 2016. In recent decades, cities around Vancouver merged to form one continuous urban developed area, a conurbation process that also took place in the San Francisco Bay Area (USA).

Although the overall aim of SDG 11 is valid for all cities, different city types have different approaches towards implementation. For example, cities experiencing fast growth and sprawl in areas of low (or medium) affluence are putting major pressure on surrounding areas, while emphasising the development of urban infrastructure such as housing and transport networks. Sprawl can enhance vulnerability to climate-change impacts as people are pushed into areas of greater risk, such as coastal areas, floodplains or mountainsides. Sprawl and densification can also occur at the same time. Densification is a complex phenomenon (Haaland and Konijnendijk van den Bosch 2015) where existing urban green spaces can fall victim, affecting local resilience and quality of life. Densification can drive up property prices and reduce the wider urban footprint on adjacent nature. However, this process can also make cities increasingly unaffordable for the less affluent.

URBANISATION OF MINDS AND ATTITUDES

The conceptualisations of urban or rural are determined by space, representation and culture. Recent conceptualisations see rurality and urbanity as cultural constructs rather than geographically bound places (Dymitrow et al. 2016). From this perspective, the urbanisation of minds is linked to a cultural hegemony, i.e. the control of culture through the domination of social groups via social institutions. In the urban world, the urban culture became rapidly hegemonic, heavily influencing minds and behaviours, also of people (or at least a part of them) living in rural settings.

One of the fundamental questions reflecting the changes in the relationship between urban and rural is the attitude concerning the temporality and transformability of the landscape. Urban thinking tends to perceive fixity in landscape, a sort of freezing of the built-up status quo. In urban community perceptions, the concept of landscape very often represents an immutable, static referential component (Bonnes et al. 2010). In contrast, rural perception of landscape is often very dynamic, related to detailed ecological and vegetation knowledge (Campos et al. 2012).

The urbanisation of minds might often influence attitudes towards forests. Urban dwellers may feel more disconnected from nature. In other instances, urbanites may see forests as an icon of nature that needs protection. This may hinder decision-making that aims at changing (even slightly) the structure of forests, thereby also affecting sustainable forest management and impacting the livelihoods of people who depend on access to and commercialisation of forest products and services. Allowing a more dynamic and legitimate change in forests in and around cities calls for inclusiveness in urban forest planning, accounting for the multiple social perceptions of and interactions with forests.

URBAN DEVELOPMENT, LOSS OF URBAN FOREST AND TRADITIONAL CULTURAL LANDSCAPES

Cities of all types (Table 11.2) face different challenges in relation to urban planning that can successfully and sustainably steer growth. In all cities, maintaining (or creating) green space, including urban forests, is facing pressure from grey infrastructure development.

While urbanisation can reduce direct pressure on forests via rural migration, the expansion of urban and residential areas can cause significant forest loss due to construction and land conversion. Sprawl is commonplace in many Global South cities, such as Kinshasa (Democratic Republic of Congo). Decades ago, Kinshasa was a small town; now it has boomed to 17 million residents, with extensive urban sprawl. Uncontrolled sprawl leads to the decrease of surrounding forests and other ecosystems and a concurrent decrease in the important ecosystem services these provide.

Many cities in Southeast Asia and China are also fast-growing and sprawling, leading to forest fragmentation, habitat loss and degradation, and land-use change. Even in more affluent cities, governments struggle to control growth. In Hangzhou, now the tenth largest city in China, rapid population and surface area growth have led to large-scale land cover changes, loss of suburban forests and green space, and forest fragmentation. This has resulted in a series of environmental and social issues, such as habitat loss, air-pollution and urban health concerns (Byrne et al. 2015, Yue et al. 2013). Urban expansion also frequently takes prime agricultural land out of production, making it increasingly necessary to use marginal lands for cropland and pastures.

Urban sprawl is commonplace in many Latin American cities, although this trend may be decelerating in some cities that are proactively planning for green spaces. Medellín (Colombia) has reinvented itself from a troubled past (related to drug crime) as one of today's most innovative cities in Latin America (Mendieta 2011). With enhanced stability and liveability has come the need to house a growing population (Mendieta 2011). Despite this, the city has been more successful than many other cities in the region in controlling sprawl, and is currently developing an extensive network of urban parks. Curitiba (Brazil) is also globally known as an example of sustainable and green city development, but it has yet to address sprawl or the fact that an increasing part of its population lives in slums (Atlas of Human Development in Brazil 2013).

The growth of some cities occurs through urban densification rather than sprawl, posing its own challenges in relation to green space planning. In Ho Chi Minh City (Vietnam), urban densification was largely unplanned, due to speculation and other factors (Zhu 2012), resulting in a major loss of tree canopy and green spaces (Thanhnien New 2016). Following settlement pressures – residential and production space allocation – intensified landscapes occurred on the urban fringe. This resulted in the unplanned intensification of functions and dramatic landscape change within confined boundaries. This kind of densification, as in the case of Ho Chi Minh, can lead to major land modification near the urbanised zones, often oriented towards industrialised agriculture with mechanised monoculture systems.

Even in affluent Vancouver (Canada), with strong environmental policies in place, fast growth through densification of the urban core has impacted tree canopy cover. At the city level, canopy cover has decreased from 22.5 per cent to less than 18 per cent since 1995 (City of Vancouver 2018). Vancouver's densification of the urban core is considered necessary, as the Coast Mountain Range and Pacific Ocean restrict development. Less affluent Tehran (Iran) is already very dense, with an average of 11 800 residents per square kilometre (World Population Review 2018). The limits to urban expansion, mobility and the city's location on two major earthquake fault lines led the Iranian government to explore plans to relocate the capital, which never materialised (Madanipour 2006). Currently, Tehran's residents have only an average of 3 m^2 of green space per capita at their disposal (van den Bosch and Sang 2017, Kabisch et al. 2017). In comparison, Vienna has 120 m^2 of green space per capita, Singapore 66 m^2 and London 27 m^2 (Baharash Architecture 2018).

Proximity to cities also causes changes in rural lifestyles and employment. In post-industrial societies, urbanisation has gradually induced desertion of rural areas, particularly among younger generations. Rural abandonment has led to intensification of the most productive lands and abandonment of marginal lands (Barbero-Sierra et al. 2013), as well as loss of traditional cultural landscapes in many cases (Van Eetvelde and Antrop 2004). Rural abandonment can also spur outright conversion of forest by industrial farmers and ranchers, especially in areas suitable for large-scale agriculture. For example, at the end of the nineteenth century, Milan's agro-industry transformed the rural area around the city, followed by the heavy industrial boom of the twentieth century. Currently, Milan's efforts have reoriented to renew the character of the rural landscape even in the metropolitan area. An associated phenomenon is the daily commuting from the rural suburbs to industrial/commercial occupations in cities.

THE IMPORTANCE OF ECOSYSTEM SERVICES PROVIDED BY URBAN FORESTS

The important contributions of forests and trees to healthy and resilient cities are increasingly recognised. Vancouver's Urban Forestry Strategy highlights the many ecosystem services provided by the city's urban forests, including for culture, recreation and social cohesion, all with potential positive influence on human health (City of Vancouver 2014, 2018). Many fast-growing, affluent cities in Asia, including Hangzhou, also stress these ecosystem services. In Medellín, where growth has slowed but sprawl needs to be controlled, the role of urban forests in improving air quality, stormwater runoff and carbon sequestration is highlighted (Mendieta 2011). Urban forests and trees in the city are considered important for combatting pollution and reducing erosion, as well as for improving public health (van den Bosch and Sang 2017, Kabisch et al. 2017). In Ho Chi Minh City, expanding the urban green–blue network is considered an important way to combat both flooding and the urban heat island effect (C40 Cities 2016).

All of the world's cities still have some form of dependency on forests and the ecosystem services they provide; however, the focus on specific services differs (Ferrini et al. 2017). While cities in affluent areas of the Global North may stress the importance of forests for recreation, tourism, water protection and biodiversity conservation, many cities in the developing world are still focusing on provisioning ecosystem services of forests. For example, dependency on fuelwood may be greatest in sub-Saharan Africa. In Kinshasa, annual household fuelwood consumption increased from 12.9 million to 14.5 million m^3, while industrial consumption increased from 66 080 to 71 066 m^3 between 2001 and 2005 (Samndong and Nhantumbo 2015). Initially, fuelwood was obtained from nearby forests, but with the complete disappearance of forestland around Kinshasa, fuelwood and charcoal are now imported from 400 to 500 km away. In other cities, forests are considered a source of food and fodder, and agroforestry and foraging systems in urban areas help supplement urban diets. Nonetheless, cities are not only dependent on surrounding rural areas for their food; they often derive food from areas much further away, impacting remote landscapes (Seto et al. 2013).

INCREASED ACCESS TO GREEN SPACE AND INTEGRATION OF URBAN FORESTS IN URBAN PLANNING

An integrative approach to planning and managing green space calls for even distribution of vegetation and equitable access to urban forests, particularly for low-income communities. The recognition of multiple perspectives on and relationships with urban forests demands inclusiveness and equity in urban forest governance. This is critical when planning for Target 11.7 (universal access to safe, inclusive and accessible, green and public spaces) and overcoming the prevalent uneven distribution of green space, whereby more affluent residents have better access (Nesbitt and Meitner 2016, Salbitano et al. 2016). The long-term sustainability of green spaces is largely dependent on local community action, inclusivity with respect to multiple perspectives and cultural connections to forests, and more integrated governance models.

Creating green, healthy, sustainable cities requires balancing urbanisation pressures with institutional capacity in models that reflect and are relevant to local realities. More affluent cities such as Vancouver, Hangzhou and Curitiba have adopted an urban forestry and green urban planning approach. As part of its Urban Forestry Strategy, Vancouver aims to plant 150 000 new public trees by 2020, while strengthening the legal protection of all urban trees (City of Vancouver 2014, 2018). The regional Metro Vancouver government has developed a series of interconnected regional management plans around its Sustainability Framework to help achieve various SDGs (Kanuri et al. 2016).

Hangzhou is recognised as one of the national Forest Cities, based on a set of criteria for sound urban forest planning and management in China. Accordingly, the Hangzhou government is protecting and rebuilding its urban forest, for example through large-scale afforestation. The city aims to regain its reputation as 'heaven on earth', partly based on green and blue spaces such as the West Lake and the Xixi Wetland (Wolch et al. 2014). In Curitiba, the role of the city's interconnected network of parks and woodlands in providing ecosystem services, such as stormwater management, is well recognised (Adler 2016). Curitiba is often highlighted as a global leader in urban sustainability. After facing rapid urban growth, a period of drastic urban-planning reform started under the leadership of Mayor Jaime Lerner during the 1970s. The city created an extensive and affordable public transport system and enhanced the infrastructure for soft traffic (i.e. biking and walking). Urban growth was restricted along a few major transport corridors. Protecting urban forests became an integral part of urban growth, and green space per capita increased from 2 m^2 to 50 m^2 since the 1970s. To protect its main river (the Iguazu) and regulate flooding, a large river park was created. Green–blue infrastructure replaced otherwise 'hard engineering' solutions such as channelling the river with concrete walls. Curitiba also recognises the importance of public places for pedestrians, as well as place-making, in terms of creating meaningful environments for socialisation, social cohesion and community building (Adler 2016, Atlas of Human Development in Brazil 2013).

Less affluent cities have also started to consider urban forestry and greening more proactively, seeing green space as an integral part of urban planning and resilience. Medellín is developing an extensive network of large and small urban parks, under a philosophy of social urbanism, with smaller urban parks and forests facilitating natural flows and assisting with biodiversity and water management (Mendieta 2011). A large-scale afforestation effort is underway to design the Metropolitan Green Belt on the slopes of the Aburrá Valley. When completed, this Green Belt will extend 75 km into the city's peri-urban hillsides. Various urban forest services are in focus, including recreation, community building, employment opportunities for residents and local food production in terraced gardens. Crucially, the Green Belt will contain urban sprawl and landslides on the steep hillsides, as well as help protect important watersheds, create more inclusive spaces and aim for increased green equity.

11.4. The Governance of a Green Urban SDG

11.4.1 The International Outlook: Building on Decentralised Partnerships

The roles of urban forests and green spaces are increasingly recognised within wider sustainable city discourses. The debate is not just about sustainable energy or transport systems in cities, but also about how these should be balanced with improved access to urban green spaces of varying sizes in proximity to where urban citizens live and interact. International and national organisations, including UN agencies, city networks and research programmes, are supporting the integration of green spaces in cities. A description of a few of these initiatives follows.

City networks and support programmes – such as C40 Cities, Local Governments for Sustainability (ICLEI), 100 Resilient Cities, Smart Cities Connect and WWF's One Planet Cities – support cities and local governments to address climate change, technological transformation and sustainability challenges. They do so by gathering data on cities and their inhabitants, facilitating exchange among cities, enhancing city capacity via training and tools, and representing cities in national or international forums, including how cities can better address urban green spaces. Although these networks do not focus directly on urban forests, they help enhance cities' engagement on urban biodiversity. For example, ICLEI set up its BiodiverCity and Cities With Nature programmes. Together with the UN Convention on Biological Diversity, ICLEI and partners coordinate the City Biodiversity Index. The World Urban Parks is an umbrella organisation connecting cities, NGOs and research organisations dedicated to expansive parks in cities and improved tree cover. Cities4Forests[9] and the Mantova Challenge 'Tree Cities of the World'[10] launched in September and December 2018, respectively, are initiatives that directly address the connection between cities and forests in and

[9] http://pilot-projects.org/projects/project/global-cities-global-forests

[10] www.wfuf2018.com/public/file/CS-MantovaChallengeENG_WFUF2018.pdf

around urban areas. The purpose of these two initiatives is manifold, including human well-being, improved management of forests, protection of biodiversity and watersheds, and combatting climate change.

Research programmes are also investigating the role and status of urban green spaces within cities. For example, the Naturvation project[11] links nature, innovation and cities, focusing on how NBS can address climate change, including an Urban Nature Atlas with some 100 city cases in Europe (Naturvation 2018). The Green Surge project[12] tested and implemented ways to connect green spaces, biodiversity, people and the green economy in cities to address land-use conflicts, climate change adaptation and human health and well-being. In March 2018, global researchers investigating the role of NBS in cities to combat climate change gathered at the Cities IPCC Conference[13] in Edmonton, Canada, proposing to set up a more global urban NBS research network. While an important network, it will most likely examine the contribution of NBS to urban climate strategies, without an explicit examination of urban forestry (Cities IPCC Conference 2018).

Communication tools and mapping are also important to consider. The i-Tree tools and Treepedia initiatives to map a city's street trees, ecosystem services and Green View Index help increase awareness of the importance of a city's tree canopy in lowering urban temperatures, creating more comfortable microclimates and mitigating air pollution or intensive rains during flooding events. Treepedia currently collects data in some 30 global cities – however, few case-study cities are in the Global South.

International policy processes are also crucial to put urban forestry considerations on the policy agenda. Cognisant of rapid global urbanisation and the unsustainability of most cities' growth, FAO formed the Urban and Peri-Urban Forestry (UPF) programme, in growing collaboration with UN agencies (e.g. UNEP, UNDP, UN-Habitat), city networks (e.g. Cities Alliance, United Cities and Local Governments), the Green Belt Movement, the International Society of City and Regional Planners, the World Resources Institute and the WWF, as well as with regional and national stakeholder groups. The UPF programme aims to raise awareness and build knowledge about urban forestry by producing normative tools, sharing policy expertise and building a knowledge network. It recognises the role of urban forestry to support food and nutrition security, provide livelihoods, alleviate poverty, reduce disaster risk, support climate-change adaptation and mitigation strategies, and facilitate recreational, cultural and social opportunities.

---

[11] https://naturvation.eu

[12] https://cordis.europa.eu/project/rcn/110888_en.html

[13] https://citiesipcc.org

Accordingly, in global policy processes and among stakeholder groups, there is growing interest in and attention to examining the importance of urban green spaces, including urban forests. This is also reflected in the NUA (Section 11.2.1), with signatories committed to promote safe, inclusive, accessible, green and quality public spaces, including access to gardens and parks to support civil engagement. Global interest comes from UN agencies, city networks, research institutes, civil society organisations and NGOs operating at local and regional levels. Nonetheless, coordination across stakeholder groups is lacking, and important lessons are often not diffused quickly or effectively enough across diverse stakeholder groups or geographic regions. Overall, an emphasis on capacity building is needed to ensure that research and practice also reach small and medium-sized cities, especially those in the Global South, where urban green spaces and urban forestry are under threat. Ultimately, municipal and regional budgetary allocations will be critical to successful implementation of urban forestry initiatives, and research and practice research communication and public-awareness raising can help make the case for such allocations.

11.4.2 Creating Multi-Scale Bridges for Collective Stewardship

The case studies of different city types discussed in Section 11.3 illustrate that governance of forests and the provision of essential forest ecosystem services is challenging. Urban areas often face a so-called scale mismatch: sustaining and enhancing ecosystem services requires the resolution of mismatches between ecological processes, on the one hand, and social processes of governance, on the other (Ernstson et al. 2010). All cities face issues of forest and tree loss and increasing pressure on surrounding forest landscapes. Few cities have managed to set up more effective, multi-scale governance structures.

Adaptation to the impacts of climate change provides potential for collective stewardship of forests. Many cities across the world, including in the Global South, have started to develop more comprehensive climate strategies (e.g. C40 Cities 2016). The implementation of these strategies is often obstructed, however, by a lack of cohesive governance and the involvement of a large range of government and other actors. Similar challenges are faced when focusing on the role of forests and forestry in providing other ecosystem services, such as food, fuelwood, construction material and settings for recreation. However, successful examples of integrated climate adaptation and collective stewardship in the Global South exist. For example, the Marikina Watershed Integrated Resources Development Alliance includes seven cities in the Manila metropolitan region (Philippines) that are working with NGOs, the private sector and civil society to rehabilitate and restore the Marikina Watershed to reduce disaster risk and improve urban resilience.

Ernstson et al. (2010) use the case of ecosystem governance in Stockholm (Sweden) to highlight the importance of social networks and network governance. They argue that substantial governance gaps exist, which need to be filled by mid-scale managers and 'scale brokers' who can operate across and link between different levels of governance. These scale brokers can help link the many bottom-up community initiatives related to forests and trees. In the USA, urban challenges such as budget limitations, ageing urban infrastructure and the impacts of natural and human-made disasters often linked to climate change have helped spur the creation of community-based environmental stewardship groups across cities and urban areas (USDA Northern Research Station 2017). Similar trends can be noted in developing-country cities such as Cochabamba (Bolivia), Guayaquil (Ecuador) and Bangkok (Thailand), where community groups have taken responsibility for urban forestry in the absence of strong government commitment (BIGTrees n.d., Fundación Pro-Bosque 2018, Konijnendijk et al. 2018). Mapping environmental governance is also an important step to develop a baseline for better stewardship. The US Forest Service embarked on an information-gathering process called STEW-MAP to identify and quantify stewardship in several US cities and internationally, including information such as organisational characteristics, geographic areas of influence and connections with other civic, private and governmental organisations (USDA Northern Research Station 2017).

In recent decades, recognition of the importance of public and civil society actors in decision-making process has increased. In Vancouver, for example, one of the city's most famous urban parks, Stanley Park, is co-managed by the Vancouver Parks and Recreation Board (VPRB) and the Stanley Park Ecology Society (SPES). SPES is primarily responsible for conservation and education in the park, with support from VPRB, while conducting research to inform VPRB's decision-making (Stanley Park Ecology Society n.d.). Other cases demonstrate governance without, or even in opposition to, formal governments. The Big Tree Project, an environmental and advocacy group in Bangkok, unites local communities to protest the government's decisions to remove large trees for commercial development (BIGTree n.d.). In other cases, businesses provide leadership in urban forest creation and stewardship, sometimes accelerating environmental action because of economic capacity. In Tokyo, several large businesses initiated and funded urban greening efforts – for example, an entire secondary woodland was established near Otemachi Tower. The woodland was established and grown outside of Tokyo, after which it was transported, piece by piece, to the downtown area (Konijnendijk et al. 2018).

11.4.3 Integrated Governance and Territorial Planning

As this chapter has shown, cities remain highly dependent on forests and trees, both within their boundaries and in their periphery. However, urban and peri-urban forests, and the many ecosystem services they provide to cities, face multiple challenges due to gaps in governance, planning and management. In the twentieth century, urban planning moved towards metropolitanism, encompassing everything from the central city to its periphery, while rural planning faded in importance, potentially marginalising rural voices in urban and peri-urban planning considerations (Dandekar et al. 2016). Today, urban–rural landscapes are highly heterogeneous, tensions exist around land use, social and economic changes happen rapidly and capacity for ecological renewal is limited (Ernstson et al. 2010).

Nonetheless, the need to bridge the urban–rural divide is unquestionable. The NUA (UN-Habitat 2016) brings out this responsibility, encouraging governance styles that integrate both urban and rural priorities. A decisive challenge is committing to integrated and inclusive policies for territorial planning. The International Guidelines on Urban and Territorial Planning (UN-Habitat 2015) promote, among other issues, integrated urban and territorial planning to improve urban–rural complementarities and food security. Moreover, Target 11.A calls for an integrated approach to planning by 'supporting positive economic, social and environmental links between urban, peri-urban and rural areas by strengthening national and regional development planning'.

Adopting integrated, comprehensive governance and stewardship approaches at the regional, landscape level requires overcoming asymmetric power dynamics between cities and surrounding communities. Efforts of this kind are illustrated by cities such as Hangzhou, which have advanced urban–rural ecological development through a strategy known as 'one theme, two goals'. The theme is 'bringing forests into cities and letting cities embrace forests'. The two goals are planting trees and growing green minds among citizens. Another example of a landscape approach to interconnect cities and forests is that of the community forests in the UK. Since the 1990s, large-scale forest establishment and enhanced woodland stewardship in some of the UK's major agglomerations has resulted in important improvements to local quality of life and environment. In most cases, local community forest teams have acted as 'scale brokers', bringing together different local government actors, businesses, and community groups (Konijnendijk 2018). Especially in Europe, but also in Canada and the USA, the green infrastructure planning concept has been instrumental in promoting regional landscape governance and stewardship (Davies et al. 2015). Similar integrative approaches could be adopted in many other parts of the world to enhance the positive implications of SDG 11 on forests and associated socio-economic systems.

References

Adler, D. 2016. Story of cities #37: how radical ideas turned Curitiba into Brazil's green capital. *The Guardian*. Available at: www.theguardian.com/cities/2016/may/06/story-of-cities-37-mayor-jaime-lerner-curitiba-brazil-green-capital-global-icon (Accessed 9 May 2018).

Akbari, H., Huang, J., Martien, P. et al. 1988. The impact of summer heat islands on cooling energy consumption and global CO_2 concentrations. In Diamond, R. C. and Goldman, C. A. (eds.) *Proceedings of ACEEE 1988 summer study in energy efficiency in buildings*. Washington DC: American Council for an Energy-Efficient Economy, vol. 5, pp. 11–23.

Arnfield, A. J. 2003. Two decades of urban climate research: A review of turbulence, exchanges of energy and water, and the urban heat island. *International Journal of Climatology* 23(1):1–26.

Atlas of Human Development in Brazil 2013. RM Curitiba. Available at: www.atlasbrasil.org.br/2013/en/perfil_rm/curitiba (Accessed 9 May 2018).

Baharash Architecture 2018. *Liveable cities: How much green space does your city have*? Available at: www.baharash.com/liveable-cities-how-much-green-space-does-your-city-have/ (Accessed 20 September 2018).

Barbero-Sierra, C., Marques, M. J. and Ruíz-Pérez, M. 2013. The case of urban sprawl in Spain as an active and irreversible driving force for desertification. *Journal of Arid Environments* 90:95–102.

Bettencourt, L. and West, G. 2010. A unified theory of urban living. *Nature* 467(7318):912–13.

BIGTree n.d. About. Available at: https://bigtreesworld.wordpress.com/about/ (Accessed 20 September 2018).

Bonnes, M., Passafaro, P. and Carrus, G. 2010. The ambivalence of attitudes toward urban green areas: Between proenvironmental worldviews and daily residential experience. *Environment and Behavior* 43(2):207–232.

Buijs, A. E., Mattijssen, T. J. M., Van der Jagt, A. P. N. et al. 2016. Active citizenship for urban green infrastructure: Fostering the diversity and dynamics of citizen contributions through mosaic governance. *Current Opinion in Environmental Sustainability* 22:1–6.

Byrne, J. A., Lo, A. Y. and Yang, J. 2015. Residents' understanding of the role of green infrastructure for climate change adaptation in Hangzhou, China. *Landscape and Urban Planning* 138:132–43.

C40 Cities 2016. *C40 Good practice guides: Ho Chi Minh City – Triple-A strategic planning*. Available at: www.c40.org/case_studies/c40-good-practice-guides-ho-chi-minh-city-triple-a-strategic-planning (Accessed 15 May 2018).

Campos, M., Velázquez, A., Bocco Verdinelli, G. et al. 2012. Rural people's knowledge and perception of landscape: A case study from the Mexican Pacific Coast. *Society and Natural Resources* 0:1–16. https://doi.org/10.1080/08941920.2011.606458.

Cities IPCC Conference 2018. Cities and Climate Change Science Conference. Available at: https://citiesipcc.org/ (Accessed 23 August 2018).

City of Vancouver 2014. *City of Vancouver Urban Forest Strategy*. Vancouver BC, Canada: City of Vancouver, Greenest City 2020 and Vancouver Board of Parks and Recreation.

City of Vancouver 2018. *Vancouver's Urban Forest Strategy*. Vancouver BC, Canada: City of Vancouver and Vancouver Board of Parks and Recreation.

Cohen, A. J., Brauer, M., Burnett, R. et al. 2017. Estimates and 25-year trends of the global burden of disease attributable to ambient air pollution: An analysis of data from the Global Burden of Diseases Study 2015. *The Lancet 389*(10082):1907–18.

Cook, E. M., Hall, S. J. and Larson, K. L. 2012. Residential landscapes as social-ecological systems: A synthesis of multi-scalar interactions between people and their home environment. *Urban Ecosystems 15*(1):19–52.

Dandekar, H. C. and Hibbard, M. 2016. Rural issues in urban planning: Current trends and reflections. *International Planning Studies* 21(3):225–9.

Davies, C., Hansen, R., Rall, E. et al. 2015. *Green infrastructure planning and implementation: The status of European green space planning and implementation based on an analysis of selected European city regions*. Deliverable 5.1 of the GREEN SURGE project. Available at: https://greensurge.eu/working-packages/wp5/files/D_5.1_Davies_et_al_2015_Green_Infrastructure_Planning_and_Implementation_v2.pdf (Accessed 7 June 2018).

Dickson, E., Baker, J. L., Hoornweg, D. and Tiwari, A. 2012. *Urban risk assessments: Understanding disaster and climate risk in cities. Urban Development Series*. Washington, DC: World Bank.

Donovan, G. H. and Butry, D. 2009. The value of shade: Estimating the effect of urban trees on summertime electricity use. *Energy and Buildings* 41(6):662–8.

Drigo, R. and Salbitano, F. 2008. *WISDOM for cities. Analysis of wood energy and urbanization using WISDOM methodology. Woodfuels Integrated Supply/Demand Overview Mapping*. Rome: FAO.

Dymitrow, M., Biegańska, J. and Grzelak-Kostulska, E. B. 2016. Deprivation and the rural-urban trap. *Tijdschrift voor Economische en Sociale Geografie* 109(1):87–108.

Ernstson, H., Barthel, S., Andersson, E. and Borgström, S. T. 2010. Scale-crossing brokers and network governance of urban ecosystem services: The case of Stockholm. *Ecology and Society* 15(4):28.

FAO 2009. *Stratégie de développement et plan d'action pour la promotion de la foresterie urbaine et périurbaine de la Ville de Bangui*. Prepared by Salbitano, F. Urban and Peri-Urban Forestry Working Paper No. 3. Rome: FAO.

FAO 2012. *Plateforme WISDOM pour N'Djaména, Tchad. Diagnostic et cartographie de l'offre et de la demande en combustible ligneux. Foresterie urbaine et périurbaine*. Forest Working Paper 8. Rome: FAO.

FAO 2018. *The State of the World's Forests 2018. Forest Pathways to sustainable development*. Rome: FAO.

Fensom, A. 2015. Asia's urbanization just beginning. Available at: http://thediplomat.com/2015/01/asias-urbanization-just-beginning/ (Accessed 8 June 2018).

Ferrini, F., Konijnendijk van den Bosch, C. and Fini, A. (eds.) 2017. *Routledge handbook of urban forestry*. London: Routledge.

Fields, B. 2009. From green dots to greenways: Planning in the age of climate change in post-Katrina New Orleans. *Journal of Urban Design* 14(3):325–44.

Fundación Pro-Bosque. 2018. *Bosque Cerro Blanco*. Available at: http://bosquecerroblanco.org/es/ (Accessed 18 September 2018).

Ghofrani, Z., Sposito, V. and Faggian, R. 2017. A comprehensive review of blue-green infrastructure concepts. *International Journal of Environment and Sustainability* 6(1).

Haaland, C. and Konijnendijk van den Bosch, C. 2015. Challenges and strategies for urban green space planning in cities undergoing densification: A review. *Urban Forestry & Urban Greening* 14(4):760–71.

Hiner C. C. 2016. Beyond the edge and in between: (Re)conceptualizing the rural–urban interface as meaning–model–metaphor. *The Professional Geographer* 68(4):520–32.

Hoornweg, D., Sugar, L., and Trejos Gomez, C. L. 2011. Cities and greenhouse gas emissions: Moving forward. *Environment and Urbanization* 23(1):207–227.

ICLEI 2018. *ICLEI – Local governments for sustainability*. Available at: https://iclei.org/en/Home.html (Accessed September 2018).

i-Tree 2018. *i-Tree: Tools for assessing and managing forests and community trees*. Available at: www.itreetools.org/ (Accessed 8 August 2018).

Johnson, E. A. J. 1970. *The organization of space in developing countries*. London: Oxford University Press.

Kabisch, N., van den Bosch, M. and Lafortezza, R. 2017. The health benefits of nature-based solutions to urbanization challenges for children and the elderly – A systematic review. *Environmental Research* 159:362–73.

Kanuri, C., Revi, A., Espey, J. and Kuhle, H. 2016. *Getting started with the SDGs in cities: A guide for stakeholders*. Sustainable Development Solutions Network. Available at: http://unsdsn.org/wp-content/uploads/2016/07/9.1.8.-Cities-SDG-Guide.pdf (Accessed 6 May 2018).

Konijnendijk, C. C. 2018. *The city and the forest: The cultural landscape of urban woodland*. Berlin: Springer.

Konijnendijk, C. C., Rodbell, R., Salbitano, F. et al. 2018. The changing governance of urban forests. *Unasylva* 69(250):37–42.

Laukkonen, J., Blanco, P. K., Lenhart, J. et al. 2009. Combining climate change adaptation and mitigation measures at the local level. *Habitat international* 33(3):287–92.

Lenhart, J., Van Vliet, B. and Mol, A. P. 2015. New roles for local authorities in a time of climate change: The Rotterdam Energy Approach and Planning as a case of urban symbiosis. *Journal of Cleaner Production* 107:593–601.

Lohr, V. I., Pearson-Mims, C. H., Tarnai, J. and Dillman, D. A. 2004. How urban residents rate and rank the benefits and problems associated with trees in cities. *Journal of Arboriculture* 30:28–35.

Madanipour, A. 2006. Urban planning and development in Tehran. *Cities* 26(6):433–38.

McBride, J.R. and Mossadegh, A. 2012. Tree-lined canals and the urban forest of Tehran. *Arboricultural Journal* 24(2–3):155–73.

Mendieta, E. 2011. Medellín and Bogotá: The global cities of the other globalization. *City* 15(2):167–80.

MIT Senseable City Lab 2018. *Treepedia*. MIT Senseable City Lab. Available at: http://senseable.mit.edu/treepedia (Accessed 5 September 2018).

Moraci, F., Errigo, M. F., Fazia, C., Burgio, G. and Foresta, S. 2018. Making less vulnerable cities: Resilience as a new paradigm of smart planning. *Sustainability* 10(3):755.

Naturvation 2018. *The Urban Nature Atlas*. Available at: https://naturvation.eu/ (Accessed September 2018).

Nesbitt, L. and Meitner, M.J. 2016. Exploring relationships between socioeconomic background and urban greenery in Portland, OR. *Forests* 7(8):162.

Nilsson, M., Griggs, D. and Visbeck, M. 2016. Policy: Map the interactions between Sustainable Development Goals. *Nature* 534(7607):320–3.

Potapov, P. V., Turubanova, S. A., Hansen, M. C. et al. 2012. Quantifying forest cover loss in Democratic Republic of the Congo, 2000–2010, with Landsat ETM+ data. *Remote Sensing of Environment* 122:106–116.

Raymond, C. M., Frantzeskaki, N., Kabisch, N. et al. 2017. A framework for assessing and implementing the co-benefits of nature-based solutions in urban areas. *Environmental Science & Policy* 77:15–24.

Revi, A., Satterthwaite, D. E., Aragón-Durand, F. et al. 2014. Urban areas. In Field, C. B., Barros, V. R., Dokken, D. J. et al. (eds.) *Climate change 2014: Impacts, adaptation, and vulnerability. Part A: Global and sectoral aspects*. Contribution of Working Group II to the Fifth Assessment Report of the Intergovernmental Panel on Climate Change New York: Cambridge University Press, pp. 535–612.

Salbitano, F., Borelli, S., Conigliaro, M. and Chen, Y. 2016. *Guidelines on urban and peri-urban forestry*. FAO Forestry Paper No.178. Rome: FAO.

Salbitano, F., Borelli, S. and Sanesi, G. 2015. Urban forestry and agroforestry. In de Zeeuw, H. and Drechsel, P. (eds.) *Cities and agriculture: Developing resilient urban food systems*. New York: Routledge, pp. 285–311.

Samndong, R. A. and Nhantumbo, I. 2015. *Natural resources governance in the Democratic Republic of Congo: Breaking sector walls for sustainable land use investments*. IIED Report. Available at: http://doi.org/10.13140/RG.2.1.2821.9285 (Accessed 23 May 2018).

Schwarz, K., Fragkias, M., Boone, C. G. et al. 2015. Trees grow on money: urban tree canopy cover and environmental justice. *PLoS ONE* 10(4):1–17.

Scott, A. J., Gilbert, A. and Gelan, A. 2007. *The urban-rural divide: Myth or reality?* SERG Policy Brief No. 2. The Macaulay Institute, Aberdeen, UK.

Seto, K. C., Parnell, S. and Emlqvist, T. 2013. A global outlook on urbanization. In Elmqvist, T., Fragkias, M., Goodness, J. et al. (eds.) *Urbanization, biodiversity and ecosystem services: Challenges and opportunities*. New York: Springer, pp. 1–12.

Stanley Park Ecology Society n.d. Overview of SPES. Available at: http://stanleyparkecology.ca/about-us/our-history/ (Accessed September 2018).

Swanwick, C., Dunnett, N. and Woolley, H. 2003. Nature, role and value of green spaces in towns and cities: an overview. *Built Environment* 29(2):94–106.

Tacoli, C. 1998. *Bridging the divide: Rural-urban interactions and livelihood strategies*. Gatekeeper Series No. 77. London: IIED.

Taylor, L. and Hurley, P. T. (eds.) 2016. *A comparative political ecology of exurbia: Planning, environmental management, and landscape change*. Berlin: Springer Verlag.

Thanhnien News. 2016. *Ho Chi Minh City announces massive tree removal plan for metro station*. Available at: www.thanhniennews.com/society/ho-chi-minh-city-announces-massive-tree-removal-plan-for-metro-station-60521.html (Accessed 5 June 2018).

UN 2014. *World urbanization prospects: The 2014 revision*. New York: UN Department of Economic and Social Affairs, Population Division.

UN 2015. *Sustainable Development Goals. Goal 11: Make cities inclusive, safe, resilient and sustainable*. Available at: www.un.org/sustainabledevelopment/cities/ (Accessed 4 May 2018).

UN 2016. *Progress of Goal 11 in 2016*. Available at: https://sustainabledevelopment.un.org/sdg11 (Accessed 5 May 2018).

UN 2017. *Progress towards the Sustainable Development Goals. Report of the Secretary-General*. United Nations Economic and Social Council. Available at: www.un.org/ga/search/view_doc.asp?symbol=E/2017/66&Lang=E (Accessed 5 May 2018).

UN-Habitat 2003. *The challenge of slums – Global report on human settlements 2003*. London: Earthscan Publications on Behalf of UN-Habitat.

UN-Habitat 2015. *International guidelines on urban and territorial planning. HS/059/15E*. Nairobi: UN Human Settlements Programme.

UN-Habitat 2016. *Urbanization and development: Emerging futures. World Cities Report 2016*. Nairobi: UN Human Settlements Programme.

UN-Habitat 2017. *The New Urban Agenda. A/RES/71/256*. Nairobi: UN Human Settlements Programme.

UNDP 2018. *Goal 11 targets*. Available at: www.undp.org/content/undp/en/home/sustainable-development-goals/goal-11-sustainable-cities-and-communities/targets.html (Accessed 3 April 2018).

UN DESA Population Division. 2015. *World urbanization prospects: The 2014 revision*. New York: United Nations.

UNISDR 2015. *Making development sustainable: The future of disaster risk management. Global assessment report on disaster risk reduction*. Geneva: UNISDR.

USDA Northern Research Station 2017. STEW-MAP: Amplifying the power of urban environmental stewardship groups. *Current Urban Field Station Topics* 3:2017.

van den Bosch, M. and Sang, Å. O. 2017. Urban natural environments as nature-based solutions for improved public health – A systematic review of reviews. *Environmental Research* 158:373–84.

Van Eetvelde, V. and Antrop, M. 2004. Analyzing structural and functional changes of traditional landscapes: Two examples from Southern France. *Landscape and Urban Planning* 67(1):79–95.

VNR 2017. *Voluntary National Reviews. Synthesis Report*. New York: UN, High-Level Political Forum on Sustainable Development.

Weitz, N., Carlsen, H., Nilsson, M. and Skånberg, K. 2017. Towards systemic and contextual priority setting for implementing the 2030 Agenda. *Sustainability Science*:1–18.

WHO 2016. *Ambient air pollution: A global assessment of exposure and burden of disease*. World Health Organization. Available at: http://apps.who.int/iris/bitstream/handle/10665/250141/9789241511353-eng.pdf?sequence=1 (Accessed 7 December 2018).

Wolch, J. R., Byrne, J. and Newell, J. P. 2014. Urban green space, public health, and environmental justice: The challenge of making cities 'just green enough'. *Landscape and Urban Planning* 125:234–44.

World Bank/WWF Alliance for Forest Conservation and Sustainable Use 2003. *Running pure: The importance of forest protected areas to drinking water.* Compiled by Nigel Dudley and Sue Stolton. Available at: assets.panda.org/downloads/runningpurereport.pdf (Accessed 8 June 2018).

World Population Review 2018. *Tehran population 2018*. Available at: http://worldpopulationreview.com/world-cities/tehran-population/ (Accessed 20 May 2018).

World Urban Parks 2013. *World Urban Parks: The organisation for open space and recreation*. Available at: www.worldurbanparks.org/en/2-content/1-welcome (Accessed 18 September 2018).

Yue, W., Liu, Y. and Fan, P. 2013. Measuring urban sprawl and its drivers in large Chinese cities: The case of Hangzhou. *Land Use Policy* 31:358–70.

Zhu, J. 2012. Development of sustainable urban forms for high-density low-income Asian countries: The case of Vietnam: The institutional hindrance of the commons and anticommons. *Cities* 29(2):77–87.

Chapter 12 SDG 12: Responsible Consumption and Production – Potential Benefits and Impacts on Forests and Livelihoods

Patrick Schröder*, Alexander S. Antonarakis, Jana Brauer, Abu Conteh, Ryo Kohsaka, Yuta Uchiyama and Pablo Pacheco

Key Points

- Although forests are not explicitly mentioned in SDG 12, achieving the targets will result in positive contributions towards forest conservation and support forest-dependent livelihoods.
- SDG 12 targets can contribute to reducing trade-offs between other SDGs; in particular, Target 12.3 (aimed at reducing food waste and food losses) can limit trade-offs between SDG 2 and SDG 15.
- SDG 12 has its limitations, including the lack of absolute limits to consumption of forest products or products that place pressures on forests leading to deforestation and forest degradation.
- The main players for achieving SDG 12 targets with positive outcomes for forests will comprise national governments, large companies and consumers involved in global value chains.

A thorough, integrative SCP approach that addresses systemic issues is required to achieve sustainable forest management and land use associated with responsible consumption.

12.1 Introduction

This chapter identifies and analyses the potential benefits, impacts and contributions of efforts to achieve SDG 12 – Sustainable Consumption and Production (SCP) – on forests and forest-dependent livelihoods. While SCP has been part of the international policy discourse for more than four decades, its uptake has not been smooth. A bias has tended towards relatively weak measures. Currently, SDG 12 has no specific direct reference to forests or forest-dependent people among its targets or indicators, despite linkages between sustainable forest management and agricultural commodity supply. These linkages have implications for deforestation and forest degradation.

* Lead author.

The significant links between SCP and forests have yet to receive sufficient attention among the expert and policy communities. Although SDG 12 is considered a major contributor to the protection and enhancement of natural resources, including forests (FAO 2018), and is seen to be particularly relevant to the supply of forest products (Brack 2018), 'progress towards this goal has so far been very limited' (Brack 2018: 5).

This chapter's overall findings align with the above assessments by FAO and the United Nations Forum on Forests (UNFF). Achieving SDG 12 targets can contribute to improving forest conservation and sustainable management by reducing pressures from forest-risk commodities (e.g. palm oil, soy, cacao, beef) and incentivising sustainable supply of forest products (e.g. timber, pulp and paper), leading to slowing or reducing current impacts.

The SDG 12 targets in Table 12.1 do not suggest any direct trade-offs that could emerge between achieving the SDG 12 targets and protecting forest ecosystem functions and services. However, depending on how governments and the private sector address SDG 12 in the forests and forest-risk commodities' supply and value-chain governance, there can be issues of leakage and indirect social impacts (e.g. exclusion of smallholders and forest-dependent communities), resulting in trade-offs with SDG 10 (Reduced Inequalities).

SDG 12 can enable conditions for advancing a more sustainable supply of forest commodities, notably timber and pulp and paper, as well as expanding the adoption of more sustainable practices in the supply of forest-risk commodities. These enabling conditions are primarily linked to the expansion of more responsible sourcing strategies downstream in the value chain, which may translate into improved standards and incentives for making upstream production more sustainable. The targets and indicators per se do not provide assurance that SDG 12 will effectively support forest conservation and sustainable forest management. To achieve the full potential of SCP approaches for forests and forest-dependent people, explicit criteria on land use and forests must be included in SDG 12 targets, accounting for leakage and spatial spillover effects (Lambin and Meyfroidt 2011).

SCP is a well-established interdisciplinary research field with a wide variety of practical life-cycle approaches, including life-cycle analysis (LCA), cleaner production, eco-efficiency, changes in consumption patterns, using less resource-intensive products, the 3Rs (reduce, reuse and recycle), moving from material products to immaterial services, energy conservation, sharing the use of products and using higher-quality products with longer lifespans (Lebel and Lorek 2008). The academic discourse on SCP differentiates between

Table 12.1 SDG 12 targets and indicators with links to forests

Targets	Indicators	Links to forests
12.1 Implement the 10YFP on SCP, all countries taking action, with developed countries taking the lead, taking into account the development and capabilities of developing countries	12.1.1 Number of countries with SCP national action plans or mainstreamed as a priority or a target into national policies	National SCP action plans can include mechanisms to support forest commodities and reduce pressures from forest-risk commodities, depending on country-specific priorities and resource base
12.2 By 2030, achieve the sustainable management and efficient use of natural resources	12.2.1 Material footprint, material footprint per capita and material footprint per GDP 12.2.2 Domestic material consumption, domestic material consumption per capita and domestic material consumption per GDP	Forest commodities and forest-risk commodities as part of materials footprints and domestic material consumption Increased material efficiency in primary and secondary processing of forest products
12.3 By 2030, halve per capita global food waste at the retail and consumer levels and reduce food losses along production and supply chains, including post-harvest losses	12.3.1 Global food loss index	Reduced food waste and losses would mean reduced need for new agricultural land, leading to reduced deforestation for agricultural supply
12.4 By 2020, achieve the environmentally sound management of chemicals and all wastes throughout their life cycle, in accordance with agreed international frameworks, and significantly reduce their release to air, water and soil in order to minimize their adverse impacts on human health and the environment	12.4.1 Number of parties to international multilateral environmental agreements on hazardous waste, and other chemicals that meet their commitments and obligations in transmitting information as required by each relevant agreement 12.4.2 Hazardous waste generated per capita and proportion of hazardous waste	Prevent forest areas being used as illegal dumping grounds for industrial and hazardous waste products Less soil and water pollution through avoidance of waste and careful management of chemicals in harvesting areas

12.5 By 2030, substantially reduce waste generation through prevention, reduction, recycling and reuse	12.5.1 National recycling rate, tons of material recycled	Reduction/improved use of harvesting and industrial residues Recycling of paper and wood products
12.6 Encourage companies, especially large and transnational companies, to adopt sustainable practices and to integrate sustainability information into their reporting cycle	12.6.1 Number of companies publishing sustainability reports	Sustainable practices in supply chains of forest commodities and forest-risk commodities, reported by companies through their sustainability reports
12.7 Promote public procurement practices that are sustainable, in accordance with national policies and priorities	12.7.1 Number of countries implementing SPP policies and action plans	Certification schemes for forest commodities and agricultural commodities used in government SPP policies and related initiatives and practices
12.8 By 2030, ensure that people everywhere have the relevant information and awareness for sustainable development and lifestyles in harmony with nature	12.8.1 Extent to which (i) global citizenship education and (ii) education for sustainable development (including climate change education) are mainstreamed in (a) national education policies; (b) curricula; (c) teacher education; and (d) student assessment	Information and environmental education about sustainable forest production, forest conservation, and life-cycle assessments of forest products
12.A Support developing countries to strengthen their scientific and technological capacity to move towards more sustainable patterns of consumption and production	12.A.1 Amount of support to developing countries on research and development for sustainable consumption and production and environmentally sound technologies	Improved technological capabilities to reduce impacts on forests, e.g. improved harvesting and processing technologies with improved materials efficiency

Table 12.1 (cont.)

Targets	Indicators	Links to forests
12.B Develop and implement tools to monitor sustainable development impacts for sustainable tourism that creates jobs and promotes local culture and products	12.B.1 Number of sustainable tourism strategies or policies and implemented action plans with agreed monitoring and evaluation tools	Nature-based and forest-based tourism strategies as part of tourism strategies; tools to monitor the impact of tourism on forest resources
12.C Rationalise inefficient fossil fuel subsidies that encourage wasteful consumption by removing market distortions, in accordance with national circumstances, including by restructuring taxation and phasing out those harmful subsidies, where they exist, to reflect their environmental impacts, taking fully into account the specific needs and conditions of developing countries and minimising the possible adverse impacts on their development in a manner that protects the poor and the affected communities	12.C.1 Amount of fossil fuel subsidies per unit of GDP (production and consumption) and as a proportion of total national expenditure on fossil fuels	Linked to subsidies for bioenergy, potential trade-offs between reduction in fossil fuel subsidies and growth in bioenergy subsidies, with both positive and negative impacts on forests

Source for targets and indicators: https://sustainabledevelopment.un.org/sdg12

improvements in technical-managerial resource efficiency and green supply chains (Rajeev et al. 2017). It also comprises systemic approaches considering social and behavioural change and sufficiency approaches (Cohen et al. 2014). Sufficiency is defined as the consumption of commodities and services in amounts just enough for ideal health (Boulanger 2010). Sufficiency principles include individual and societal restraint in consumption, precaution and 'polluter pays'. It considers planetary risks, not just in the short term for immediate beneficiaries, but also in the long term for the under-represented and future generations (Princen 2003). In the context of public health and sustainable food systems, sufficiency approaches regarding voluntary reduction and possible policy targets are important to address the overconsumption of meat (Allievi et al. 2015), one of the main drivers of tropical deforestation.

Efficiency measures can encourage demand and consumption, offsetting gains; sufficiency approaches are necessary to address these rebound effects, also known as Jevons' Paradox (Jevons 1865). There is evidence of rebound effects in the energy and transport sectors and the manufacturing industries; it has been observed in the agricultural sector, when improvements in water productivity and irrigation efficiency resulted in increases in total water use (Song et al. 2018). Regarding climate change, relying solely upon energy efficiency to reduce carbon emissions is misguided (Herring and Sorrell 2009). Likewise for forests: relying solely on efficiency measures in the plantation, harvest and supply of forest products or forest-risk commodities could result in increases in demand and higher consumption levels. Environmental rebounds could also emerge from conservation policies that decrease the use of tropical hardwood, leading to a consumption shift towards materials using chemicals, toxic components or higher CO_2 emissions (Maestre-Andrés et al. 2012).

The need for complementarity between efficiency and sufficiency approaches can be illustrated by the example of pulp and paper. The efficiency approach aims at improving the supply-chain and industrial-production processes of mills by improving material and energy efficiency through new technologies (Griffin et al. 2018) and increasing the use of recycled fibre. A systemic approach to SCP is more comprehensive, focusing on sustainable uses of paper, behavioural changes of institutional and industrial users and private consumers, and finding solutions to address the growing global demand for paper and paper products. While efficiency aspects are included under SDG 12 and are linked to SDG 9 (Industry, Innovation and Infrastructure), the more systemic ones are largely missing – with the exception of Target 12.8, aiming to provide more consumer information for sustainable lifestyles.

An analysis of SDG 12's overall effectiveness for achieving SCP patterns (Bengtsson et al. 2018) concludes that the current conception of targets mainly relies on efficiency approaches rather than a systemic approach considering sufficiency and inequality in consumption patterns. SDG 12 represents a partial and inadequate conceptualisation of SCP, hampering effective implementation and progress towards sustainability. The paper provides some suggestions on how governments and other actors involved in operationalisation of SDGs could more effectively pursue SCP from a systemic standpoint. While not specifically discussing the impact on forest resources and livelihoods, the findings are directly relevant for forests.

An efficiency-based approach, while potentially positive, does not automatically guarantee significant positive outcomes for forest conservation and sustainable management, nor for forest livelihoods. One of the main issues for forests is how SDG 12 can contribute to SCP for forest products, as well as reduce the negative impacts from forest conversion to meet growing demand for forest-risk commodities (e.g. soy, palm oil, beef and cocoa). Most SCP approaches implement voluntary sustainability standards in corporate sourcing or public procurement, aiming to increase efficiency and reduce environmental impacts in the production phase. While this approach is an important element for forest conservation efforts and reducing deforestation from large-scale agri-food production, what is less clear is how the efficiency-based approach will impact forest-based livelihoods. Small-scale producers may be excluded from value chains, if they are unable to comply with more stringent criteria, mainly due to uptake costs and market access.

Our analysis focuses in particular on the 10-Year Framework Plan (10YFP) on SCP (Target 12.1), food waste reduction (Target 12.3), the role of the private sector in adopting sustainable corporate practices (Target 12.6), sustainable public procurement (Target 12.7), sustainable tourism (Target 12.B) and rationalising inefficient fossil fuel subsidies (Target 12.C). The analysis features selected country-specific case studies on how adopting SCP principles and practices can provide a positive contribution to forests and forest livelihoods by achieving specific SDG 12 targets, along with associated issues arising from the process.

12.2 The 10YFP on SCP

Target 12.1 concerns in particular the 10-Year Framework Plan (10YFP) on SCP, an outcome of the Rio+20 conference, coordinated by the United Nations Environment Programme (UNEP). Various commentators have criticised the 10YFP for defaulting to 'weak' forms of sustainable consumption intervention, focusing on efficiency and technological innovation (Hobson et al. 2013). We review projects and initiatives implemented under the 10YFP that relate to forest protection and conservation. The 10YFP has 6 programmatic areas: consumer information, sustainable tourism, sustainable lifestyles and education, sustainable buildings and construction, sustainable food systems and sustainable public procurement (SPP). In 2018 the 10YFP was renamed and rebranded as the One Planet Network.

The 10YFP involves more than 500 stakeholders, including governments and implementing partners (UN bodies, civil society and private sector organisations). The 10YFP programmes are closely related to several SDG 12 targets; while links to forest and forest actors are not explicit, a number of international and national forest actors are involved (Table 12.2), the most active and visible of which is the Forest Stewardship Council (FSC).The FSC is

Table 12.2 The 6 programmes of the 10YFP and selected forest stakeholders

10YFP programme area	Forest-related stakeholders involvement	Links with SDG 12 targets
Consumer Information	Indonesia's Ministry for Environment and Forestry ISEAL Alliance	12.1, 12.8
Sustainable tourism	Rainforest Alliance IUCN WCPA Tourism and Protected Areas Specialist Group	12.1, 12.2, 12B
Sustainable lifestyles and education	Japan's Ministry of Environment France's Ministry of Ecology, Sustainable Development and Energy Sweden's Ministry of Environment	12.1, 12.3
Sustainable buildings and construction	Finland's Ministry of the Environment Various green building initiatives	12.1, 12.2, 12.7
Sustainable food systems	IFOAM – Organics International FAO	12.1, 12.3, 12.8
Sustainable public procurement	UNEP (lead) FSC Germany's Ministry for Environment, Nature Conservation, Building and Nuclear Safety (BMUB) Various national ministries and organisations dealing with SPP policies on forest products	12.1, 12.7, 12.8

devoted to the certification of forests and forest product value chains. For the 10YFP to become more relevant to forests, it has to connect to other important initiatives such as the New York Declaration on Forests, a partnership of governments, multinational companies, civil society and Indigenous peoples that strives to halve deforestation by 2020 and end it by 2030.

The 10YFP has developed a complex institutional structure involving numerous stakeholders. For example, the 10YFP Sustainable Tourism Programme has established a multi-stakeholder advisory committee of committed institutions from different geographic regions and categories (governmental agencies, non-governmental organisations (NGOs), private sector businesses, intergovernmental organisations as well as academia and UN agencies) and acts as a forum for consultation, advice and support to the Lead, Co-Leads and Coordination Desk for its implementation. Similarly, the SPP programme set up a multi-stakeholder advisory committee that includes some forest actors, such as the FSC and Germany's Ministry for Environment, Nature Conservation, Building and Nuclear Safety.

The 10YFP has established a global database on SCP initiatives (SCP Clearinghouse). Of the 1000+ initiatives and programmes registered, less than 10 focus explicitly on forests, indicating the weak link between the 10YFP and forests; however, there are many other forest initiatives that have not linked up or reported to the SCP Clearinghouse. Additionally, some of the 10YFP initiatives working on sustainable food systems, lifestyles and healthy diets have positive indirect effects on forests.

Target 8.4 also refers to the 10YFP, establishing a link between the 10YFP and the SDG 8 on economic growth and decent work. The wording about resource efficiency and decoupling of growth indicates that the 10YFP is largely situated within the neoliberal growth paradigm. Nevertheless, it provides a starting point for an alternative development model that aims for absolute reductions in resource use while also addressing inequality of existing consumption and production patterns.

12.3 Reducing Food Waste

The total global food wastage in 2017 involved almost 1.4 billion ha of land – about 28 per cent of agricultural land worldwide (FAO 2018). The food waste related to meat and dairy production is a major contributor to land conversion – here there are direct links to Targets 12.3 and 12.8 concerning sustainable lifestyles and healthy nutrition. Meat and dairy occupies about 78 per cent of the total land area in agricultural production, whereas their contribution to total food wastage makes up 11 per cent of all food (FAO 2013). This ratio indicates a high 'land intensity' of this commodity group's food waste compared with other commodity groups, even if wastage volumes in all regions are comparatively low. For ruminants, the share of concentrated feed (e.g. grains, soymeal), its constituents (such as maize or soy) and the yields in the originating regions of these crops influences the arable-land occupation intensity (FAO 2013). By region, the major areas where food losses occur are North Africa, Western Asia and Central Asia, accounting for 27 per cent of food-loss areas globally (FAO 2013).

Target 12.3 – 'by 2030, halve per capita global food waste at the retail and consumer levels and reduce food losses along production and supply chains, including post-harvest losses' – is positively related to forest protection and SDG 2 (Zero Hunger). By preventing food waste and food losses, the conversion of forest areas to agricultural land for food production, which remains the main driver of deforestation, can be reduced or avoided. Target 12.3 can support forest conservation and avoid the potential trade-off between SDG 2 and SDG 15 (Life on Land), one of the major trade-offs identified in the SDG framework (ICSU 2016).

Forests are fundamental to food security and improved livelihoods. Agricultural development, governance of land-use changes and active policy interventions are key factors affecting forest conversion (FAO 2016). Globally, one-third of food produced for human consumption is lost or wasted each year (FAO 2013). This indicates an opportunity to improve global food security and food-chain resource use, which would significantly lower environmental impacts from greenhouse gas (GHG) emissions and land conversion, thereby reducing threats to forest resources. Preventing food waste and food losses reduces demand for agriculture, forestry and other land-use products, thereby decreasing not only inputs (fertiliser, energy, machinery), but also demand for land (Smith et al. 2014) and generating benefits for other SDGs, such as 13 (Climate Action).

12.4 Tackling Waste through the 3Rs

Target 12.5 aims to 'by 2030, substantially reduce waste generation through prevention, reduction, recycling and reuse'. The reduce–reuse–recycle, or 3Rs, as part of the circular economy is a SCP approach that combines both the technical-managerial efficiency aspect of recycling and the systemic approach to reduce overall consumption and waste generation. Since the circular economy is increasingly recognised as an alternative to the conventional economic model of 'take–make–throw away', Target 12.5 has direct links with SDG 8 (Decent Work and Economic Growth) and SDG 9 (Industry, Innovation and Infrastructure).

In the forest context, Target 12.5 is relevant for three stages in the value chains of paper and wood products: (1) reducing process waste in the production phase of paper and wood products; (2) reducing the amounts of paper and wood products used during the consumption phase; and (3) reducing waste generation at the end-of-life stage through reuse of wood products and recycling of wood and fibre. The world consumes about 300 million tons of paper annually. Globally, paper production accounts for about 35 per cent of industrial roundwood (Martin 2011). Although much progress has been made in recovering and recycling paper – 230 million tonnes were recovered in 2016, a 354 per cent increase since 1980 (FAOSTAT 2016) – most paper today is still made from virgin pulp, while recycled paper accounts for only 38 per cent of the global fibre supply. There is much potential to increase recycling rates; however, 100 per cent recovery rates are impossible to achieve and the recycled fibre needs to be complemented with virgin fibre for technical reasons. Its maximum level depends on the product mix of the paper industry, which limits the potentials of paper recycling.

The third largest producer and consumer of paper and cardboard, Japan (RISI 2017), provides an example of paper recycling trends: increasing ratios in paper recycling and used paper collection, and continually growing recycled paper use since 2000 (Figure 12.1).

It is necessary to enhance the efficiency of 3R approaches on paper in large economies to contribute to global sustainable management of forests. Notably, impacts of 3R approaches on paper vary depending on the origin,

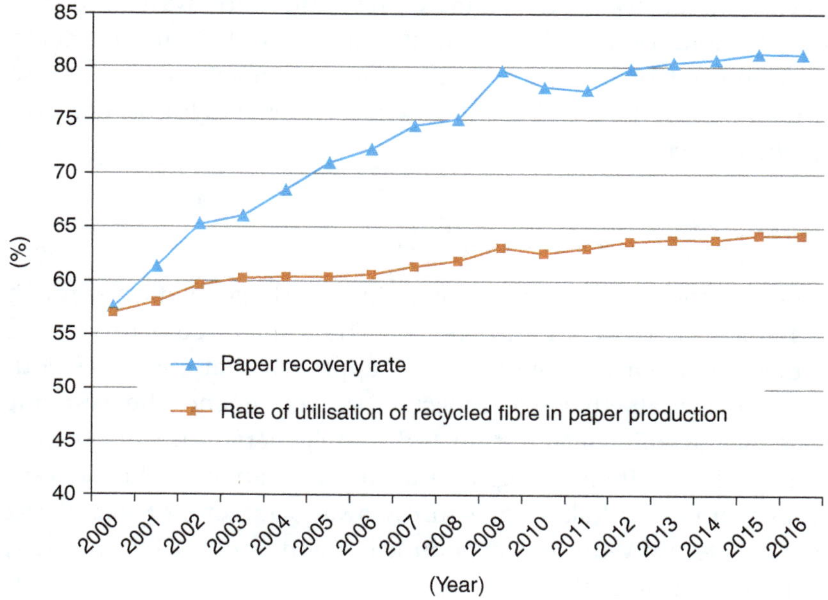

Figure 12.1 Trends of used paper collection ratio and paper recycling ratio in Japan. Source: Paper Recycling Promotion Center, Japan.

type and status of trees used for material and the processes used in production and recycling (Čabalová et al. 2011). In the case of Japan, efforts of industrial sectors related to paper production and facilitation of collective collection of used paper supported by municipalities contributed to the continuous enhancement in efficiency of paper recycling since the 1980s (synergies with SDG 9). Policies and institutions exist to facilitate recycling of resources, including the Act on Promoting Green Procurement, in force since 2001, that regulates the ratios of used paper content in paper-production materials (see also Box 12.3).

12.5 The Role of Business as an Actor in Forest Resource Management

Business is a main actor in global forest-commodity and agricultural value chains that place pressure on forest resources. Its active participation and contribution will be crucial to achieving SDG 12 and the wider goals of sustainability of forest resources and livelihoods. Target 12.6 specifically addresses the private sector: 'encourage companies, especially large and transnational companies, to adopt sustainable practices and to integrate sustainability information into their reporting cycle'. A widespread practice of businesses in the forest industry is to prepare sustainability reports, which often rely on forest certification schemes, among others, in accounting for their sustainability performance.

In addition to reporting, there are several approaches companies of all sizes can use to help achieve SCP targets, including multinationals, local small and medium-sized enterprises and smallholders and communities controlling forest resources. In relation to global forest commodity chains, in the following section we focus on forest certification schemes, and zero-deforestation initiatives, especially those linked to palm oil. Regarding forest livelihoods, we discuss analogue forestry models and Indigenous practices.

12.5.1 Forest Certification Schemes

Several targets of SDG 12 are connected to the implementation of international and national forest certification schemes that seek to help manage natural and planted forests resources more sustainably and efficiently (Target 12.2). Such schemes can enable businesses to improve sustainability performance and adopt sustainable practices (Target 12.6); they can help the public sector to promote and adopt SPP practices (Target 12.7); and, by tracking forest product production and manufacturing via chains of custody, they can inform consumers about sustainable development (Target 12.8).

The two major international certification systems are the FSC and the Programme for the Endorsement of Forest Certification (PEFC). These two systems use internationally agreed upon principles of sustainable forest management and adopt mechanisms that favour their voluntary, credible, transparent, cost-efficient and non-discriminatory nature with respect to forest types and owners. The FSC and PEFC dominate forest certification worldwide in terms of land area. The FSC is a globally applied system, while the PEFC has criteria and rules to endorse national certification systems.

The FSC promotes a triumvirate of environmentally responsible, socially beneficial and economically viable forest management. By 2018, FSC-certified forest area exceeded 200 million ha (UNECE/FAO 2018). The PEFC originally worked mostly in Europe and North America but has expanded to South America and East Asia, with certification in Central and South America carried out since the 1990s (Romero et al. 2013). Working to overcome high certification costs, both the PEFC and the FSC developed and popularised group certifications, supporting community forests, small-scale private forest owners and small rural communities to become incorporated into sustainable forest management efforts.

The total global certified forest area, subtracting double counting, was 431.4 million ha in mid-2017, up by 3.7 million ha from December 2016. The uptake of forest certification, including the FSC and PEFC, has been particularly strong in the UNECE region, which hosts 85 per cent (about 365 million ha) of the forest area certified globally. In 2016, about one-third of the global industrial roundwood supply originated in certified forests (UNECE / FAO 2018). Most certified forests are located in high-income countries, and 90 per cent in boreal and temperate forests (MacDicken et al. 2015). Of the total area certified, 47 per cent is in North America, 38 per cent in Europe (of which, 53 per cent is in the EU, and 31.5 per cent in Russia), 5 per cent in Oceania, 4.5 per cent in Asia (half of which is in East Asia), almost 4 per cent in South America and the Caribbean (half of which, Brazil) and 1.4 per cent in Africa.

Forest certification schemes were created to advance sustainable forest management, including environmental, social and economic aspects (Romero et al. 2013). Certification in tropical forests is associated with social and environmental benefits (Burivalova et al. 2017). Nevertheless, in developing countries it is often not financially sustainable in the short term without external subsidies, as good management practices tend to imply higher costs, even if price premiums on products are sometimes obtained. Benefits from price premiums and market access are limited, but certification yields social benefits including learning, governance, community empowerment and reputational benefits (Carlson and Palmer 2016).

Other studies focus on individual regions and countries. For example, in Indonesia, Miteva et al. (2015) suggest that forest certification schemes have reduced deforestation by 5 per cent, air pollution by 31 per cent, firewood dependence by 33 per cent and respiratory infections by 32 per cent, but they had no effect on fire incidence and increased forest perforation. In Brazil, McDermott et al. (2015) indicate that forest certification initiatives have favoured large firms and external markets, while smallholders' disadvantaged position due to tenure insecurity, complex forest registration, planning and management requirements has led to less success in reducing forest degradation and generating local benefit. In the Congo basin, Cerutti et al. (2017) argue that forest certification has resulted in better working and living conditions for workers and their families, better negotiation institutions between the local population and logging companies, more effective benefit-sharing mechanisms, and respect for customary rights in forest areas for agriculture, hunting and collection of non-timber forest products. Certification in Mexico had no clear impacts on deforestation (Blackman et al. 2018), but it did help community forest enterprises and small carpentry workshops, with support of NGOs and government agencies, to promote sound forest management, rural development and SCP practices (Klooster 2011).

An analysis in the EU by Gómez-Zamalloa et al. (2011) shows that ecologically (i.e. biodiversity, structure and function of forests), the impacts of certification schemes are positive; economically (i.e. price of wood, price of certified products and market access), consumers would be willing to pay slightly more for certified products, but forest owners would have to bear the certification costs themselves; and socially, certification schemes would increase public knowledge of certification and increase social integration into forest management. In the EU, forest certification challenges included fragmented forest ownership and rural abandonment, but not excessive legislation (Gómez-Zamalloa et al. 2011). In North America, Moore et al. (2012) found that certified organisations increased forest inventory programmes, established geographic inventory systems, controlled exotic invasive species, monitored chemical use and best management practices, and increased natural heritage planning. They also increased social and community efforts through releasing management plans, reporting their programmes and consulting with the public, and they promoted outreach and extension activities. Forest certification effects in Estonia, Germany, Latvia, Russia, Sweden and the UK showed significant improvements within certified forests (Hirschberger 2005): ecologically (protection of natural reserves and key habitats, increase in deadwood, restoration of threatened forest types), economically (marketing of forest products, income, recreational use) and socially (health and safety legislation, equipment and training, worker qualifications,

Box 12.1 Biodiversity Impacts of Swedish and Russian Forest Certification

Sweden and Russia are the two biggest FSC-certified countries in Europe and have a high opportunity for biodiversity and conservation (Elbakidze et al. 2011). Russia holds the largest area of forest on the planet (815 million ha) and the largest area of primary forest (273 million ha); it certifies 60 million ha through the FSC and PEFC (FAO 2015). Sweden holds the second largest area of certified forest (28 million ha) and certifies 23 million ha under the FSC and PEFC (FAO 2015). Forest certification schemes deliver biodiversity conservation through maintaining/improving ecologically valuable forest land, preventing conversion of forests to cropland, and reducing logging pressure on high-conservation-value forests (Gulisson 2003). A study on forest sites in Sweden and Russia (Elbakidze et al. 2011) investigates biodiversity conservation at spatial scales from individual trees to stands to landscapes, and finally to eco-regions. This analysis shows that the Russian standard implied higher ambitions related to biodiversity conservation than the Swedish standard at various spatial scales – mainly larger ones. The reasons are that Russia contains three times more land set aside for focal species and ecosystem connectivity; Russia has less history of forest management and thus more abundant naturally dynamic forests with high biodiversity value; also, Russian forests are state-owned, while in Sweden there is a mix of public, state, industrial and private ownership. The role of national FSC standards for biodiversity in these cases is positive, but the standards need to be integrated with formal forest protection at various spatial scales, and require the collaboration of different stakeholders.

employment rights, local stakeholder decision-making). The case study in Box 12.1 looks at the impacts of forest certification schemes on biodiversity in more detail by comparing the cases of Sweden and Russia, the two biggest FSC-certified countries in Europe.

12.5.2 Deforestation-Free Supply in Forest-Risk Commodities

Companies are increasingly committing to embrace sustainable supply and, specifically, to delink their supply chains from deforestation in response to increasing consumer pressure. Zero-deforestation commitments have been adopted by consumer goods companies, traders and retailers in the context of the Consumer Goods Forum in 2010 and the New York Declaration of Forests in 2014. These companies have been putting pressure on their suppliers to embrace similar types of commitments. Private sector commitments aimed at eliminating deforestation from a company's operations or supply chain involve objectives with different levels of ambitions (Jopke and Schoneveld

2018). Lambin et al. (2018) suggest that these commitments take different forms (e.g. codes of conduct, individual and collective pledges) and often fall short in achieving their targets. Company pledges vary in the degree to which they include time-bound interventions and the definitions and criteria to achieve verifiable outcomes. According to these same authors, in terms of short- and long-term sustainability for forest resources, 'zero-deforestation policies by companies may be insufficient to achieve broader impact on their own due to leakage, lack of transparency and traceability, selective adoption and smallholder marginalisation' (Lambin et al 2018: 109).

In the specific case of palm oil, the major traders and corporate groups producing it (Wilmar, Musim Mas Group, Golden Agri Resources, Cargill) in Malaysia and Indonesia, which are the main producer countries, have adopted No Deforestation, No Peat and No Exploitation policies. While these groups have placed pressure on their third-party suppliers to comply with zero-deforestation commitments, this has proven difficult to achieve in practice, particularly with independent smallholders, a portion of which also supply to independent mills (Skye and Paoli 2015). Some of the main palm oil corporate groups had already certified part of their operations based on the Roundtable on Sustainable Palm Oil (RSPO) standards (Box 12.2). However,

Box 12.2 The Roundtable on Sustainable Palm Oil (RSPO)

The RSPO was established in 2004 as a non-profit, industry-led trade organisation to promote the production and use of sustainable palm oil (Oosterveer et al. 2014). The RSPO provides certification for sustainably produced palm oil (Certified Sustainable Palm Oil-CSPO) based on its principles and criteria, which include conservation of biodiversity and natural resources and reductions in greenhouse gases. The RSPO emerged through a multi-stakeholder process, including private sector and civil society organisations. Since its inception, the RSPO has seen a slow but steady uptake of the standard. As of March 2018, 19 per cent of global palm oil production was RSPO-certified (RSPO 2018). Recent research suggests RSPO certification has reduced losses of primary forests in certified plantations; yet, it also found that certified operations were those with more consolidated plantations, likely with few forests under pressure (Carlson et al. 2018). Certification of palm oil plantations has been slower in new frontier areas.

Transnational processes, such as the EU Renewable Energy Directive, have included RSPO as part of the different certification mechanisms to ensure compliance with sustainable palm oil supply for the EU biodiesel market. There is an ongoing debate about whether the EU should apply more stringent criteria for

Box 12.2 (cont.)

the procurement of palm oil, including a ban on palm oil for biodiesel. A long-awaited EU Action Plan on Deforestation and Forest Degradation – suggested in 2017, discussion to be initiated in 2019 – may eventually consider options for strengthening EU regulations on supply chains of agro-commodities and other forest-risk commodities as a way to reduce their negative effects in producer countries. The Amsterdam Declaration – a group of seven European countries (Denmark, France, Germany, the Netherlands, the UK, Norway and Italy) – committed to 100 per cent sustainable sourcing of palm oil by 2020, relying on RSPO standards.

Initiatives formulated under zero-deforestation frameworks by major corporate groups, while relying on sustainable palm oil certification, are trying to move beyond the uptake of voluntary sustainability standards such as RSPO to make a stronger case for more explicit criteria on zero deforestation. RSPO, through RSPO Next, is considering the inclusion of explicit zero-deforestation criteria in order to improve the accountability of company commitments (Jopke and Schoneveld 2018).

voluntary certification was questioned for failing to reduce the expansion of oil palm plantations into forest lands (Pirard et al. 2015). An alternative approach to the High Conservation Value concept that is part of the RSPO criteria adopted by companies is the High Carbon Stock approach, which provides clearer criteria for ensuring compliance with zero-deforestation commitments (HCSA 2016).

The main issues for companies have been tracing their suppliers, particularly independent farmers, and the complex local market networks dominated by intermediaries (Jelsma and Schoneveld 2016). In this same vein, Lyons-White and Knight (2018) identify a number of barriers to the realisation of zero-deforestation supply chains in the palm oil sector. Barriers include perceived incompatibility of no-deforestation commitments and development priorities and the complexity of the supply chain – hindering the implementation of no-deforestation commitments by obscuring palm oil traceability and hindering engagement with indirect customers or suppliers. The existing model 'in which companies adopt unilateral no-deforestation commitments is unsuited to the complexity of the palm oil supply chain and is, therefore, likely to fail' (Lyons-White and Knight 2018: 311).

In order to tackle these challenges, new initiatives are emerging to foster partnerships between the public and private actors in the palm oil sector, particularly in Indonesia. These initiatives, often orchestrated by NGOs, aim at supporting private efforts to clean supply chains, mainly through improved traceability systems and delivery of technical services to independent farmers while also promoting more active government engagement, mainly in land-use planning and land-tenure regularisation – particularly at the provincial level (Luttrell et al. 2018). Pacheco et al. (2018) argue that these initiatives constitute experimentalist approaches with potential to overcome main performance gaps in the palm oil sector (i.e. land-tenure conflict, yield differences between large- and small-scale producers and carbon debt). These initiatives have three broad objectives: (1) to refine and harmonise sustainability regulations, standards and tools; (2) to implement business models that increase productivity while overcoming the challenges of involving smallholders; and (3) to reconcile value-chain interventions with territorial perspectives by adopting jurisdictional approaches. These approaches are receiving increasing attention and are increasingly orchestrated by provincial-level governors and facilitated by NGOs, which tend to operate as intermediaries.

12.5.3 Forest Livelihoods, Analogue Forestry and Non-Timber Forest Products

Forests provide food, wood energy, medicines, fodder, fibre, income and employment opportunities, as well as ecosystem services such as biodiversity and climate regulation. An example of how the objectives of forest-based livelihoods and forest conservation can be aligned is through approaches such as analogue land-management techniques to create biodiverse agroforestry systems inspired by mature forest ecology (and, hence, 'analogous' in structure and function). Furthermore, many Indigenous practices can be linked to analogue forestry.

Analogue forests differ from traditional agroforestry in their emphasis on mimicking native forest structure and in their biodiversity. This increases their utility, both in diversity of production and in the ecosystem services they provide. They are more effective as biological corridors since their structure more closely approximates a natural forest and the remnant forest patches they seek to connect (Dickinson 2014). The inherent species richness in analogue forestry systems can provide diversified income streams by providing multiple products that offer a variety of processing and marketing opportunities. For example, systems that produce non-timber forest products (NTFPs) such as spices (e.g. vanilla, nutmeg, cinnamon and black pepper) can also produce cut flowers, animal fodder, rice, beans, bamboo and plants for essential oils such as patchouli (Aguilar and Gates 2013). The viability of this approach from an economic perspective depends on markets for the products. Value can be added by using a label or brand associated with a set of production standards, such as third-party or participatory guarantee systems (Aguilar and Gates 2013). The International Federation of Organic Agriculture Movements (IFOAM) standard to certify forest garden products (FGP) is an example of how analogue forestry can be linked to international forest value chains. The certification aims to develop markets for products such as teas, spices, tamarind, guaraná, nuts and other products from countries including Sri Lanka, India, Vietnam, Thailand and Brazil (IAFN 2014).

An example of a sustainable NTFP from Brazil is the açai berry, which can be harvested through sustainable agroforestry practices that support local livelihoods and forest conservation efforts (Tunçer and Schroeder 2010). Another example is the Brazil nut, a crucial non-timber product in the Amazon – a very tall (up to 50 m), slow-growing, carbon-rich species that can live for over 500 years and thrives in dense, undisturbed rainforests. Sustainable exploitation and regeneration of such species is crucial to secure this practice and product into the future (Wadt et al. 2008). Analogue forestry is a recent concept that needs to be tested in specific local conditions; SCP and SDG 12 could be an approach to enhance the economic viability of its broader application.

12.6 Sustainable Public Procurement Policies

Target 12.7 – 'promote public procurement practices that are sustainable, in accordance with national policies and priorities' – is closely related to forest certification schemes. Public procurement represents 5–20 per cent of total forest product consumption; the indirect impact on the sector is much larger. Certification schemes often serve procurement policies as a tool to verify and demonstrate that products come from sustainably managed forests. There is a large number of countries that have specific public procurement policies for forest products. This section looks at sustainable public procurement (SPP) schemes and their links to certification schemes such as the FGP for the international market, the BES 6001 Responsible Sourcing of Construction Products and other standards. SPP schemes are expected to enable regime shifts towards sustainable societies with socio-economically and environmentally harmonised production and consumption cycles (Trindade et al. 2018). In 2017, the International Organisation for Standardisation (ISO) issued a new standard, ISO 2040, to guide sustainable procurement. This new standard can be used to facilitate the introduction of SPP in public sectors. The development of information technology is another factor that can support SPP. In the academic discourses on SPP, the contribution of E-procurement and communication with suppliers to facilitate SPP is discussed (Walker and

Brammer 2012). SPP practices are implemented in different regions of the world; however, the concept of SPP varies significantly among sectors and regions (Walker and Brammer 2009, 2011). For example, while local authorities tend to focus on buying from local and small-sized suppliers to facilitate production and consumption cycles in the given region, education sectors emphasise environmental aspects (Walker and Brammer 2009). These kinds of variations may cause conflicts among stakeholders, hindering national and international collaborations on SPP.

Appropriate timing is required when introducing SPP, considering the specific markets of the countries involved (Crespin-Mazet and Dontenwill 2012). Individual regions, with different market characteristics, need to develop their own concepts and schemes of SPP. Finding similarities among SPP regions, as well as individual uniqueness, and sharing knowledge and experience regarding SPP can contribute to global collaboration for facilitating such schemes. Furthermore, considering the cultural background of regions and implementing educational activities can facilitate the implementation of SPP (Aragão and Jabbour 2017, Delmonico et al. 2018).

In recent academic discourses, SPP has been discussed in the framework of green public procurement (Lundberg and Marklund 2018). The need for scientific evidence on effects of SPP to support decision-making on SPP schemes in relation to forest sectors has been emphasised (Cerutti et al. 2018). One of the major challenges is to link global forest certification systems such as the PEFC and FSC with SPP initiatives. An example of standards for building and construction within SPP policies is the BES 6001 Responsible Sourcing of Construction Products (UNECE / FAO 2016). It requires a range of life-cycle criteria to be met as part of the procurement process for construction materials. If mainstreamed across public procurement systems, it has the potential to influence the entire life-cycle of materials. However, as it stands, it is likely to mainly impact the production and consumption phases. At the end of 2014, the total number of valid BES 6001 certificates stood at 89, covering 76 companies. Recently, the FGP was approved by IFOAM-Organics International. FGP is the first set of organic standards developed in countries of the Global South and emphasises the place-based and sustainable production activities to produce forest products. Choosing the products with appropriate certificates to circulate them through SPP can enhance the effects of SPP on the aspects of environment, economy and society in individual countries and regions. Several national PEFC standards have been established in developing countries, such as the Philippine Sustainable Forest Certification System or the Pan-African Forest Certification (covering Cameroon, Congo and Gabon).

Box 12.3 Sustainable Public Procurement in Japan – Impacts and Benefits for Forests

While the Japanese government has not yet conducted SPP in all three aspects – environment, economy and society – the 2001 Act on Promoting Green Procurement emphasises the environmental ones. Based on this Act, the national and local governments and independent administrative institutions introduced green procurement practices. The government and institutions have an obligation to procure products and services that can reduce environmental impacts (Ministry of the Environment 2001). The definition of eco-friendly is updated once a year; the 2017 guidelines promoted the procurement of 274 products and services in 21 sectors (Ministry of the Environment 2017).

Regarding SPP in forestry sectors, Japan has yet to develop regulatory frameworks to prohibit and reduce illegal logging. The above Act is only a policy to facilitate the procurement of legal timber and wood products (Shimamoto 2014). The SPP guidelines include specific guidelines for forest products, including legality and sustainability certification of timber and wood products, and credit for procurement of certified timber and for use of lumber and wood chips from thinning. Nevertheless, other countries and environmental NGOs argue that Japan provides a market for illegal forest products because they lack relevant regulatory frameworks (Shimamoto 2014). Furthermore, public procurement of wood products is estimated to account only for 2–3 per cent of the national demand (Morita 2007). To promote green procurement, the 'Eco Mark' label is used to help users find the promoted products and services under the Act. However, uptake of green procurement among private sectors and improvement of the schemes to mainstream SPP among the government remain challenging tasks (CSO Network Japan 2017).

12.7 Sustainable Tourism Plans and Monitoring

Target 12.B – 'develop and implement tools to monitor sustainable development impacts for sustainable tourism that creates jobs and promotes local culture and products' – has relevance to forest conservation and sustainable forest management, and has potentially positive impacts on livelihoods. Although Target 12.B only asks for tools to monitor impacts, there is a link with the programmatic area of Sustainable Tourism of the 10YFP under Target 12.1. Some limited evidence indicates that establishing closer links between sustainable tourism and forest management would potentially benefit and support forest conservation.

We have explored the potential benefits and possible trade-offs between sustainable tourism (e.g. ecotourism, community-based tourism) and forests in different country contexts. However, the empirical evidence base is not solid as there is only limited research on the impact of nature-based tourism on forests. IUFRO recognises the potential of sustainable tourism and nature-based tourism for forest conservation efforts, while at the same time acknowledging the need for more research to examine how land-management agencies can use innovative techniques to incorporate local community residents into tourism decision-making (IUFRO 2017). IUFRO also emphasises that forest managers and policymakers have yet to recognise the potential benefits that quality nature-based tourism planning and management can bring to local communities and forest conservation.

Forest landscape management is extremely complex, involving a wide range of factors and institutions; integrating nature-based tourism considerations into the management plans increases both complexity and the number of stakeholders, posing challenges to implementation. According to Rizio and Gios (2014), the difficulties involved in the coordination of such a complex system stem from the multiple interests, stakeholders and utility flows. The whole set of forest and tourism management tools and mechanisms needs to be integrated into a collaborative management approach in order to achieve long-term sustainability through nature-based tourism. There are opportunities to align nature-based tourism and sustainable forest management in a collaborative approach.

The focus on operating an economically viable tourism destination that also conserves forests could be a successful approach to forest conservation for local governments, financing the conservation of the forest area and funding forest reserves through tourism. In addition, designation of land and forests as sustainable tourism destinations can be a way to protect them from invasions and illegal deforestation activities. A prominent example is Costa Rica, which has successfully pursued this approach (Bien 2010); the integration of farming communities and Indigenous communities into ecotourism has had positive livelihood impacts through improving marginal sources of incomes. At the same time, this approach can support official acknowledgement of the ecological importance of the forest region by locals, government and tourists (which links to Target 12.8).

The variety of forest landscapes, in terms of wildlife and the ecosystems more generally, offer numerous opportunities for tourism. Forest peoples often play a role in making a place tourist-friendly, both in terms of interesting cultural features and in terms of their role in managing services. Forest landscapes often serve at least two functions within the local tourism context: they are both a good in their own right and a background for the pursuit of sustainable recreational activities. Furthermore, these tourism uses provide potential income sources for the local population. To conclude, key issues for Target 12.B are to monitor (and control) overuse of protected areas for tourism and promote benefit sharing with local populations.

12.8 Fossil Fuel Subsidy Phase-Out and National Bioenergy Policies and Strategies

Target 12.C is to

> rationalise inefficient fossil fuel subsidies that encourage wasteful consumption by removing market distortions, in accordance with national circumstances, including by restructuring taxation and phasing out those harmful subsidies, where they exist, to reflect their environmental impacts, taking fully into account the specific needs and conditions of developing countries and minimising the possible adverse impacts on their development in a manner that protects the poor and the affected communities.

It relates directly to SDG 7 (Affordable and Clean Energy) and SDG 13 (Climate Action).

According to the International Monetary Fund, fossil fuel subsidies amounted globally to USD 233 billion in 2015 – more than four times the subsidies promoting renewable energy. Fossil fuel subsidies have a negative impact on the development of renewable energy (Bridle and Kitson 2014). The elimination of fossil fuel subsidies would not only be beneficial from a climate change perspective (creating synergies with SDG 7 and 13), since these subsidies work in practice as a negative carbon price, but would also help eliminate a significant market distortion that encourages inefficient consumption of fossil fuels (Sampedro et al. 2018). The overall environmental benefit of phasing out fossil fuel subsidies will be positive, especially for the global climate in terms of reducing CO_2 emissions from sectors dependent on fossil fuels.

It is unclear how the phase-out of subsidies would affect forests. The concern is that it could increase the use of forest resources for energy generation (e.g. traditional biomass, fuelwood, pellets). Although the issue of subsidies for the production and use of biomass, bioenergy and biofuels is not included in Target 12.C, we consider these interlinked issues of fossil fuel phase-out important in relation to national strategies for bioenergy and biofuels. Production, processing and use of forest-based biomass for energy is not necessarily harmful to forests, but it does require careful policy design to avoid trade-offs and unintended consequences.

EU member states have committed to phase out environmentally harmful studies – including fossil fuels – by 2020. European governments have made parallel pledges to end inefficient fossil fuel subsidies under the G7 and the G20. Despite the commitments, 11 European countries and the EU provided at least EUR 112 billion annually during the period 2014–2016 in subsidies towards fossil fuel production and consumption, with EUR 4 billion from the EU itself (Whitley et al. 2017). Fiscal support and other measures for subsidies included, as in the case of Italy, a reduction on tax applied to diesel used in the agricultural sub-sectors of farming, horticulture, forestry and aquaculture, as well as VAT concessions to petroleum products for use in agriculture, forestry and inland fisheries (Whitley et al. 2017). According to research by Chatham House (Brack 2017), many policies intended to boost biomass use are 'not fit for purpose' because they inadvertently increase pollution by ignoring emissions from burning wood in power stations and failing to account for changes in forest carbon stocks. The UK rules for bioenergy and those recently revised by the EU are inadequate for managing and monitoring emissions from burning biomass. Global electricity generation from biomass, including wood pellets, more than doubled from 2005 to 2015, and the EU has emerged as the world's biggest user of biomass for electricity generation: bioenergy is expected to contribute 57 per cent of the EU's total renewable energy by 2020 (Brack 2017). The implications of these growth trends for forests, forest-based products and the climate have yet to be systematically analysed.

Optimisation of biomass-flow cascades increase resource-use efficiency and may reduce competition; however, there may be trade-offs, again between SDG 2 (Zero Hunger) and SDG 15 (Life on Land). For example, using crop residues for bioenergy or roughage supply may leave less carbon and nutrients on cropland, reduce soil quality and carbon storage in soils and increase the risk of losses of carbon through soil erosion. Residues are also often used as forage, particularly in the tropics (Smith et al. 2014). In conclusion, Target 12.C's impacts on forests can be positive, but due to complex interlinkages between sectoral fuel and biomass subsidies, safeguarding measures are needed for mitigating and avoiding possible negative outcomes.

12.9 Conclusions

Although SDG 12 targets do not explicitly mention forests and forest-based livelihoods, and even though they mainly take a so-called efficiency approach to SCP, our assessment shows that the targets are important for forest conservation and forest livelihoods. Absolute limits to consumption of resources such as forest products or meat are not part of SDG 12, and the targets do not address systemic issues of deforestation and overconsumption of forest resources and agricultural products that drive deforestation. Therefore, the overall benefits of SDG 12 to long-term sustainability of forests and forest livelihoods are limited to slowing down impacts, rather than reversing unsustainable trends. Unless mainstreamed throughout the economy, it is unlikely that the important Target 12.2 (by 2030, achieve the sustainable management and efficient use of natural resources) will be met through the approaches presented under this SDG, such as consumer education, SPP, voluntary sustainability initiatives and reporting by the private sector, food waste reduction, eco-labelling and sustainable tourism. Therefore, measures to be taken under other SDGs, notably SDG 15, are necessary. Regarding SDG 12 target implementation, the main programme covering various issues is UNEP's 10YFP on SCP. Our analysis shows that the 10YFP does not specifically focus on forests or forest livelihoods; however, some of the initiatives, such as SPP, consumer information and eco-labelling, are consistent with and linked to ongoing efforts to further SCP of forest commodities. There are potentials for the 10YFP to initiate more specific activities and programmes to directly address forest protection and livelihoods.

Linking national and sub-national deforestation analyses with national SCP plans could be a promising strategy, especially SPP schemes or national sustainable tourism development plans and policies. SPPs are useful tools to promote the uptake of forest certification schemes to achieve SDG 12 with potentially positive impact on forests.

National governments and large companies involved in global value chains of forest products and agricultural commodities will be the main players for achieving SDG 12 targets. How SDG 12 can be leveraged to advance conservation efforts for forests is, however, not straightforward. We have attempted to provide a number of examples of how synergies between SDG 12 and forest protection and livelihoods can be created. SDG 12 can possibly also address and lessen trade-offs between other SDGs such as SDG 2 and SDG 15 (food production versus biodiversity and forest conservation) and SDG 9 and SDG 15 (industrialisation versus biodiversity and forest conservation) (See Table 12.3). To enhance forest conservation through SCP, a more integrative approach addressing systemic issues needs to be adopted. This would include efforts to address industrial drivers, poverty, food security and other underlying causes of deforestation, such as increasing levels of consumption driven by consumerism.

Table 12.3 Synergies and trade-offs associated with SDG 12 in relation to forests

Target	Synergies and Trade-offs	Chapter Section
12.1	***Synergies*** Improve resource efficiency in production and consumption and decouple economic growth from environmental degradation in accordance with the 10YFP (8.4) Mobilise resources to finance sustainable forest management (15.B)	12.2
12.2	See SDG 12.6	12.5
12.3	***Synergies*** Ensure sufficient food and end hunger (2.1) Ensure sustainable food production that maintains ecosystems and improve soil and land quality (2.4) Improve diversity of terrestrial landscapes and reduce the need for deforestation (15.1, 15.2)	12.3
12.4	***Synergies*** Reduction of waste and toxic chemicals dumped in forest areas (15.1, 15.2)	12.4
12.5	***Synergies*** Foster innovation to tackle waste through environmentally sound technologies (9.4) Strive for a land degradation-neutral world (15.3) Integrate ecosystem values into planning (15.9) Mobilise finance and provide incentives for sustainable forest management (15.B) Improve multi-stakeholder partnerships (16.7, 17.16)	12.4
12.6	***Synergies*** Improved biodiversity (15.2) Reduced deforestation (15.1) Reduced GHGs (13.2) Improved welfare and access to land (1.4) Improved working conditions and safety (8.4, 8.8) Improved negotiation between actors (16.7, 17.16)	12.5

Table 12.3 (cont.)

Target	Synergies and Trade-offs	Chapter Section
	Trade-offs Can increase forest canopy perforation (15.3) Can reduce biodiversity in monodominant stands (15.1) Can reduce profitability and productivity (8.2) Can disadvantage smallholders and forest communities in their access to markets and multi-stakeholder partnerships (1.1, 8.4, 8.8, 16.7, 17.6) Focus on large transnationals may limit economic, social and environmental links between rural to urban areas (11.A)	
12.7	***Synergies*** Promote uptake of forest certification programmes (15.1) Reduce corruption between local stakeholders, national authorities and international organisations and increase transparency (16.5, 16.6) May be strengthened through considering cultural background of regions and education (11.4, 4.4)	12.6
12.8	***Synergies*** Promote uptake of forest certification programmes (15.1)	12.5
12.A	***Synergies*** Improve technical and statistical capacity to implement sustainable forest management (15.1)	
12. B	***Synergies*** Increase local income through diversification and innovation (8.2) Raise local, government and tourist awareness of the value of nature heritage (11.4, 12.8) Promote collaborative management approach between all actors (16.7, 17.16)	12.7
12.C	***Synergies*** Increase the global share of renewable energy (7.2) Reduce greenhouse gas emissions (13.2) Improve economic and social viability of sustainable forest management (15.2)	12.8

Table 12.3 (cont.)

Target	Synergies and Trade-offs	Chapter Section
	Trade-offs Bioenergy subsidies may reduce forest cover and biodiversity (15.1, 15.2) May affect land that is currently used for agriculture (2.1, 2.4)	

References

Aguilar, E. and Gates, C. 2013. Creating better opportunities: PGS and analog forestry. *Farming Matters* 6:24–26

Allievi, F., Vinnari, M. and Luukkanen, J. 2015. Meat consumption and production – Analysis of efficiency, sufficiency and consistency of global trends. *Journal of Cleaner Production* 92:142–151.

Aragão C. and Jabbour, C. J. 2017. Green training for sustainable procurement? Insights from the Brazilian public sector. *Industrial and Commercial Training* 49:48–54.

Bengtsson, M., Alfredsson, E., Cohen, M., Lorek, S. and Schroeder, P. 2018. Transforming systems of consumption and production for achieving the Sustainable Development Goals: Moving beyond efficiency. *Sustainability Science*. https://doi.org/10.1007/s11625-018-0582-1.

Bien, A. 2010. Forest-based tourism in Costa Rica as a driver for positive social and environmental development. *Unasylva* 61(236):49–53.

Blackman, A., Goff, L. and Planter, M. R. 2018. Does eco-certification stem tropical deforestation? Forest Stewardship Council certification in Mexico. *Journal of Environmental Economics and Management* 89:306–33.

Boulanger, P. 2010. Three strategies for sustainable consumption. *Sapiens* 3(2):1–10.

Brack, D. 2017. *Woody biomass for power and heat: Impacts on the global climate*. Chatham House Research Paper. February 2017. London: Chatham House.

Brack, D. 2018. *Sustainable consumption and production of forest products*. Background study prepared for the thirteenth session of the United Nations Forum on Forests. Available at: www.un.org/esa/forests/wp-content/uploads/2018/04/UNFF13_BkgdStudy_ForestsSCP.pdf (Accessed 29 July 2019).

Brammer, S. and Walker, H. 2011. Sustainable procurement in the public sector: An international comparative study. *International Journal of Operations & Production Management* 31(4):452–76.

Bridle, R. and Kitson, L. 2014. *The Impact of fossil-fuel subsidies on renewable electricity generation*. GSI Report. Winnipeg: IISD.

Burivalova, Z., Hua, F., Koh, L. P., Garcia, C. and Putz, F. 2017. A critical comparison of conventional, certified, and community management of tropical forests for timber in terms of environmental, economic, and social variables. *Conservation Letters* 10(1):4–14.

Čabalová, I., Kačík, F., Geffert, A. and Kačíková, D. 2011. The effects of paper recycling and its environmental impact. In Broniewicz, E. (ed.) *Environmental Management in Practice* (open access), pp. 329–350. Available at: www.intechopen.com/books/environmental-management-in-practice/the-effects-of-paper-recycling-and-its-environmental-impact (Accessed 29 July 2019).

Carlson, A. and Palmer, C. 2016. A qualitative meta-synthesis of the benefits of eco-labeling in developing countries. *Ecological Economics* 127:129–45.

Carlson, K., Heilmayr, R., Gibbs, H. K. et al. 2018. Effect of oil palm sustainability certification on deforestation and fire in Indonesia. *Proceedings of the National Academy of Sciences* 115(1):121–126.

Cerutti, A. K., Ardente, F., Contu, S., Donno, D. and Beccaro, G. L. 2018. Modelling, assessing, and ranking public procurement options for a climate-friendly catering service. *The International Journal of Life Cycle Assessment* 23(1):95–115.

Cerutti, P. O., Lescuyer, G., Tacconi, L. et al. 2017. Social impacts of the Forest Stewardship Council certification in the Congo basin. *International Forestry Review* 19(S2):1–14.

Cohen, M., Brown, H. and Vergragt, P. (eds.) 2014. *Innovations in sustainable consumption: New economics, socio-technical transitions, and social practices*. Northampton, MA: Edward Elgar Publishing.

Crespin-Mazet, F. and Dontenwill, E. 2012. Sustainable procurement: Building legitimacy in the supply network. *Journal of Purchasing and Supply Management* 18(4):207–17.

CSO Network Japan 2017. *Report of sustainable public procurement in the era of SDGs*. Available at: www.csonj.org/images/reportbook001.pdf (in Japanese) (Accessed 11 April 2018).

Delmonico, D., Chiappetta Jabbour, C. J., Pereira, S. et al. 2018. Unveiling barriers to sustainable public procurement in emerging economies: Evidence from a leading sustainable supply chain initiative in Latin America. *Resources, Conservation and Recycling* 134:70–9.

Dickinson, A. K. 2014. Analog Forestry – Creating productive landscapes. *European Tropical Forest Research Network* 56:103–10.

Elbakidze, M., Angelstam, P., Andersson, K., Nordberg, M. and Pautov, Y. 2011. How does forest certification contribute to boreal biodiversity conservation? Standards and outcomes in Sweden and NW Russia. *Forest Ecology and Management* 262(11):1983–95.

FAO 2013. *Food Waste Footprint: Impact on Natural Resources*. Summary Report. Rome: FAO.

FAO 2015. *Global forest resources assessment 2015*. FAO Forestry Paper No. 1. Rome: FAO.

FAO 2016. *State of the world's forests 2016. Forests and agriculture: Land-use challenges and opportunities*. Rome: FAO.

FAO 2018. *Transforming food and agriculture to achieve the SDGs. 20 interconnected actions to guide decision-makers*. Rome: FAO.

FAOSTAT 2016. *Forestry Database*: Global production and trade of forest products in 2016. Available at: www.fao.org/faostat/en/#data/FO (Accessed 29 July 2019).

Fauset, S., Johnson, M. O., Gloor, M. et al. 2015. Hyperdominance in Amazonian forest carbon cycling. *Nature Communications* 6:6857.

Gómez-Zamalloa, M. G., Caparrós, A. and Ayanz, A. S. M. 2011. 15 years of forest certification in the European Union. Are we doing things right? *Forest Systems* 20(1):81–94.

Griffin, P., Hammond, G. and Norman, J. 2018. Industrial decarbonisation of the pulp and paper sector: A UK perspective. *Applied Thermal Engineering* 134:152–62.

Gulisson, R. 2003. Does forest certification conserve biodiversity? *Oryx* 37(2):154–65.

HCSA (High Carbon Stock Approach) 2016. *Agreement on Unified Approach to Implementing No Deforestation Commitments*. HCS Convergence Working Group Public Statement. Available at: http://d2ouvy59p0dg6k.cloudfront.net/downloads/hcs_convergence_public_statement_7_november_2016.pdf (Accessed 17 July 2019).

Herring, H. and Sorrell, S. (eds.) 2009. *Energy efficiency and sustainable consumption: The rebound effect*. London: Palgrave Macmillan.

Hirschberger, P. 2005. *The Effects of FSC-certification in Estonia, Germany, Latvia, Russia, Sweden and the United Kingdom: An analysis of corrective action requests*. WWF European Forest Programme.

Hobson, K. 2013. 'Weak' or 'strong' sustainable consumption? Efficiency, degrowth, and the 10 year framework of programmes. *Environment and Planning C: Politics and Space* 31(6):https://doi.org/10.1068/c12279.

IAFN (International Analog Forestry Network) 2014. *Forest garden product standards approved!* Available at: www.analogforestry.org/forest-garden-standards-approved/ (Accessed 8 February 2019).

ICSU (International Council for Science) 2016. *A draft framework for understanding SDG interactions*. Working Paper. Paris: ICSU.

IUFRO 2017. Forest tourism can mean billions in economic benefits. *IUFRO 125th Anniversary Congress Spotlight*, 8 July 2017. International Union of Forest Research Organisations.

Jelsma, I. and Schoneveld, G. C. 2016. *Towards more sustainable and productive independent oil palm smallholders in Indonesia: Insights from the development of a smallholder typology*. Working Paper 210. Bogor, Indonesia: CIFOR.

Jevons, W. S. 1865. *The coal question; An enquiry concerning the progress of the nation, and the probable exhaustion of our coal-mines*. 2nd ed. London: Macmillan

Jopke, P. and Schoneveld, G. C. 2018. *Corporate commitments to zero deforestation: An evaluation of externality problems and implementation gaps*. Occasional Paper 181. Bogor, Indonesia: CIFOR.

Klooster, D. 2011. The local instrumentality of global standards: How Mexican indigenous communities use FSC certification to foster a furniture production network. In Ponte, S., Gibbon, P. and Vestergaard, J. (eds.) *Governing through standards: Origins, drivers and limitations*. London: Palgrave McMillan, pp. 266–88.

Lambin, E., Gibbs, H., Heilmayr, R. et al. 2018. The role of supply-chain initiatives in reducing deforestation. *Nature Climate Change* 8:109–16.

Lambin, E. and Meyfroidt, P. 2011. Global land use change, economic globalization, and the looming land scarcity. *Proceedings of the National Academy of Sciences* 108(9):3465–3472. https://doi.org/10.1073/pnas.1100480108.

Lebel, L. and Lorek, S. 2008. Enabling sustainable production-consumption systems. *Annual Review of Environment and Resources* 33:241–75.

Lundberg, S. and Marklund, P. O. 2018. Green public procurement and multiple environmental objectives. *Economia e Politica Industriale* 45:37–53.

Luttrell, C., Komarudin, H., Mike, M. et al. 2018. *Implementing sustainability commitments for palm oil in Indonesia: Governance arrangements of sustainability initiatives involving public and private actors*. Working Paper 241. Bogor, Indonesia: CIFOR.

Lyons-White, J. and Knight, A. 2018. Palm oil supply chain complexity impedes implementation of corporate no-deforestation commitments. *Global Environmental Change* 50:303–13.

MacDicken, K. G., Sola, P., Hall, J. E., et al. 2015.Global progress toward sustainable forest management. *Forest Ecology and Management* 32:476.

Maestre-Andrés, S., Calvet Mir, L., van den Bergh, J., Ring, I. and Verburge, P. 2012. Ineffective biodiversity policy due to five rebound effects. *Ecosystem Services* 1(1):101–110.

Martin, S. 2011. *'Paper Chase'*. Ecology Global Network. Available at: www.ecology.com/2/10/paper-chase/ (Accessed 21 August 2018).

McDermott, C. L., Irland, L. C. and Pacheco, P. 2015. Forest certification and legality initiatives in the Brazilian Amazon: Lessons for effective and equitable forest governance. *Forest Policy and Economics* 50:134–42.

Ministry of the Environment (Japan) 2001. *Guidance of the Act on Promoting Green Procurement.* Available at: www.env.go.jp/en/policy/economy/pdf/gpp_pamphlet.pdf (Accessed 1 April 2018).

Ministry of the Environment (Japan) 2017. *The guideline for promoting green procurement.* Available at: www.env.go.jp/policy/hozen/green/attach/gpp%20pamphlet_eng.pdf (Accessed 11 April 2018).

Miteva D. A., Loucks C. J., Pattanayak S. K. 2015. Social and environmental impacts of forest management certification in Indonesia. *PLoS ONE* 10(7):e0129675. https://doi.org/10.1371/journal.pone.0129675.

Moore, S. E., Cubbage, F. and Eicheldinger, C. 2012. Impacts of Forest Stewardship Council (FSC) and Sustainable Forestry Initiative (SFI) forest certification in North America. *Journal of Forestry* 110(2):79–88.

Morita, K. 2007. *Issues of illegal logging and challenge of Japan: Towards legal timber trade, Council of Solution for Illegal Logging Issues and Legal Timber Trade*, Available at: http://goho-wood.jp/event/event5/c2-out-jp.pdf (in Japanese) (Accessed 11 April 2018).

Oosterveer, P., Adjei, B. E. Vellema S. and Slingerland, M. 2014. Global sustainability standards and food security: Exploring unintended effects of voluntary certification in palm oil. *Global Food Security* 3:220–226. https://doi.org/10.1016/j.gfs.2014.09.006

Pacheco, P., Schoneveld, G., Dermawan, A., Komarudin, H. and Djama, M. 2018. Governing sustainable palm oil supply: Disconnects, complementarities, and antagonisms between state regulations and private standards. *Regulation & Governance*. doi:10.1111/rego.12220. Available at: https://onlinelibrary.wiley.com/doi/full/10.1111/rego.12220 (Accessed 29 July 2019).

Pirard, R., Fishman, A., Snych, S., Kbidzinski, K. and Pacheco, P. 2015. *Deforestation-free commitments: The challenge of implementation – An application to Indonesia*. CIFOR Working Paper 181. Bogor, Indonesia: CIFOR.

Princen, T. 2003. Principles for sustainability: From cooperation and efficiency to sufficiency. *Global Environment Politics* 3(1):33–50. doi:10.1162/152638003763336374.

Rajeev, A., Pati, R., Padhi, S. and Govindan, K. 2017. Evolution of sustainability in supply chain management: A literature review. *Journal of Cleaner Production* 162:299–314.

RISI 2017. *Annual Review of Pulp and Paper Statistics*. Fastmarkets RISI. Available at: www.risiinfo .com/product/annual-review-of-global-pulp-paper-statistics/ (Accessed 8 August 2018).

Rizio, D. and Gios, G. 2014. A sustainable tourism paradigm: Opportunities and limits for forest landscape planning. *Sustainability* 6:2379–91. doi:10.3390/su6042379.

Romero, C., Putz, F. E., Guariguata, M. R. et al. 2013. *An overview of current knowledge about the impacts of forest management certification: A proposed framework for its evaluation*. CIFOR Occasional Paper 91. Bogor, Indonesia: CIFOR.

RSPO (Roundtable on Sustainable Palm Oil) 2018. *Our impact*. Available at: www.rspo.org/ about/impacts (Accessed 24 February 2019).

Sampedro, J., Arto, I. and González-Eguino, M. 2018. Implications of switching fossil fuel subsidies to solar: A Case Study for the European Union. *Sustainability* 10:50. doi:10.3390/ su10010050.

Shimamoto, M. 2014. What should we do for tropical forest management and biodiversity conservation? *Review of Environmental Economics and Policy Studies* 7(1):33–6 (in Japanese).

Skye, G. and Paoli, G. 2015. *Indonesian oil palm smallholder farmers: A typology of organizational models, needs, and investment opportunities*. Bogor, Indonesia: Daemeter Consulting.

Smith, P., Bustamante, M., Ahammad, H. et al. 2014. Agriculture, Forestry and Other Land Use (AFOLU). In *AR5 Climate Change 2014: Mitigation of Climate Change*. Contribution of Working Group III to the Fifth Assessment Report of the IPCC. Cambridge: Cambridge University Press, pp. 811–922.

Song, J., Guo, Y., Wu, P. and Sun, S. 2018. The agricultural water rebound effect in China. *Ecological Economics* 146:497–506.

Trindade, P. C., Antunes, P. and Partidário, P. 2018. SPP toolbox: Supporting sustainable public procurement in the context of socio-technical transitions. *Sustainability* 10(1):1–26.

Tunçer, B. and Schroeder, P. 2010. Sambazon: Creating environmental and social value through marketing the açai berry; sustainable agro-forestry practices in the Brazilian Amazon. In Tischner, U., Stø, E., Kjaernes, U. and Tukker, A. (eds.) *System innovation for sustainability 3: Case studies in sustainable consumption and production – food and agriculture*. Sheffield: Greenleaf Publishing, pp. 160–75.

UNECE / FAO 2016. *Promoting sustainable building materials and the implications on the use of wood in buildings: A review of leading public policies in Europe and North America*. Geneva, United Nations.

UNECE / FAO 2018. *Forest products annual market review 2017–2018*. Geneva: United Nations.

Wadt, L. H. O., Kainer, K. A., Staudhammer, C. L. and Serrano, R. O. P. 2008. Sustainable forest use in Brazilian extractive reserves: Natural regeneration of Brazil nut in exploited populations. *Biological Conservation* 141:332–46.

Walker, H. and Brammer, S. 2009. Sustainable procurement in the United Kingdom public sector. *Supply Chain Management: An International Journal* 14(2):128–37.

Walker, H. and Brammer, S. 2012. The relationship between sustainable procurement and E-procurement in the public sector. *International Journal of Production Economics* 140(1):256–68.

Whitley, S., van der Burg, L., Worrall, L. and Patel, S. 2017. *Cutting Europe's lifelines to coal. Tracking subsidies in 10 countries*. Policy Briefing. London: Overseas Development Institute.

Chapter 13 SDG 13: Climate Action – Impacts on Forests and People

Bas Louman*, Rodney J. Keenan, Daniela Kleinschmit, Stibniati Atmadja, Almeida A. Sitoe, Isilda Nhantumbo, Ronnie de Camino Velozo and Jean Pierre Morales

Key Points

- The Paris Agreement is the principal international instrument for achieving SDG 13. Current commitments under the agreement are likely to be insufficient to remain under the 2°C limit. If average global temperature increases exceed 2°C, there will be increased risks to forest area, biodiversity and species composition, and forest ecosystem services. These will most strongly impact poor and marginalised people.
- If mitigation activities are widely adopted, forests and the people depending on them will benefit through lower reductions in forest area and biodiversity and improved functions and services.
- Forest- and land-based climate action has the potential to support the adaptation of society to climate change and contribute to up to 20 per cent of the needed emission reductions to meet the 2°C goal. Despite these opportunities, only 3 per cent of climate finance is used for this purpose.
- Scaling up of such climate actions is necessary but requires stronger linkages between global climate change mitigation goals and local development and adaptation priorities. These can best be achieved through landscape-scale, locally driven, long-term approaches that engage all relevant actors and industry sectors, including agriculture. Community forest management shows promise for combining mitigation objectives with strengthening adaptive capacity.
- Evidence-based policies and regulations, education and market mechanisms that are linked to the economic benefits of forest-based climate actions can support improved decision-making by governments, communities and industries.

* Lead author.

13.1 Introduction: Climate Change, SDG 13 and Forests

SDG 13 is action to combat climate change. The most recent progress report on SDG 13[1] indicates that global temperatures have reached 1.1°C above pre-industrial levels due to increased greenhouse gas concentrations, with increasingly costly extreme weather events and rising sea levels. Urgent action is required to reduce emissions and adapt to climate change, and forests can play an important role. This chapter provides insights into the links between SDG 13, forests and the people who depend on them.

Climate change is a major challenge for societies and environments, and forests have been integral to this challenge. Forests play an important role in the global carbon cycle, and tropical deforestation and forest degradation are significant contributors to global emissions. Forests capture and store carbon dioxide in living biomass and help store carbon in soils. Above and below ground, there are 795–927 Gt of carbon stored in the soils and vegetation of the world's forests, more than half of it (55 per cent) in the tropics (Pan et al. 2011). Biomes differ in how this carbon is distributed between soil and vegetation: in boreal forests, 60 per cent of carbon is in the soils; in tropical forests, only 32 per cent. Loss of these forests has contributed an average of 2.27 Gt to global annual CO_2 emissions between 2001 and 2013 (Zarin et al. 2016). This occurred mainly in tropical forests (Liu et al. 2015).

Combined with afforestation and improved forest management, avoiding deforestation and reducing degradation can mitigate 4–20 per cent of global emission-reduction targets required to meet the +2°C limit of the Paris Agreement (Forsell et al. 2016). Such mitigation actions in the land-use sector are essential to keep global warming below 1.5°C above pre-industrial average global temperatures by the end of this century (IPCC 2018).

In addition, forests can help people and environments become less vulnerable to climate change by providing protection from floods and storms, providing food, materials or alternative income after crop failures and improving water quality and, in some cases, yield for downstream users (Osman-Elasha et al. 2009).

On the other hand, forests can be affected by climate change which will favour the survival and growth of some species, but hamper that of others. Climate change is also expected to reinforce the current threats to forests and their ecosystem services by increasing the frequency and intensity of fires, pests and diseases, extreme events (flooding and storms) and changing precipitation regimes. Achievement of Targets 13.1 and 13.2 will reduce the

[1] https://sustainabledevelopment.un.org

negative impacts of climate change and contribute to the resilience of forest ecosystems. Furthermore, forests may be affected positively or negatively by the implementation of measures under SDG 13 involving conversion of land to other forms of land cover or land use.

Humans are connected to forests via complex socio-ecological systems (Figure 13.1). In Section 13.2, we analyse how actions and measures under SDG 13 (see list of targets in Table 13.1) may influence interactions between forests and people.[2] We use documented case studies that illustrate climate actions and their impacts on forests and people at local and regional levels, focusing on two major international initiatives that link forests to climate change: forest and landscape restoration (FLR) and REDD+.[3] In Section 13.3

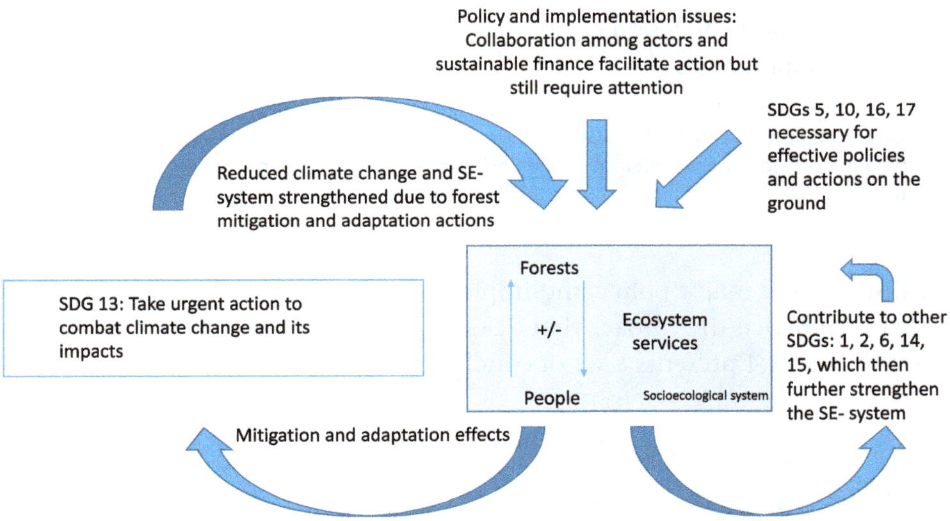

Figure 13.1 Framework of the relationship between forest-based socio-ecological (SE) systems and SDG 13. Forest-based SE systems (landscapes) have the potential to generate a virtuous cycle for the implementation of SDG 13: if climate actions consider forests in their policies, measures and actions, the mitigation and adaptation potential of forests will be enhanced and will better contribute to SDG 13 targets and the Paris Agreement's main goal of reducing the carbon concentration in the atmosphere. The latter will slow down climate change and allow forests to adapt better to new conditions. Achievement of SDGs 5, 10, 16 and 17 will reinforce the positive effects of SDG 13 on forests, while we expect that the achievement of positive effects of SDG 13 on forests will support the achievement of SDGs 1, 2, 6, 14 and 15 (Reed et al. 2015), which in turn may have positive effects on SDG 13 and forests. Greater collaboration and sustainable finance are still major challenges for SDG 13 implementation.

[2] The effects of climate on forests have been amply analysed elsewhere (e.g. Swamy et al. 2018).

[3] Reducing Emissions from Deforestation and Forest Degradation and the role of conservation, sustainable management of forests and enhancement of forest carbon stocks.

Table 13.1 SDG 13: Take action to combat climate change[4]

Target	Action
13.1	Strengthen resilience and adaptive capacity
13.2	Integrate climate change measures into national policies, strategies and planning
13.3	Improve education, awareness-raising and human and institutional capacity
13.A	Fully operationalise the Green Climate Fund through its capitalisation (USD 100 billion/year)
13.B	Support least developed countries, small island developing States, including focusing on women, youth and local and marginalised communities

Source: Adapted from https://sustainabledevelopment.un.org/sdg13

[4] Acknowledging that the United Nations Framework Convention on Climate Change (UNFCCC) is the primary international, intergovernmental forum for negotiating the global response to climate change.

we discuss how major policy and implementation issues related to these initiatives influence the impact that SDG 13 may have on forests and people, and Section 13.4 presents a set of conclusions.

13.2 Targets and Their Possible Impacts

13.2.1 Strengthen Resilience and Adaptive Capacity to Climate-Related Hazards and Natural Disasters – Target 13.1

This target focuses on reducing the potential impacts of climate change. This can be done through national and local disaster risk-reduction strategies or building resilience to more chronic, longer-term stresses due to climate change. Forests may play an important role. Well-managed forests can reduce flood or landslide risks and provide valuable safety nets for local people in times of stress, providing supplemental food sources when crops or pastures fail due to droughts, fires or pests or are impacted by extreme weather events. This role of forests has been recognised in reducing the impacts of disasters such as tsunamis, storms and typhoons. For example, in Southeast Asia, mangroves reduce future climate risks while also providing food, habitat for aquatic biodiversity and carbon storage (Murdiyarso et al. 2015).

Such hazards may occur more frequently or increase in intensity due to climate change (Dale et al. 2001) and will affect both forests and people, reducing the capacity of forests to provide benefits and mitigate climate change.

This is aggravated by unsustainable land use: human interference may exacerbate forest fires, for example, in boreal (Flannigan et al. 2009) or tropical regions (Nepstad et al. 2008), or they may increase forest fragmentation (Laurence and Williamson 2001). This vicious circle can only be interrupted through deliberate actions, such as those oriented at sustainable use, conservation and restoration of forests.

Many international and national initiatives related to forests are focused on their contribution to current biodiversity conservation objectives or forest-based climate mitigation. These initiatives can have positive impacts on both forests and people. For example, REDD+ activities often have little positive impact on the well-being of forest-dependent communities because they do not add to these communities' resilience to outside changes (Akamani and Hall 2015, Duchelle et al. 2018). If such mitigation strategies addressed local access to forests and forest resources, improving such access could contribute to the adaptive capacity of forest-dependent communities (Turnhout et al. 2017).

In other cases, mitigation may reduce the adaptive capacities of the forest or of local people. Planting large areas of fast-growing trees in degraded forest areas, for example, will increase local carbon stocks, but may decrease water availability and increase the risk of droughts. Forest-based mitigation and adaptation investments need to be carefully planned to address impacts on water resources (van Dijk and Keenan 2007).

The experiences with forest-based climate actions so far have shown that: (1) the impacts of forest-based mitigation and adaptation activities on people's climate resilience depends on their level of social organisation and their level of participation in these activities, and (2) investment in forest-based climate change action is still limited.

In summary, reaching Target 13.1 will have positive impacts on forests and people if the role of forests in climate-hazard reduction can be properly valued and is recognised. Additionally, sufficient evidence of these values should be provided in order to reduce the risk perceptions that investors have of investments in sustainable land use, while social organisation and participation of local people needs to be strengthened to ensure that the most vulnerable people can also benefit from these investments.

13.2.2 Integrate Climate Change Measures into National Policies, Strategies and Planning – Target 13.2

Forest-based adaptation and mitigation have been incorporated into global climate agreements since the United Nations Framework Convention on Climate Change (UNFCCC) was ratified in 1992. The following mechanisms and programmes are included in the framework:

- **Article 3.3 and 3.4 of the Kyoto Protocol.** This provides for Annex 1 parties to use forest-based activities (afforestation, reforestation and avoided deforestation) to meet their emission-reduction commitments.
- **Clean Development Mechanism (CDM).** Under the Kyoto Protocol, reforestation and afforestation are eligible activities.[5] Less than 1 per cent of CDM projects are in forestry due to stringent requirements for forest carbon monitoring, additionality, permanence and leakage avoidance and limitations on the scale of projects.
- **Reducing emissions from deforestation and forest degradation and the role of conservation, sustainable management of forests and enhancement of forest carbon stocks (REDD+).** REDD+ is recognised in the Paris Agreement as a climate change mitigation action. International funding mechanisms such as the Forest Carbon Partnership Facility (FCPF), the Forest Investment Program and the UN-REDD Programme were established to help countries prepare for and implement REDD+. The Green Climate Fund has made investments in developing countries to support forest-based activities. The Biocarbon Fund is one of the mechanisms for incentivising action through payment for results.
- **Nationally Appropriate Mitigation Actions (NAMAs).** A set of voluntary policies and actions that contribute to achieving a country's mitigation commitments, which differ by country. Some examples of NAMAs (plantations in Chile, livestock and coffee in Costa Rica) indicate their potential to contribute to forest and tree cover. Developing a NAMA, however, takes time, because it combines technical and policy aspects, involves many stakeholders and is country-driven, with limited resources.
- **Paris Agreement.** This Agreement sets a comprehensive long-term temperature goal for all parties, potential for increased ambition, and mechanisms for regular review of nationally determined contributions. Implementation of this Agreement is essential to achieving SDG 13.

Two other international initiatives also relate forests to climate action. These aim to support the Aichi Targets of the Convention on Biological Diversity (CBD) and the UNFCCC:

- **The Bonn Challenge.** The Bonn Challenge was launched in 2011 to restore 150 million ha of land by 2020. It has led to many pledges for FLR.
- **New York Declaration on Forests (NYDF).** Made in 2015, it expands the goals of the Bonn Challenge to 350 million ha by 2030, in line with Agenda 2030. As of September 2018, 56 countries or jurisdictions have

[5] www.cdmpipeline.org/cdm-projects-type.htm

committed to restore more than 160 million ha of forest, potentially sequestering 15.66 Gt of CO_2 by 2030.[6]

There has also been a range of bilateral investments to reduce deforestation emissions and expand forest cover, most notably by the Norwegian government in Brazil and Indonesia and more recently by the UK and German governments in a number of countries.

Of all these initiatives, REDD+ and FLR have received substantial support from governments and the private sector, and they can work in tandem to contribute to maintaining forest cover and sustaining restored forests and landscapes.

REDD+ addresses deforestation, degradation, conservation, management and restoration of carbon stocks in forests in developing countries. Countries interested in REDD+ need to go through a readiness phase, which includes strengthening policy frameworks, institutions and human resources. This is followed by implementation and results-based payments phases. However, pledged funding for REDD+ stays well behind the projected needs to reduce deforestation (Turnhout et al. 2017). A lack of human and institutional capacity among developing countries to access and use funds efficiently is limiting investment. This can be seen in the early experiences of implementing REDD+ (see Box 13.1). This has also hampered access to and equitable distribution of financial resources at local scales, particularly for small and medium-sized forest enterprises and communities (Myers et al. 2018). A further limitation is the top-down nature of global climate programmes. Global objectives (to reduce emissions from deforestation) need to be better translated to align with the needs and interests of local actors to be successfully achieved at a local scale (Sanders et al. 2017), and need to demonstrate clear benefits for local, poor people.

Box 13.1 REDD+ Progress and Issues: Challenges to Achieving Long-Term Positive Impacts

REDD+ discussions and initiatives have been successful in putting the spotlight on deforestation; early REDD+ initiatives have achieved institutional changes in the forest sector and have started intersectoral coordination, bringing representatives of the government, environmental organisations, commercial (forest and agricultural) productive organisations, civil society organisations and communities to the same table (Brockhaus et al. 2017). Important issues such as land tenure, regulations and transparency are at least partially addressed (Sunderlin

[6] www.bonnchallenge.org

Box 13.1 (cont.)

et al. 2014), and more detailed information – not all limited to REDD+ – continuously becomes available on the forests, even from lesser-known carbon pools in some of the remotest places on Earth (Webb et al. 2017).

Different authors question whether the approach taken (i.e. REDD+ related initiatives) will produce the long-term results needed for the forest–society interactions to become sustainable and climate proof (Lund et al. 2016). One of the criticisms is that current policies and measures (PAMs) are oriented towards strengthening existing relations and practices rather than promoting transformational change that allows for reducing or eliminating the underlying causes of deforestation and forest degradation. Indeed, addressing these underlying causes is one of the major challenges encountered by REDD+ practitioners, as is the unwillingness of relevant sectors to collaborate in solving these issues (Angelsen et al. 2017) and the emphasis on technical approaches rather than seeking political approaches to resolve problems of recognition and justice (Myers et al. 2018).

At the same time, REDD+ has not received the amounts of finance originally envisaged, and is not likely to (Angelsen et al. 2017). This is partially due to the less-than-expected global mitigation commitments, but it is also linked to uncertainties related to the long-term contribution of forest-based mitigation initiatives for reduced emissions and carbon storage.

These uncertainties may relate to difficulties in transparent reporting on carbon results (Enrici and Hubacek 2018), transparent mechanisms for benefit distribution that ensure forest maintenance in the long term (Lund et al. 2016, Myers et al. 2018), and the future needs for agricultural land and subsequent changes in land value, pressures on the forest and stakeholder relations (Lapola et al. 2014). These may be affected by markets and other policy areas such as trade, as well as climate change; current and future abilities to manage conflicts (Myers et al. 2018, Sunderlin et al. 2014); the need for, and difficulty of, measuring compliance with the REDD+ safeguards (Jagger et al. 2014); and the need to match the expectations of global policy with local development needs (Sanders et al. 2017). The way that REDD+ initiatives are visualised may also affect their performance. Weatherley-Singh and Gupta (2017) argue that, to be successful, REDD+ initiatives need to balance ecological, political and economic aspects. They studied a REDD+ initiative in Madagascar that was set up with a more ecological approach, seeking to benefit smallholders and preserve biodiversity. However, despite some progress in reducing deforestation, it was challenging for the initiative to address the underlying causes of deforestation, such as the lack of intersectoral coordination, national political support and stakeholder involvement at multiple governance levels.

Directly or indirectly, existing efforts have had a positive influence on forest area by contributing to reducing the rate of annual loss. How much of this is due to climate action is difficult to discern from the available information. Despite these efforts, forest loss continues (Keenan et al. 2015), with the highest rates in poorer countries in the tropics.

FLR actions have potential to link mitigation with adaptation objectives. In many countries, restoration of forest lands contributes to increased resilience and reduced vulnerability by restoring essential ecosystem services (Stanturf et al. 2015). Despite many pledges to restore forests and landscapes (Box 13.2; see also AFR100[7]), there is a big gap in meeting restoration targets under the Bonn Challenge. For upscaling implementation, the challenges that remain to be solved include governance issues (who decides on what to do where and when), regulatory frameworks that facilitate FLR, the institutional arrangements to be able to implement them and agreement on the best way to monitor progress (Mansourian et al. 2017). Additionally, further finance will be necessary to meet the Bonn and New York goals.

To make the implementation of climate actions more transparent, countries have been asked to propose contributions to the goals of the Paris Agreement through the **Nationally Determined Contributions (NDC)**. These NDCs need to have clear and fair mitigation targets, may deal with adaptation and

Box 13.2 Seeking the Implementation of International Commitments of the Bonn Challenge in Latin America: Initiative 20x20[8]

Initiative 20x20 brings international intentions to actions on the ground and combines the objectives of three international conventions (Biodiversity, Climate and Degraded Lands). Under this initiative, several countries in Latin America and the Caribbean developed a regional mechanism to restore 20 million ha of forests and landscapes by 2020. After its launch during the Conference of Parties (COP) 20 in Lima, a dialogue was set up among more than 40 technical organisations, 20 financial organisations and 17 countries, building a coalition that should reach investments of up to USD 1 billion. As of 2018, countries and organisations have committed to restore 53.2 million ha by 2020, pledged USD 2.1 billion and committed USD 1.4 billion. Approximately USD 400 million comes from investors seeking positive social and environmental impacts as well as financial returns. Initiative 20x20 aims to attract more private investments by showing the business case for restoration, including climate mitigation and adaptation benefits.

[7] https://afr100.org/

[8] https://initiative20x20.org/

show a variety of ways in which countries incorporate or propose to incorporate climate actions into their national policies and strategies. NDCs are useful to monitor progress against the commitments of each country, and countries are expected to produce biannual reports. Nevertheless, commitments in the NDCs are not sufficient to maintain global warming below the 2°C limit (Rogelj et al. 2016). Even the implementation of these commitments is limited, due to lack of capacity and financial resources.

Meeting the global warming target would reduce the risk of losing species because of climate change. Without mitigation only 11 of 33 globally significant biodiversity conservation areas (GSBCA) would maintain their functions as refuge for 75 per cent of today's' species. Current mitigation commitments would bring that up to 50 per cent, while achieving the Paris Agreement target of reduced global warming would maintain the refuge function in 67 per cent of the current GSBCA (Warren et al. 2018). The Intergovernmental Panel on Climate Change (IPCC 2018) estimates that mitigation actions that can reduce global warming by another 0.5°C – from 2°C to 1.5°C – will lower the risks for biodiversity and forest area, lowering the risk of extinction by as much as 67 per cent for insects and 50 per cent for plant and vertebrate species. Even for many species not threatened by extinction, climate change will cause changes in their ecosystem and likewise affect local forest-depending people (Pecl et al. 2017).

On the positive side, more than 100 countries include forests in their NDC, representing more than 80 per cent of the global forest area. Approximately 20 per cent of the estimated total net emission reductions to be achieved through the NDCs are expected to come from the land use, land-use change and forestry sector (Forsell et al. 2016). Forests are typically mentioned in relation to one of the instruments mentioned above (NAMA, REDD+, CDM, FLR).

Chile, for example, blended its commitments to the CBD, UNFCCC and the United Nations Convention to Combat Desertification into one single strategy: climate change and vegetational resources. This strategy operationalises its NDC and proposes to plant 100 000 ha with mainly native species and to restore and sustainably manage an additional 100 000 ha of natural forest lands (Moraga and Sartori 2017). This strategy also addresses adaptation concerns, proposing direct interventions in 267 000 ha in order to safeguard biodiversity and environmental services. Although it can be expected that if these targets are met, many people will be positively affected, the strategy itself does not make an estimate of the potential impacts on people. It remains to be seen whether Indigenous people perceive the strategy as positive for their well-being, considering that they already have serious problems with current water and land-tenure regimes. The strategy does not explicitly offer solutions to these problems. In Chile, the main challenges for climate action implementation are considered to be the participation of all stakeholders, existing human capacities and the fact that 84 per cent of the proposed budget will depend on the availability of international finance. While the first two challenges are addressed as part of the strategy, through training of staff and consultation programmes, the third challenge depends on the international preparedness to contribute to Chile's efforts to combat climate change.

While every country is different, it can be expected that even implementation of these insufficient NDCs will positively impact on forest area, diversity and ecosystem services. The benefits of these impacts, however, may not accrue equally to all stakeholders. Each country will have to overcome its own implementation challenges, many of which will relate to stakeholder participation, capacities and access to additional finance.

13.2.3 Improve Education, Awareness-Raising and Human and Institutional Capacity on Climate Change Mitigation, Adaptation, Impact Reduction and Early Warning – Target 13.3

Awareness of climate change is essential to stimulate action. Education, skills and institutional capacity should allow people to implement these actions. The forest sector has a long track record of raising awareness about environmental concerns, and for many years forest extension programmes have aimed at increasing the capacity of local people to contribute to forest management and conservation (see Chapter 4). In Costa Rica, education and awareness were found to be major triggers for conservation of trees on farmlands (Louman et al. 2016). Within the context of REDD+ and the fight against illegal logging, strengthening institutional capacity has received major attention. Most of the recent REDD+ financing, estimated to be nearly USD 2 billion, has been spent on strengthening national capacities in preparation for REDD+. However, as the example of Chile shows, education, skills and institutional capacity remain major concerns for the successful implementation of forest-based mitigation and adaptation initiatives. These challenges must be addressed for Targets 13.1 and 13.2 to be fully achieved.

13.2.4 Mobilising Jointly USD 100 Billion Annually by 2020 – Target 13.A

Currently, USD 10.3 billion has been pledged to the Green Climate Fund (GCF).[9] However, overall costs of climate change actions are estimated to be about USD 2.4 trillion for the energy sector alone (IPCC 2018). In comparison

[9] Of this, USD 10.2 billion has been signed (www.greenclimate.fund/how-we-work/resource-mobilization).

to some climate actions in the energy, transport and industry sectors, investing in the forest sector is a relatively low-cost contribution to the overall goals of reduced atmospheric carbon concentrations and increased resilience, in particular of forest-dependent people.

Looking at past assignment of climate funds to forests (Bird et al. 2017a, 2017b, 2017c), we could expect that by 2020 about USD 3 billion per year will become available for emission reduction in forests if Target 13A is met. In theory, USD 3 billion annually could cover 12 per cent of forest emissions (Douglas and Simula 2010). Whether this will occur, however, requires further analysis of how these funds are being used. A large part of this money will still need to be spent on preparing the conditions for emission reductions to take place and be measured. It may take many years for most countries to reach such readiness. In addition, if allocation policies of climate funds do not change, this money would have to pay for FLR as well. Clearly, current fund allocations are not sufficient to make maximum use of the mitigation and adaptation potential of forests.

13.2.5 Raising Capacity for Effective Climate Change-Related Planning and Management in Least Developed Countries and Small Island Developing States, Including Focusing on Women, Youth and Local and Marginalised Communities – Target 13.B

Climate action PAMs may affect marginalised groups in different ways. Issues such as tenure clarity and security, conflict resolution, transparency in government decisions and empowerment of marginalised groups must be elements of any climate action to reduce further marginalisation of these groups (White et al. 2010). Eighty-five per cent of farmers worldwide are smallholder farmers, many of whom belong to marginalised groups and who utilise unsustainable land-use practices due to insufficient resources and/or knowledge. They contribute to greenhouse gas emissions and are often among the most vulnerable people due to their reliance on rain-fed agriculture. Women, youth and local marginalised groups also require special attention, as their level of vulnerability to climate change may differ from others. Women in the Yucatan Peninsula, for example, are more vulnerable to extreme events such as hurricanes because they communicate through different networks than the one usually used for warning systems (Soares Moraes et al. 2011). However, it has been argued that generalisations about women's vulnerability and virtuousness in reference to climate change might lead to an increase in women's responsibility without corresponding rewards (Arora-Jonsson 2011).

The relatively new global agreements on forests (the NYDF) and climate change (Paris Agreement) put even more strain on marginalised people: now they are not only required to meet their own needs with few resources, but are also expected to contribute to global needs and priorities.

The meaningful participation of these marginalised groups in planning and management of climate actions is crucial for both successfully implementing climate action and achieving positive impacts on the people who need it most. This requires strengthening the planning and management capacities of these groups.

Australia's experience in reducing emissions by paying Aboriginal communities to reintroduce traditional burning practices on customary lands is a good example of an approach that meets climate policy goals while providing community income and social benefits and improving biodiversity conservation. In this case it also involves younger community members with their land and intergenerational transfer of land management skills (Russell-Smith et al. 2017).

The forest sector has several decades of experience in working with local marginalised groups and in less-developed countries, on which climate action can build. Within the REDD+ context, for example, safeguards have been agreed upon, including benefit-sharing mechanisms to stimulate greater participation, upholding the rights of these different groups and incentivising climate mitigation actions undertaken by them. As seen in Box 13.1, however, the implementation of such safeguards beyond policy statements still lags behind, and they do not yet ensure that marginalised groups will benefit from REDD+ actions (Tehan et al. 2017).

13.3 Policy and Implementation Issues

Both FLR and REDD+ are forest-related climate actions receiving much attention in current international and national policies. For both, large sums of money have been pledged, but implementation in the field has not met expectations. While for each of these international initiatives there are a series of factors that explain (lack of) progress, two issues are common to both: they require collaboration between actors from different sectors and at different scales (local-national-international), and they require more and better financing.

13.3.1 Collaboration between Public, Private and Civic Society Actors and Their Organisations

A wide variety of actors, institutions and organisations are related to forest-based climate actions. Globally, agreements have evolved that deal with relations among countries in the framework of climate change (UNFCCC), forests (UN Forum on Forests, Bonn Challenge and NYDF) and the SDGs, all of which consider both climate and forests. Climate actions, however, are implemented locally, with or without the support of national authorities and (inter)national civil society organisations and enterprises. The implementation of the negotiated agreements poses new challenges to the countries: (1) in the short and medium term, global development priorities may not necessarily coincide with national priorities, and (2) implementation requires the collaboration of a range of global, national and local actors.

LINKING PRIORITIES OF GLOBAL, NATIONAL AND LOCAL ACTORS

Often, contradictions exist between global and local priorities related to forests that need to be resolved. Globally, climate change mitigation receives the highest priority, since it poses serious risks for everyone and requires a concerted global action early enough to prevent future problems. Adaptation is also important and – especially in the case of small island nations – extremely urgent, but until recently it was largely a national or local problem. Although slowly changing, as reflected in the Paris Agreement text, the provision and distribution of finance to date, and the generation of knowledge and technology within the framework of the UNFCCC and SDG 13, focus more on mitigation than adaptation.

In most tropical and developing countries, adaptation is important, although their NDCs often emphasise mitigation actions to capture additional international finance, which thus far has prioritised mitigation. Until recently, the Adaptation Fund was the major source of international finance for adaptation. Two per cent of the proceeds from certified emission reductions issued from projects in the CDM of the Kyoto Protocol were set aside for this fund. This is changing with the establishment of the GCF, whose mandate is to have a 50/50 balance between investments on mitigation and adaptation.

Such contradictions are particularly relevant for global initiatives such as FLR and REDD+, which emphasise mitigation goals but have possible synergies and trade-offs with national and local adaptation goals and other SDGs. Synergies and trade-offs may differ according to the local contexts (see Chapters 2, 6–9, 12, 15–17).

A key challenge is to link international goals for emission reductions to local priorities for economic development and poverty alleviation, which have often been drivers of deforestation and degradation. Efforts to effectively translate these global goals have been challenged by local interests and issues associated with land allocation and resource development. Those seeking to implement forest conservation measures must deal with the history of land allocation decisions and conflict between traditional land users, recent smallholder immigrants and larger capital investors in agricultural development (Sanders et al 2017).

In the case of REDD+, for example, actors on the ground need to address the underlying causes from technical, social, economic and policy perspectives. Achieving that through a carbon-centric approach may be less effective than seeking multipurpose forest and tree-based management systems (MFTMS), with mitigation as a secondary result (Ellison et al. 2017). MFTMS have the advantage that they allow the addressing of locally felt needs. This is also valid for FLR initiatives. Vergara et al. (2016) show, for example, that under current carbon prices, the net present value of carbon sequestration services from landscapes is only about 25 per cent of the total present value of average landscape restoration activities in Latin America, the other 75 per cent being attributable to activities that provide locally more appreciated production and regulation services.

The potential beneficiary effects of considering more than one objective in the context of climate change are also highlighted by Locatelli et al. (2015). They found both synergies and trade-offs between agricultural and forest management strategies oriented to adaptation and mitigation. In particular, they found trade-offs where mitigation strategies caused negative impacts on people's adaptive capacity, for example through negative impacts on local land-use opportunities or the availability of water tied to large-scale biofuel plantations. Improving the integration of adaptation and mitigation can be achieved by (1) integrating climate, forest and agricultural policies and strategies across multiple scales (national to local), (2) improving the metrics for monitoring synergies and trade-offs, (3) greater international recognition of the synergies between adaptation and mitigation and the need to treat them together rather than separately (policies, finance) and (4) generating more empirical evidence of the synergies and trade-offs between adaptation and mitigation (Duguma et al. 2014a). The way these issues are addressed may greatly influence the impact of SDG 13 on forests: if PAMs continue to address mitigation and adaptation separately, many opportunities will be lost for optimising synergies and reducing trade-offs.

In the context of both FLR and REDD+, the potential synergies between adaptation and mitigation can be leveraged as a common denominator for linking global, national and local priorities. To be able to do so, however, it is important that FLR and REDD+ focus on the process as well as on the quality and dimension of the final results.

LANDSCAPE APPROACHES TO FOSTER ACTOR COLLABORATION FOR FLR AND REDD+

The benefits of sustainable forest management (SFM), FLR and REDD+ programmes depend on the level of social organisation that existed prior to, or that developed during programme activities (Akamani and Hall 2015, Duchelle et al. 2018). For example, in Honduras, those in a community forestry programme responded quicker to post-hazard recovery assistance after

Hurricane Mitch than communities outside the programme.[10] Whether such benefits reach the people that need it most depends on their access to the programme. In Ghana, access was linked to previous ties with the people and institutions that set up a community forest programme and not necessarily linked directly with the needs of the local people (Akamani and Hall 2015).

A landscape approach is a learning process that fosters collaboration and social organisation, allowing landscape managers to adapt to changing circumstances, involve stakeholders and achieve multiple objectives within defined geographic spaces (Sayer et al. 2013). Experiences have shown that landscape approaches have the potential to facilitate the process of linking global to local priorities (Minang et al. 2014). Numerous landscape initiatives are being implemented. Among the major benefits perceived by the stakeholders in these initiatives are improved institutional planning and coordination, greater collaboration among sectors, and forest planning that serves national and local needs (Hart et al. 2015).

Embracing multiple objectives is a lesson learned from landscape approaches relevant for the implementation of SFM, FLR and REDD+. This is an important factor that determines success, allows for coordination among various programmes and opens financing possibilities from different sources (Hart et al. 2015, see also Box 13.2). Leadership, trust, vision, common concern and the existence of bridging organisations are essential elements for successful governance of socio-ecological systems (Hayes and Persha 2010, Lorenzo et al. 2014). In an analysis of FLR in South Asia these same factors have also been mentioned as success factors (IUFRO et al. 2018). These factors, however, rarely form part of formal planning cycles.

The main challenges of landscape approaches are a lack of political coherence, sustainable finance and, despite increased collaboration, full stakeholder engagement (Hart et al. 2015). To overcome some of these challenges, jurisdictional approaches are being applied, in particular in relation to REDD+ and zero-deforestation initiatives. Such approaches align landscape planning initiatives to (sub)national administrative areas and their authorities, seeking political backing of the initiatives. While these address the issue of political coherence, their success also depends on sustainable finance and full engagement. In Mozambique, such a jurisdictional approach is seen as a strategy to access finance from different sources (Box 13.3).

Since landscape approaches typically bring multiple disciplines together, they can be useful processes to coordinate the implementation of other SDGs. They offer opportunities to maximise synergies and reduce trade-offs among SDGs. By focusing on sustainable use of natural resources, these approaches are important for the achievement of SDGs 1 (No Poverty), 2 (Zero Hunger),

[10] Personal observation.

Box 13.3 A Landscape Approach in Practice: The Zambezia Integrated Landscape Management Programme

In Mozambique, the main drivers of deforestation and forest degradation are agriculture, illegal logging and wood extraction for domestic use. During the 2015 COP21 in Paris, Mozambique signed a Letter of Intent (LOI) with the FCPF Carbon Fund on the purchase of carbon emission reductions up to a value of USD 50 million. Under this LOI, the country committed to implementing REDD+ and established the jurisdictional Zambezia Integrated Landscape Management Programme (ZILMP). It promotes innovative and decentralised governance arrangements at the provincial and district levels and coordinates activities implemented with impacts on emissions from deforestation and forest degradation. The programme covers nine districts in the Zambezia Province – an area of 53 000 km^2, of which 66 per cent is covered by forests with an estimated current annual deforestation rate of 0.62 per cent.

The Zambezia Sustainable Development Platform is a new model for multistakeholder engagement in which, for the first time in Mozambique, civil society, the private sector, academia and government have an adequate forum to discuss issues related to natural resource management. The platform creates opportunities for broad discussions and learning among stakeholders.

The ZILMP works with several government initiatives financed by the World Bank: (1) the Conservation Areas for Biodiversity and Development Project (USD 46.3 million), (2) the Agriculture and Natural Resources Landscape Management Project (USD 40 million) and (3) the Mozambique Forest Investment Project (USD 47 million). Despite different entry points and focuses, all the projects have a common goal: to enhance the living conditions of communities through the sustainable use of forests and other natural resources. Activities to reduce the emissions from deforestation and forest degradation include: (1) conservation agriculture with the communities surrounding the Gilé National Reserve, (2) introduction of 4000 clean cook stoves and (3) spatial analysis to prioritise activities across the landscape.

Adapted from: The World Bank (2015).

6 (Clean Water and Sanitation), 13 (Climate Action), 14 (Life below Water) and 15 (Life on Land). This is particularly relevant since these SDGs are interrelated. Poor people are in general more vulnerable to climate change, so reducing poverty will decrease vulnerability, while at the same time reducing climate change through mitigation measures will reduce the climate-related hazards to which these people may become exposed. Hunger is often exacerbated by climate-related events, such as droughts or floods. Water resources

are directly affected by climate change, as is life below water and life on land. Any level of achievement of the SDG 13 goals, even if insufficient to remain under the 2°C limit, will have a positive influence on moving towards the other goals, compared to not implementing any climate action at all. Applying landscape approaches will facilitate making such linkages at the local level.

On the other hand, SDGs 5 (Gender Equality), 10 (Reduced Inequalities), and 16 (Peace, Justice and Institutions) are instrumental to the implementation of landscape approaches.

Landscape approaches are not a panacea for the local implementation of climate action; due to their relatively recent nature, little evidence can yet be presented on how they contribute to the success of such implementation. However, they build on lessons learned from past experiences with similar approaches, and when they adhere to implementation principles (Sayer et al. 2013) they can offer great opportunities to address the challenges of better implementation for SDG 13 targets having positive impacts on forests and people.

13.3.2 Finance for Implementation

Estimating the amount of funding currently invested in climate change is challenging. In 2013 some USD 331 billion was invested globally for climate-related activities (Buchner et al. 2014), although not all of this was classified as official climate funds. These funds fall well short of the needs for mitigation and adaptation: the same authors suggest, based on their review of other publications, that to transform only the energy sector to a low-carbon sector requires about USD 1 trillion a year until 2030. Currently, investors and international finance organisations are focused on this sector since it promises more straightforward links to emission reductions than those in the land-use sector (Duguma et al. 2014b). However, the forest-based offset market is growing (Forest Trends 2018), and is likely to increase as emission-reduction targets become tougher and low-cost energy transition options are taken up.

Of the USD 61 billion official climate fund monies spent between 2011 and 2015, only 3 per cent was dedicated to forests (Bird et al. 2017a, 2017b, 2017c, MDB 2017, OECD 2015), and most of that for mitigation purposes. If this trend continues, and the USD 100 billion annual target (Target 13.A) is reached, after 2020 about USD 3 billion annually would become available. This is well below the estimated opportunity cost of avoided deforestation (Douglas and Simula 2010) – USD 25 billion USD per year – and would also need to cover adaptation investments. Investment levels, particularly for forest and forest-based adaptation, lag behind needs. There are several reasons for this: (1) it is difficult to determine specific adaptation needs of forests and forest-dependent people due to the uncertainty about long-term climate changes and the variability of responses for different species and forest types to changing climatic conditions; (2) because of lack of experience with the analysis of climate-change impacts and management responses, there is no continuous learning process that can catalyse the adaptation of forest management to address climate change impacts (Lawrence and Haasnoot 2017); and (3) forest managers are generally reluctant to implement new practices without clear evidence of benefits.

Being able to show evidence of the full costs and benefits of trees in the landscape should leverage private finance from a range of different actors (Vergara et al. 2016). This is particularly relevant if you want to capture more private investment that contributes to both mitigation and adaptation. In the context of the Bonn Challenge and the NYDF, it will be difficult to reach the target to restore 350 million ha of land by 2030 without such private finance.

To achieve the climate goals, small and medium-sized enterprises and communities also need to be involved: they are affected by climate change and need to adapt; they contribute to the problem of deforestation and degradation, yet can also provide important contributions to solutions (de Jong et al. 2018). Their access to climate finance, however, is limited (MacQueen et al. 2014), and is often negatively influenced by unclear land-tenure regimes (van Dijk and Savenije 2009). In the forest sector of many countries, there is also a failure to apply existing legislation, along with weak control mechanisms, weak institutions and a lack of political will to make the necessary transformational changes (van Dijk and Savenije 2009).

Another issue that calls for careful attention is the potential in climate finance for both synergies and trade-offs across different sectors and different scales. Synergies exist between the agricultural and forest sectors: protection of forests is essential for regular and clean water flows (Ellison et al. 2017); in agroforestry systems, ecosystem services are provided as well as agricultural crops and tree products; and in the restoration of degraded landscapes, trees and other conservation measures can contribute to future agricultural crop production. Trade-offs may also occur. Forest programmes focusing only on carbon, for example, may reduce local capacities for adaptation (Duguma et al. 2014a, Locatelli et al. 2015). Finance that facilitates synergies will be more efficient than finance generating trade-offs.

Reducing direct finance and other incentives (e.g. tax deductions) for land uses that negatively affect forests and their ecosystem services is as important for forest-based mitigation and adaptation as increasing the availability and sustainable and inclusive use of finance. More than USD 380 billion is invested annually in the land-use sectors (FAO 2017), and only a small proportion of this considers the negative impacts of land use on forests and their functions and services. Many financed land-use activities still threaten the quality or quantity of forests. For example, in many tropical countries, the expansion of extensive livestock management and large-scale monocultures of agro-commodities impinges on forest areas. At the same time, inadequate or overbearing legislation and regulation provide opportunities for corruption and cause high transaction costs for legal forest use (Navarro and Bermudez 2006). Transport infrastructure development, mining and hydropower are having increasing impacts on forests, with associated effects on the rights, lives and livelihoods of local people. The stakes for these projects are often much higher than for agricultural development, and it often requires well-organised political and social movements to counter them (Bebbington et al. 2018).

Experiences with climate finance show that much of the assigned money (1) does not enter the forest sector and, in some cases, may be detrimental to the forest sector (e.g. by promoting large-scale commercial agriculture through subsidies or facilitating finance), (2) is not used efficiently in that it addresses mitigation and adaptation issues separately rather than addressing them jointly from the outset, and (3) if invested in the forest sector, is not used efficiently since it is invested in conditions where existing policy frameworks, institutional settings and available human resources cause high transaction costs that discourage the management of forests and trees.

Many opportunities lie in the conversion of business-as-usual investments into investments that improve local climate resilience of forests and people while contributing to low-carbon development. Considering that at the global level, USD 380 billion is invested annually in land uses that impact forests or otherwise increase emissions, investors should seriously consider reducing the climate risks of their investments. To help change the mindset of the investors, it is necessary to have a good understanding of who finances what and why beneficiaries prefer one type of finance over another. In addition, it is necessary to identify which locally relevant policy, regulatory, planning and development arrangements could facilitate directing that mindset towards greater investment in resilience and low-carbon development.

13.4 Conclusions

Climate change affects all people, land uses and land covers. Implementation of SDG 13 is critical for forests and for the people that interact with and depend on them. Successful achievement of the SDG 13 targets alone will not be sufficient to ensure conservation and better management of existing forests and restoration of degraded forest lands. Coordination is necessary because SDG 13 is closely interrelated with other SDGs, especially SDG 5 (Gender Equality), SDG 10 (Reduced Inequalities), SDG 12 (Responsible Consumption and Production), SDG 16 (Peace, Justice and Institutions) and SDG 17 (Partnerships for the Goals). Their implementation and awareness of their impact on forests are prerequisites for achieving SDG 13. This requires a high degree of intersectoral coordination and collaboration among stakeholders at all levels, a clear focus on ecosystem services generated by forests (e.g. climate regulation, hydrological services) and the identification of the policy and practice linkages between adaptation and mitigation in forest ecosystems and the people who depend on them.

Reaching the Paris goals and SDG 13 targets also depends on the development of new programmes and funding models. Our analysis finds two cross-cutting issues that offer potential to facilitate such transformational change: (1) new types of relationships among the various actors involved, and (2) alternative approaches to land-use investments. Considerable evidence and experience provide a foundation on which to build these new relationships and investment approaches.

For example, activities that aim to support Targets 13.3 and 13.B can improve local participation in integrated land-use planning approaches with multiple development and conservation goals, which will increase the willingness of local people to implement climate actions. When local stakeholders have the capacity to negotiate and the results of this process are guaranteed to be implemented by local, provincial, national or international authorities, there is a clear indication that improved outcomes for forests, people and climate can be achieved. While many of these landscape-level initiatives are relatively new, they show promise in bringing together diverse interests to identify shared goals and build more sustainable forest management solutions. Lessons can be learned from these initiatives to strengthen SFM, FLR and jurisdictional approaches to REDD+ that will not only support wider implementation of SDG 13 and the Paris Agreement, but can also contribute to achieving the synergies and balance the trade-offs that may result from the application of adaptation and mitigation actions.

Though there is significant investment in climate finance, little of it is going to the forest sector, while at the same time billions of dollars are invested in ways that are detrimental to forests, such as subsidies or finance to increase the scale of commercial agriculture. Another important factor is that funding allocated to forests is not used efficiently to address the requirements of both mitigation and adaptation. In other cases, current policies, laws and regulations can cause high transaction costs that discourage the improved management of forests and trees. With good design, conditions associated with international funds for REDD+ can improve forest governance and transparency, and international and national funds are becoming more aware of the need for integrated finance to address the role of forests and climate change in a multifaceted way. However, current efforts are not sufficient to (1) avoid the continuing forest loss or the detrimental impacts that climate change is having on forest health, vitality and diversity or (2) increase the rate of FLR and its contributions to both mitigation and adaptation. Many are now looking to the private sector as a new actor in climate governance and as the primary source of funds for climate actions related to forests and forest landscapes. This requires appropriate regulations and strong business cases built on clear evidence of the positive impacts of investment on climate, economic development and financial returns.

Increased forest-based climate actions will increase the capacity of forests to contribute to mitigation and reduce the vulnerability of forest-dependent people. Embedding forests into the climate policies, measures and actions under SDG 13 can generate a virtuous cycle, deriving greater benefits from forests for people, reducing the future rate of climate change and ensuring that forests and forest-dependent people are more resilient and have the capacity to anticipate and adapt to future change.

References

Akamani, K. and Hall, T. E. 2015. Determinants of the process and outcomes of household participation in collaborative forest management in Ghana: A quantitative test of a community resilience model. *Journal of Environmental Management* 147:1–11.

Angelsen, A., Brockhaus, M., Duchelle, A. E. et al. 2017. Learning from REDD+: a response to Fletcher et al. *Conservation Biology* 31(3):718–20.

Arora-Jonsson, S. 2011. Virtue and vulnerability: Discourses on women, gender and climate change. *Global Environmental Change* 21:744–51.

Avelino, J., Cristancho, M., Georgiou, S. et al. 2015. The coffee rust crises in Colombia and Central America (2008–2013): Impacts, plausible causes and proposed solutions. *Food Security* 7(2):303–21.

Bebbington, A. J., Bebbington, D. H., Sauls, L. A. et al. 2018. Resource extraction and infrastructure threaten forest cover and community rights. *Proceedings of the National Academy of Sciences* 115(52):13164–73.

Bird, N., Watson, C. and Schalatek, L. 2017a. *The global climate finance architecture. Climate funds update.* Washington, DC: Heinrich Boll Stiftung.

Bird, N., Watson, C., Schalatek, L. and Keil, K. 2017b. *Climate finance thematic briefing: REDD+ finance. Climate funds update.* Washington, DC: Heinrich Boll Stiftung.

Bird, N., Watson, C., Schalatek, L. and Keil, K. 2017c. *Climate finance thematic briefing: Adaptation finance. Climate funds update.* Washington, DC: Heinrich Boll Stiftung.

Brockhaus, M., Korhonen-Kurki, K., Sehring, J. et al. 2017. REDD+, transformational change and the promise of performance-based payments: A qualitative comparative analysis. *Climate Policy* 17(6):708–30.

Buchner, B., Abramskiehn, D., Stadelmann, M. et al. 2014. *The global landscape of climate finance 2014.* Climate Policy Initiative. Available at: https://climatepolicyinitiative.org/wp-content/uploads/2014/11/The-Global-Landscape-of-Climate-Finance-2014.pdf (Accessed 29 July 2019).

Dale, V. H., Joyce, L. A., McNulty, S. et al. 2001. Climate change and forest disturbances: Climate change can affect forests by altering the frequency, intensity, duration, and timing of fire, drought, introduced species, insect and pathogen outbreaks, hurricanes, windstorms, ice storms, or landslides. *BioScience* 51(9):723–34.

De Jong, W., Pokorny, B., Katila, P., Galloway, G. and Pacheco, P. 2018. Community forestry and the Sustainable Development Goals: A two way street. *Forests* 9(6):331.

Douglas, J. and Simula, M. 2010. *The future of the world's forests: Ideas vs ideologies* (vol. 7). Berlin: Springer Science and Business Media.

Duchelle, A. E., Seymour, F., Brockhaus, M. et al. 2018. *REDD+: Lessons from national and subnational implementation* (Ending Tropical Deforestation Series). Washington, DC: World Resources Institute.

Duguma, L. A., Minang, P. A. and van Noordwijk, M. 2014a. Climate change mitigation and adaptation in the land use sector: From complementarity to synergy. *Environmental Management* 54(3):420–32.

Duguma, L. A., Wambugu, S. W., Minang, P. A. and van Noordwijk, M. 2014b. A systematic analysis of enabling conditions for synergy between climate change mitigation and adaptation measures in developing countries. *Environmental Science & Policy* 42:138–48.

Ellison, D., Morris, C. E., Locatelli, B. et al. 2017. Trees, forests and water: Cool insights for a hot world. *Global Environmental Change* 43:51–61.

Enrici, A. and Hubacek, K. 2018. Challenges for REDD+ in Indonesia: A case study of three project sites. *Ecology and Society* 23(2):7.

FAO 2017. *The future of food and agriculture. Trends and challenges*. Rome: FAO.

Flannigan, M., Stocks, B., Turetsky, M. and Wotton, M. 2009. Impacts of climate change on fire activity and fire management in the circumboreal forest. *Global Change Biology* 15(3):549–60.

Forest Trends 2018. *Voluntary carbon markets insights: 2018 outlook and first-quarter trends*. Available at: www.forest-trends.org/publications/voluntary-carbon-markets/ (Accessed 30 January 2019).

Forsell, N., Turkovska, O., Gusti, M. et al. 2016. Assessing the INDCs' land use, land use change and forest emission projections. *Carbon Balance and Management* 11:26. http://doi .org/10.1186/s13021-016–0068-3.

Hart, A. K., Milder, J. C., Estrada-Carmona, N. et al. 2015. Integrated landscape initiatives in practice: Assessing experiences from 191 landscapes in Africa and Latin America. In Minang, P., van Noordwijk, M., Freeman, O. E. et al. (eds.) *Climate-smart landscapes: Multifunctionality in practice*. Nairobi: World Agroforestry Centre (ICRAF), pp. 89–101.

Hayes, T. and Persha, L. 2010. Nesting local forestry initiatives: Revisiting community forest management in a REDD+ world. *Forest Policy and Economics* 12(8):545–53.

IPCC 2018. *Global warming of 1.5°C. Summary for policy makers*. Available at: www.ipcc.ch/report/sr15/ (Accessed 30 January 2019).

IUFRO, Global Partnership on Forest and Landscape Restoration and Bonn Challenge 2018. *International knowledge-sharing workshop, a contribution to the implementation of the Bonn Challenge. Best practices for the implementation of forest landscape restoration in South Asia*. Summary report. Available at: www.iufro.org/fileadmin/material/science/spps/spdc/Sri_Lanka_2018/FLR-Knowledge-sharing_WS-Sri_Lanka-Summary_Report.pdf (Accessed 30 January 2019).

Jagger, P., Brockhaus, M., Duchelle, A. et al. 2014. Multi-level policy dialogues, processes, and actions: Challenges and opportunities for national REDD+ safeguards measurement, reporting, and verification (MRV). *Forests* 5(9):2136–2162.

Jindal, R., Swallow, B. and Kerr, J. 2008. Forestry-based carbon sequestration projects in Africa: Potential benefits and challenges. *Natural Resources Forum* 32(2):116–30.

Keenan, R. J., Reams, G. A., Achard, F. et al. E. 2015. Dynamics of global forest area: Results from the FAO Global Forest Resources Assessment 2015. *Forest Ecology and Management* 352:9–20.

Lapola, D. M., Martinelli, L. A., Peres, C. A. et al. 2014. Pervasive transition of the Brazilian land-use system. *Nature Climate Change* 4(1):27.

Laurance, W. F. and Williamson, G. B. 2001. Positive feedbacks among forest fragmentation, drought, and climate change in the Amazon. *Conservation Biology* 15(6):1529–35.

Lawrence, J. and Haasnoot, M. 2017. What it took to catalyse uptake of dynamic adaptive pathways planning to address climate change uncertainty. *Environmental Science & Policy* 68:47–57.

Liu, Y. Y., van Dijk, A. I., de Jeu, R. A. et al. 2015. Recent reversal in loss of global terrestrial biomass. *Nature Climate Change* 5(5):470.

Locatelli, B., Pavageau, C., Pramova, E. and Di Gregorio, M. 2015. Integrating climate change mitigation and adaptation in agriculture and forestry: opportunities and trade-offs. *Wiley Interdisciplinary Reviews: Climate Change* 6(6):585–98.

Lorenzo, J, Carrera, F., de camino, R. and Villalobos, R. 2014. *Participatory forest governance in Ibero America: Social and political impacts of the Model Forests approach*. Salt Lake City, Utah, USA. IUFRO XXIV World Congress.

Louman, B., Gutiérrez, I., Le Coq, J. F. et al. 2016. Avances en la comprensión de la transición forestal en fincas costarricenses. *Revibec: Revista de la Red Iberoamericana de Economia Ecológica* 26:0191–206.

Lund, J. F., Sungusia, E., Mabele, M. B. and Scheba, A. 2016. Promising change, delivering continuity: REDD+ as conservation fad. *World Development* 89:124–39.

Macqueen, D., Andaya, E., Begaa, S. et al. 2014. *Prioritising support for locally controlled forest enterprises*. London: IIED. Available at: https://pubs.iied.org/pdfs/13572IIED.pdf (Accessed 30 January 2019).

Mansourian, S., Dudley, N. and Vallauri, D. 2017. Forest landscape restoration: progress in the last decade and remaining challenges. *Ecological Restoration* 35(4):281–8.

MDB 2017. *Joint report on multilateral development banks' 2016 climate finance*. Available at: https://publications.iadb.org/bitstream/handle/11319/8505/2016_joint_report_on_mdbs_climate_finance.pdf?sequence=1&isAllowed=y (Accessed 30 January 2019).

Minang, P. A., van Noordwijk, M., Freeman, O. E., Mbow, C., de Leeuw, J. and Catacutan, D. (eds.) 2014. *Climate-smart landscapes: Multifunctionality in practice*. ASB Partnership for The Tropical Forest Margins.

Moraga, J. and Sartori, A. 2017. *Estrategia Nacional de Cambio Climatico y Recursos Vegetacionales*. Santiago, Chile. CONAF. Available at: www.conaf.cl/cms/editorweb/ENCCRV/ENCCRV-3a_Edicion-17mayo2017.pdf (Accessed 29 July 2019).

Murdiyarso, D., Purbopuspito, J., Kauffman, J. B. et al. 2015. The potential of Indonesian mangrove forests for global climate change mitigation. *Nature Climate Change* 5(12):1089.

Myers, R., Larson, A. M., Ravikumar, A. et al. 2018. Messiness of forest governance: How technical approaches suppress politics in REDD+ and conservation projects. *Global Environmental Change* 50:314–324.

Navarro, G. and Bermudez, G. 2006. *Rentabilidad del manejo de bosques naturales y su competitividad respecto a otros usos de la tierra en Costa Rica*. Serie de Cooperación Técnica Economía y Gobernanza Forestal, vol. 4. Turrialba, Costa Rica: Tropical Agricultural Research and Higher Education Center (CATIE).

Nepstad, D. C., Stickler, C. M., Soares-Filho, B. and Merry, F. 2008. Interactions among Amazon land use, forests and climate: Prospects for a near-term forest tipping point. *Philosophical Transactions of the Royal Society B: Biological Sciences* 363(1498):1737–46.

OECD 2015. *Climate Fund Inventory: Report and database*. Available at: www.oecd.org/env/cc/database-climate-fund-inventory.htm (Accessed 30 January 2019).

Oliver, C. D., Nassar, N. T., Lippke, B. R. and McCarter, J. B. 2014. Carbon, fossil fuel, and biodiversity mitigation with wood and forests. *Journal of Sustainable Forestry* 33(3):248–75.

Osman-Elasha, B., Adger, N., Brockhaus, M. et al. 2009. Future socio-economic impacts and vulnerabilities. In Seppala, R., Buck, A. and Katila, P. (eds.) *Adaptation of forests and people to climate change. A Global Assessment Report*. IUFRO World Series Vol. 22. Helsinki: IUFRO, pp. 101–22.

Pan, Y., Birdsey, R. A., Fang, J. et al. 2011. A large and persistent carbon sink in the world's forests. *Science*: 333(6045):988–93.

Pecl, G. T., Araújo, M. B., Bell, J. D. et al. 2017. Biodiversity redistribution under climate change: Impacts on ecosystems and human well-being. *Science*: 355 (6332):eaai9214.

Reed, J., van Vianen, J. and Sunderland, T. 2015. *From global complexity to local reality: Aligning implementation pathways for the Sustainable Development Goals and landscape approaches* (vol. 129). Bogor, Indonesia: CIFOR.

Rogelj, J., den Elzen, M., Hohne, N. et al. 2016. Paris Agreement climate proposals need a boost to keep warming well below 2°C. *Nature Climate Change* 534:631–9.

Russell-Smith, J., Monagle, C., Jacobsohn, M. et al. 2017. Can savanna burning projects deliver measurable greenhouse emissions reductions and sustainable livelihood opportunities in fire-prone settings? *Climatic Change* 140(1):47–61.

Sanders, A. J. P., da Silva Hyldmo, H., Prasti H. et al. 2017. Guinea pig or pioneer: Translating global environmental objectives through to local actions in Central Kalimantan, Indonesia's REDD+ pilot province. *Global Environmental Change* 42:68–81.

Sayer, J. A., Margules, C., Boedhihartono, A. K. et al. 2017. Measuring the effectiveness of landscape approaches to conservation and development. *Sustainability Science* 12(3):465–76.

Sayer, J., Sunderland, T., Ghazoul, J. et al. 2013. Ten principles for a landscape approach to reconciling agriculture, conservation, and other competing land uses. *Proceedings of the National Academy of Sciences* 110(21):8349–56.

Soares Moraes, D., Gutiérrez Montes, I., Romero Pérez, R. et al. 2011. *Capitales de la comunidad, medios de vida y vulnerabilidad social ante huracanes en la costa yucateca: Un acercamiento a través de la experiencia de San Felipe, Yucatán*. Serie técnica. Informe técnico no. 385. Turrialba, Costa Rica: CATIE, IICA, IMTA.

Stanturf, J. A., Kant, P., Lillesø, J.-P. B. et al. 2015. *Forest landscape restoration as a key component of climate change mitigation and adaptation*. World Series, vol. 34. Vienna, Austria: International Union of Forest Research Organizations (IUFRO).

Stringer, L. C., Dougill, A. J., Fraser, E. et al. 2006. Unpacking 'participation' in the adaptive management of social–ecological systems: A critical review. *Ecology and Society* 11(2):39.

Sunderlin, W. D., Larson, A. M., Duchelle, A. E. et al. 2014. How are REDD+ proponents addressing tenure problems? Evidence from Brazil, Cameroon, Tanzania, Indonesia, and Vietnam. *World Development* 55:37–52.

Swamy, L., Drazen, E., Johnson, W. R. and Bukoski, J. J. 2018. The future of tropical forests under the United Nations Sustainable Development Goals. *Journal of Sustainable Forestry* 37(2):221–56.

Tehan, M. F., Godden, L. C., Young, M. A. and Gover, K. A. 2017. *The impact of climate change mitigation on indigenous and forest communities: International, national and local law perspectives on REDD+*. Cambridge: Cambridge University Press.

Turnhout, E., Gupta, A., Weatherley-Singh, J. et al. 2017. Envisioning REDD+ in a post-Paris era: Between evolving expectations and current practice. *Wiley Interdisciplinary Reviews: Climate Change* 8(1):e425.

Van Dijk, A. I. and Keenan, R. J. 2007. Planted forests and water in perspective. *Forest Ecology and Management* 251:1–9.

Van Dijk, K. and Savenije, H. 2009. Towards national financing strategies for sustainable forest management in Latin America. Overview of the present situation and the experience in selected countries. *Forestry Policy and Institutions Working Paper 21*. Rome: FAO.

Vergara, W., Gallardo-Lomeli, L., Ríos, A. R. et al. 2016. *The economic case for landscape restoration*. Washington, DC: World Resources Institute.

Warren, R., Price, J., VanDerWal, J., Cornelius, S. and Sohl, H. 2018. The implications of the United Nations Paris Agreement on climate change for globally significant biodiversity areas. *Climatic Change* 147(3–4):395–409.

Weatherley-Singh, J. and Gupta, A. 2017. An ecological landscape approach to REDD+ in Madagascar: Promise and limitations? *Forest Policy and Economics* 85:1–9.

Webb, E. E., Heard, K., Natali, S. M. et al. 2017. Variability in above-and below ground carbon stocks in a Siberian larch watershed. *Biogeosciences* 14(18):4279–94.

White, A., Hatcher, J., Khare, A. et al. 2010. Seeing people through the trees and the carbon: Mitigating and adapting to climate change without undermining rights and livelihoods. In Mearns, R. and Norton, A. (eds.) *The social dimensions of climate change: Equity and vulnerability in a warming world*. World Bank, pp. 277–301. Available at: https://openknowledge.worldbank.org/bitstream/handle/10986/2689/520970PUB0EPI11C010disclosed0Dec091.pdf?sequence=1&isAllowed=y (Accessed 29 July 2019).

World Bank. 2015. *Zambézia Integrated Landscape Management Programme*. Brochure. Maputo, Mozambique.

Zarin, D. J., Harris, N. L., Baccini, A. et al. 2016. Can carbon emissions from tropical deforestation drop by 50% in 5 years? *Global Change Biology* 22(4):1336–47.

Chapter 14 SDG 14: Life below Water – Impacts on Mangroves

Daniel A. Friess*, Toe Toe Aung, Mark Huxham, Catherine Lovelock, Nibedita Mukherjee and Sigit Sasmito

Key Points

- SDG14 focuses on fisheries, though coastal forests such as mangroves are indirectly linked as they support fisheries and associated human coastal populations.
- SDG 14 benefits coastal forests, but negative impacts are also envisaged. These include (but are not limited to) encouraging new deforestation drivers, reducing environmental justice and encouraging governance recentralisation. SDG 14 may also encourage the creation of very large marine protected areas (often in the open ocean) that do not cover coastal forests or cannot be adequately enforced without concomitant increases in funding.
- Considering coastal forests more explicitly during the planning of SDG 14 targets may anticipate or ameliorate some of these negative impacts.
- With the exception of Target 14.1, the below-water focus of most SDG 14 targets means that terrestrial–marine linkages (e.g. sediment, nutrients, pollution, financial flows) are not strongly acknowledged; it is in this transition zone where coastal forests are found.
- Governance challenges increase the likelihood that SDG 14 will have negative impacts on coastal forests. Coastal forests often fall through policy gaps between terrestrial and marine legislation and between different governance levels. Governance decentralisation (itself threatened by some SDG 14 targets) and community management may negate some impacts.
- Other SDGs are likely to impact coastal forests and SDG 14. Conflicting objectives identified in particular include SDGs 1 (No Poverty), 2 (Zero Hunger), 11 (Sustainable Cities and Communities) and 15 (Life on Land).

---

* Lead author.

14.1 Introduction

The SDGs provide multiple opportunities for coastal and marine areas (Szabo et al. 2015) by addressing coastal poverty, prioritising conservation and explicitly recognising climate change. Coastal and marine environments are relevant to most SDGs, but are explicitly considered under SDG 14, Life below Water: Conserve and sustainably use the oceans, seas and marine resources for sustainable development. SDG 14 aims to increase the protection and sustainable management of coastal and marine ecosystems and their resources while addressing threats such as pollution and ocean acidification. National policymakers have been criticised for not prioritising SDG 14 to the same degree as other SDGs (Custer et al. 2018). However, SDG 14 was a particular focus at the recent 2017 High Level Political Forum on Sustainable Development, where 17 out of 43 countries explicitly stated in their Voluntary National Reviews how they were working towards SDG 14 (UN DESA 2017).

While development and environmental concerns are not always in conflict, the inherent development focus of the SDGs means that these goals may themselves have negative environmental impacts (Kopnina 2016). The coastal and marine focus of SDG 14 means that it may have impacts on coastal forested ecosystems, such as intertidal mangrove forests, beach dune forests and tidal freshwater forested wetlands. With population densities significantly higher in the coastal zone than interior areas (Neumann et al. 2015), the negative impacts of SDG 14 on coastal forests discussed here are also expected to impact the hundreds of millions of people who directly or indirectly derive benefits from coastal forest ecosystems.

This chapter outlines the potential positive contributions of mangroves to SDG 14 and the negative impacts of SDG 14 implementation on (1) mangrove forests and (2) the local communities that derive direct and indirect livelihood benefits from them. We differentiate between these because SDG 14 may affect human systems differently from natural systems. Teasing out such interactions and complexities is key to understanding the myriad impacts that SDG 14 may have on coastal forests.

14.2 Mangrove Forests as a Lens to Analyse SDG 14

14.2.1 Why Focus on Mangrove Forests?

Multiple ecosystems come under the definition of coastal forests, including beach forests and tidally influenced freshwater forested wetlands. In this study, mangrove forests have been chosen as a proxy for coastal forests for the following reasons:

1. Mangroves cover 83 500–137 000 km^2 (Giri et al. 2011, Hamilton and Casey 2016) across the tropics, subtropics and warm temperate zones. Mangroves are potentially relevant to the SDG aspirations of at least 118 countries and territories (Giri et al. 2011), spanning a gradient of economic development across the Global North and Global South.

2. Potentially hundreds of millions of people rely directly on mangroves and their ecosystem services.

3. Mangroves are strongly linked to fisheries (Carrasquilla-Henao and Juanes 2017), a key component of SDG 14, due to the role of the forest as a spawning and nursery ground for commercially important fish species.

4. As mangroves are located between terrestrial and marine zones, they provide strong synergies between SDG 14 and other SDGs. For example, mangrove conservation (Target 14.2) provides coastal protection benefits, strengthening coastal community resilience to climate-related hazards (Target 13.1).

5. Increased international policy attention around mangroves and high levels of scientific knowledge compared with other coastal forest types provides more case studies and literature to discuss potential SDG 14 impacts.

Our focus on mangroves precludes a global analysis, though we are still able to make comparisons between the Global North and the Global South since mangrove-holding countries span a gradient of economic development. Several countries in the Global North have subtropical mangrove resources, including the USA, Australia, New Zealand and the overseas territories of several European countries.

A focus on mangroves excludes terrestrial forests located along the coast that are *not* coastal forests. While we make links to these forest types in relevant instances, we generally do not consider them here because terrestrial forests are supratidal, so may only be intermittently flooded compared to mangroves. Thus, they are not as strongly linked to coastal fisheries as mangroves, and thus may not be immediately covered by SDG 14. Instead, they are more likely to be managed under SDG15 (Life on Land). Splitting SDGs 14 and 15 into water and land misses key linkages between these spheres.

14.2.2 The Relevance of Mangrove Forests to SDG 14

SDG 14 strongly focuses on fisheries, which is one ecosystem service provided by mangroves, with many coastal communities across the tropics directly using mangrove forests as fishing grounds and nursery areas (Carrasquilla-Henao and Juanes 2017). Complex root systems shelter juvenile fish from predators, and mangroves provide food and nutrients for fishes. Mangroves provide additional ecosystem services to coastal communities, including storm protection, pollutant trapping and a variety of cultural ecosystem services, which can all contribute in some form to most of the SDG 14 targets (Table 14.1). Most recently, mangroves have been placed high on the policy agenda of many international bodies due to their role in carbon sequestration and storage. Mangroves are an example of a blue carbon ecosystem, with an ability to store carbon at densities three to five times that of other tropical forests (Donato et al. 2011). This makes mangroves a useful tool to help offset the fossil-fuel emissions of a number of countries under the Paris Agreement (Taillardat et al. 2018).

Table 14.1 Contribution of mangrove ecosystem services to SDG 14

Ecosystem service	Contribution	SDG target(s)
Provisioning services Fish production	Positive linkages exist between fish production and mangrove extent (Whitfield 2017)	Various
Fuel (wood, charcoal)	High-calorific mangrove wood can be used through mangrove harvesting (Sillanpää et al. 2017), a potentially sustainable resource.	14.2, 14.7
Non-timber forest products (e.g. honey, waxes, tannins, non-fish foods)	Numerous provisioning ecosystem services can be extracted under sustainable management (Uddin et al. 2013)	14.2, 14.7
Regulating services Carbon storage and sequestration	Carbon storage provides financial incentives to protect and sustainably manage mangroves (Alongi 2011)	14.2, 14.5, 14.7
Coastal protection	Roots and topography reduce wave energy through friction; coastal protection is a strong driver of mangrove restoration (Spalding et al. 2014)	14.2
Waste processing	Mangroves can assimilate pollutants in their soils and biomass (Ouyang and Guo 2016)	14.1

Table 14.1 (cont.)

Ecosystem service	Contribution	SDG target(s)
Ocean acidification regulation	Mangroves can increase water alkalinity (Sippo et al. 2016); however, mangroves only influence pH at local scales	14.3
Cultural services Tourism	Mangrove tourism provides livelihoods and a financial incentive for conservation (Foucat 2002)	14.1, 14.5
Recreation	Recreation provides well-being, livelihoods and a financial incentive for conservation (Ahmad 2009)	14.1, 14.5
Education	Traditional ecological knowledge about mangroves and their resources can complement scientific knowledge	14.A

14.2.3 A Framework to Understand the Impact of SDG 14 on Mangrove Forests

While mangroves may contribute to achieving SDG 14, this goal does not explicitly focus on mangrove forests. Ecosystems not explicitly considered by an SDG may be more likely to be negatively impacted. In an analysis of the 2017 VNRs to the UN, only 11 of 118 countries and territories with mangroves mentioned SDG 14 in their executive summaries. Of these, 10 focused predominantly on the role of fisheries, reflecting the dominant focus of SDG 14. Only 5 mentioned mangroves, most doing so in a single sentence. For example, Bangladesh's review states that mangrove afforestation could protect the coastal zone and islands (Government of the People's Republic of Bangladesh 2017). In the VNRs, mangroves are likely to be implicit within SDG targets that vaguely describe marine and coastal ecosystems (Target 14.2) or coastal and marine areas (Target 14.5).

Some SDG targets and their indicators may indeed be positive for mangroves and others may not. Impacts may also be positive or negative depending on whether the impact accrues on the ecosystem or the local communities that rely on them (Table 14.2). In Section 14.3 we consider each SDG 14 target and its indicator, and the potential negative impacts each may have on (1) the mangrove ecosystem; and (2) local communities reliant on mangrove resources.

Table 14.2 The possible effects of the SDG 14 targets on the mangrove ecosystem and associated local human communities [yellow = potential positive benefit; orange = potential for both mixed impacts]

Target	Indicator	Relevance to mangrove ecosystem	Relevance to local communities
14.1 By 2025, prevent and significantly reduce marine pollution of all kinds, in particular from land-based activities, including marine debris and nutrient pollution	Index of coastal eutrophication and floating plastic debris density	POSITIVE – Reduction in pollution is positive for forest health and macrobenthic biodiversity	MIXED – Positive: local communities benefit from reduced pollution over the long term – Negative: pollution controls could have negative short-term economic impacts for local industries, with knock-on impacts on local employment
14.2 By 2020, sustainably manage and protect marine and coastal ecosystems to avoid significant adverse impacts, including by strengthening their resilience, and take action for their restoration in order to achieve healthy and productive oceans	Proportion of national exclusive economic zones managed using ecosystem-based approaches	MIXED – Positive: sustainable management and restoration increases mangrove area, health and ecosystem services – Negative: it could promote large-scale monoculture planting in unsuitable areas, leading to failed restoration	MIXED – Positive: increased ecosystem services for communities to use – Negative: international donor-supported restoration can lead to community dependency – Negative: sustainable management activities – e.g. forestry or Payments for Ecosystem Services can lead to land grabs and conflicts within communities

14.3 Minimise and address the impacts of ocean acidification, including through enhanced scientific cooperation at all levels	Average marine acidity (pH) measured at agreed suite of representative sampling stations	POSITIVE – Positive impacts for calcified organisms such as shellfish	POSITIVE – Positive impacts for livelihoods linked to shellfish fisheries
14.4 By 2020, effectively regulate harvesting and end overfishing, illegal, unreported and unregulated fishing and destructive fishing practices and implement science-based management plans, in order to restore fish stocks in the shortest time feasible, at least to levels that can produce maximum sustainable yield as determined by their biological characteristics	Proportion of fish stocks within biologically sustainable levels	MIXED – Positive: for fisheries related to mangroves – Negative: could lead to increased aquaculture in mangrove areas to compensate for reduced wild-caught fishing	MIXED – Positive: fisheries may become more sustainable over the long term – Negative: stronger regulation of overfishing and unregulated fishing may have negative short-term economic impacts – especially if enforcement efforts focus on local communities rather than on large industrial players (which have a bigger ecological impact but may be more politically sensitive to regulate)

Table 14.2 (cont.)

Target	Indicator	Relevance to mangrove ecosystem	Relevance to local communities
14.5 By 2020, conserve at least 10 per cent of coastal and marine areas, consistent with national and international law and based on the best available scientific information	Coverage of protected areas in relation to marine areas	MIXED – Positive: a greater percentage of mangroves will be protected – Negative: 'leakage' pushes deforestation pressures to neighbouring unprotected mangroves – Negative: potential for 'paper parks' that look good on paper but are not enforced or resourced – Negative: mangroves may be excluded from this target because it is easier to achieve such large targets in open-ocean areas	MIXED – Positive: more protected mangroves mean more ecosystem services for communities – Negative: communities can be excluded or removed from certain types of marine protected areas – Negative: governments and industry can use protected areas as a land-grabbing tool – Negative: potential equity issues (gender, ethnicity, class)

14.6 By 2020, prohibit certain forms of fisheries subsidies which contribute to overcapacity and overfishing, eliminate subsidies that contribute to illegal, unreported and unregulated fishing and refrain from introducing new such subsidies, recognising that appropriate and effective special and differential treatment for developing and least-developed countries should be an integral part of the World Trade Organization fisheries subsidies negotiation	Progress by countries in the degree of implementation of international instruments aiming to combat illegal, unreported and unregulated fishing	MIXED – Positive: could remove perverse onshore aquaculture subsidies and incentives that encourage mangrove deforestation – Negative: could encourage shift to other agriculture types, if aquaculture no longer financially attractive	MIXED – Positive: if linked to perverse aquaculture subsidies and incentives that encourage mangrove deforestation; communities may retain land or access to mangroves – Negative: may cause reduced employment opportunities – Negative: can change or remove local economic structures

Table 14.2 (cont.)

Target	Indicator	Relevance to mangrove ecosystem	Relevance to local communities
14.7 By 2030, increase the economic benefits to small island developing States and least-developed countries from the sustainable use of marine resources, including through sustainable management of fisheries, aquaculture and tourism	Sustainable fisheries as a percentage of GDP in small island developing States, least-developed countries and all countries	MIXED – Positive: should reduce environmental impacts if forest resources used/ harvested sustainably – Negative: increased economic benefits could lead to unregulated development and cause environmental harm e.g., tourism and aquaculture can cause disturbance and mangrove loss – Negative: increased livelihoods due to sustainable management can increase demand for forest products, causing further environmental harm	MIXED – Positive: sustainable management protects ecosystem services that communities use – Positive: increased local livelihoods – Negative: economic uses can exclude certain parts of the community, e.g. communities restricted access to REDD+ sites or can no longer extract certain resources

14.A Increase scientific knowledge, develop research capacity and transfer marine technology, taking into account the Intergovernmental Oceanographic Commission Criteria and Guidelines on the Transfer of Marine Technology, in order to improve ocean health and to enhance the contribution of marine biodiversity to the development of developing countries, in particular small island developing States and least-developed countries	Proportion of total research budget allocated to research in the field of marine technology	MIXED – Positive: if improvement of aquaculture techniques increase efficiency and reduce demand for converting mangrove forests – Positive: if technology for restoration is enhanced – Negative: technological innovation could lead to increased pressure to clear mangroves for aquaculture	MIXED – Positive: improvement of aquaculture techniques increases profitability – Negative: technical innovation could reduce labour requirements
14.B Provide access for small-scale artisanal fishers to marine resources and markets	Progress by countries in the degree of application of a legal/regulatory/policy/institutional framework which recognises and protects access rights for small-scale fisheries	MIXED – Positive: if it encourages more sustainable practices – Negative: increased access could have negative environmental impacts if not regulated adequately	POSITIVE – Increases livelihoods – Increases access to resources and environmental justice, especially if institutional frameworks can promote gender issues alongside access rights

Table 14.2 (cont.)

Target	Indicator	Relevance to mangrove ecosystem	Relevance to local communities
14.C Enhance the conservation and sustainable use of oceans and their resources by implementing international law as reflected in UNCLOS [UN Convention on the Law of the Sea], which provides the legal framework for the conservation and sustainable use of oceans and their resources, as recalled in paragraph 158 of The Future We Want	Number of countries making progress in ratifying, accepting and implementing through legal, policy and institutional frameworks, ocean-related instruments that implement international law, as reflected in the UNCLOS, for the conservation and sustainable use of the oceans and their resources	POSITIVE – Positive impacts on migratory species that use mangroves for some part of their lifecycle (birds, turtles) – Positive for reducing impacts of oil or other chemical spills, if UNCLOS extends to mangroves – Positive indirect impact if offshore fisheries management is improved, reducing pressure on near-shore resources.	POSITIVE – Less-polluted mangroves will better provide ecosystem services such as fisheries

14.3 SDG 14 Targets, Actors and Potential Impacts

14.3.1 Prevent and Significantly Reduce Marine Pollution – Target 14.1

Pollution, marine litter and eutrophication are significant issues in tropical coastal areas (Todd et al. 2010). Marine debris stems from improper solid-waste management practices, poor behavioural choices made by consumers and fishers (e.g. plastic pollution from fluvial sources, dumping of fishing gear or ship-generated waste) and poor waste disposal facilities or lack of access to them. Nutrient sources contributing to eutrophication include aquaculture outflows, factory discharges into rivers and coastal areas, and agricultural and urban runoff, all of which are expected to be exacerbated by global change (Rabalais et al. 2009). These issues are caused by numerous diffuse sources, making their management and reduction particularly difficult.

To address the issue of marine litter, the UN Environment Programme (UNEP) launched the Global Partnership on Marine Litter in June 2012. Known as the Honolulu Strategy, by 2025 it aims to reduce the impacts of litter, enhance cooperation and coordination through multi-stakeholder platforms, promote knowledge sharing and monitoring, promote economic development through resource efficiency and waste prevention, increase awareness, and assess emerging health and ecological issues of plastic waste (UNEP 2012). The G20 group of governments has also created an *Action Plan on Marine Litter* (G20 2017). Most EU countries have national policies to reduce marine litter (IUCN 2017), and similar national and regional strategies exist for nutrient and sediment pollution (Chen 2015).

ANTICIPATED IMPACTS ON MANGROVES

Protecting mangroves from pollution is important; a healthy mangrove can help buffer other coastal ecosystems from pollution and eutrophication because of its ability to trap heavy metals and excess nutrients (Valiela and Cole 2002). Reducing nutrient pollution also increases mangrove tolerance to extreme weather events (Feller et al. 2015). Mangroves are exposed to terrestrial and marine pollution sources because they exist at the transitional margin between land and sea. Thus, initiatives such as the Honolulu Strategy have great potential to protect mangroves through Target 14.1. Negative impacts can only be foreseen if Target 14.1 drives a focus on eutrophication and litter to the exclusion of other pollution types, such as noise or light pollution (which may affect mangrove fauna).

ACTORS INVOLVED OR AFFECTED

Addressing pollution may have unintended negative consequences if implementation does not adequately consider equity issues, e.g. who pays for or bears the burden of pollution and its control and who makes decisions about pollution prevention. Are local communities involved and, if so, in what capacity? Marginalised populations and lower-income countries can bear a disproportionate cost of the negative impacts of pollution, while wealthy polluters can pay off their pollution debt (Torras and Boyce 1998). However, overregulation may also have short-term negative economic impacts for industries, with indirect impacts for the local people they employ. Thus, aspects of the Honolulu Strategy that emphasise the economic benefits of resource efficiency and waste prevention are particularly welcome.

Risk transfer is another key issue. In the absence of adequate land-based waste disposal, marine litter that is collected may eventually return to the coast. When planning ecosystem-based adaptation approaches to reduce pollution, we must consider the adequacy of pollution remediation mechanisms at source and whether we are shifting the problem to other ecosystems.

14.3.2 Sustainably Manage and Protect Coastal Ecosystems, Including Their Restoration – Target 14.2

Sustainably managing natural resources is a formidable challenge in any ecosystem due to weak governance, limited resources, corruption and conflicting pressures for short-term economic growth. Mangrove management shares these and comes with additional complications. Mangroves suffer from cross-sectoral conflicts and lack of communication, as they frequently fall under both marine and terrestrial jurisdictions, or neither (Friess et al. 2016a, Primavera 2000). Mangroves are also not included in many international forestry conservation and management initiatives – e.g. the initial lack of protocols and standards for including mangroves in international Payments for Ecosystem Services (PES) schemes such as REDD+.

Target 14.2 could be achieved by strengthening legislation that promotes sustainable management. Many mangrove-holding countries have national laws that aim to protect them either under specific mangrove laws or generic environmental protection policies (Slobodian et al. 2018). Hence, the challenge is not to advocate for new legislation but to enforce existing laws and build strong governance. Improved governance requires greater cooperation between sectors and actors. Ecosystem-based approaches that recognise the importance of ecosystem services and can adapt when faced with change are also important. Engaging with and empowering local communities in forest governance, often building on existing customary institutions, has been proven to ensure equity and effectively provide resources that are not available to national institutions (Friess et al. 2016a). Market-based solutions may also be effective if regulated effectively. These may involve traditional sustainable-production forestry, such as the Matang Mangrove Forest Reserve in Malaysia, which provides wood for charcoal to local and regional markets (Goessens et al. 2014), or the BUMWI forestry concession in West Papua, Indonesia, the world's only Forest Stewardship Council (FSC)-certified mangrove plantation, providing wood chips to international buyers (Sillanpää et al. 2017). Alternatively, other ecosystem services, such as carbon, may be marketed through PES (e.g. Mikoko Pamoja in Kenya). However, market-based solutions come with their own potential issues, such as governance recentralisation (Phelps et al. 2010), inadequate benefit sharing (Phelps et al. 2013) and other challenges to environmental justice (Locatelli et al. 2014).

Target 14.2 could also be achieved through mangrove rehabilitation, which is increasingly being proposed over large scales. The 2011 Bonn Challenge aims to restore 150 million ha of degraded and deforested land by 2020 and 350 million ha by 2030. Of the countries that have pledged, 27 have mangroves within their national borders. The Bonn Challenge is promoted as a vehicle to achieve existing international commitments, including the CBD[1] Aichi Target 15 (relating to ecosystem resilience and restoration), UNFCCC[2] REDD+ goals and the Rio+20 land degradation neutrality goal. A mangrove-specific initiative is the Global Mangrove Alliance (GMA) – a partnership between major conservation NGOs – that aims to stop deforestation and increase current mangrove area by 20 per cent by 2030 (GMA 2017).

ANTICIPATED IMPACTS ON MANGROVES

Given current threats, establishing ambitious, well-funded and centrally co-ordinated targets for mangrove restoration seems an obvious and necessary goal. However, large-scale mangrove restorations often fail. For example, 97 per cent of replanting efforts surveyed in Sri Lanka showed partial or complete mortality of trees (Kodikara et al. 2017), and thousands of hectares of mangrove plantations in the Philippines have been established on inappropriate areas, which reduces mangrove survival and damages and replaces valuable habitats such as seagrasses (Primavera and Esteban 2008). These failures occur even though several practitioner manuals and best-practice documents exist (Lewis and Brown 2014, ZSL 2015). There is a disconnect between the recommendations of these manuals and the target setting, coupled with low monitoring requirements of international donors.

Even when healthy trees are established, planting can result in ecologically impoverished monocultures that are less resistant to ecological stress and disturbance (Villamayor et al. 2016). For these reasons, approaches based

[1] Convention on Biological Diversity.
[2] UN Framework Convention on Climate Change.

on hydrological restoration, such as Ecological Mangrove Restoration (EMR) (Lewis 2005), should be used whenever possible. These approaches take time, expertise and extensive on-site pre-planning, but rushing to reach inflated artificial planting targets is likely to result in failures and increased cynicism about rehabilitation potential.

ACTORS INVOLVED OR AFFECTED

Mangroves are complex socio-ecological systems, with numerous actors at different hierarchical levels involved in their restoration and sustainable management. Large-scale restorations are often promoted by international donors and NGOs, as is seen with the Bonn Challenge and the GMA. National and local government agencies are often involved in granting permission for or implementing restoration, though jurisdictions among agencies may be blurred (Primavera 2000).

Large-scale restoration projects are conducted in degraded areas where human influence is high, affecting neighbouring communities. Environmental justice often demands the involvement of local communities in restoration. Practical arguments for involving local people are also strong since resources for external management may be limited. However, paying communities to undertake restoration, as is the norm, can set up a cycle of dependency. This includes economic dependency, with communities potentially encouraging rehabilitation failure so that they can get paid to replant multiple times; planting payments can become a substantial part of the local economy, providing little incentive for restoration to be successful (Thompson 2018). However, dependency can be broken by using other metrics of success (e.g. focusing on per cent of survival, as opposed to per cent planted), paying communities for tasks other than planting (e.g. digging creeks for hydrological restoration), or co-funding restoration projects in conjunction with local communities (Thompson 2018). Some of these approaches have been applied with great success, such as hydrological restoration through community-based EMR in Indonesia, Thailand and Latin America (Brown et al. 2014).

Other methods of achieving Target 14.2 may also impact local communities. Monetised incentives for mangrove management, such as the demarcation of forestry concessions or PES schemes, may challenge environmental justice by risking elite capture and land grabs by powerful stakeholders (Beymer-Farris and Bassett 2012) or exacerbate existing gender inequalities (e.g. where male heads of households receive payments). Explicit commitments and safeguards to environmental justice and gender equality as part of project design and execution are essential to mitigate these risks.

Mangroves are complex socio-ecological systems. Large-scale global restoration and sustainable management ambitions required under Target 14.2 must account for this complexity by involving local as well as national and international stakeholders. Nested governance, responsive to local needs and guided by environmental justice and ecosystem management, may seem complex when set against the stirring simplicity of the SDG visions. Realising the promises of those visions for mangroves will mean the careful use of what we already know about successful mangrove management, restoration and governance.

14.3.3 Regulate Harvesting and End Overfishing – Target 14.4

Fishing is critical to the food security and livelihoods of hundreds of millions of people. Unfortunately, 89.5 per cent of fisheries are either fully exploited or overfished (FAO 2016), up from 70 per cent a decade earlier (FAO 2006). Target 14.4 aims to reduce overfishing through improved fisheries management, reduction of harmful fishing practices, stronger monitoring and enforcement, reduced by-catch and eco-labelling (Vierros and Buonomo 2017). The Sundarbans – one of the largest mangrove forests in the world – provides a good example of how Target 14.4 could be implemented. This area faced over-exploitation of fisheries by local communities in the 1990s, resulting in declining yields. The Bangladesh government, supported by the Asian Development Bank, strengthened legislation to deal with this issue, including the banning of fishing for certain species, minimum size limits for harvesting and the introduction of minimum net sizes (Hoq et al. 2007).

ANTICIPATED IMPACTS ON MANGROVES

Target 14.4 may be focused primarily on offshore fisheries, as this is where a large amount of fishing effort is based. However, if there is a stronger focus on inshore fisheries, then this target is anticipated to positively impact faunal components of the mangrove ecosystem by reducing harvesting pressure on fish and shellfish stocks. For example, traditional regulation of fisheries in Okinawa's mangroves and other inshore areas has successfully sustained fish stocks over centuries (Akimichi and Ruddle 1984). Similarly, the sasi system – a local traditional coastal resource management system in eastern Indonesia – is associated with increased effectiveness of environmental protection (McLeod et al. 2009). Linking existing traditional and government regulations with further business-sector demand may be an alternative way to ensure the sustainability of wider fisheries value chains and the mangroves that support them. However, the success of measures such as these relies on sufficient enforcement, whether by government or the community, and the provision of alternative livelihoods and food sources if access to fisheries is limited. Of particular concern is whether stronger enforcement in capture fisheries pushes food and economic security more towards aquaculture, the dominant driver of mangrove deforestation across the tropics (Hamilton 2013, Richards and Friess 2016).

ACTORS INVOLVED OR AFFECTED

State actors are often heavily involved in the implementation and enforcement of Target 14.4 and frequently face challenges and limitations in resources for operational costs and enforcement. The actors contributing most to heavy fishing pressure are likely to be commercial operators, especially in open-sea fisheries. It is less clear whether this is the case in mangrove-related fisheries, where local communities contribute to localised overfishing. Regulations and enforcement of overuse are thus expected to impact local communities and their economic pathways (Silva-Cavalcanti and Costa 2009). Impacts include increased transaction costs of fishing (if particular fishing equipment is banned or changed) and decreased food security (if certain fish species or fishing areas are banned). In these cases, alternative food and livelihood sources need to be considered. The implementation of Target 14.4 requires rules designed to control large-scale commercial fishery operations that do not place unintended restrictions on local fish harvesting and consumption.

14.3.4 Conserve at Least 10 Per Cent of Coastal and Marine Areas – Target 14.5

Protected areas are a traditional method of habitat conservation, and marine protected areas (MPAs) have increased in the tropics, driven by national and international policy concerns. Target 14.5 further promotes protected areas as a conservation tool by pushing for an increase in the proportion of coastal and marine areas to be protected to 10 per cent by the year 2020, aligning closely with Aichi Target 11 (Rees et al. 2017). MPAs cover about 3.25 million km^2 globally (Roberts et al. 2018) and some countries have already achieved SDG Target 14.5, such as Belize protecting 21 per cent of its national waters (Government of Belize 2017). Yet the targets set by some countries are currently not strong enough. For example, only 6.3 per cent of Bangladesh's coastal and marine protected area is scheduled to be protected by 2020 (Government of the People's Republic of Bangladesh 2017), though the already protected Sundarbans mangroves do make up a proportion of this.

MPAs are often focused on reefs (Edgar et al. 2014) or open-water ecosystems, such as those important for pelagic fisheries. Mangroves have not traditionally been the main focus of MPAs (Friess et al. 2016a), therefore we would expect the proportion of mangroves under protection to lag behind what is required of SDG Target 14.5. If SDG Target 14.5 is to protect 10 per cent of coastal and marine areas, and particularly mangroves, this will require the substantial strengthening of national and international protected-area legislation, including the establishment of new mangrove MPAs and proper enforcement for existing areas. The proportion of mangroves covered in protected areas should be reported by member states in their annual National Biodiversity Strategic Action Plans to the Convention on Biological Diversity to monitor progress.

ANTICIPATED IMPACTS ON MANGROVES

While the establishment of new MPAs would be expected to result in positive environmental outcomes, several negative impacts are envisaged. Firstly, ambitious targets to increase protected area extent are achieved through the establishment of very large marine protected areas (VLMPAs) that cover hundreds of thousands of square kilometres (Jones and De Santo 2016). Because of their size, VLMPAs are primarily established in the open ocean. It is harder to achieve large-scale protection in the coastal zone due to the many conflicting stakeholders and the need for resource access. Target 14.5 may promote the creation of VLMPAs in the open ocean at the expense of smaller coastal MPAs that would incorporate coastal forests and benefit the local communities that rely on them.

Secondly, MPAs will only have positive environmental outcomes if sufficiently resourced and enforced so that infringements such as local mangrove deforestation can be stopped. Setting up MPAs without adequate resourcing often leads to the establishment of 'paper parks' that technically meet Target 14.5 requirements but show low success on the ground. For example, the establishment of protected areas in Indonesia has not necessarily reduced mangrove deforestation (Miteva et al. 2015), where oil palm has been found encroaching into protected areas in Langkat Regency, north Sumatra, and other protected areas have suffered from aquaculture encroachment. We should move beyond simple area targets and focus instead on protected area quality, not quantity (Barnes et al. 2018).

Finally, protected areas can drive leakage, where deforestation and degradation are stopped within the protected area but shifted off-site into neighbouring unprotected mangroves or other unprotected ecosystems. Leakage is a common issue in forested ecosystems; it is hard to monitor and remains a major challenge in mangroves (Locatelli et al. 2014). Leakage can be reduced by increasing monitoring around the protected area, increasing the scale of the protected area to cover locations particularly at risk or creating protected area networks. However, it is debatable whether these solutions solve leakage or push it even further off-site.

ACTORS INVOLVED OR AFFECTED

MPAs are most successful when they have clear objectives and strong enforcement (Edgar et al. 2014). These are most often in the control of the state actors who have the capacity to manage and monitor large areas, e.g. government national park agencies. Centralised resource management goes against decades of governance decentralisation in the tropics, with the potential for local community exclusion (Phelps et al. 2010) and the reassertion of state control over community lands that can lead to land grabs (Beymer-Farris and Bassett 2012). This has been highlighted as a potential consequence of Aichi Target 11 (and, by extension, SDG Target 14.5), which explicitly requires equitable management of MPAs (Rees et al. 2017).

Environmental justice has important implications for MPA success because MPA performance is not determined solely by top-down processes. When national governments implement Target 14.5, they should consider how to incorporate communities and local practices into MPA design. The International Union for Conservation of Nature (IUCN) promotes seven categories of protected areas, some of which allow community use or interaction (IUCN 2012). MPAs that have strong community support can be more successful in achieving conservation objectives compared to those without (Francis et al. 2002). Similarly, community-managed mangrove forests can have better conservation outcomes than state-managed forests (Sudtongkong and Webb 2008). The equity and economic security of local communities must always be considered, particularly because mangrove protected areas can be associated with short-term livelihood losses, while the economic benefits may accrue over the long term (McNally et al. 2011).

14.3.5 End Certain Forms of Fisheries Subsidies – Target 14.6

Industrial fisheries politically and administratively overwhelm concerns about mangrove conservation, as nations exploit fisheries for income and food security. Their importance means that fisheries are subsidised in most nations, with global fishing subsidies valued at USD 25–29 billion annually (Sumaila et al. 2010). In Global North countries containing mangroves (e.g. Japan, Taiwan, USA) subsidies overwhelmingly support catch fisheries (European Commission 2017). In many Global South nations subsidies also support onshore aquaculture activities through fuel subsidies, tax exemptions and aquaculture extension. For example, the Government of India subsidises pond construction, input costs, hatcheries and monitoring costs (DAHD 2016). Target 14.6 is currently ambiguous as to whether such farmed fishing practices fall under its remit.

Industrial aquaculture, supported by subsidies, has been the predominant cause of mangrove loss in recent decades, representing one of the major threats to mangrove-dominated coastal and delta areas (e.g. Richards and Friess 2016). The importance of the aquaculture sector for economic development means that fishery subsidies have been made in both direct and indirect ways: such subsidies have contributed to many of the environmental problems seen with the aquaculture industry (Neiland et al. 2001), such as deforestation (Barbier and Cox 2004). Achieving Target 14.6 requires changes to national legislation that currently provides aquaculture subsidies and incentives in many tropical countries.

ANTICIPATED IMPACTS ON MANGROVES

Removing aquaculture subsidies should have immediate positive benefits for mangrove ecosystems by making it more expensive and less appealing to expand aquaculture operations into mangrove areas. The concerns are that it may shift people's attention to other economic activities that impact mangroves, especially if they are also the focus of government subsidies. In Southeast Asia we are already seeing the replacement of mangroves and abandoned aquaculture by oil palm, particularly in Indonesia and Malaysia (Richards and Friess 2016). This may become an even greater threat in the future, as oil palm is now the focus of substantial national financial subsidies in order to achieve ambitious production targets to secure food and economic security in these countries. Thus, in the medium to long term, mangrove deforestation will persist in many countries even as aquaculture subsidies are phased out.

ACTORS INVOLVED OR AFFECTED

Target 14.6 signals to actors that we need to change our expectations that we can increase fishery production through mangrove clearance in unsustainable ways. As this involves changes to national legislation and economic priorities, state actors are key to Target 14.6 implementation. Significant political will at the country level is required to end fishery subsidies and the allocation of mangrove areas to aquaculture.

Commercial actors and community actors will be affected by Target 14.6. At the local scale, conflicts between two parts of society and Target 14.6 will occur. Investing in aquaculture is often expensive, so only affluent stakeholders are able to invest in such a business or secure appropriate loans (Barbier and Cox 2004, Primavera 1997). These investments will be impacted by changes in aquaculture subsidies. Local communities who are employed by commercial stakeholders may face unemployment or reduced employment if fisheries subsidies are removed. The need for further economic gains and employment opportunities is likely to drive actors into other, potentially unsustainable industries, as noted above. Target 14.6 needs to take a broader view than solely aquaculture subsidies, to include alternative subsidised income streams and the impacts they may have on the mangrove ecosystem.

14.3.6 Increase Economic Benefits from the Sustainable Use of Marine Resources – Target 14.7

Target 14.7 focuses on improving the economies of developing nations and small island states by increasing the sustainable use of marine resources, primarily through sustainable fisheries. This could be achieved in many ways, promoted by technological, commercial and governance influences. Target 14.7 suggests that aquaculture and tourism can play a role in achieving Target 14.7, though these do not match with the SDG indicator for this target. It may be unclear how current forms of intensive aquaculture in mangrove areas could match the sustainable-use focus of Target 14.7. Mangrove ecotourism fits under the broad description of Target 14.7 and is a popular management activity in many mangrove areas across the tropics, particularly in Central America and some Southeast Asian countries (such as the Philippines and Malaysia). To increase economic benefits in the coastal zone, it is good that Target 14.7 looks beyond fisheries. However, we see that there are more opportunities because mangroves could be a key contributor to Target 14.7 if the sustainable use of other marine resources could be incorporated. This can be done most obviously with provisioning ecosystem services provided by mangroves (timber, fuelwood, non-timber forest products, food resources, pharmaceuticals) and some regulating services (carbon, nutrients). When broadened beyond fisheries, Target 14.7 has strong synergies with Target 14.2.

Standards and certifications can promote sustainable management practices and they do exist in fishing activities (e.g. Marine Stewardship Council) and mangrove aquaculture (e.g. in Vietnam). This can be expanded to other mangrove ecosystem services. For example, FSC certification requires strict adherence to environmental and social standards for logging operations, with certified products generally attracting a premium on the market, particularly in the Global North (Hoang et al. 2015). There is huge scope to expand this: currently, only one FSC-certified mangrove concession exists worldwide, in West Papua, Indonesia (Sillanpää et al. 2017). PES (see Target 14.2) is another method of promoting a switch from exploitative to sustainable management: it requires stakeholders to change land-use practices to protect and/or increase the ecosystem service of interest.

ANTICIPATED IMPACTS ON MANGROVES

Target 14.7 is expected to positively impact fisheries, since they are the main focus. We also expect positive impacts on other mangrove ecosystem services, if the target expands to them. However, as the following examples demonstrate, 'sustainable' management can have adverse and unintended consequences on ecosystem quality.

Ecotourism is an increasingly common method of mangrove management in many countries. While ecotourism can increase local livelihoods and educate visitors, it is frequently associated with negative environmental impacts. For example, mangrove ecotourism in Langkawi, Malaysia, has been associated with pollution and erosion, as speedboats disobey speed limits during tours (Lee 2013).

Matang Mangrove Forest Reserve is an approximately 30 000 ha charcoal production area in Malaysia that has been managed through rotational harvesting for some 100 years. As such, it is often held up as a leading example of long-term sustainable logging (Shaharuddin et al. 2005). However, these practices have progressively turned this area from a biodiverse mangrove into a monoculture, as *Rhizophora* spp. are strongly preferred for charcoal. Its long-term sustainability has also been brought into question due to lower propagule production and natural regeneration (Goessens et al. 2014) and potential declines in plant productivity.

Sustainable management may increase local livelihoods but can have perverse impacts on environmental resources. For example, sustainable sea cucumber fisheries have been promoted by an NGO in Madagascar to create more secure livelihoods for local communities. These communities previously conducted small-scale selective logging for poles and house-frame construction. However, increased disposable income in these communities means that many households have upgraded their homes to lime render, which increases durability and is a status symbol. Lime render requires the collection of a large volume of gastropod shells, which are burned down in kilns fuelled by mangrove wood (Figure 14.1). This has forced a transition from selective logging to larger-scale mangrove clearance when lime orders come in (Scales et al. 2018).

ACTORS INVOLVED OR AFFECTED

Target 14.7 should have positive economic impacts for commercial parties and local communities who adhere to sustainable management, if a product has a suitable premium and an existing market. Local communities stand to benefit especially if sustainable management certifications have built in strong social safeguards, such as FSC certification.

Social safeguards are important because sustainable management practices can have negative consequences for local communities if production rates are lowered. This is particularly so if sustainable resource use requires high levels of initial investment, which can lead to bigger companies pushing out smaller local companies. For example, the promotion of mangrove ecotourism by a government development agency in Langkawi, Malaysia, has encouraged larger national companies to come into the area to invest in the operation, with negative consequences for small local operators who

Figure 14.1 Clearcutting of mangroves in the Bay of Assassins, Southwest Madagascar, in response to market demands for lime kilns. Photo by Dan Friess.

had been conducting ecotourism activities but lack the financial resources or political connections to compete (Thompson et al. 2018). Negative social consequences are often seen in PES schemes (Pascual et al. 2014, Pouyal et al. 2016), where sustainable management and conservation of the ecosystem service (e.g. carbon) excludes local communities from lands in order to minimise impact. In Tanzania efforts to prepare for a REDD+ project led to land grabs and state protectionism at the expense of community environmental justice (Beymer-Farris and Bassett 2012).

14.3.7 Increase Scientific Knowledge, Capacity and Marine Technology – Target 14A

Target 14.A could be achieved by the transfer of capacity and marine technology related to onshore aquaculture – for example, by sharing knowledge and technology that improves aquaculture yields or diversifies aquaculture products such as algae without increasing the current aquaculture footprint. Knowledge sharing may be difficult in some cases as aquaculture is a commercial endeavour, so there is little incentive to share information with competitors. Aquaculture extensionists and new networks of communication (such as efforts to communicate market prices or mobile technology to alert shrimp disease) play an important role in achieving this target.

Capacity and knowledge could also be shared on the topic of mangrove restoration. As noted under Target 14.2, mangrove restoration is notoriously unsuccessful (Kodikara et al. 2017, Lewis 2005, Primavera and Esteban 2008), most often because inappropriate species are planted in inappropriate locations. Several practitioner manuals exist in multiple languages (Lewis and Brown 2014) to facilitate knowledge transfer from successful projects, so it is not an issue of creating more materials and manuals. Rather, it is an issue of communication, translation and knowledge sharing. Target 14.A could bolster this through international platforms such as the IUCN Mangrove Specialist Group and the GMA.

ANTICIPATED IMPACTS ON MANGROVES

Target 14.A may positively benefit mangroves if improvement of aquaculture techniques increases yields within existing ponds and reduces pollution, thereby reducing demand for converting neighbouring mangrove forests and other adjacent ecosystems. However, this has just as much chance of negatively impacting mangroves, as technological innovation leads to increased profitability, increasing pressure to clear more mangroves as aquaculture expands to take advantage of commercial opportunities. There is huge debate in the conservation field about whether increased productivity results in land sparing or agricultural expansion, with several studies suggesting the latter to be true, to the extent that conservation costs actually increase (Carrasco et al. 2014, Phelps et al. 2013).

ACTORS INVOLVED OR AFFECTED

Several actors are involved in marine technology and its transfer, including the commercial actors creating such technology and the national and international platforms facilitating knowledge transfer (e.g. the GMA could play a key role in communicating proper standards for mangrove restoration).

Local communities and businesses will be impacted by any increases in knowledge and technology, creating a positive impact on livelihoods if the improvement of aquaculture techniques increases profitability and if it trickles down to local workers. However, technological innovation may be just as likely to reduce workforce requirements, with negative impacts on employment and livelihoods potentially pushing people into more destructive practices. Target 14.A suggests that knowledge and technology can only be created through research by key gatekeepers and then transmitted to those on the ground; this disregards the huge contribution of traditional ecological knowledge and the capacity of local and Indigenous communities to improve fisheries and coastal management.

14.3.8 Provide Access for Small-scale Artisanal Fishers to Marine Resources and Markets – Target 14.B

Target 14.B is an important way to improve local livelihoods and increase environmental justice by securing access to marine resources (such as fish) and markets that allow monetising those ecological benefits. As reflected in Indicator 14.B.1, this is most likely to be achieved in a top-down manner by legislative reform that encourages the transfer of marine resource rights to local communities. Improving market access requires more transparency in supply chains and technology investments that give local communities access to market information (e.g. mobile phone platforms to communicate market prices).

ANTICIPATED IMPACTS ON MANGROVES

Community-based mangrove management can have positive benefits for mangrove area and health since management is placed in the hands of resource users who know the local context and communities have an incentive to conserve their own resources (Friess et al. 2016a, Sudtongkong and Webb 2008). So, increasing regulatory and institutional frameworks that recognise community access rights are welcome, especially if such frameworks are able to incentivise and enforce sustainable management by local communities. Target 14.B may have mixed small-scale impacts on mangrove-related fisheries, as increasing resource access may promote more people to undertake fishing activities. This could lead to (shell)fish population declines if not properly and equitably managed. The broader mangrove ecosystem can be negatively impacted if trees are harvested for poles and other artisanal fishing infrastructure.

ACTORS INVOLVED OR AFFECTED

The hope is that this SDG target will have the intended positive impacts on local community actors by increasing livelihoods through more direct connections to markets. Increasing resource access and market access are also positive outcomes for environmental justice, giving local communities more control over their local resources and their livelihoods. However, success is most likely when there are strong local–state relationships and co-management (see examples presented by Defeo et al. 2016 in Latin America), so that appropriate, equitable and enforceable frameworks can be drafted. For frameworks to be equitable, they need to address benefit sharing and gender issues, as women are key collectors of (shell)fish and forest products as well as being heavily involved in their processing (Lau and Scales 2016). However, frameworks that focus on fishing practices alone may ignore these contributions.

14.3.9 Enhance Conservation and Sustainable Management through UNCLOS – Target 14.C

Articles 61 and 62 of the UN Convention on the Law of the Sea (UNCLOS) describe countries' obligations for the conservation and use of living marine resources within the exclusive economic zone. The conservation of biodiversity and protection of marine habitats are obligations under Article 194. Articles 117–19 describe obligations for the conservation of living resources in the high seas; Articles 207–12 address pollution from the land, sea and atmosphere. Migratory and straddling fish stocks are also considered under UNCLOS.

As per the indicator for Target 14.C, most mangrove-holding countries have already ratified UNCLOS (Cambodia, Colombia, El Salvador, Iran, the United Arab Emirates and the USA have signed but not ratified; Eritrea, Peru and Venezuela have done neither). Thus, this target is most likely to have an impact by focusing on the implementation of UNCLOS principles that promote incorporation of mangrove conservation into national laws.

ANTICIPATED IMPACTS ON MANGROVES

Mangroves are not explicitly considered under UNCLOS, but may be indirectly addressed since UNCLOS serves as a framework for the Convention on Biological Diversity and also the Ramsar Convention, both of which cover mangrove forests. UNCLOS has been used to enforce conservation of connected populations. While this has largely focused on globally connected fisheries (e.g. tuna), mangroves and the organisms that reside within them are also connected by sea and could fall under this remit. Due to these connections, UNCLOS can be used as a framework for regional cooperation on conservation (Ramesh et al. 2017). Articles that consider the impacts of pollution are particularly relevant, such as oil pollution, which can cause considerable damage within mangroves (Duke 2016), and other pollutants, whether from sea or land. Thus, we consider that this SDG target will have largely positive impacts on mangroves if mangroves and their inshore areas are explicitly considered.

ACTORS INVOLVED OR AFFECTED

Target 14C involves actors at multiple hierarchical levels, particularly national and international stakeholders involved in the implementation of UNCLOS principles into national policies. Though these stakeholders are not impacted, they play an important implementation role. Conversely, local communities may not play an important implementation role, but they are impacted by these national policies. Local communities will benefit from a less-polluted coastal environment; however, pollution controls should not inadvertently restrict or harm local community activities along the coast. Local communities may also benefit from the fisheries protection aims of UNCLOS, though in reality these are most appropriate for high-seas fisheries outside of a country's exclusive economic zone, and may have little relevance to coastal communities.

14.4 General Themes Regarding the Impact of SDG 14 on Coastal Forests

Three strong themes emerge across all SDG 14 targets that increase the risk of adverse consequences to coastal forested ecosystems.

14.4.1. Lack of Focus on Coastal Forested Ecosystems

Coastal forested ecosystems need explicit consideration within SDG 14. The focus on fisheries leans towards marine rather than coastal ecosystems. A focus on fisheries is rightly important, because fisheries are heavily threatened globally, with important implications for development and the environment. However, this chapter shows examples of how this focus may have unintended consequences on forested ecosystems that directly and indirectly support fisheries. Target 14.6 provides a pertinent example: this target is solely focused on fisheries, and the Indicator of 'implementation of international instruments aiming to combat illegal, unreported and unregulated *fishing*' suggests a focus on fishing. However, subsidies for land-based fish production (e.g. tax breaks for aquaculture) are huge drivers of mangrove deforestation. The removal of aquaculture subsidies may encourage other forms of agriculture that are now more financially viable. We are already seeing this in Southeast Asia, with the emergence of oil palm as a driver of mangrove deforestation (Richards and Friess 2016). Other SDG targets implicitly bias management efforts towards fisheries, such as the creation of VLMPAs (Target 14.5) at the expense of smaller, coastal MPAs that incorporate mangroves, or focusing on sustainable fisheries management (Target 14.7), though the latter has huge potential application to other mangrove ecosystem services. Mangroves can make an important contribution to SDG 14, but they have to be included.

14.4.2 A Marine Focus of SDG 14 Misses Important Linkages

Only Target 14.1 considers terrestrial–marine linkages, through upstream pollution. This is a problem because it is more than just pollution that connects the terrestrial and marine zones; they are also connected by sediments, financial flows, community uses and governance arrangements. This disconnection also means that some forest types are not considered; for example, terrestrial forests co-located along the coast are excluded, because they would be managed under SDG 15 (Life on Land), though they may be impacted by SDG 14. Of all the SDGs, SDG 14 and 15 (alongside SDG 11 Sustainable Cities and Communities) stand out because they are the only explicit system- or location-based SDGs. The other SDGs are sectoral (e.g. governance, poverty, education, gender equality), and as such can be incorporated more widely into the other SDGs.

14.4.3. The Importance of Multi-Stakeholder Engagement

SDGs are most likely to be implemented by national-level state actors, as they are the signatories to the SDGs and associated conventions and are well-placed to mainstream the SDGs into national development and environmental conservation planning. However, many of the impacts of SDG 14 discussed here are exacerbated by the dominance of top-down governance. Multi-stakeholder collaboration and engagement may make it more likely to anticipate negative impacts of SDG 14 on coastal communities. Many tropical countries have decentralised natural resource governance over the past few decades, which has increased the success of conservation and sustainable management actions, particularly for mangroves (Sudtongkong and Webb 2008). Linked to this, environmental justice for local communities is another theme that emerges from this chapter, with clear links to decentralisation. Policies and management actions that ensure equitable access to ecosystem benefits are likely to discourage some of the negative and unintended impacts of SDG 14 outlined here, indicating the importance of incorporating principles of SDG 16 (Peace, Justice and Institutions) into all the SDGs.

14.5 Synergies and Trade-Offs

While a full analysis is beyond the scope of this chapter, it is clear that coastal forests can contribute to all SDGs (Ramsar 2018). Likewise, many SDGs can have positive, synergistic effects on SDG 14 targets and mangroves. However, they are also likely to be antagonistic to mangrove conservation and management in some settings.

SDG14's focus on fisheries can contribute to achieving SDG1 (No Poverty) (e.g. Coulthard et al. 2011). However, trade-offs are envisaged, because increasing livelihoods can lead to greater environmental degradation, as in the Madagascar example given earlier (Scales et al. 2018). Other examples exist of environmental degradation driven by projects that increase coastal livelihoods, for example through aquaculture or linking fisheries to international markets (e.g. Armitage and Johnson 2006). As such, efforts to increase livelihoods to achieve SDG 1 can go against sustainable resource use required by SDG 14.2.

Mangroves and the goals of SDG 14 can also contribute to achieving SDG 2 (Zero Hunger) due to their important role in the food security of coastal communities. However, there is potential for conflicts among SDG 2 and SDG 14 targets such as 14.2, 14.5 and 14.6 because mangroves continue to be converted to rice agriculture and aquaculture, incentivised by government tax breaks, subsidies and production targets for food security. Most recently, Indonesia and Malaysia have ambitious short- and medium-term plans to increase palm oil production. This is expected to expand current mangrove deforestation frontiers into new areas such as Papua, Indonesia (Richards and Friess 2016).

A push to achieve SDG 2 will indirectly impact mangroves and fisheries. Soil erosion related to land-cover change increases sediment load in water courses that drain into the coastal zone, which at high magnitudes is detrimental to mangroves due to pneumatophore smothering (Sidik et al. 2016). Increased suspended sediment concentrations have negative impacts on mangrove-associated fisheries. Agricultural intensification to achieve SDG 2 is likely to increase eutrophication in downstream coastal waters, with low dissolved oxygen affecting coastal flora and fauna.

SDG 11 (Sustainable Cities and Communities) can positively impact urban mangroves. Some of the largest cities in the tropics occur in mangrove-rich deltas, such as the Chao Phraya (Bangkok, Thailand) and the Mekong (Ho Chi Minh, Vietnam) (Tessler et al. 2015). Other large coastal cities, such as Mumbai, Singapore and Hong Kong, have substantial mangroves. Maintaining mangroves within urban landscapes has multiple benefits, including coastal protection, flood control, nutrient processing, carbon storage and cultural services (Everard et al. 2014, Friess et al. 2016b). However, degraded coastal wetlands within city landscapes can also be sources of contaminated seafood (Dsikowitzky et al. 2011) and habitats for mosquitoes (Claflin and Webb 2017), which may reduce their amenity within urban landscapes.

SDG 15 (Life on Land) is highly synergistic with coastal forests, located in the transition between the terrestrial and marine realms, including a transitional space for many terrestrial faunal species. Mangroves rely on terrestrial connections; restoring rivers and their freshwater and sediment fluxes to the coastal oceans enhances mangrove resilience to sea-level rises (Lovelock et al. 2015). However, their position between the terrestrial and marine spheres means that mangroves may fall through the policy gap in many countries (Friess et al. 2016a, Primavera 2000), with some government agencies considering them neither terrestrial (SDG 15) nor marine (SDG 14).

14.6 Conclusions

Coastal forested ecosystems such as mangrove forests are strongly linked to poverty and development since they provide ecosystem services to potentially hundreds of millions of people. An SDG focused on coastal and marine ecosystems – the life below water – is therefore encouraging. However, SDG 14 is likely to have indirect and unintended consequences for the very ecosystems it aims to protect and the local communities that rely on them. Anticipating negative consequences requires thinking and planning at multiple scales and a multidisciplinary view of SDG 14 implementation that incorporates multiple stakeholders at different hierarchical levels. Ultimately, many SDG 14 targets require increasing local environmental justice and resource management.

We need to more explicitly consider coastal forested ecosystems within SDG 14, as not doing so may explain the potential for unintended consequences on coastal forests. Ultimately, coastal and marine ecosystems face challenges when forced into one SDG, so there is high potential for conflict. A stronger recognition of the unique challenges of the coastal zone, and coastal forested ecosystems in particular, throughout all SDGs may raise their profile so that they can be more strongly considered in conservation and development planning.

Acknowledgements

This chapter benefited from discussions and feedback at the FAO-WFSE workshop and the Asia Research Institute, National University of Singapore. DAF was supported by the Singapore Social Sciences Research Council's project 'Sustainable governance of transboundary environmental commons in Southeast Asia' (MOE2016-SSRTG-068). Katie Arkema (Stanford University) contributed to some of the ideas discussed in this chapter.

References

Ahmad, S. 2009. Recreational values of mangrove forest in Larut Matang, Perak. *Journal of Tropical Forest Science* 21:81–7.

Akimichi, T. and Ruddle, K. 1984. The historical development of territorial rights and fishery regulation in Okinawan inshore waters. *Senri Ethnological Studies* 17:7–88.

Alongi, D. M. 2011. Carbon payments for mangrove conservation: Ecosystem constraints and uncertainties of sequestration potential. *Environmental Science and Policy* 14:462–70.

Armitage, D. and Johnson, D. 2006. Can resilience be reconciled with globalization and the increasingly complex conditions of resource degradation in Asian coastal regions? *Ecology and Society* 11(1):2.

Barbier, E. B. and Cox, M. 2004. An economic analysis of shrimp farm expansion and mangrove conversion in Thailand. *Land Economics* 80:389–407.

Barnes, M. D., Glew, L., Wyborn, C. and Craigie, I. D. 2018. Prevent perverse outcomes from global protected area policy. *Nature Ecology & Evolution* 2:759–62.

Beymer-Farris, B. A. and Bassett, T. 2012. The REDD menace: Resurgent protectionism in Tanzania's mangrove forests. *Global Environmental Change* 22:332–41.

Brown, B., Fadillah, R., Nurdin, Y., Soulsby, I. and Ahmad, R. 2014. Community based ecological mangrove restoration in Indonesia. *S.A.P.I.E.N.S* 7(2). Available at: http://journals.openedition.org/sapiens/1589 (Accessed 29 July 2019).

Carrasco, L. R., Larrosa, C., Milner-Gulland, E. J. and Edwards, D. P. 2014. A double-edged sword for tropical forests. *Science* 346:38–40.

Carrasquilla-Henao, M. and Juanes, F. 2017. Mangroves enhance local fisheries catches: A global meta-analysis. *Fish and Fisheries* 18:79–93.

Chen, C.-L. 2015. Regulation and management of marine litter. In Bergman, M., Gutow, L. and Klages, M. (eds.) *Marine Anthropogenic Litter*. Berlin: Springer: pp. 395–428.

Claflin, S. B. and Webb, C. E. 2017. Surrounding land use significantly influences adult mosquito abundance and species richness in urban mangroves. *Wetlands Ecology and Management* 25:331–44.

Coulthard, S., Johnson, D. and McGregor, J. A. 2011. Poverty, sustainability and human wellbeing: A social wellbeing approach to the global fisheries crisis. *Global Environmental Change* 21:453–63.

Custer, S., DiLorenzo, M., Masaki, T., Sethi, T. and Harutyunyan, A. 2018. *Listening to leaders 2018: Is development cooperation tuned-in or tone-deaf?* Williamsburg: AidData at the College of William & Mary.

DAHD 2016. *Centrally sponsored scheme on development of inland fisheries and aquaculture.* Department of Animal Husbandry, Dairying and Fisheries, Government of India. Available at: http://dahd.nic.in/related-links/centrally-sponsored-scheme-development-inland-fisheries-and-aquaculture (Accessed 30 July 2018).

Defeo, O., Castrejón, M., Pérez-Castañeda, R. et al. 2016. Co-management in Latin American small-scale shellfisheries: assessment from long-term case studies. *Fish and Fisheries* 17:176–92.

Donato, D. C., Kauffman, J. B., Murdiyarso, D. et al. 2011. Mangroves among the most carbon-rich forests in the tropics. *Nature Geoscience* 4:293–97.

Dsikowitzky, L., Nordhaus, I., Jennerjahn, T. C. et al. 2011. Anthropogenic organic contaminants in water, sediments and benthic organisms of the mangrove-fringed Segara Anakan Lagoon, Java, Indonesia. *Marine Pollution Bulletin* 62:851–62.

Duke, N. C. 2016. Oil spill impacts on mangroves: Recommendations for operational planning and action based on a global review. *Marine Pollution Bulletin* 109:700–15.

Edgar, G. J., Stuart-Smith, R. D., Willis, T. J. et al. 2014. Global conservation outcomes depend on marine protection areas with five key features. *Nature* 506:216–20.

European Commission 2017. *Study on the subsidies to the fisheries, aquaculture, and marketing and processing subsectors in major fishing nations beyond the EU.* MARE/2011/01 Lot 2, European Commission, Brussels.

Everard, M., Jha, R. R. and Russell, S. 2014. The benefits of fringing mangrove systems to Mumbai. *Aquatic Conservation: Marine and Freshwater Ecosystems* 24:256–74.

FAO 2006. *The state of the world's fisheries and aquaculture*. Rome: FAO.

FAO 2016. *The state of the world's fisheries and aquaculture*. Rome: FAO.

Feller, I. C., Dangremond, E. M., Devlin, D. J. et al. 2015. Nutrient enrichment intensifies hurricane impact in scrub mangrove ecosystems in the Indian River Lagoon, Florida, USA. *Ecology* 96:2960–72.

Foucat, V. S. 2002. Community-based ecotourism management moving towards sustainability, in Ventanilla, Oaxaca, Mexico. *Ocean & Coastal Management* 45:511–29.

Francis, J., Nilsson, A. and Waruinge, D. 2002. Marine protected areas in the Eastern African region: How successful are they? *Ambio* 31:503–11.

Friess, D. A., Thompson, B. S., Brown, B. et al. 2016a. Policy challenges and approaches for the conservation of mangrove forests in Southeast Asia. *Conservation Biology* 30:933–49.

Friess, D. A., Richards, D. R. and Phang, V. X. H. 2016b. Mangrove forests store high densities of carbon across the tropical urban landscape of Singapore. *Urban Ecosystems* 19:795–810.

G20 2017. *G20 Action Plan on Marine Litter*. Group of 20. Available at: www.g20.utoronto .ca/2017/2017-g20-marine-litter-en.pdf (Accessed 8 January 2019).

Giri, C., Ochieng, E., Tieszen, L. L. et al. 2011. Status and distribution of mangrove forests of the world using Earth observations satellite data. *Global Ecology and Biogeography* 20:154–9.

GMA 2017. *The Global Mangrove Alliance*. The Global Mangrove Alliance. Available at: www .mangrovealliance.org/wp-content/uploads/2017/08/global-mangrove-alliance_strategy.pdf (Accessed 8 January 2019).

Goessens, A., Satyanarayana, B., Van der Stocken, T. et al. 2014. Is Matang Forest in Malaysia sustainably rejuvenating after more than a century of conservation and harvesting management? *PLoS ONE* 9:e0105069.

Government of Belize 2017. *Belize's Voluntary National Review for the Sustainable Development Goals*. VNR Submission to Department of Economic and Social Affairs, United Nations.

Government of the People's Republic of Bangladesh 2017. *Eradicating poverty and promoting prosperity in a changing world voluntary national review (VNR), 2017*. VNR Submission to Department of Economic and Social Affairs, United Nations.

Hamilton, S. E. 2013. Assessing the role of commercial aquaculture in displacing mangrove forest. *Bulletin of Marine Science* 89:585–601.

Hamilton, S. E. and Casey, D. 2016. Creation of a high spatio-temporal resolution global database of continuous mangrove forest cover for the 21st century (CGMFC-21). *Global Ecology and Biogeography* 25:729–38.

Hoang, H. T., Hoshino, S. and Hashimoto, S. 2015. Forest stewardship council certificate for a group of planters in Vietnam: SWOT analysis and implications. *Journal of Forest Research* 20:35–42.

Hoq, M. E. 2007. An analysis of fisheries exploitation and management practices in Sundarbans mangrove ecosystem, Bangladesh. *Ocean & Coastal Management* 50:411–27.

IUCN 2012. *Guidelines for applying the IUCN protected area management categories to marine protected areas*. Gland, Switzerland: International Union for Conservation of Nature.

IUCN 2017. *National Marine Plastic Litter Policies in EU Member States: an Overview*. Gland, Switzerland: International Union for Conservation of Nature.

Jones, P. L. and De Santo, E. M. 2016. Is the race for remote, very large marine protected areas (VLMPAs) taking us down the wrong track? *Marine Policy* 73:231–4.

Kodikara, K. A., Mukherjee, N., Jayatissa, L. P., Dahdouh-Guebas, F. and Koedam, N. 2017. Have mangrove restoration projects worked? An in-depth study in Sri Lanka. *Restoration Ecology* 25:705–16.

Kopnina, H. 2016. The victims of unsustainability: A challenge to Sustainable Development Goals. *International Journal of Sustainable Development & World Ecology* 23:113–21.

Lau, J. D. and Scales, I. R. 2016. Identity, subjectivity and natural resource use: How ethnicity, gender and class intersect to influence mangrove oyster harvesting in The Gambia. *Geoforum* 69:136–46.

Lee, M. 2013. The reality of balancing tourism development and protecting the nature heritage of Langkawi Island, Malaysia. *Journal of Ecotourism* 12:197–203.

Lewis, R. R. 2005. Ecological engineering for successful management and restoration of mangrove forests. *Ecological Engineering* 24:403–18.

Lewis, R. R. and Brown, B. 2014. *Ecological mangrove rehabilitation: A field manual for practitioners*. Mangrove Action Project, Canadian International Development Agency and Oxfam.

Locatelli, T., Binet, T., Kairo, J. G. et al. 2014. Turning the tide: How blue carbon and payments for ecosystem services (PES) might help save mangrove forests. *Ambio* 43:981–5.

Lovelock, C. E., Cahoon, D. R., Friess, D. A. et al. 2015. The vulnerability of Indo-Pacific mangrove forests to sea-level rise. *Nature* 526:559–63.

Mazarrasa, I., Olsen, Y. S., Mayol, E., Marbà, N. and Duarte, C. M. 2013. Rapid growth of seaweed biotechnology provides opportunities for developing nations. *Nature Biotechnology* 31:591.

McLeod, E., Szuster, B. and Salm, R. 2009. Sasi and marine conservation in Raja Ampat, Indonesia. *Coastal Management* 37:656–76.

McNally, C. G., Uchida, E. and Gold, A. J. 2011. The effect of a protected area on the tradeoffs between short-run and long-run benefits from mangrove ecosystems. *Proceedings of the National Academy of Sciences* 108:13945–50.

Miteva, D. A., Murray, B. C. and Pattanayak, S. K. 2015. Do protected areas reduce blue carbon emissions? A quasi-experimental evaluation of mangroves in Indonesia. *Ecological Economics* 119:127–35.

Neiland A. E., Soley, N., Varley, J. B. and Whitmarsh, D. J. 2001. Shrimp aquaculture: economic perspectives for policy development. *Marine Policy* 25:265–79.

Neumann, B., Vafeidis, A. T., Zimmermann, J. and Nicholls, R. J. 2015. Future coastal population growth and exposure to sea-level rise and coastal flooding – a global assessment. *PLoS ONE* 10:e0131375.

Ouyang, X. and Guo, F. 2016. Paradigms of mangrove in treatment of anthropogenic wastewater pollution. *Science of the Total Environment* 544:91–7

Pascual, U., Phelps, J., Garmendia, E. et al. 2014. Social equity matters in Payments for Ecosystem Services. *BioScience* 64:1027–36.

Phelps, J., Carrasco, L. R., Webb, E. L., Koh, L. P. and Pascual, U. 2013. Agricultural intensification escalates conservation costs. *Proceedings of the National Academy of Sciences* 110:7601–7.

Phelps, J., Webb, E. L. and Agrawal, A. 2010. Does REDD+ threaten to recentralise forest governance? *Science* 328:312–13.

Poudyal, M., Ramamonjisoa, B. S., Hockley, N. et al. 2016. Can REDD+ social safeguards reach the 'right' people? Lessons from Madagascar. *Global Environmental Change* 37:31–42.

Primavera, J. H. 1997. Socio-economic impacts of shrimp culture. *Aquaculture Research* 28:815–27.

Primavera, J. H. 2000. Development and conservation of Philippine mangroves: Institutional issues. *Ecological Economics* 35:91–106.

Primavera, J. H. and Esteban, J. M. 2008. A review of mangrove rehabilitation in the Philippines: Successes, failures and future prospects. *Wetlands Ecology and Management* 16:345–58.

Rabalais, N. N., Turner, R. E., Diaz, R. J. and Justić, D. 2009. Global change and eutrophication of coastal waters. *ICES Journal of Marine Science* 66:1528–37.

Ramesh, R., Purvaja, R., Krishnan, P. et al. 2017. Conservation of coastal wetlands: An appraisal of the policy and legal framework in South Asian nations. In *Wetland Science*. New Delhi: Springer, pp. 515–44.

Ramsar 2018. *Wetlands and the SDGs: Scaling up wetland conservation, wise use and restoration to achieve the Sustainable Development Goals*. Ramsar Convention on Wetlands.

Rees, S. E., Foster, N. L., Langmead, O., Pittman, S. and Johnson, D. E. 2017. Defining the qualitative elements of Aichi Biodiversity Target 11 with regard to the marine and coastal environment in order to strengthen global efforts for marine biodiversity conservation outlined in the United Nations Sustainable Development Goal 14. *Marine Policy* 93:241–50.

Richards, D. R. and Friess, D. A. 2016. Rates and drivers of mangrove deforestation in Southeast Asia, 2000–2012. *Proceedings of the National Academy of Sciences* 113:344–9.

Roberts, K. E, Valkan, R. S. and Cook, C. N. 2018. Measuring progress in marine protection: A new set of metrics to evaluate the strength of marine protected area networks. *Biological Conservation* 219:20–7.

Scales, I. R., Friess, D. A., Glass, L. and Ravaoarinorotsihoarana, L. A. 2018. Rural livelihoods and mangrove degradation in south-west Madagascar: Lime production as an emerging threat. *Oryx* 52:641–5.

Shaharuddin, M. I., Azahar, M., Razani, U. et al. 2005. *Sustainable management of Matang Mangroves: 100 years and beyond*. Forestry Department Peninsular Malaysia, Kuala Lumpur.

Sidik, F., Neil, D. and Lovelock, C. E. 2016. Effect of high sedimentation rates on surface sediment dynamics and mangrove growth in the Porong River, Indonesia. *Marine Pollution Bulletin* 107:355–63.

Sillanpää, M., Vantellingen, J. and Friess, D. A. 2017. Vegetation regeneration in a sustainably harvested mangrove forest in West Papua, Indonesia. *Forest Ecology and Management* 390:137–46.

Silva-Cavalcanti, J. S. and Costa, M. F. 2009. Fisheries in protected and non-protected areas: Is it different? The case of *Anomalocardia brasiliana* at tropical estuaries of northeast Brazil. *Journal of Coastal Research* SI56:1454–8.

Sippo, J. Z., Maher, D. T., Tait, D. R., Holloway, C. and Santos, I. R. 2016. Are mangroves drivers or buffers of coastal acidification? Insights from alkalinity and dissolved inorganic carbon export estimates across a latitudinal transect. *Global Biogeochemical Cycles* 30:753–66.

Slobodian, L. N., Rodriguez Chaves, M., Nguyen, L.T. and Rakotoson, L. N. 2018. *Legal Frameworks For Mangrove Governance, Conservation and Use: Assessment Summary*. IUCN, Geneva, Switzerland, and WWF Germany, Berlin, Germany.

Spalding, M. D., Ruffo, S., Lacambra, C. et al. 2014. The role of ecosystems in coastal protection: a climate change and coastal hazards. *Ocean & Coastal Management* 90:50–7.

Sudtongkong, C. and Webb, E. L. 2008. Outcomes of state- vs. community-based mangrove management in southern Thailand. *Ecology and Society* 13:27.

Sumaila, U. R., Khan, A. S., Dyck, A. J. et al. 2010. A bottom-up re-estimation of global fisheries subsidies. *Journal of Bioeconomics* 12:201–25.

Szabo, S., Renaud, F. G., Hossain, M. S. et al. 2015. Sustainable Development Goals offer new opportunities for tropical delta regions. *Environment: Science and Policy for Sustainable Development* 57:16–23.

Taillardat, P., Friess, D. A. and Lupascu, M. 2018. Mangrove blue carbon strategies for climate change mitigation are most effective at the national scale. *Biology Letters* 14:20180251.

Tessler, Z. D., Vörösmarty, C. J., Grossberg, M. et al. 2015. Profiling risk and sustainability in coastal deltas of the world. *Science* 349:638–43.

Thompson, B. S. 2018. The political ecology of mangrove forest restoration in Thailand: Institutional arrangements and power dynamics. *Land Use Policy* 78:503–14.

Thompson, B. S., Gillen, M. J. and Friess, D. A. 2018. Challenging the principles of ecotourism: insights from entrepreneurs on environmental and economic sustainability in Langkawi, Malaysia. *Journal of Sustainable Tourism* 26:257–76.

Todd, P. A., Ong, X. and Chou, L. M. 2010. Impacts of pollution on marine life in Southeast Asia. *Biodiversity and Conservation* 19:1063–82.

Torras M. and Boyce J. K. 1998. Income, inequality and pollution: a reassessment of the environmental Kuznets Curve. *Ecological Economics* 25:147–60.

Uddin, M. S., de Royter van Steveninck, E., Stuipm, M. and Shah, M. A. 2013. Valuation of provisioning and cultural services of a protected mangrove ecosystem: A case study on Sundarbans Reserve Forest, Bangladesh. *Ecosystem Services* 5:88–93.

UN DESA 2017. *Voluntary National Reviews*. High Level Political Forum on Sustainable Development, Department of Economic and Social Affairs, United Nations.

UNEP 2012. *The Honolulu Strategy: A Global Framework for Prevention and Management of Marine Debris*. United Nations Environment Programme.

Valiela, I. and Cole, M. L. 2002. Comparative evidence that salt marshes and mangroves may protect seagrass meadows from land-derived oxygen loads. *Ecosystems* 5:92–102.

Vierros, B. and Buonomo, R. 2017. *In-depth analysis of ocean conference voluntary commitments to support and monitor their implementation.* Department of Economic and Social Affairs, United Nations.

Villamayor, B. M., Rollon, R. N., Samson, M. S., Albano, G. M. and Primavera, J. H. 2016. Impact of Haiyan on Philippine mangroves: Implications to the fate of the widespread monospecific *Rhizophora* plantations against strong typhoons. *Ocean & Coastal Management* 132:1–14.

Whitfield, A. K. 2017. The role of seagrass meadows, mangrove forests, salt marshes and reed beds as nursery areas and food sources for fishes in estuaries. *Reviews in Fish Biology and Fisheries* 27:75–110.

ZSL 2015. *Community-based mangrove rehabilitation training manual.* Zoological Society of London, Philippine Tropical Forest Conservation Foundation Inc.

Chapter 15 SDG 15: Life on Land – The Central Role of Forests in Sustainable Development

Jeffrey Sayer*, Douglas Sheil, Glenn Galloway, Rebecca A. Riggs, Gavyn Mewett, Kenneth G. MacDicken, Bas Arts, Agni K. Boedhihartono, James Langston and David P. Edwards

Key Points

- There will be trade-offs between SDG 15 and other SDGs resulting from competition for land, but there are also synergies and opportunities.
- The principal opportunity of SDG 15 is that it will be recognised and integrated, along with the other SDGs, in all developments.
- The main risk is that short-term priorities and a 'business as usual' approach will undermine this opportunity for integration and synergy and SDG 15 will often be overlooked.
- The scale, and complexity, of challenges for conserving life on land, versus the limited resources available, pose many challenges.
- Greater cross-sectoral integration, not just sectoral policy reform, is essential to advancing SDG 15.
- We encourage conservation and development professionals to engage with those responsible for all the Agenda 2030 targets to ensure that SDG 15 is a priority in all SDG related processes.

15.1 Introduction

The claim that stewardship of terrestrial ecosystems, particularly forests and their rich biodiversity, is essential for sustainable development has achieved broad recognition. The conservation of life on land is recognised as Goal 15 of the 17 Sustainable Development Goals (SDGs). When the SDGs were adopted by the UN in 2015, all indicators showed that life on Earth was in decline (Schipper et al. 2008, Tittensor et al. 2014), eroding the ability to meet human needs (Pimm et al. 2014). SDG 15 asks for the protection, restoration and sustainable use of terrestrial ecosystems along

* Lead author.

with the sustainable management of forests, combating desertification, halting and reversing land degradation and halting biodiversity loss. Clearly, the pressures of population growth, economic development and increased consumption will only intensify the challenges for the maintenance of life on land. In this chapter we examine the opportunities and challenges that merit particular attention if we are serious about SDG 15 and reversing the decline of life on Earth. In principle, SDG 15 (Life on Land) is recognised as having equal prominence alongside other SDGs. As there are widespread calls for integration and explicit attention to synergies and trade-offs among SDG outcomes (Le Blanc 2015, Stafford-Smith et al. 2017), the existence of SDG 15 should result in conservation concerns and issues having a raised profile within the wider decision-making community. Unfortunately, this appears far from assured. Many people, especially in the biodiverse tropics, view conservation as a concern of rich Western nations with little relevance to immediate local needs. Western conservation agencies that focus on charismatic species and the concerns and preferences of Western audiences do little to dispel these preconceptions (Meijaard and Sheil 2008). At the same time, most people worldwide, including most politicians, policymakers and experts, now live in cities, where they are disconnected from nature and forests (Soga et al. 2016). A recent survey by Price Waterhouse Coopers found that in many countries SDG 15 and SDG 14 (Life below Water) are getting less attention than other SDGs (Scott and McGill 2018). Development and conservation are viewed as being in competition – and development is the more immediate priority.

This chapter summarises recent forest and biodiversity trends and the diverse contexts in which progress towards SDG 15 will play out. We focus on the factors and conditions likely to influence achievement of the goals and targets. We then examine three broad areas: (1) forest cover and management, (2) biodiversity and (3) financial and policy instruments. We examine the implementation of SDG 15 in Indonesia, where challenges over forest and natural resource stewardship will influence the attainment of all SDGs. We advocate a more inclusive, proactive and outcome-based approach to SDG 15, firmly rooted in realistic scenarios. We hope to see transformational change rather than business as usual. (See Table 15.1 for SDG 15 targets and indicators.)

Table 15.1 SDG 15 Goal, targets and indicators

SDG 15: Protect, restore and promote sustainable use of terrestrial ecosystems, sustainably manage forests, combat desertification, and halt and reverse land degradation and halt biodiversity loss

Targets	Indicators
15.1 By 2020, ensure the conservation, restoration and sustainable use of terrestrial and inland freshwater ecosystems and their services, in particular forests, wetlands, mountains and drylands, in line with obligations under international agreements	15.1.1 Forest area as a proportion of total land area 15.1.2 Proportion of important sites for terrestrial and freshwater biodiversity that are covered by protected areas, by ecosystem type
15.2 By 2020, promote the implementation of sustainable management of all types of forests, halt deforestation, restore degraded forests and substantially increase afforestation and reforestation globally	15.2.1 Progress towards sustainable forest management
15.3 By 2030, combat desertification, restore degraded land and soil, including land affected by desertification, drought and floods, and strive to achieve a land degradation-neutral world	15.3.1 Proportion of land that is degraded over total land area
15.4 By 2030, ensure the conservation of mountain ecosystems, including their biodiversity, in order to enhance their capacity to provide benefits that are essential for sustainable development	15.4.1 Coverage by protected areas of important sites for mountain biodiversity 15.4.2 Mountain Green Cover Index
15.5 Take urgent and significant action to reduce the degradation of natural habitats, halt the loss of biodiversity and, by 2020, protect and prevent the extinction of threatened species	15.5.1 Red List Index

15.6 Promote fair and equitable sharing of the benefits arising from the utilisation of genetic resources and promote appropriate access to such resources, as internationally agreed	15.6.1 Number of countries that have adopted legislative, administrative and policy frameworks to ensure fair and equitable sharing of benefits
15.7 Take urgent action to end poaching and trafficking of protected species of flora and fauna and address both demand and supply of illegal wildlife products	15.7.1 Proportion of traded wildlife that was poached or illicitly trafficked
15.8 By 2020, introduce measures to prevent the introduction and significantly reduce the impact of invasive alien species on land and water ecosystems and control or eradicate the priority species	15.8.1 Proportion of countries adopting relevant national legislation and adequately resourcing the prevention or control of invasive alien species
15.9 By 2020, integrate ecosystem and biodiversity values into national and local planning, development processes, poverty reduction strategies and accounts	15.9.1 Progress towards national targets established in accordance with Aichi Biodiversity Target 2 of the Strategic Plan for Biodiversity 2011–2020
15.A Mobilise and significantly increase financial resources from all sources to conserve and sustainably use biodiversity and ecosystems	15.A.1 Official development assistance and public expenditure on conservation and sustainable use of biodiversity and ecosystems
15.B Mobilise significant resources from all sources and at all levels to finance sustainable forest management and provide adequate incentives to developing countries to advance such management, including for conservation and reforestation	15.B.1 Official development assistance and public expenditure on conservation and sustainable use of biodiversity and ecosystems
15.C Enhance global support for efforts to combat poaching and trafficking of protected species, including by increasing the capacity of local communities to pursue sustainable livelihood opportunities	15.C.1 Proportion of traded wildlife that was poached or illicitly trafficked

Source: https://sustainabledevelopment.un.org/sdg15

15.2 Trends and Contexts

SDG 15 endorses priorities already established through previous conventions and agreements, such as the Convention on Biological Diversity (CBD) and its Aichi Biodiversity Targets and Nagoya Protocol, the UN Convention to Combat Desertification and the Convention on International Trade in Endangered Species. Most SDG 15 indicators draw on existing data and measures. This is potentially both a strength and a weakness. Using existing metrics allows for long-term analysis but runs the danger of perpetuating business as usual and not encouraging attention to emerging needs (see Steffen et al. 2015).

15.2.1 Forest Cover and Management

Forest cover and condition are influenced by various drivers – direct and indirect. A growing population, trending towards 9 billion, is increasing demand for food and other commodities and placing more pressure on land while also affecting livelihoods and health. Climate change may curtail food production in some locations and create new options in others (Lobell et al. 2011). Competing demands on land require the management of trade-offs, which will raise many challenges (Laurance et al. 2013). Achieving SDG 15 will require addressing these drivers and trade-offs so as to bolster life on land and accommodate wider societal values.

Many studies document trends and changes in the extent and condition of the world's forests and biodiversity (FAO 2010, 2012, Keenan et al. 2015, Sloan and Sayer 2015). For example, FAO's Forest Resources Assessments have found that between 1990 and 2015, global forest cover decreased by 3.1 per cent, to 30.6 per cent of global land area. Total forest area declined from more than 4.1 billion ha to below 4 billion ha. The rate of loss has slowed since 2010, but there are significant regional variations. From 2010 to 2015, boreal and subtropical forests suffered virtually no net loss while temperate forests gained over 2 million ha per year. Deforestation remained higher in the tropics, with 5–6 million ha lost annually in the same period, a reduction from nearly 10 million ha per year in the 1990–2000 period (FAO 2015).

Using different definitions and methods, Global Forest Watch (2019) shows continued, if uneven, decline in tropical tree cover since 2001. Agriculture has been responsible for about 80 per cent of tropical and subtropical deforestation (Kissinger et al. 2012). Over this same period, large-scale commercial agriculture has replaced subsistence agriculture as the most important driver of change across the tropics and subtropics. In recent years commercial agriculture has been responsible for 70 per cent of deforestation in Latin America, compared to 30–35 per cent in tropical Asia and Africa (Kissinger et al. 2012).

In recent years the FAO's FRA studies have reported on changes in the condition, management and production of forests (FAO 2012). Though forest area has increased in many parts of the world, much of this is a result of expansion of industrial tree plantations. Planted forest area increased in all regions between 1990 and 2015, while natural forest area declined in all regions except Europe and Oceania (FAO 2015). Plantations constitute about 7 per cent of the world's forests (FAO 2015) and will likely continue to expand beyond 2030.

Another trend concerns sources of wood and timber. The area of timber-production forest in low-income countries is declining, though it remains stable in higher-income countries. Meanwhile, the volume of wood harvested is increasing in every region except Europe and North America (FAO 2015). Addressing growing demand for wood in the tropics may ultimately require more intensive forest management or greater timber imports from temperate and boreal forests.

15.2.2 Biodiversity

Current global species extinction rates are estimated to be about three orders of magnitude above those of the prehuman world (Pimm et al. 2014). Habitat loss, unsustainable hunting, introduced invasive species and other factors have contributed to 322 recorded extinctions of terrestrial vertebrates since 1500. Meanwhile, the populations of most remaining species have declined (Dirzo et al. 2014, Pimm et al. 2014). Detailed assessments by the International Union for Conservation of Nature (IUCN) have identified 24 307 species that are currently facing significant threat of extinction (IUCN 2019).

Forests, in particular tropical forests, harbour most of Earth's species and most of the threatened species (Vira et al. 2015). In IUCN's most recent global assessment of mammal species, 25 per cent (1139) were judged 'threatened with extinction' while another 15 per cent (836) were 'data deficient' (Pimm et al. 2014, Schipper et al. 2008). These figures are underestimates, as we still know little about the overall diversity of many tropical taxa. New species are still being discovered in even the best-known groups; for example, 85 new primates were described between 2000 and 2016 (Estrada et al. 2017) and a new great ape (an orangutan, *Pongo tapanuliensis*) was described in 2017 (Nater et al. 2017).

The risk of species loss reflects multiple factors – including habitat loss, modification and fragmentation, over-exploitation, interactions with other species and climate change – and combinations of all these factors (Selwood et al. 2015). When a species is lost from a forest, the ecological processes that depend upon that species are lost with it; this reduces community resilience and can provoke further species losses (Dirzo et al. 2014, Redford 1992).

These threats and changes are ongoing worldwide. For example, terrestrial ecosystems are increasingly fragmented by infrastructure (Laurance et al. 2014), posing new challenges for those seeking to protect life on land. Climate change is a major emerging threat. The Intergovernmental Panel on Climate Change (IPCC) shows that the Earth warmed by an estimated 0.74°C over the last century and is forecast to warm by another 1.5–4.5°C this century (IPCC 2013). Places where forest can grow, and in which species can persist, will change as climate changes. While the worst impacts likely lie beyond 2030, significant changes are now underway. Various local extinctions have already been attributed to changes in climate (Cahill et al. 2013, Wiens 2016,). Predicted changes mean that many species will not persist unless they move to more favourable locations, but many species face barriers to such movement or appear unable to move rapidly enough to find and track suitable climates (Corlett and Westcott 2013). Even common species appear at risk (Warren et al. 2013). While outcomes remain uncertain, many authoritative accounts based on IPCC scenarios anticipate numerous extinctions (Cahill et al. 2013). Even if forests persist, climate will influence their growth and productivity and intensify threats, such as drought and fire that impact their ecological and economic viability.

15.2.3 Social–Economic Systems

For several billion people, wild plants and animals remain essential elements of daily life. Forests and biodiversity underpin subsistence, providing goods and services important for health, fuel and income. Many farmers still depend on wild resources. Food security and nutrition are bolstered and sustained by wild species to a much greater extent than is often widely recognised (Colfer et al. 2006). Many million people lack ready access to modern healthcare and depend on wild plants and animals for medicines. Forest-dependent people are among the world's poorest and most marginalised. In many cases, wild resources provide a crucial safety net, allowing people to find food and manage after crises (Liswanti et al. 2011). As these dependent populations grow and access to wild resources declines, there will be major implications for people's lives and livelihoods. If these wild resources are no longer available, difficult lives will be made even harder.

As challenges to development and sustainability become increasingly evident (climate change, water, food security, inequality etc.), bottom-up commitments may be stronger drivers of change than top-down discourses and policy. SDG 15 conveys a sense that central government is the predominant driver of moves to conserve life on land, but recent practice suggests that in many countries moves to achieve conservation through decentralised governance systems are gaining traction (Agrawal et al. 2008, Ribot et al. 2006). Decentralised and community resource management are being pursued in many countries and in diverse contexts, but they are not always the panacea that their advocates claim (Agrawal and Gibson 1999, Boedhihartono 2017, Robinson et al. 2014). Some communities that have maintained strong control over their lands and resources remain effective in achieving desirable conservation outcomes and are willing to see large tracts of land set aside in perpetuity: an example is the protection of Papua's Foja Mountains and large areas of the Mamberamo Valley that are held to be sacred (Sheil et al. 2015). Nonetheless, in many cases there is a tension between the management of resources for local goals and the need to conserve public goods values (Sayer et al. 2017).

The growth of economies, a major target of many other SDGs, will have major impacts on terrestrial biodiversity. Human populations will move, cities will grow, agricultural technologies will allow for producing more on less land (Sayer and Cassman 2013). If these trends continue, the supply and demand of forest goods and services in 2030 and beyond will be primarily determined by indirect drivers outside the forest sector (Bruinsma 2009). The planet as a whole may be advancing through a forest transition in which nations experience a period of protracted forest loss, followed by reaching a low point and then a stage of widespread reforestation and forest recovery (Rudel et al. 2005). Different regions are advancing at different speeds, with some areas still suffering dramatic forest loss, such as Southeast Asia, and others showing substantial gain, such as the tropical Andes. Those who seek to implement SDG 15 need to do so with these likely changes in mind. Maintaining forest-dependent biodiversity through the low point in the transition, and the rapid expansion of suitable habitat in the post-transition era, will be important (see Wright and Muller-Landau 2006 and subsequent discussions, e.g. Melo et al. 2013, Rudel et al. 2009). Examples include the widespread protection even of young or degraded forests and control of damaging practices such as hunting, over-exploitation and fire that may reduce the conservation value of human impacted forests (Chazdon et al. 2009). The implication is that there are many domains where policymakers can make a major difference to the maintenance or loss of life on land. Those seeking to address SDG 15 should identify the right policy signals to ensure the best outcomes for biodiversity and for societies' future needs. These are long-term objectives: 2030 should not be seen as an end point – the changes that are under way in 2030 will influence life on land for millennia.

Aside from the many sectoral challenges, those addressing SDG 15 will also confront long-standing issues of legal enforcement. Consider the challenges posed by the illegal trade in wildlife and timber – worth, respectively, an estimated USD 8–10 billion (Haken 2011) and USD 7 billion per year – with links to powerful organised crime systems and insurgencies (Milner-Gulland 2018). In many regions military interests manage and protect these activities. Trends suggest that 'regulatory approaches are being overwhelmed by rising prices and growing relative poverty between areas of supply and centers of demand. Aggressive enforcement of trade controls, in particular bans, can increase profits for traffickers and lead to the involvement of organized criminals with the capacity to operate even under increased enforcement effort' (Challender and MacMillan 2014: 484). Foresters, conservationists and enforcement agencies are ill-equipped to confront such forces. More light needs to be shed on these issues, which will prove challenging. According to Global Witness, in 2016 there were 200 confirmed murders of environmental activists, wildlife rangers and Indigenous leaders trying to protect their land (Watts and Vidal 2017).

15.2.4 What Are the Future Challenges for Conservation of Life on Land?

What challenges will forests and biodiversity face in the coming decade? How will the context of conservation and sustainable use change from that which existed in 2015? How will progress on the other SDGs impact the attainment of SDG 15? Several major trends are already apparent, all of which will impact on life on land. If economies continue to grow as they have in recent decades, then many more people will escape from extreme poverty – including many who currently depend upon forest resources for their livelihoods. The dependency of these people on forests will decrease. If economies stagnate, then the rural poor will suffer disproportionately and achieving conservation goals will be more difficult. Under optimistic scenarios, people in rural areas will have better health and education and, consequently, smaller families. The ability to provide services, peace and opportunities to those who survive in and around forests, rangelands and wetlands will be a major determinant of many of the SDGs, including SDG 15. A major unknown is the fate of the 2 billion-plus subsistence farmers in the tropics: will economies and education provide attractive alternative livelihoods on or off the lands? If progress is made on the other SDGs, then we see the following scenario unfolding between now and 2030:

- People will move from rural areas to cities.
- Agriculture will continue to intensify and move towards larger, more mechanised farms. Industrial agriculture, especially tree crops, will continue to expand into forested areas. Productivity gains may reduce demand for land, but in some regions expansion may displace farmers into other frontier areas (Laurance et al. 2013).

- A shift will occur in the demand for agricultural commodities – for example an increase in meat and dairy consumption – as peoples' purchasing power increases.
- Infrastructure will continue to expand into forest areas to access mineral resources and new land for agriculture.
- Demand for near-natural forests for their biodiversity values and other ecosystem services will increase. Demand for forests for recreational use, especially in peri-urban areas, will also grow (Tyrväinen et al. 2005).
- Processes of decentralisation of natural resources management to communities and local governments will continue. Integrated management practices may become more widely adopted.
- Countries will move towards a green economy or bio-economy and increasing forest extent and health will be supported by this trend.
- Timber harvesting from natural forests will continue to decline as more timber is produced at less cost from plantations, on farms and along roadsides.
- Capacity to assess and monitor a wide range of forest attributes, particularly with improved remote sensing and application of the 'internet of things', will increase.
- Attempts to address some of the world's climate concerns through forest conservation and restoration will intensify.
- Climate-related stresses on forests and the associated risks from fire and invasive species will intensify.

The SDGs are part of a greater process. The SDGs will not be entirely met by 2030, with inevitable disparity in the progress made in different countries. Conserving life on land will remain a work in progress and SDG 15 will play out in a variety of contexts. Contexts will strongly influence the approaches that would be appropriate to pursue SDG 15 targets. Those implementing the SDGs will need the awareness, flexibility and understanding to adapt to the nuances and opportunities of their specific situations.

15.3 Achieving SDG 15 Targets

15.3.1 Targets for Forest Cover and Management

The measurement and interpretation of forest resource changes is challenging. Even measuring changes in forest area remains beyond the capacity of many countries – though increasingly easy access to satellite-based imagery is helping address this. For example, *Terra-i*[1] and *Global Forest Watch*[2] both provide access to land-cover data. FAO's FRA already collects data applicable to SDG 15 – including forest area as a proportion of land area and country compliance with requirements to report on policies relevant to sustainable forest management. The FRA also reinforces concerns about SDG 15. Forest area change means different things to different countries. For many forest-rich tropical countries, economic growth is seen as requiring increased forest conversion to agriculture and other uses. A case in point is Bolivia, which in its submission to the FAO Forest Resources assessment for 2015 stated the intention to reduce forest area by some 75 per cent, presumably in order to spur economic growth. This demonstrates the conundrum governments face in planning forest area. Many governments will choose the path of economic growth over forest conservation, but they seldom make this de facto policy explicit.

Indicator 15.2.1 requires measurement of progress towards sustainable forest management. What defines and determines sustainable forest management remains unclear for many. The topic is contentious, and achievements are difficult to quantify. This is demonstrated in how the FRA has handled the topic since FRA 2010. In the 2010 reporting year, countries were simply asked how much forest area was under sustainable forest management. Guidelines were lacking and the reported values were not taken seriously outside the FAO process. In the most recent FRA (FAO 2015), countries were asked to report a range of values including areas under management plans – inclusive of those that had community inputs, monitoring of management plans and public forest resource reporting (MacDicken 2015). This approach allows users to select indicators that best fit their understanding of sustainable forest management. It is a pragmatic solution, but there are still concerns that the selected indicators neglect crucial factors such as the contributions of forests – and how they are managed – to people.

By focusing on forest cover and protected forest extent, SDG 15 may underestimate the complexity of forest land governance and the differing values of forest types. Forests where timber has been harvested often retain significant biodiversity values and protecting these values is important, especially when the other option is forest clearance (Edwards et al. 2014). For centuries, tropical foresters sought to protect and manage natural forests in a sustainable manner (Dawkins and Philip 1998, Ghazoul and Sheil 2010, Wiersum 1995). Forestry's destructive image arose in the second half of the twentieth century when large-scale industrial timber harvesting was widely promoted

[1] www.terra-i.org

[2] www.globalforestwatch.org/

without regard for sustainability. Nonetheless, though overshadowed, efforts to protect and sustainably harvest tropical forests have continued into the modern era. Two large tropic-wide meta-analyses found that selectively harvested production forests retain a species richness of animals, insects and plants similar to that found in undisturbed forests (Gibson et al. 2011, Putz et al. 2012), and far higher species richness than competing non-forestland uses (Gibson et al. 2011). Production forests also harbour an array of IUCN red-listed species (Edwards et al. 2010), underscoring their potential to protect critical biodiversity. Managed timber concessions are often better protected than poorly staffed protected areas (Meijaard and Sheil 2007). However, harvesting timber at higher intensities (Burivalova et al. 2014), failing to retain patches of undisturbed forest within production landscapes (Edwards et al. 2014) or using more destructive conventional rather than reduced-impact logging techniques (Bicknell et al. 2015) devalues the conservation potential of production forests. Intensively managed tropical forests tend to lose certain species and become more homogeneous (Alroy 2017).

Production forests should be seen as an addition to rather than a replacement for more strictly protected areas. While there are many uncertainties and unknowns, when we look at the big picture we do largely know which practices improve conservation outcomes (Dale et al. 2000, IUCN and ITTO 2019, Meijaard et al. 2005). Achieving SDG 15 requires an open-minded approach that seeks opportunities to promote improved conservation outcomes not only through the protection of large-scale intact landscapes devoid of people, but also through the small-scale trade-offs that arise in more densely settled regions.

15.3.2 Targets for Biodiversity

Achieving SDG 15 requires an intensification of efforts to implement the Aichi Biodiversity Targets and a range of other plans and strategies elaborated by intergovernmental processes. The Aichi targets encompass more ambitious goals than those originally agreed on by the CBD in 2002. Available data from the 55 different biodiversity indicators compiled by the World Conservation Monitoring Centre suggest that the Aichi targets, set by the CBD in 2010, will not be achieved by the 2020 target date (Tittensor et al. 2014). By 2010, 31 indicators had not been achieved and biodiversity was still declining with no substantial reductions in the rate of loss. Pressures on biodiversity were still increasing. There were some successes, including improvements in the nominal coverage of protected areas (Butchart et al. 2010), but most of the increase was in protected landscapes whose biodiversity values may not be high.

By 2016, some 14.7 per cent of the world's terrestrial surface was officially protected (217 155 areas): coverage is generally higher in the tropics (Brooks et al. 2009). Yet, as protected area targets are often achieved at the cost of reduced management standards and abilities, this alone is not a sufficient proxy for improved biodiversity outcomes (Dudley et al. 2016, Watson et al. 2014,). Many species occur primarily, or exclusively, outside formal protected areas or require much larger habitat areas to ensure viability (Brooks et al. 2009, Ricketts et al. 2005, Rodrigues et al. 2004). This reflects the tendency for governments to protect areas that are economically marginal and under only limited threat. Such approaches do not yield the best conservation outcomes. Protected areas are effective only if adequately resourced, yet this is often not the case (Brooks et al. 2009, Bruner et al. 2001, Inamdar et al. 1999). These shortfalls in funding tend to be most severe in the poorest countries – the places where investments in conservation can likely make the biggest direct contributions. A simple national-scale conservation-status index based on changes in the IUCN red-listing of birds and mammals can predict with remarkable accuracy the positive impacts of the financial investments made towards achieving conservation versus benefits from economic, agricultural and population growth (Waldron et al. 2017). This model indicates that the relative benefits of conservation spending are greatest in the poorest, most biodiverse nations (Waldron et al. 2017).

McCarthy et al. (2012) examined global conservation needs and estimated that the conservation of terrestrial species would require an annual expenditure of USD 76.1 billion, equivalent to 0.1 per cent of global GDP, or less than 5 per cent of global defence spending (Sheil 2017). Current expenditures average about half of these requirements in higher-income countries and less than one-third (31 per cent) in lower-income countries. While McCarthy et al. (2012) are doubtless correct that conservation should obtain increased financial support, their approach neglects local costs and consequences (Sheil et al. 2013), including past injustices, widespread evictions and mistreatment of local populations (Agrawal and Redford 2009, Dowie 2011). These are not just a 'colonial legacy'; many communities living near protected areas suffer from the presence of wild animals: crops are destroyed, domestic animals are killed and people are attacked (Hill 2015, Naughton-Treves et al. 2011). With current laws often not offering any provisions for direct compensation, victims bear the costs.

Such human costs of protected areas matter both as an ethical concern in their own right, not least because they can erode advances being made on other SDGs such as ending poverty (SDG 1), access to affordable energy (SDG 7) and justice (SDG 16), and because they undermine community support and escalate threats. They pose a challenge to local support for conservation (Sheil et al. 2013). Indeed, failure to satisfactorily address such costs and concerns has often turned communities against conservation (Baker et al. 2012, Boissière et al. 2009, Sharpe 1998, Temudo 2012), with much conservation expenditure required to counter the ensuing conflicts (Roe 2008). We already see politicians speaking against conservation and promising that if they are elected, they will ensure that people will get their lands back (Sassen et al. 2013). In a world in which conservation requires societal support, the manner in which local costs are dealt with appears crucial. While the biophysical measures of SDG 15 do not address these trends and their drivers, those looking to achieve these goals should.

15.3.3 Targets for Financial and Policy Instruments

The SDG 15 targets related to governance, particularly Targets 15.9–15.12, cover policy integration, resource mobilisation and capacity-building for the conservation, sustainable use and fair and equitable sharing of benefits from forests, biodiversity and landscapes. Of these, the call for policy integration is the most innovative. Indeed, the loss of terrestrial biodiversity is favoured by existing sectoral approaches that underlie land-use decisions and their hierarchical ordering. Mining and infrastructure are higher priorities than agriculture, which in turn is higher than forestry and conservation. Integrating biodiversity concerns into all land-use policies and decisions would be an advance. Thus, agriculture, infrastructure and forestry could become much more nature-inclusive, which might then substantially reduce their impacts on biodiversity. The absence of reference to natural resource issues in most of the SDGs is symptomatic of a fundamental trade-off among the priorities of the different sectors.

The advancement of nature-inclusive land-use policies will in many cases be opposed by vested interests. The decisions that drive the political economies of biodiversity loss and deforestation, including illegal practices and organised crime, are shaped by these economic-interests. Greater cross-sectoral integration, not just sectoral policy reform, is essential. Transformational change of political economies is surely essential, but such change appears unlikely. Along with policy changes, the behaviour of producers and consumers needs to be fundamentally changed. Working with the best integrative practices at the landscape level can be a first step forward (Sayer et al. 2013).

The other two institutional targets under SDG 15 are resource mobilisation and capacity-building. External forest finance has relatively little impact in most cases because the value of forest production or alternative land uses is typically much higher than development assistance funding. Simula (2008) provided an estimate of total official development assistance (ODA) to the forest sector at nearly USD 2 billion per year, of which about USD 1.3 billion is to forestry and some USD 700 million to forest conservation. In contrast, wood removals from tropical Africa, Asia, Oceania and South America were reported to be approximately 1803 x 10^9 cubic metres in 2011. Using a 2011 value of USD 150 per cubic metre this comes to about USD 270 billion, or more than 100 times the value of total official ODA flows to forests.[3] Simula's estimate of USD 1.3 billion of ODA to forestry is less than 0.5 per cent of estimated value of wood production. This does not include the value of non-timber forest products, or the value of domestic public-sector expenditure. In addition to external ODA flows, Whiteman et al. (2015) report some USD 7.3 billion in public-sector forest expenditure in tropical and subtropical countries in 2010. As another comparison, consider palm oil: in 2016, the value of Indonesian palm oil exports was USD 18.1 billion (GAPKI 2017). The financial drivers impacting forestlands are vastly larger than government allocations to the forest sector. ODA contributions to the forest sector remain important to support research and to provide examples of sustainable forestry, but many of the problems of unsustainable resource use will only be resolved with widespread changes in the behaviour of both producers and consumers and their governments.

15.4 SDG 15 in Indonesia: A Case in Point

Indonesia is a major player in the SDG process. President Susilo Bambang Yudhiono was co-chair of the UN committee that developed the SDGs. Indonesia is striving to be a leader in the pursuit of the SDG concept.

Indonesia officially classes 91 million ha (49.8 per cent of its land area) to forest. However, much of this land is no longer forested, and there are multiple overlapping claims on the land. Government figures state a net annual loss of forest of 0.7 per cent from 2010 to 2015 (684 000 ha) (FAO 2015). Relatively intact old-growth forests represent at most 50 per cent of total forest area and are stated to be declining by 800 000 ha per year. Much deforestation occurs in areas previously degraded by swidden agriculture or logging, in lowland areas, although recently some expansion of deforestation has occurred in the uplands (Margono et al. 2014). Approximately 45 per cent of recent deforestation has occurred within industrial concessions, mainly for oil palm (Abood et al. 2015).

Indonesia is home to 10 per cent of the world's flowering plant species, of which 55 per cent are endemic; 12 per cent of mammal species; 17 per cent

[3] Calculated based on ITTO Market Report December 2011 prices with an unweighted approximation of USD 100 per cubic metre of domestic logs and USD 200 per cubic metre of export logs.

of bird species; 16 per cent of reptile species; and 35 per cent of primate species (CBD Secretariat 2018). Habitat loss, deforestation, fragmentation and degradation means that 1259 species are threatened. This is despite protected areas covering 15 per cent of total sea and land area (von Rintelen et al. 2017) and 41 per cent of forests (Abood et al. 2015). Lowland forests contain the most biodiversity, but are the most threatened due to increasing pressure from population growth, infrastructure development, fires and conversion to industrial estate crops (CBD Secretariat 2018).

As an emerging economy, development in Indonesia is rapidly expanding into areas rich in terrestrial and marine biodiversity. About 10 per cent of Indonesians live below the national poverty line, a figure in annual decline. Population growth means that average farm size is becoming smaller (McCarthy and Robinson 2016). Human development in the eastern provinces lags behind the islands closer to the administrative centre. Government investments in infrastructure to improve the lives of the poorest provinces, such as West Papua and Maluku, will increase the pressure on forests as access becomes easier for extractive industries. Some 50–70 million Indonesians have self-identified as 'Indigenous', and many of these people rely on natural resources for their livelihoods (AMAN 2013). Many Indigenous Indonesians, particularly those living in forests without legal rights, are among those most affected by forest degradation and loss. Integration of forest policy decisions with local realities is increasing, and the government is working towards recognising the rights of Indonesia's rural poor, particularly the Indigenous forest-dependent peoples.

Decisions on use of forestlands lie with the Ministry of Environment and Forestry (MoEF). Indonesia's constitution acknowledges the concept of traditional ownership, but also declares that the state has responsibility for the nation's natural resources (Wrangham 2002). This ambiguity permitted post-independence governments to assert ever-greater control over forest areas, leading to a complex situation where different state institutions have overlapping, and seemingly incompatible, rights and responsibilities for the same lands. For decades, campaigners and others have urged the Indonesian government to return control of forests to local communities. Indonesia's Constitutional Court finally agreed in its decision No. 35/PUU-X/2012. Consequently, President Jokowi pledged that local rights would be respected and that 12.7 million ha of forestland would be returned to communities by 2019. This transfer of rights is continuing, but progress has been slow, inhibited by legislative and political hurdles.

The pledged redistribution of 12.7 million ha of forestland to community and Indigenous groups could significantly contribute to conserving life on land in Indonesia. While most areas will be under community forestry schemes, an increasing proportion of forestland will now come under the private ownership of Indigenous groups. Considerable uncertainty remains as to how the behaviour of communities will change in response to these new land-tenure arrangements (Sayer et al. 2017). More than 40 million ha of the forest estate is licenced to concessionaires: 21.49 million ha for timber exploitation and 19.4 million for oil palm plantations (McCarthy and Robinson 2016). These industries drive rural economies and contribute significantly to the national economy. Current investments in the service sector and tourism may reduce economic reliance on natural resource extraction and exports. Agricultural and forestry land-management standards will be critical to maintaining Indonesia's life on land in a future period of economic transition. The tensions between drivers of the globalised economy and the rights of rural and Indigenous populations in Indonesia echo similar tensions in many developing tropical countries.

15.4.1 SDG Implementation in Indonesia

The SDGs in Indonesia are overseen by an SDG coordination team (Figure 15.1) that integrates targets and indicators within the medium-term plans of central and provincial governments. The national SDG action plan is the template for all provincial action plans for the period up until 2020.

The top-down nature of the SDGs presents challenges for the application of SDG 15 in many countries, including Indonesia. The SDGs prioritised in the national SDG action plan must pass through the national and subnational development plan and budgets before implementation at the local level. In this process, the Indonesian government identifies which goals align

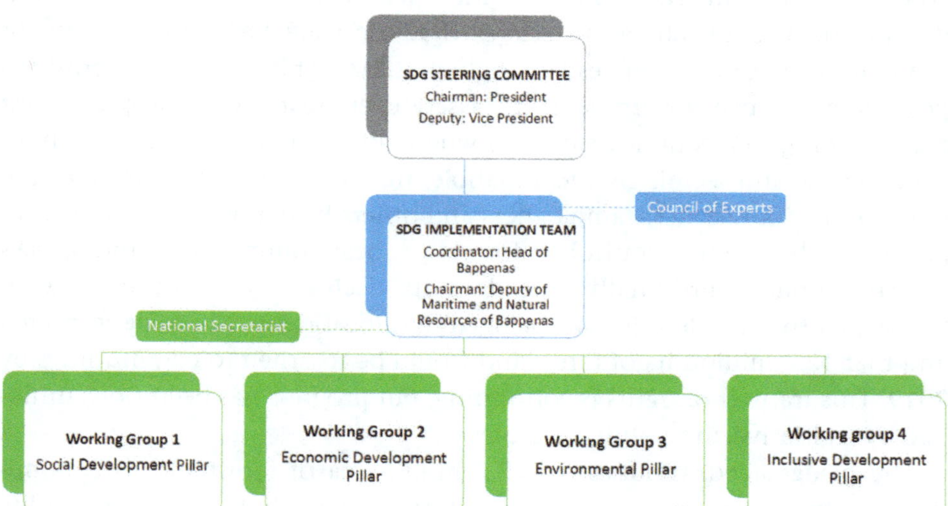

Figure 15.1 Organisational structure of SDG coordination in Indonesia.

with national and regional priorities and allocates resources towards these priorities. As a result, some SDGs receive higher budget allocations, particularly SDG1 (No Poverty) and SDG2 (Zero Hunger). Targets 15.4, 15.A and 15.B are not identified in the national development plan, so we would expect to see no investment in, for instance, mountain conservation. Considering the trend for increased deforestation in upland areas, the omission of Target 15.4 is a concern for protection of biodiversity and preservation of forested land.

In Riau, the first province to complete its SDG action plan, SDG 15 receives little attention: just three targets – 15.1, 15.2 and 15.3 – are included. Unlike the national SDG action plan, Targets 15.5–15.9 are excluded. For a province that has extensive industrial forestry concessions, it is surprising to see that key biodiversity and invasive species targets are not addressed. An acknowledged value of the SDG process – to encourage integration and to make trade-offs more explicit – is not being addressed in the Riau Provincial Action Plan. Interestingly, some of the estate crop industries operating in Riau are using the SDGs as a framework for guiding their own activities in ways that promote environmental and social sustainability.

Indonesia is prioritising the SDG Target 15.2 for the restoration of degraded land. The MoEF aims to identify 500 000 ha of forest for restoration by 2019, and to have actually restored 100 000 ha by that date. The way this is done will have important consequences for other SDGs. In the past, degraded lands have often been allocated for estate crop development. There will still be pressure from local governments to convert degraded land to estate crops in order to drive local economic development. If MoEF restores natural forest, then a significant contribution to SDG 15 could be made. Further expansion of estate crops could lead local communities to lose their land and suffer economically (SDG 1) and culturally (SDG 2 Zero Hunger, SDG 10 Reduced Inequalities). People could be forced to move and could degrade forests in other areas. Estate crop development could provoke land conflicts and run counter to SDG 16 (Peace, Justice and Institutions).

A harmonised approach to the SDGs requires a full understanding of the social–ecological systems where change is happening. The danger is that SDG 15 is only used as a measuring and communication tool, unable to address the myriad of interconnected impacts that result from focusing only on achieving the headline indicator measurements.

15.4.2 Conclusions for Indonesia

Indonesia is taking a strong position on implementing the SDGs, but, like many other countries, is challenged to fully understand how the SDGs interact. The establishment of 12.7 million ha of forestlands under community management has implications for SDGs 1–3 and many more, but currently there is no framework for capturing this information and measuring the impacts of such policies on the other SDGs. Government policies tend to favour the pursuit of individual goals and may overlook trade-offs. As in many countries, SDG implementation is the responsibility of sectoral institutions. It is proving difficult to deal with the interconnected web of targets and indicators that Agenda 2030 envisages.

In many cases, the only options for Indonesia's rural population to escape poverty require converting more forestland to agriculture or for people to move to cities. Local and provincial governments are understandably sympathetic to the needs of the rural poor. Reclassification of land as 'other land use' allows more land-clearing by smallholders but also by estate crop companies. Estate crops provide employment, drive local economies and are favoured by local government. The political economy of land-based investments in Indonesia has been widely debated (Barr and Sayer 2012). Lessons learned from REDD+ demonstrate the difficulty of coordinating action against deforestation and greenhouse gas emissions on a national scale. Navigating synergies and trade-offs of SDGs will have to be context-based, recognising that forests in different parts of Indonesia are at different stages of transition. The diversity of Indonesia means SDGs cannot take a one-size-fits-all approach. As tenure arrangements continue to evolve, the challenge of conserving life on land will have to be met in a context of contested land claims and greatly increased pressures on forests. The dilemma facing the achievement of the SDGs in Indonesia is symptomatic of the situation found in many tropical countries under extreme development pressures.

15.5 Synergies and Trade-offs

The achievement of SDG 15 will have impacts on and be decided by the other SDGs. Context is everything, and the responsibility for each country to establish national priorities for implementation of the SDGs will determine the extent of synergies and trade-offs.

Life on land constitutes the essential environmental underpinning for human existence, and hence for all other SDGs. The degree to which other goals are given higher priority than SDG 15 will determine the extent of potentially large negative trade-offs. This is obvious when there are competing demands for land: for example, converting forests to agriculture could mean that SDG 2 competes with SDG 15. However, if SDG 2 is achieved through increased agricultural efficiency and more food is produced on less land, then there is potential for a positive synergy. A major potential synergy exists between attainment of SDG 15 and Target 16.6, the development of effective institutions. There is an urgent need in many countries to reform the institutions responsible for forests so that they can better deal with the multiple values of forests and the diverse stakeholders concerned by these values. Many other potential synergies with forests and forest biodiversity have an impact on people's lives, such as the maintenance of pollinators required for various crops (IPBES 2016). We know from advances over the past decade that, in many regions, accessible groundwater (Ilstedt et al. 2016) and the maintenance of reliable rainfall depends on forests and tree cover to a much greater degree than most policymakers and their advisors yet realise (Ellison et al. 2017, Sheil 2018). A continuing dialogue around these potential synergies, both among experts and among the general population, would ensure more informed decision-making in which synergies leverage balanced sustainable development to achieve lasting benefits for life on land.

15.6 Discussion

SDG 15 is welcome in directing high-level attention to the need for conservation; but without an emphasis on integration, wide political and public engagement and greater responsiveness to local needs, SDG 15 risks perpetuating a sectoral, top-down approach. Decision-makers are now seeking to break down sectoral barriers, develop new institutional arrangements for managing landscapes and focus attention on the underlying causes of ecological decline rather than the symptoms. Integrated approaches to managing life on Earth have been recognised as essential for several decades (Brundtland 1987, IUCN et al. 1980). Many countries are experimenting with decentralised institutions for managing natural resources (Sayer et al. 2005). Experiences gained through these initiatives can create opportunities for further transformation. Integrated, landscape-scale governance arrangements are now being widely pursued (Sayer and Collins 2012). Much of the success of the SDGs will depend on whether SDG 15 encourages or impedes these ongoing changes.

The future of life on Earth depends largely on actions taken in support of all the SDGs. One might expect numerous references to environmental constraints in the indicators for all SDGs, but such references are few: there is only one mention of the word 'forests' in the entire SDG text outside SDG 15, in Target 6.6. Conservation and development are interdependent. Life on land, and forests in particular, is in general being husbanded better in countries that score well on the development indicators included in the other SDGs. Forests continue to decline in poorer countries with weak institutions. The fate of life on land really depends upon progress on SDGs 1–14 and 16–17.

In the introduction to this chapter we noted that SDG 15 could raise the profile of conservation within the broader community. We also noted the tendency for SDG 15 to be viewed as a second-tier goal behind the others. We must combat this: it is not the intention of Agenda 2030 to select and privilege some SDGs over others. This points to the need for SDG 15 advocates to reach out to those responsible for the other SDGs to show how and where the greatest synergies and most modest trade-offs can be achieved. We need to show that conservation can work with development rather than against it.

Many questions advanced by the conservation and development community over recent years and decades remain contentious. Many of these are central to the advancement of SDG 15. For example, are intensified (land-sparing) farming approaches preferable to low intensity (land-sharing) systems (Kremen 2015, Phalan et al. 2011)? We now know the answers depend on circumstances, and the fate of any 'spared land' is key. Continuation of such debates and further technical advances remain necessary to ensure better informed decisions.

15.7 Conclusions

If global economies continue to grow and people become more prosperous, then attitudes to forests and biodiversity will change. People may be more receptive to conservation, and moves towards green growth and bio-based economies will alleviate pressures on life on land. If economies stagnate between now and 2030, then the opposite may occur: people will be more concerned about their short-term well-being and may resist conservation measures. These trends may manifest themselves differently in poorer and richer countries. The degree to which SDG 15 and all the other SDGs are attained will be highly dependent on economic, social and political trends from now until 2030.

The targets and goals indicate what Agenda 2030 aspires to achieve, but they do not tell us how to get there. Since progress will be needed in highly divergent local contexts, there is need for the SDG process to be based upon locally learned lessons so that these can feed into SDG and forest policies at higher levels. Many decisions will be taken at the landscape level through a continued process of experimentation and adaptation. Progress in any of the SDG targets may involve fits and starts, setbacks and failures; learning must be an integral part of the agenda. Navigating an optimal path for development in both the short term and the longer term is a major challenge. Integrating SDG 15 into all the processes guided by the other SDGs will be essential to support life on land to 2030 and beyond.

References

Abood, S. A., Ser Huay Lee, J., Burivalova, Z., Garcia-Ulloa, J. and Koh, L. P. 2015. Relative contributions of the logging, fiber, oil palm, and mining industries to forest loss in Indonesia. *Conservation Letters* 8(1):58–67. doi.org/10.1111/conl.12103.

Agrawal, A., Chhatre, A. and Hardin, R. 2008. Changing governance of the world's forests. *Science* 320:1460–2.

Agrawal, A. and Gibson, C. 1999. Enchantment and disenchantment: The role of community in natural resource conservation. *World Development* 27(4):629–49.

Agrawal, A. and Gibson, C. C. 2001. *Communities and the environment: ethnicity, gender, and the state in community-based conservation*. New Brunswick: Rutgers University Press.

Agrawal, A. and Redford, K. 2009. Conservation and displacement: An overview. *Conservation and Society* 7(1):1–10.

Alroy, J. 2017. Effects of habitat disturbance on tropical forest biodiversity. *Proceedings of the National Academy of Sciences* 114:6056–61.

AMAN (Aliansi Masyarakat Adat Nusantara) 2013. *Indonesia and the denial of Indigenous peoples' existence*. Available at: www.aman.or.id/indonesia-and-the-denial-of-Indigenous-peoples-existence/ (Accessed 16 February 2019).

Baker, J., Milner-Gulland, E. and Leader-Williams, N. 2012. Park gazettement and integrated conservation and development as factors in community conflict at Bwindi Impenetrable Forest, Uganda. *Conservation Biology* 26:160–70.

Barr, C. M. and Sayer, J. A. 2012. The political economy of reforestation and forest restoration in Asia–Pacific: Critical issues for REDD+. *Biological Conservation* 154:9–19.

Bicknell, J. E., Struebig, M. J. and Davies, Z. G. 2015. Reconciling timber extraction with biodiversity conservation in tropical forests using reduced-impact logging. *Journal of Applied Ecology* 52(2):379–88.

Boedhihartono, A. K. 2017. Can community forests be compatible with biodiversity conservation in Indonesia? *Land* 6(1):21.

Boissière, M., Sheil, D., Basuki, I., Wan, M. and Le, H. 2009. Can engaging local people's interests reduce forest degradation in Central Vietnam? *Biodiversity and Conservation* 18(10):2743–57.

Brooks, T. M., Wright, S. J. and Sheil, D. 2009. Evaluating the success of conservation actions in safeguarding tropical forest biodiversity. *Conservation Biology* 23:1448–57.

Bruinsma, J. 2009. *The resource outlook to 2050*. Paper presented at Expert Meeting on How to Feed the World. Rome: FAO.

Brundtland, G. H. 1987. *Report of the World Commission on Environment and Development: 'Our Common Future'*. New York: United Nations.

Bruner, A. G., Gullison, R. E., Rice, R. E. and Da Fonseca, G. A. 2001. Effectiveness of parks in protecting tropical biodiversity. *Science* 291:125–8.

Burivalova, Z., Şekercioğlu, Ç. H. and Koh, L. P. 2014. Thresholds of logging intensity to maintain tropical forest biodiversity. *Current Biology* 24(16):1893–8.

Butchart, S. H., Walpole, M., Collen, B. et al. 2010. Global biodiversity: Indicators of recent declines. *Science* 328(5982):1164–8.

Cahill, A. E., Aiello-Lammens, M. E., Fisher-Reid, M. C. et al. 2013. How does climate change cause extinction? *Proceedings of the Royal Society B: Biological Sciences* 280:1–10.

CBD Secretariat 2018. *Indonesia Country Profile*. Available at: www.cbd.int/countries/profile/default.shtml?country=id (Accessed 26 January 2019).

Challender, D. W. and Macmillan, D. C. 2014. Poaching is more than an enforcement problem. *Conservation Letters* 7:484–94.

Chazdon, R. L., Peres, C. A., Dent, D. et al. 2009. The potential for species conservation in tropical secondary forests. *Conservation Biology* 23:1406–17.

Colfer, C., Sheil, D., Kaimowitz, D. and Kishi, M. 2006. Forests and human health in the tropics: Some important connections. *UNASYLVA* 57(2):3.

Corlett, R. T. and Westcott, D. A. 2013. Will plant movements keep up with climate change? *Trends in Ecology & Evolution* 28:482–8.

Dale, V. H., Brown, S., Haeuber, R. A. et al. 2000. Ecological principles and guidelines for managing the use of land. *Ecological Applications* 10:639–70.

Dawkins, H. C. and Philip, M. S. 1998. *Tropical moist forest silviculture and management: A history of success and failure*. New York: CAB International.

Dirzo, R., Young, H. S., Galetti, M. et al. 2014. Defaunation in the Anthropocene. *Science* 345:401–6.

Dowie, M. 2011. *Conservation refugees: The hundred-year conflict between global conservation and native peoples*. Cambridge, MA: MIT Press.

Dudley, N., Phillips, A., Amend, T., Brown, J. and Stolton, S. 2016. Evidence for biodiversity conservation in protected landscapes. *Land* 5(4):38.

Edwards, D. P., Larsen, T. H., Docherty, T. D. et al. 2010. Degraded lands worth protecting: The biological importance of Southeast Asia's repeatedly logged forests. *Proceedings of the Royal Society B: Biological Sciences* 278(1702):82–90.

Edwards, D. P., Tobias, J. A., Sheil, D., Meijaard, E. and Laurance, W. F. 2014. Maintaining ecosystem function and services in logged tropical forests. *Trends in Ecology & Evolution* 29(9):511–20.

Ellison, D., Morris, C. E., Locatelli, B. et al. 2017. Trees, forests and water: Cool insights for a hot world. *Global Environmental Change* 43:51–61.

Estrada, A., Garber, P. A., Rylands, A. B. et al. 2017. Impending extinction crisis of the world's primates: Why primates matter. *Science Advances* 3(1):e1600946. doi:10.1126/sciadv.1600946.

FAO 2010. Global forest resources assessment 2010: Main report. *FAO Forestry Paper* 163. Rome: FAO.

FAO 2012. FRA 2015: Terms and definitions. *Forest Resource Assessment Working Paper* 180. Rome: FAO.

FAO 2015. *Global forest resources assessment 2015: Desk reference*. Rome: FAO.

GAPKI (Gabungan Pengusaha Kelapa Sawit Indonesia) 2017. *Refleksi Industri Kelapa Sawit 2016 and Prospek 2017*. Available at: https://gapki.id/news/1848/refleksi-industri-kelapa-sawit-2016-prospek-2017 (Accessed 27 July 2019).

Ghazoul, J. and Sheil, D. 2010. *Tropical rain forest ecology, diversity, and conservation*. New York: Oxford University Press.

Gibson, L., Lee, T. M., Koh, L. P. et al. 2011. Primary forests are irreplaceable for sustaining tropical biodiversity. *Nature* 478:378–81.

Global Forest Watch 2019. *Forest monitoring designed for action*. Available at: www.globalforestwatch.org (Accessed 26 January 2019).

Haken, J. 2011. Transnational crime in the developing world. *Global Financial Integrity* 12.

Hill, C. M. 2015. Perspectives of 'conflict' at the wildlife–agriculture boundary: 10 years on. *Human Dimensions of Wildlife* 20(4):296–301.

Ilstedt, U., Tobella, A. B., Bazié, H. et al. 2016. Intermediate tree cover can maximize groundwater recharge in the seasonally dry tropics. *Scientific Reports* 6.

Inamdar, A., De Jode, H., Lindsay, K. and Cobb, S. 1999. Capitalizing on nature: protected area management. *Science* 283(5409):1856–7.

IPBES 2016. *Summary for policymakers of the assessment report on pollinators, pollination and food production*. Potts, S. G., Imperatriz-Fonseca, V. L., Ngo, H. T. et al. (eds.) Bonn, Germany: IPBES.

IPCC 2013. *Climate change 2013: The physical science basis. Contribution of working group I to the fifth assessment report of the Intergovernmental Panel on Climate Change*. Stocker, T. F., Qin, D., Plattner, G.-K. et al. (eds.) Cambridge: Cambridge University Press.

IUCN 2019. *The IUCN red list of threatened species. Version 2018–2*. Available at: www.iucnredlist.org (Accessed 26 January 2019).

IUCN and ITTO (International Tropical Timber Organization) 2019. *Guidelines for the conservation and sustainable use of biodiversity in tropical timber production forests*. Yokohama: ITTO And IUCN.

IUCN, UNEP and WWF 1980. *World Conservation Strategy: Living Resource Conservation for Sustainable Development*, Gland, Switzerland: IUCN.

Keenan, R. J., Reams, G. A., Achard, F. et al. 2015. Dynamics of global forest area: Results from the FAO Global Forest Resources Assessment 2015. *Forest Ecology and Management* 352:9–20.

Kissinger, G., Herold, M. and De Sy, V. 2012. *Drivers of deforestation and forest degradation: A synthesis report for REDD+ Policymakers*. Lexeme Consulting, Vancouver Canada, August 2012.

Kremen, C. 2015. Reframing the land-sparing/land-sharing debate for biodiversity conservation. *Annals of the New York Academy of Sciences* 1355(1):52–76.

Laurance, W. F., Clements, G. R., Sloan, S. et al. 2014. A global strategy for road building. *Nature* 513:229–32.

Laurance, W. F., Sayer, J. and Cassman, K. G. 2013. Agricultural expansion and its impacts on tropical nature. *Trends in Ecology & Evolution* 29(2):107–16.

Le Blanc, D. 2015. Towards integration at last? The Sustainable Development Goals as a network of targets. *Sustainable Development* 23:176–87.

Liswanti, N., Sheil, D., Basuki, I., Padmanaba, M. and Mulcahy, G. 2011. Falling back on forests: How forest-dwelling people cope with catastrophe in a changing landscape. *The International Forestry Review* 13(4):442–55.

Lobell, D. B., Schlenker, W. and Costa-Roberts, J. 2011. Climate trends and global crop production since 1980. *Science* 333:616–20.

MacDicken, K. G. 2015. Global Forest Resources Assessment 2015: What, why and how? *Forest Ecology and Management* 352:3–8.

Margono, B. A., Potapov, P. V., Turubanova, S., Stolle, F. and Hansen, M. C. 2014. Primary forest cover loss in Indonesia over 2000–2012. *Nature Climate Change* 4(8):730–5.

McCarthy, D. P., Donald, P. F., Scharlemann, J. P. W. et al. 2012. Financial costs of meeting global biodiversity conservation targets: Current spending and unmet needs. *Science* 338(6109):946–9.

McCarthy, J. F. and Robinson, K. 2016. Land, economic development, social justice and environmental management in Indonesia: The search for the people's sovereignty. In McCarthy, J. F. and Robinson, K. (eds.) *Land and development in Indonesia: Searching for the people's sovereignty*. Singapore: ISEAS–Yusof Ishak Institute, pp. 1–32.

Meijaard, E. and Sheil, D. 2007. A logged forest in Borneo is better than none at all. *Nature* 446:974.

Meijaard, E. and Sheil, D. 2008. The persistence and conservation of Borneo's mammals in lowland rain forests managed for timber: Observations, overviews and opportunities. *Ecological Research* 23:21–34.

Meijaard, E., Sheil, D., Nasi, R., et al. 2005. *Life after logging: Reconciling wildlife conservation and production forestry in Indonesian Borneo*. Bogor, Indonesia: CIFOR.

Melo, F. P., Arroyo-Rodríguez, V., Fahrig, L., Martínez-Ramos, M. and Tabarelli, M. 2013. On the hope for biodiversity-friendly tropical landscapes. *Trends in Ecology & Evolution* 28(8):462–8.

Milner-Gulland, E. 2018. Documenting and tackling the illegal wildlife trade: Change and continuity over 40 years. *Oryx* 52(4):597–8.

Nater, A., Mattle-Greminger, M. P., Nurcahyo, A. et al. 2017. Morphometric, behavioral, and genomic evidence for a new orangutan species. *Current Biology* 27(22):3487–98.

Naughton-Treves, L., Alix-Garcia, J. and Chapman, C. A. 2011. Lessons about parks and poverty from a decade of forest loss and economic growth around Kibale National Park, Uganda. *Proceedings of the National Academy of Sciences* 108(34):13919–24.

Phalan, B., Onial, M., Balmford, A. and Green, R. E. 2011. Reconciling food production and biodiversity conservation: Land sharing and land sparing compared. *Science* 333:1289–91.

Pimm, S. L., Jenkins, C. N., Abell, R. et al. 2014. The biodiversity of species and their rates of extinction, distribution, and protection. *Science* 344(6187):1246752.

Putz, F. E., Zuidema, P. A., Synnott, T. et al. 2012. Sustaining conservation values in selectively logged tropical forests: The attained and the attainable. *Conservation Letters* 5(4):296–303.

Redford, K. H. 1992. The empty forest. *BioScience* 42(6):412–22.

Ribot, J. C., Agrawal, A. and Larson, A. M. 2006. Recentralizing while decentralizing: How national governments reappropriate forest resources. *World Development* 34:1864–86.

Ricketts, T. H., Dinerstein, E., Boucher, T. et al. 2005. Pinpointing and preventing imminent extinctions. *Proceedings of the National Academy of Sciences of the United States of America* 102:18497–18501.

Robinson, B. E., Holland, M. B. and Naughton-Treves, L. 2014. Does secure land tenure save forests? A meta-analysis of the relationship between land tenure and tropical deforestation. *Global Environmental Change* 29:281–93.

Rodrigues, A. S., Akcakaya, H. R., Andelman, S. J. et al. 2004. Global gap analysis: Priority regions for expanding the global protected-area network. *Bioscience* 54:1092–1100.

Roe, D. 2008. The origins and evolution of the conservation-poverty debate: A review of key literature, events and policy processes. *Oryx* 42(04):491–503.

Rudel, T. K., Coomes, O. T., Moran, E. et al. 2005. Forest transitions: Towards a global understanding of land use change. *Global Environmental Change* 15(1):23–31.

Rudel, T. K., Defries, R., Asner, G. P. and Laurance, W. F. 2009. Changing drivers of deforestation and new opportunities for conservation. *Conservation Biology* 23(6):1396–405.

Sassen, M., Sheil, D., Giller, K. E. and Ter Braak, C. J. 2013. Complex contexts and dynamic drivers: Understanding four decades of forest loss and recovery in an East African protected area. *Biological Conservation* 159:257–68.

Sayer, J. and Cassman, K. G. 2013. Agricultural innovation to protect the environment. *Proceedings of the National Academy of Sciences* 110(21):8345–8.

Sayer, J. A. and Collins, M. 2012. Forest governance in a changing world: Reconciling local and global values. *The Round Table* 101:137–46.

Sayer, J. A., Maginnis, S. and Laurie, M. 2005. *Forests in landscapes: Ecosystem approaches to sustainability*. London and Sterling, VA: Earthscan and IUCN.

Sayer, J., Margules, C. and Boedhihartono, A. K. 2017. Will biodiversity be conserved in locally-managed forests? *Land* 6(1):6.

Sayer, J., Sunderland, T., Ghazoul, J. et al. 2013. Ten principles for a landscape approach to reconciling agriculture, conservation, and other competing land uses. *Proceedings of the National Academy of Sciences* 110(21):8349–56.

Schipper, J., Chanson, J. S., Chiozza, F. et al. 2008. The status of the world's land and marine mammals: Diversity, threat, and knowledge. *Science* 322:225–30.

Scott, L. and McGill, A. 2018. *From promise to reality: Does business really care about the SDGs?* United Kingdom: Price Waterhouse Coopers.

Selwood, K. E., Mcgeoch, M. A. and Mac Nally, R. 2015. The effects of climate change and land-use change on demographic rates and population viability. *Biological Reviews* 90:837–53.

Sharpe, B. 1998. 'First the forest': Conservation, 'community' and 'participation' in south-west Cameroon. *Africa* 68(1):25–45.

Sheil, D. 2017. Exploring local perspectives and preferences in forest landscapes: Towards democratic conservation. In Sanz, N., Lewis, R. C., Mata, J. P. and Connaughton, C. (eds.) *Tropical forest conservation: Long-term processes of human evolution, cultural adaptations and consumption patterns*. Mexico City: Unesco, pp. 262–83.

Sheil, D. 2018. Forests, atmospheric water and an uncertain future: The new biology of the global water cycle. *Forest Ecosystems* 5:1–22.

Sheil, D., Boissière, M. and Beaudoin, G. 2015. Unseen sentinels: Local monitoring and control in conservation's blind spots. *Ecology and Society* 20(2):39.

Sheil, D., Meijaard, E., Angelsen, A., Sayer, J. and Vanclay, J. K. 2013. Sharing future conservation costs. *Science* 339(6117):270–1.

Simula, M. 2008. *Financing flows and needs to implement the non-legally binding instrument on all types of forests*. Washington, DC: World Bank.

Sloan, S. and Sayer, J. A. 2015. Forest resources assessment of 2015 shows positive global trends but forest loss and degradation persist in poor tropical countries. *Forest Ecology and Management* 352:134–45.

Soga, M., Gaston, K., Koyanagi, T., Kurisu, K. and Hanaki, K. 2016. Urban residents' perceptions of neighbourhood nature: Does the extinction of experience matter? *Biological Conservation* 203:143–50.

Stafford-Smith, M., Griggs, D., Gaffney, O. et al. 2017. Integration: The key to implementing the Sustainable Development Goals. *Sustainability Science* 12:911–19.

Steffen, W., Broadgate, W., Deutsch, L., Gaffney, O. and Ludwig, C. 2015. The trajectory of the Anthropocene: The great acceleration. *The Anthropocene Review* 2(1):81–98.

Temudo, M. P. 2012. 'The white men bought the forests': Conservation and contestation in Guinea-Bissau, Western Africa. *Conservation and Society* 10(4):354–66.

Tittensor, D. P., Walpole, M., Hill, S. L. L., et al. 2014. A mid-term analysis of progress toward international biodiversity targets. *Science* 346(6202):241–4.

Tyrväinen, L., Pauleit, S., Seeland, K. and de Vries, S. 2005. Benefits and uses of urban forests and trees. In Konijnendijk, C., Nilsson, K., Randrup, T. and Schipperijn, J. (eds.) *Urban Forests and Trees*. Berlin: Springer, pp. 81–114.

Vira, B., Agarwal, B., Jamnadass, R. et al. 2015. Forests, trees and landscapes for food security and nutrition. In Vira, B., Wildburger, C. and Mansourian, S. (eds.) *Forests and food: Addressing hunger and nutrition across sustainable landscapes*. Cambridge: Open Book Publishers, pp. 9–27.

von Rintelen, K., Arida, E. and Häuser, C. 2017. A review of biodiversity-related issues and challenges in megadiverse Indonesia and other Southeast Asian countries. *Research Ideas and Outcomes* 3(1):e20860.

Waldron, A., Miller, D. C., Redding, D. et al. 2017. Reductions in global biodiversity loss predicted from conservation spending. *Nature* 551:364–7.

Warren, R., Vanderwal, J., Price, J. et al. 2013. Quantifying the benefit of early climate change mitigation in avoiding biodiversity loss. *Nature Climate Change* 3:678–82.

Watson, J. E., Dudley, N., Segan, D. B. and Hockings, M. 2014. The performance and potential of protected areas. *Nature* 515:67–73.

Watts, J. and Vidal, J. 2017. Environmental defenders being killed in record numbers globally, new research reveals. *The Guardian* (UK) 13 July. Available at: www.theguardian.com/environment/2017/jul/13/environmental-defenders-being-killed-in-record-numbers-globally-new-research-reveals (Accessed 27 January 2019).

Whiteman, A., Wickramasinghe, A. and Piña, L. 2015. Global trends in forest ownership, public income and expenditure on forestry and forestry employment. *Forest Ecology and Management* 352:99–108.

Wiens, J. J. 2016. Climate-related local extinctions are already widespread among plant and animal species. *PLOS Biology* 14(12):e2001104.

Wiersum, K. F. 1995. 200 years of sustainability in forestry: Lessons from history. *Environmental Management* 19:321–9.

Wrangham, R. 2002. Changing policy discourses and traditional communities. In Colfer, C. and Resosudarmo, D. A. P. (eds.) *Which way forward? People, forests, and policymaking in Indonesia*. Washington, DC: Resources for the Future, Center for International Forestry Research (CIFOR) and Institute of Southeast Asian Studies (ISEAS), pp. 20–35.

Wright, S. J. and Muller-Landau, H. C. 2006. The uncertain future of tropical forest species. *Biotropica* 38(4):443–5.

Chapter 16 SDG 16: Peace, Justice and Strong Institutions – A Political Ecology Perspective

Constance L. McDermott*, Emmanuel Acheampong, Seema Arora-Jonsson, Rebecca Asare, Wil de Jong, Mark Hirons, Kaysara Khatun, Mary Menton, Fiona Nunan, Mahesh Poudyal and Abidah Setyowati

Key Points

- Understanding the impacts of SDG 16 on forests and people requires attention to the power dynamics that shape how all 17 SDGs are interpreted and implemented across the Global North and South.
- As SDGs were agreed upon by nation states, SDG 16 places a strong emphasis on state power and the rule of law.
- Yet inclusive governance requires the involvement of diverse actors, and consideration for customary laws and other non-state forms of rule-making at global to local scales.
- Many national laws governing forests and land use favour political elite, large-scale industry actors and international trade.
- The development and strengthening of legal frameworks that support all of the SDGs – including those relevant to human rights, income inequalities, land tenure, gender and environmental protection – requires equal or greater priority than law enforcement. Otherwise, law enforcement will reinforce inequities and unsustainable practices.
- SDG 16 provides an opportunity to overcome the stereotypes of the Global North as the referential role model for peace and democracy, by highlighting the role of the North in fostering market inequalities and global conflicts, and drawing attention to barriers to democratic and inclusive participation within the Global North.
- How transparency, accountability and justice are conceived and prioritised shapes their impact on forests, as well as the degree to which their achievement either empowers forest-dependent peoples or excludes them from meaningful and informed engagement.

* Lead author.

16.1 Introduction

SDG 16 highlights core principles of governance relevant to the design and implementation of all 17 SDGs. Yet despite the seemingly universal appeal of principles such as 'peace' and 'justice', negotiations over SDG 16 were heavily contested, revealing a complex geopolitical landscape of competing international priorities (Slotin and Elgin-Cossart 2013). Developing countries have raised concerns that the SDGs might be hi-jacked by the UN's peace and security agenda, perceived as prioritising securitisation while ignoring the links between peace and broader goals such as social equity, climate change mitigation and development (Slotin and Elgin-Cossart 2013). These political battles help to explain the arguably mixed messages embedded in SDG 16's 12 targets (see Table 16.1). While some targets on inclusiveness and justice suggest a pluralist and 'bottom up' approach to governance, many other targets are consistent with Westphalian notions of a strong state[1] reliant on legal formalisation, and legitimised through representative democracy. The goal's 23 indicators reinforce this emphasis on the state, and are backed by extensive demands for quantitative and qualitative data.

The implications of implementing SDG 16 for forests and forest-dependent peoples depend on how the principles, targets and indicators are interpreted, prioritised, monitored and reported. Implementation may vary, for example, according to whether emphasis is placed on (1) a strong and centralised nation state and the enforcement of state laws and regulations or (2) a more pluralistic or decentralised notion of good governance where power is distributed across scales and institutions, and authority exercised by states, market-based initiatives and/or ground-up, locally driven processes. Likewise, it will depend on whether states internalise their own visions and commitments to good governance, or whether most of their efforts focus on gathering data and reporting on the 23 more narrowly defined indicators.

The importance of interpretation reflects the socially constructed nature of core governance concepts such as 'justice' and 'accountability', and differing perceptions of the legitimacy and appropriateness of formal versus informal governance, and state-based, market-based and customary institutions. Hence, while the SDGs may represent international consensus on overarching principles of good governance, in practice their implementation will be strongly influenced by the political priorities and power dynamics that continue to unfold at global to local scales. This, in turn, will hold different implications for forests and people in different country contexts.

[1] The 1648 treaties of Westphalia were instrumental in enshrining the sovereignty and authority of nation states in international law. See, for example, Cutler 2001: 1024.

Table 16.1 SDG 16 targets
16.1 Significantly reduce all forms of violence and related death rates everywhere
16.2 End abuse, exploitation, trafficking and all forms of violence against and torture of children
16.3 Promote the rule of law at the national and international levels and ensure equal access to justice for all
16.4 By 2030, significantly reduce illicit financial and arms flows, strengthen the recovery and return of stolen assets and combat all forms of organized crime
16.5 Substantially reduce corruption and bribery in all their forms
16.6 Develop effective, accountable and transparent institutions at all levels
16.7 Ensure responsive, inclusive, participatory and representative decision-making at all levels
16.8 Broaden and strengthen the participation of developing countries in the institutions of global governance
16.9 By 2030, provide legal identity for all, including birth registration
16.10 Ensure public access to information and protect fundamental freedoms, in accordance with national legislation and international agreements
16.A Strengthen relevant national institutions, including through international cooperation, for building capacity at all levels, in particular in developing countries, to prevent violence and combat terrorism and crime
16.B Promote and enforce non-discriminatory laws and policies for sustainable development
Source: https://sustainabledevelopment.un.org/sdg16

This chapter outlines these complexities and identifies key challenges, opportunities and trade-offs in implementing SDG 16 in a way that is good for forests and people across very different environmental, social and economic contexts. Section 16.2 examines how three inter-related themes articulated in the SDG 16 text intersect with trends in forest governance: (1) peace and the reduction of armed conflict; (2) rule of law, accountability, transparency and access to justice; (3) inclusiveness and participation. These themes are discussed in a general sense, and then case studies illustrate key challenges and trade-offs. Section 16.3 synthesises lessons learned on how context shapes the impacts of SDG 16 implementation, and the implications for promoting inclusive and environmentally effective governance. This conclusion speaks to the two 'operationalising targets' of SDG 16 (16A and 16B), focused on institutions and non-discriminatory laws and policies.

Consistent with several other authors in this book, we are less concerned with hypothetical discussions of what it would mean to achieve the *ends* of universal peace, justice and strong institutions, and more focused on the importance of the *means* by which states responsible for SDG implementation attempt to achieve these, and how this shapes outcomes. This allows for a more critical, 'political ecology' perspective to unpack how power dynamics shape human-nature interactions and their outcomes. Such an approach recognises the power dynamics inherent in the design and implementation of the SDGs, and the dynamic and contested nature of governance. It highlights how the SDGs are a product of a particular UN process, agreed on by national governments and interpreted and used in the context of international processes, including bi-lateral and multi-lateral finance. This could in part explain the strong implied focus within SDG 16 on state-associated institutions, with much reference to laws, legal enforcement, legal recognition, public officials and other state-centric language. It also sparks broader questions about the relative roles of the Global North and South in deciding how peace and justice are defined and operationalised within the SDGs.

16.2 Governance under SDG 16

The term 'governance' has gained increasing traction over the last few decades, reflecting changes in the locus and nature of power and authority under globalisation (Rosenau 1995). Global economic growth and the expansion of international trade have generated new challenges – from deforestation and climate change to social inequalities and civil unrest– that have catalysed new forms of social coordination at multiple scales, involving diverse state, private and civil society actors (Castells 2008, Gunningham 2009).

In this context, 'governance' may be broadly defined as 'the formation and stewardship of the formal and informal rules that regulate the public realm, the arena in which state as well as economic and societal actors interact to make decisions' (Hyden et al. 2004: 16). With specific reference to natural resources, Campese (2016: 7) defines natural resource governance as 'the norms, institutions, and processes that determine how power and responsibilities over natural resources are exercised, how decisions are taken and how citizens – including women, men, youth, Indigenous peoples and local communities – secure access to, participate in, and are impacted by the management of natural resources'. Such an understanding reveals how the governance of 'sustainable development' and associated norms within international policy instruments such as the SDGs is not simply a technical challenge involving state actors implementing universally agreed-upon norms; it is an inherently political process involving contestations over who should govern what and for whom (Ferguson 1994).

The globalisation of social and economic relations and the territorial expansion of commercial agriculture and extractive industries, along with associated information technologies and telecommunications, have fostered the expansion of inter-governmental agreements, market-based instruments such as sustainability certification, and multi-stakeholder platforms, all of which interact in complex ways with existing state, local and/or traditional sources of formal and informal authority. The resulting power dynamics are often unequal, with the 'resource-strong' – donors, private entities or governments, often external to the sites of implementation – dictating the terms of trade and creating new governance structures impacting local decision-making. This has spurred conflicts over the appropriate nature and scale of authority, and the relative legitimacy of various institutions to govern decision-making (Bodansky 1999).

The following examination of three themes under SDG 16 will be viewed within this context. In each case we consider the literature on their relation to forests and livelihoods, relevant governance trends, and implications for the implementation of SDG 16 and the SDGs more generally.

16.2.1 Peace and the Reduction of Violence and Armed Conflict

SDG 16 covers all forms of violence and abuse, at multiple scales, both organised and un-organised. The strongest emphasis is on physical violence, but psychological violence is also mentioned (Indicator 16.1.3). The literature addressing the impacts of violence on forests and forest-dependent peoples falls into one of two relatively narrow categories: violence against civilians directly related to land and resource conflicts; and indirect impacts on forests and local communities from armed conflicts, such as political rebellion or organised crime.

Most of this literature focuses more on the ends than the means: how conflict does or does not impact forests and people, rather than on how the purpose of the conflict, or how it is suppressed, influences resulting impacts. This is perhaps congruent with a dominant discourse that peace is the natural outcome of the rule of law and democratic participation. Yet, as observed by Paret (2015: 107), the reality is much more complex: 'violent practices may become tools of liberation, promoting democracy by empowering marginalized groups ... [or] ... democracy may become a tool of domination, undermining dissent by constituting as violent those persons and actions that deviate from formal institutional channels'.

The realisation of SDG 16 requires that citizens be free to defend their land and resource rights and to engage in environmental advocacy in a nonviolent manner and without fear of intimidation or reprisal. A growing body of evidence suggests this is currently not the case across many regions worldwide. The NGO Global Witness has developed a database of murders of 'environmental defenders': activists or local populations engaged in defending land or natural resources. According to this database, an estimated global average of four environmental defenders are murdered every week (Global Witness 2017). For 2016 and 2017, 16 per cent of these murders were associated with logging and a further 20 per cent with wildlife poaching, often linked to forested areas. These rates are believed to be significantly under-reported, particularly in countries with limited free press, and they exclude much larger numbers of people who are injured, threatened, intimidated or criminalised.

In the broader literature on organised armed conflict, Melander et al. (2016) identify several types: (1) state-based armed conflict, involving at least one government actor; (2) non-state conflicts between rebel groups and militias and (3) one-sided killing of unarmed civilians, by states or formally organised non-state groups. Following World War II, the largest number of deaths are attributed to internal, state-based armed conflict, along with a recent rapid rise in 'internationalised' (internationally supported) intra-state conflict (especially within Syria, where state actors also play a significant role) (Melander et al. 2016). Of significance here, state governments are tasked with SDG implementation when they themselves are sometimes the leading perpetrators of violence. Furthermore, as highlighted by Scott(1998) and Rudel et al. (2009), state actors have historically encouraged the settlement and clearance of forest frontiers as a strategy to solidify their control over rural populations or territories.

Concerning the underlying drivers of armed conflict, Collier and Hoeffler (2000) distinguish between 'grievance', which relates to the acuteness of the professed cause of conflict, and 'greed', denoting the presence of political and economic opportunity. Donovan et al. (2007) and de Koning et al. (2008), looking specifically at forest and conflict linkages, argue for the importance of both grievance and greed in driving conflict and associated forest change. They highlight how people living in or near forests are in many parts of the world at the margins of state support and services. For instance, de Koning (2007) finds that political and economic inequalities were the underlying causes of several of the civil wars that were taking place in West Africa's forests during the first decade of the twenty-first century. Armed conflicts themselves exacerbate causes for grievance because of their negative impact on human capital, especially in rural settings – including forest-dependent communities (Chamarbagwala and Morain 2011). However, other contemporary conflict researchers argue that grievance and greed are more correlates of conflict than its drivers, and variously emphasise the causal roles of weak and illegitimate state institutions (e.g. Fearon 2011) or sectarian inequalities (e.g. Stewart 2009).

International actors and large-scale industry also play a role in forest conflict, for example when agricultural and mining firms compete with local communities for land and resources. At the same time, the rise of international governance aimed at forest conservation and reducing forest emissions (themes of SDG 15 and SDG 13, respectively) involves yet another competing set of interests in the forest frontier, focused on carbon, biodiversity and conservation (Fairhead et al. 2012, Kaag and Zoomers 2014, Scheidel and Work 2018).

The international emphasis on conservation has, in some cases, led to 'green militarisation' or the use of military or paramilitary personnel or techniques to protect national parks and limit poaching of wildlife (see Lunstrum 2014). Büscher and Ramutsindela (2015) write of 'green violence', which expands the concept of green militarisation to include broader concepts of violence, both material and non-material, and encompassing discursive and social violence used to promote environmental protection. Green violence includes forced removal of local people from parks, killing of poachers and discourse in favour of such killings (Büscher and Ramutsindela 2015). In some cases, 'shoot-on-sight' policies have been advocated (Messer 2010).

Whatever the causes of violent conflict, its impacts on forests are variable. In terms of biodiversity, the literature reports both negative (Gaynor et al. 2016, Ordway 2015) and positive impacts, the latter largely because the presence of conflict discourages forest access (Álvarez 2003, Burgess et al. 2015, Harwell 2010, McNeely 2003). Forest impacts may increase when the conflict dissipates (Blom and Yamindou 2001) as priority is placed on generating peace dividends, including construction and the resettlement of ex-combatants. Demand for wood and agricultural land may surge during post-conflict years (Harwell 2010). Additionally, negative impacts on forest governance linger, through the limited effectiveness of public administration and poor implementation of forest and nature policies (Staver et al. 2007).

Likewise, conflict has variable effects on forest-dependent communities. Armed conflicts in forest regions often involve interventions by government forces or their allies, which reduces community access to forests, or diminishes the availability of resources (Álvarez 2003, Harwell 2010). Insurgent groups impose demands on communities for food, forest products and other goods and services (Harwell 2010). Under armed conflict conditions, land grabbing is more readily condoned, resulting in communities losing access to forests and their goods and services. Displaced groups, especially refugees, turn to forests when other livelihood sources have almost evaporated or livelihoods need to be rebuilt. In such contexts, international sanctions imposed on so-called conflict timber can undermine local economic welfare (Price et al. 2007).

Two reviews on conflict timber (de Jong et al. 2007, Price 2003) synthesise findings across numerous cases of armed conflicts involving forests and forest peoples worldwide, including the two countries we explore in our case studies: Colombia and Peru. Many of these cases are classified as civil wars, although for each the nature of the conflict varied. In all case countries, except perhaps the Democratic Republic of Congo (DRC) and Colombia, the organised armed conflicts have stopped or have transformed into other types of conflict. Many regions remain unstable, characterised by violence, low compliance with state law and ineffective state governments. In Peru and Colombia remnants of insurgence groups continue as organised crime (Pettersson and Wallensteen 2015). In the coming decades, it is possible that organised crime will dominate forest-based armed conflicts, as has happened in Peru and Colombia where Mexican drug cartels have taken over from insurgence groups. This has negative consequences for local people's well-being, although not necessarily for their incomes.

These findings highlight how the challenge of reducing violence and armed conflict must be viewed in the broader context of multi-scale governance and environmental and social welfare. A reliance on state actors to quell violence and promote peace could, in some contexts, have negative environmental and social outcomes (UN 2015), including the conversion of forest frontiers to commercial land use as a means to exert state control, and the repression of political dissent (Scott 1998).

CASE STUDY 16.1 PEACEBUILDING IN COLOMBIA AND PERU

Both Colombia and Peru have undergone extended periods of armed conflict with widespread implications for forest cover and the livelihoods of forest-dependent people. In Peru, the Maoist group the Shining Path concentrated their activities in the highlands and the capital city of Lima from the 1980s to the 1990s. The conflicts and instability linked to the violence led to internally displace people (IDPs) (estimates range from 310 000 to 600 000), many of whom migrated either to Lima or to forested areas in the Peruvian Amazon. Population growth in the Peruvian Amazon, and subsequent deforestation, is often attributed to this displacement (Ravikumar et al. 2017).

In Colombia, a 2016 peace agreement ended more than 60 years of armed conflict led by the Revolutionary Armed Forces of Colombia (FARC), which concentrated much of its activities in the Amazonian region. As of 2014, there were 5.8 million IDPs in Colombia (Hojen 2015), with some displaced as a direct result of the armed conflict, and others due to drug trade and other conflicts.

In both countries, the conflicts were linked to inequality and calls for land reforms. Insecurity of land tenure has been shown to have a negative impact on forest cover in conflict areas in Colombia (Dávalos 2001). Castro-Nuñez et al. (2017a) find further evidence for the importance of unequal land distribution and land grabbing in the conflict, as well as evidence that the presence of forest commons was associated with lower rates of conflict.

Timber may serve as an important means to finance armed conflicts (Castro-Nuñez et al. 2017a), particularly in forest frontier areas where there is little state presence and a high concentration of commercially valuable species. It is important to note, however, that the presence of forest cover is itself not an indicator of the presence of armed conflict (Harwell 2010, Rustad et al. 2008).

Regarding the net effect of conflict on forests in Colombia and Peru, several main effects have been observed. In both countries, there have been cases of 'gunpoint conservation' where productive activities are curtailed by the violence or enforcement of curfews (Dávalos 2001). There has also been pressure on forest lands due to cultivation of illicit crops (coca) in regions where armed conflicts exclude the presence of law enforcement (Castro-Nuñez et al. 2017a). In Colombia, armed groups actively conserved forest cover as a means to avoid surveillance (McNeely 2003). In Peru, some regions saw displacement to areas with high forest cover in search of safer, more stable living environments (Shanee and Shanee 2016).

Importantly, the impact of armed conflict on forests is not always consistent at national or regional levels. In some cases in Colombia, the FARC used forests as cover for covert operations and/or funded their activities by selling forest resources, thereby maintaining forest cover but contributing to forest degradation (Álvarez 2003). In contrast, in the San Lucas mountain range, the FARC and the National Liberation Army (ELN) actively enforced bans on hunting and logging, thereby protecting some forests from degradation (Dávalos 2001). In cases where the local populations chose to remain, the conflicts significantly impacted land-use and livelihood strategies, thereby influencing forest cover. In other regions, the FARC and the Shining Path encouraged coca cultivation, leading to deforestation (Álvarez 2003). While the FARC were actively promoting coca cultivation, the ELN supported coca eradication and instead promoted mining, which also had negative impacts on forested lands (Dávalos 2001). In areas with gold or lands appropriate for cattle, Sánchez-Cuervo and Aide (2013) find that the presence of armed groups decreased forest cover.

The effect of migration on forests also varies. In the case of Peru, conflict led to migration to cities and to regions of the Amazon with high forest cover. In

Colombia, where conflict areas were often in regions with high forest cover, migration did not have such a clear impact on forests at the regional scale.

These examples highlight the importance of (1) the particularities of the armed group's activities (e.g. FARC sometimes used forests for cover and other times promoted deforestation for coca cultivation) and (2) the conflict's location relative to areas of high forest cover (e.g. Peru's conflicts in low-forest areas led to migration to forested areas, whereas the reverse is true for Colombia).

While Peru has had almost two decades of peace, Colombia has only recently begun the peacebuilding process. It is important to distinguish between measures that lead to decreases in direct violence (e.g. institutions that limit violence) and those that address the structural drivers of violence (e.g. securing land tenure as a means to address land conflict). Colombia and Peru have gone through different stages in the transition towards lasting peace, and these distinctions are important for the SDGs and links to forests and land-use more broadly. Peacemaking includes processes of negotiation and dialogue to shift away from violent conflict and reach a peace agreement. Peacekeeping usually involves military interventions that help transitions towards peace by separating the fighting parties and actively preventing violence from erupting. Peacebuilding is intended to promote transitions towards lasting peace by addressing the structural causes of conflict and strengthening the capacity of local actors to overcome conflicts (Doyle and Sambanis 2000). Reinstating the 'rule of law' and maintaining it (Peacekeeping and Peacemaking) are not sufficient to address long-held and legitimate grievances (Peacebuilding), nor do they ensure justice (Corntassel and Holder 2008).

Peru's Truth and Reconciliation Commission (CVR) worked in areas affected by the Shining Path to support local communities in their search for peace and reconciliation. A report (CVR 2003) outlining the depth of the conflict, the responsible parties and the long-term impacts recommended reparation actions. Some of these were put into law, but many have not been implemented. The underlying issues of inequality, land conflicts and structural racism towards Indigenous peoples have not been resolved (Corntassel and Holder 2008). This failure's impact on forests remains to be studied.

In Colombia, peacebuilding approaches aim to reduce the causes of conflict (land-related grievances) and to curb access to resources for funding armed groups (e.g. coca). Castro-Nuñez et al. (2017b) found that the current programmes underway in areas emerging from the conflict, including land-tenure programmes, conditional payments for production of alternative crops and forest conservation, are compatible with both development aims and forest conservation. Clearly more research will be needed to understand how these programmes unfold in practice.

16.2.2 The Rule of Law, Accountability, Transparency and Access to Justice

The rule of law is a major focus of SDG 16, consistent with the wider trend in international governance to promote legality and legal enforcement as pathways to sustainability (McDermott 2014). This trend is particularly evident in the forest sector, having gained major prominence with the rise of illegal logging initiatives in the early 2000s. For example, the EU Forest Law Enforcement, Governance and Trade (FLEGT) Action Plan (EC 2003) has spurred a number of initiatives to eradicate illegal wood from EU supply chains, following the logic that law enforcement is a necessary stepping stone to good governance and sustainability (EC 2003). FLEGT mechanisms include the EU Timber Regulation, which prohibits the import of wood into the EU that was produced in violation of the laws of the country of origin (EC 2010), and Voluntary Partnership Agreements (VPAs) with partner countries in the Global South, for the development of 'legality licensing systems' (EC 2005). Once these systems are in place, all wood imported into the EU from a partner country must bear a special 'legality license' that verifies the wood was legally produced.

This push for legality has spilled into other normative texts and strategies of major relevance to forests. The New York Declaration on Forests, signed by 27 national governments and numerous corporations and NGOs, includes a claim that policies and measures to address illegal logging have 'proven very effective in reducing forest loss' (UN Climate Summit 2014). No evidence is provided to back this statement, but its presence highlights the normative prominence of legality in international discourse. The harnessing of EU markets to demand legal verification is likewise a key recommendation listed in an EU-commissioned feasibility study for an action plan to combat deforestation and degradation, where the scope of legality verification is extended to the production of agricultural crops associated with deforestation (COWI 2018).

SDG 16 couples this emphasis on the rule of law with an approach to accountability and transparency that prioritises the eradication of state-based corruption (Targets 16.5, 16.6). Specifically, Indicators 16.5.1 and 16.5.2 require data quantifying and reporting on levels of individual and corporate bribery of public officials. This aligns, at least in part, with trends in international governance to promote global transparency as a means to facilitate a 'level playing field' for international trade. Such transparency is based on a principle of external surveillance and control through standardisation and external monitoring, reporting and verification (e.g. Mutersbaugh 2005, Wood 2013). Applying a universalising discourse of transparency, as a taken-for-granted social good and antidote to corruption, assumes the existence of a functioning state with clear and equitable rules and an even-handed judiciary. This overlooks how complex patron–client relations generally arise in the absence of strong state institutions with widespread legitimacy (Peluso 2018), and how accusations of corruption are frequently harnessed for political gain (Khan 1998).

A universalising, decontextualised conception of transparency is also evident, for example, in the UNFCCC[2] mechanism Reducing Emissions from Deforestation and Forest Degradation, plus carbon stock enhancement, sustainable forest management, and conservation (REDD+). REDD+ has focused singular attention on the monitoring, reporting and verification (MRV) of forest carbon, a single chemical element amenable to international standardisation and monetisation. The focus on carbon inevitably obscures many other forest values less readily measured and verified by external actors. These other non-carbon environmental and social concerns are subsumed under the REDD+ 'safeguards'.

Sikor (2013) critiques this separation of 'safeguards' from the broader 'technical' design of REDD+, arguing that MRV and other REDD+ technologies of transparency themselves play a central role in shaping the distribution of rights and benefits. Fundamentally, the implementation of REDD+ cannot be separated from the broader governance contexts in which it falls, or from the priorities of those tasked with its enactment on the ground (Khatun et al. 2015, Kull et al. 2015). This intersection of the global and the local has social impacts much less transparent and amenable to international monitoring than changes in forest cover. Decisions about which REDD+ safeguards can and should be monitored are themselves contested. REDD+ poses substantial risks to local communities if it generates new forms of knowledge for national and/or international actors which are poorly understood at local levels, yet may be used to restrict local access to resources.

The interplay of scale and transparency can also be observed in the concept of a 'global commons' inherent in international climate governance. This claiming of space by international actors enables international regulation of practices that previously had local and national provenance. Villagers are thus drawn into a shifting assemblage of international governance, and are subject to new identifications as entrepreneurs and responsible environmental citizens, meant to look after the global commons. While this may be seen, in practice, as 'global citizenship', it involves citizen responsibilities

[2] United Nations Framework Convention on Climate Change.

without corresponding rights (Arora-Jonsson et al. 2016). Meanwhile, the growing technical, bureaucratic and political complexity of international processes such as REDD+, and associated MRV, decrease their transparency to most global and local citizens, undermining citizens' abilities to independently assert whatever rights they do hold (McDermott 2014). Hence, incongruously, the transparency called for in initiatives such as REDD+ risks disempowering the people it aims to empower.

International transparency can at times strengthen the power of minority groups, particularly when those minority interests align with broader international norms and agendas. For example, some Indigenous groups have successfully aligned with international environmental or human rights organisations to stop the allocation of mining or other commercial concessions on traditional lands (Woodman 2014). The relevant lesson for SDG 16 is that the effect of international demands for transparency and accountability on forest-dependent peoples depends on what is being made transparent and accountable, and to whom.

The concept of 'justice' articulated in SDG 16 likewise raises questions of justice for what and for whom. Targets 16.3 and 16.9 appear to equate justice with promoting the rule of law and the creation of legal identities through registration (Indicator 16.9.1), while also making select reference to international human rights laws (Indicator 16.8.1). This largely bypasses the difficult question of how state definitions of justice inevitably privilege some actors and some conceptions of justice over others. In contrast, recent literature addressing REDD+ and other payments-for-nature schemes argue that justice, and associated concepts of equity, are better viewed as complex and multidimensional, involving many actors both within and beyond the state, and inevitably requiring trade-offs. For example, McDermott et al. (2013) highlight how equity and justice are socially constructed, and must be understood in the context of how they are defined and by whom, what their goals are and who counts as legitimate subjects of justice or equity. Likewise, they may involve procedural dimensions (e.g. conceptions of due process), as well as distributive (e.g. the fair distribution of benefits) and contextual dimensions. The impact on forests and people of achieving justice under the banner of SDG 16 will similarly depend on the degree to which SDG 16 implementation conceptualises and acknowledges the full assemblage of actors and institutions beyond the state, and whose justice is served or undermined.

The impact of implementing SDG 16 targets that promote the rule of law, transparency and justice will depend on a wide range of factors, including pre-existing formal and informal institutions, and the scales of the actors and markets involved in its implementation. National forest laws in many countries favour political elites and/or large-scale industry actors. The focus of international initiatives on international transparency may render local actors, and the impacts on local markets, relatively invisible or illegitimate (Hirons et al. 2018). Yet it is not inevitable that the implementation of SDG 16 will simply reinforce these trends. The following case study in Ghana draws on the examples of FLEGT VPA and REDD+ processes to illustrate how the way governance is understood and approached in any given intervention can further entrench existing power dynamics or lead to transformative change.

CASE STUDY 16.2 A COMPARISON OF THE FLEGT VPA AND REDD+ IN GHANA

Oil, gold, cocoa, timber and agriculture are the main resources underpinning the Ghanaian economy. For timber and cocoa, in particular, there are concerns over the impact of unsustainable timber harvest and expansion of low-yield cocoa into forested areas. Ghana's timber harvest is three times the annual allowable cut (Hansen et al. 2012) and, coupled with the conversion of forest to cocoa farms, may be undermining the long-term benefits flowing from intact forests, including carbon storage and sequestration and the provision of climatic conditions suitable for cocoa and biodiversity conservation (Benhin and Barbier 2004, Gockowski and Sonwa 2011, Government of Ghana 2005, Hansen et al. 2009, Obiri et al. 2007, Ruf 2011, Wade et al. 2010). Two international policy responses have emerged that align with these concerns: FLEGT's VPA programme and REDD+.[3] This case study briefly reviews the development of these two strategies in Ghana, comparing key features of their approach to legal reform so as to illustrate differences between strategies that prioritise legal formalisation, auditability and enforcement (top-down) and those that account for customary norms and locally driven governance processes (bottom-up).

It has been estimated that 70 per cent of Ghana's timber production is illegal (Hansen and Treue 2008). In 2009, Ghana and the EU signed a FLEGT VPA trade agreement to eradicate this illegality and stop the import of illegal wood into the EU (Beeko and Arts 2010). The agreement maintains the legal rights of the state over economically valuable timber, which the state may sell to private concessionaires. State control extends to native trees growing on farmlands, so that enforcing state rights effectively marginalises small-scale and local actors. Most farmers have no legal rights to timber growing on their farms, and can only access their timber through illegal harvesting.

At the heart of the VPA strategy is the idea that the legality of timber can be independently verified and audited. To deliver this accountability, the VPA established a Timber Legality Assurance System (TLAS) that includes a wood-tracking system (WTS) to provide an auditable chain of custody for

[3] The 'plus' refers to additional objectives associated with enhancing forest carbon stocks, enhancing biodiversity and ensuring the social component of sustainability is addressed.

timber. The TLAS covers imports, pre-felling processes, felling and processing, right through to point-of-sale. Once Ghana's TLAS is accepted by the EU, timber can be sold as 'FLEGT licensed' (i.e. assured legality) and can thereby be imported into the EU. To date, Ghana's efforts to implement the VPA have been focused on producing FLEGT-licensed timber for export, though no FLEGT licenses have been granted in Ghana so far. Well-documented concerns about how the current legal system in Ghana perpetuates the criminality of farmers harvesting trees on their own lands effectively remain unaddressed (Hansen et al. 2018, Hirons 2018).

In a parallel (and largely unconnected) effort, the World Bank has been supporting Ghana's proposals for REDD+ under the UNFCCC to be based on a sustainable intensification strategy for the cocoa sector. This strategy includes a 'climate-smart cocoa' production approach focused on increasing yields (productivity per unit area) through the adoption of best practices and recommended shade levels, combined with community-based landscape governance and land-use planning processes. These measures aim to reduce expansion pressure on forests and to incentivise maintenance and enhancement of carbon stocks in the landscape (Asare 2014). The strategy's success requires that cocoa farmers access critical resources and gain improved tenure. Currently, farmers have little motivation to maintain shade trees because they lack formal rights to native trees[4] and are therefore incentivised to remove shade trees to minimise the risk that the government will allocate the native trees on their farms to timber concessionaires, who will damage cocoa[5] when they remove the trees. Although there are legal provisions for farmers to be compensated for damage to their cocoa plants and for community benefits to be provided through Social Responsibility Agreements, these are often ignored, and the majority of benefits are captured by local elites.

In light of these dynamics, the REDD+ process in Ghana prioritises tree-tenure reform (Asare 2014) that would enshrine the customary norms and rights characterising much of the local decision-making on farms. This would allow farmers to negotiate their own agreements with local authorities concerning the harvest and sale of trees on their farms.[6] These reforms could be

[4] A policy reform in 2002 resulted in the amendment of the Timber Resource Management Act 547 to Act 617. The amended Act 617 makes provision for granting ownership rights to individuals who plant timber trees on farmlands. However, these rights do not apply to naturally regenerating trees, and if farmers plant native trees it is difficult for them to prove that they were planted and not naturally recurring.

[5] Timber contractors use heavy machinery to drag whole trees out of farms, damaging cocoa trees in the process. Chainsaw operators 'slice' the trees where they are felled and the lumber is carried out by hand, which is considerably less damaging to cocoa.

[6] Government revenues under a reformed tenure system would be provided for by taxation in markets, rather than stumpage fees, which are not collected by the state when felling is illegal.

tested and refined under existing proposals, such as the community resource management area (CREMA) mechanism, which authorises communities to manage their own resources. Monitoring would be largely based on indicators reflecting the CREMA development and management process; CREMA would report on outcomes to partner entities. This locally tailored approach contrasts with standardised international systems that focus on external monitoring and control. The CREMA approach potentially delivers greater rights to communities in managing the resources they steward and greater proportions of the associated benefits.

These two major forest policy initiatives in Ghana have contrasting approaches to governance and legal reform. The FLEGT VPA focuses on state control over forest resources and emphasises auditability and enforcement while the tree-tenure reforms under REDD+ propose to incorporate customary norms into statutory law. These two approaches have markedly different impacts on local groups. Under FLEGT, the local domestic producers and consumers of timber risk further criminalisation and exclusion from the benefits of timber harvest as a livelihood activity (Hansen et al. 2018, Hirons et al. 2018). Under REDD+, proposals for tree tenure improve local actors' ability to control and benefit from trees on their land (Asare 2014).

This characterisation of the difference between FLEGT and REDD+ in Ghana is somewhat simplified.[7] The findings also cannot, by themselves, be generalised to other country contexts. The intent here is to illustrate how the assumed links between legality and sustainability should be more critically and deeply investigated. Forest laws can be legal yet ecologically unsustainable and socially inequitable; illegal production can be both sustainable and fair. Approaches to legal reform that attempt to incorporate the informal rules and norms underpinning sustainable resource management are more likely to support the wider objectives outlined in the SDGs than those that benefit large private entities and states at the cost of marginalising and criminalising local actors.

16.2.3 Inclusiveness and Participation

Target 16.7 calls for 'responsive, inclusive, participatory and representative decision-making at all levels'. Like the previous themes, the indicators under this target strongly imply a focus on state institutions and national-level reporting. Trends in international discourse on governance encompass a much broader and more far-reaching understanding of participation. For the forest sector this means a push towards decentralisation in forest governance

[7] For example, tree-tenure reform has also been mooted during FLEGT discussions, but it is REDD+ that is leading the way on current efforts on this issue.

across levels of government, as well as the active involvement of non-state actors and local communities in decision-making (Arts 2014, Gilmour 2016).

In many parts of the world more participatory approaches to forest management have been formalised through community-based or collaborative forest management. It is estimated that in 62 countries, covering 65 per cent of the world's forests, 28 per cent of the forest area is community managed (Gilmour 2016). Under such arrangements, forest users are often required to set up a decision-making structure, usually based on elected community representation, such as a forest-user group or committee, with a mandate and functions prescribed in legislation. In practice, these initiatives are mostly 'top-down', unlike some customary and Indigenous governance systems which in some cases are rooted in centuries of traditional practice. While state-regulated arrangements provide opportunities for forest users to participate in forest management in ways that are recognised and supported by the state, the governance structures and systems may either conflict with existing customary practice or reinforce inequities, thereby falling short of fully inclusive or representative decision-making. It is often the case within communities that elites – those with higher status, power and/or wealth – dominate formal community management and co-management systems at the local level, steering decisions and capturing benefits.

Given sufficient time and investment, state-orchestrated participation can offer opportunity for non-elites to challenge this elite capture. For example, Lund and Saito-Jensen (2013), in their study of participatory forestry in Tanzania and India over more than a decade, observe that over time marginalised people successfully organised themselves and formed alliances with external actors to resist elite control, and made use of electoral systems to gain increased authority. Likewise, Persha and Andersson (2014) find that the involvement of external agencies, such as NGOs, can help to reduce the scale and negative effects of elite capture.

Research on the opportunities for women to be part of participatory forest structures and processes has, for example, examined which factors enable and constrain women's participation and whether women's presence has any impact on the nature and effectiveness of forest management. Agarwal (2009, 2010) finds that women's participation in community forest institutions in India and Nepal often led to stricter rules, though other factors can also be at play. Coleman and Mwangi (2013) find that women from households with higher levels of education and wealth are more likely to be influential participants in community forestry. Arora-Jonsson (2014) argues that the mandatory presence of women in forest committees may simply rubber-stamp legitimacy to external actors without changing the status quo. Women have gained greater influence over decision-making by organising women's groups outside of mainstream institutions. Arora-Jonsson advocates for acknowledging the multiple spaces in which everyday decision-making on the forests actually takes place, including informal types of social organisation (e.g. informal village groups tasked with keeping forests and villages clean, fetching water, planting herbs in the forest) that are frequently overlooked by external actors and researchers (Arora-Jonsson 2009).

Regarding the overall impacts of formal participatory forest processes, there is some evidence of positive outcomes for forest cover (Bowler et al. 2012, Gilmour 2016), but inadequate evidence of improved livelihoods (Samii et al. 2015). Bowler et al. (2012) call for capacity for evaluation to be built into participatory forest management programmes. An FAO review of 40 years of community-based forestry[8] identifies the following as key challenges for sustainable forest management: (1) the adoption of such approaches tends to be found in highly degraded forests that offer little in the way of benefits to communities; (2) communities are often not granted access to valuable resources, particularly timber; (3) the emphasis is on protection rather than management; and (4) decision-making power largely remains with the central government (Gilmour 2016). This reflects what has happened with decentralisation more generally, where the majority of decision-making power and resources remain with central government. These patterns continue, despite widespread recognition of the need for participation to be meaningful, inclusive and equitable (Nunan et al. 2018). Additionally, studies of participatory forest management tend to focus on cases where external actors are involved in conceptualising, implementing or supporting community forest activities. A myriad of traditional, community-based management strategies are often not included in these reviews or impact studies.

The meaningful inclusion of local communities in resource management depends on the strength of their underlying land tenure and use rights, and their capacity to benefit from those rights (Ribot and Peluso 2003). Multi-country synthesis studies of research on community forestry reiterate the importance of the bundles of rights communities have to land and resources, including the material benefit from forests (Baynes et al. 2015, Gnych et al. 2018). These overarching issues of rights and access are notably absent from SDG 16 targets and indicators.

The following case studies serve to situate these challenges to participatory forest management in specific country contexts. Case Study 16.3 illustrates Indonesia's high levels of ambition and investment in community forestry,

[8] The literature varies on the definitions and terms used to refer to participatory approaches to forest management. Our focus here is deliberately broad, and encompasses a range of practices which include community-based forestry, participatory forest management and other associated frames.

as well as the many challenges and trade-offs faced in realising that ambition. Case Study 16.4 provides a comparative case study of Sweden and India that challenges commonly held assumptions about the relative quality of participation in the Global North and South.

CASE STUDY 16.3 INCLUSION AND EXCLUSION IN INDONESIA'S SOCIAL FORESTRY INITIATIVE

Around 91 per cent of Indonesia's forested areas are administered by the state, and communities in an estimated 31 957 villages in and around forest areas lack secure access to forests and livelihoods (RRI 2013). In recognition of this, the government of Indonesia has recently spearheaded a major push towards social forestry, with an ambitious target to allocate 12.7 million ha of forest area to local and Indigenous communities by 2019, as stipulated by a series of social forestry policies.[9] These policies have the triple objectives of securing communities' access to forest resources, alleviating poverty and improving forest conditions. This landmark decision has been praised by many as a promising pathway for more inclusive and equitable development. It is also hoped it will clarify forest tenure, and resolve widespread and longstanding forest tenure conflicts (Eghenterr et al. 2017).

A core procedural mechanism for Indonesia's social forestry strategy has been to engage a diversity of stakeholders in social forestry working groups at the national and provincial levels. These groups currently cover 24 out of 34 provinces and engage 1118 individuals as focal points (Social Forestry Working Group 2017). The working groups are intended to facilitate the development and implementation of social forestry policies and stimulate the achievement of social forestry targets through preparing and updating digital maps to determine targeted priority areas,[10] as well as to carry out capacity-building activities and support communities with permit applications and permit verification processes. To speed up the achievement of the social forestry target, in 2016 the government issued an integrated social forestry policy that is expected to simplify procedures to obtain a permit, thereby reducing the time it takes from up to several years to a maximum of three months.

[9] Under social forestry policies, communities will be able to apply permits to manage the 'state forests' through several schemes: (a) community forestry (*hutan kemasyarakatan*); (b) village forests (*hutan desa*); (c) community plantation forests (*hutan tanaman rakyat*); (d) customary forests (*hutan adat*) and (e) forming partnership with private sectors (*kemitraan kehutanan*). Unlike other social forestry schemes, which only provide communities with rights to access forest resources, *adat* forest allows communities to have ownership rights over their customary forests. This was made possible by a constitutional court ruling that stipulates *adat* forests should no longer be considered part of state forests.

[10] *Peta Persiapan Perhutanan Sosial* or a preparation map for social forestry target.

Nevertheless, progress in reaching targets has been slow relative to ambitions, with a total of 1 065 056 ha reported as designated in September 2017 (MoEF 2017). The recognition of customary (*adat*) forests has been slower still (Myers et al. 2017). Out of the 9.1 million ha identified as *adat* forest in a map prepared by the National Indigenous People's Alliance of the Archipelago (AMAN), only 17 092 ha have been recognised by the state (Arumingtyas 2018). This slow progress is due, in part, to long and arduous government procedures, and overlapping claims over forestlands (Arumingtyas 2018, HUMA 2015).

While Indonesia's social forestry policies emphasise seemingly inclusionary politics, their implementation shows various forms of exclusion that could limit Indigenous and local communities' ability to benefit from forest resources. Despite inclusive processes at the national and provincial levels, Indigenous and local communities are at risk of being minimally involved at the village level due to the rush to achieved targeted hectares allocated for social forestry schemes. Field observations of social forestry policy implementation in several sites in Central Kalimantan suggest that local and Indigenous communities are minimally engaged in the issuance of social forestry permits. In the haste to issue permits to meet ambitious targets, and given the complexity of the permitting process, many members of communities holding such permits lack awareness of the rights and responsibilities the permits entail.[11] Some community members consider the permits additional burdens if the forestlands allocated to them are neither productive nor easily accessible. There is also evidence of elite capture (Maryudi et al. 2012), and/or the exclusion of women from forest decision-making processes and forest benefits (Setyowati 2012). Other forest-user groups remain excluded from the process entirely, including transmigrant communities, or communities making claims to lands allocated to company concessions or conservation areas, or to lands that have been removed from the category of state forestlands (Myers et al. 2017).

With the exception of *adat* forests, forest tenure reform in the social forestry policies refers to allocating usufruct rights to communities while retaining the state's ownership over the forestlands. Most of the legal rights stipulated in the forestry policies are time bound and come with use restrictions based on the classification of the allocated forest (Moeliono et al. 2017). For instance, a social forestry permit stipulated in a protected forest (*hutan lindung*) requires communities to improve forest protection and only allows the use of non-timber forest products. Moreover, the allocation is conditional

[11] Informal discussion with villagers on the community forest plantation permit (HTR), Pulang Pisau District, Central Kalimantan, November 2016.

on fulfilling responsibilities that communities often find difficult, such as submitting management plans and managing the forest in accordance with particular technical requirements. In a recent social forestry scheme which specifically targeted forest areas in Java, controlled by the state-owned enterprise, only severely degraded forest areas in need of rehabilitation could be allocated to communities.[12] Without sufficient technical and budgetary support, such allocations, rather than improving local development, limit local economic aspirations and further marginalise local communities.

CASE STUDY 16.4 THE INTERNAL AND EXTERNAL POLITICS OF PARTICIPATION IN SWEDEN

Seema Arora-Jonsson's work (2018) challenges the discourse of international development that paints the Global North as a referent for democracy and public participation, and the Global South as in need of Northern assistance and capacity-building to achieve conformity with Northern ideals. She illustrates this argument through case studies of Sweden as both a domestic and an international actor, contrasting Sweden's domestic and international policies regarding democratic decision-making in environmental governance (2018).

She points to the 1990s as an era that institutionalised international norms of decentralisation, embedded, in part, in a neoliberal agenda to shift environmental responsibility to local groups and individuals (Arora-Jonsson 2017). Within this trend, Sweden played a leading role in highlighting the importance of people's participation in decision-making, as well as serving as an international champion for gender equality. Gender forms a central theme and a priority for the Swedish Development Agency (Sida) and Sida was instrumental in supporting the World Bank to make participation of different groups central to questions of development aid. This contrasts with the relative absence of the mention of people's participation in environmental policy-making within Sweden, including few requirements for government workers to employ a gender perspective.

Her comparative case studies of two forest communities, in India and Sweden, likewise illustrate the disconnect between Sweden's international image as a gender champion and its local realities in forestry decision-making. In the Nayagarh district in Odisha, India, the openly acknowledged gender differences voiced by local actors, with support from NGOs and international actors, legitimised the formation of women's groups and collective action to address women's priorities. In contrast, in the village of Drevdagen in western Sweden, cultural perceptions of Sweden as a highly developed, modern

[12] MoEF Decree No.39/2017 on Social Forestry Schemes in Perum Perhutani Area.

country with strong norms of gender equality served to diminish the collective voice of women. In such a context, complaints of female oppression could be written off as signs of the weakness of individual women who have failed to assert themselves socially, or as an aberration in a country assumed to be gender-equal, rather than a reflection of broader societal discrimination (Arora-Jonsson 2009).

16.3 Conclusion

Understanding SDG 16 from a political ecology perspective requires attention to how states and other actors exercise their power in interpreting and implementing the SDGs. These dynamics are influenced by a range of larger governance trends, as well as by complex local particularities. States form the core authority behind the SDGs and are responsible for reporting on indicators, which at least partially explains why the language of SDG 16 places such a strong emphasis on state institutions. However, an understanding of governance as multi-nodal, networked and contested makes it impossible to ignore how reinforcing state power may undermine other governance institutions, such as traditional and local authorities.

In order to examine these issues, this chapter focuses on three different thematic areas encapsulated in SDG 16 – peace, justice and participation – to identify what existing literature can tell us about how SDG 16 implementation might impact forests and people. These analyses have also been grounded in more specific case studies to further unpack associated opportunities and challenges.

In regards to peace, our findings illustrate how peace cannot be separated from the broader context of environmental and social welfare. While SDG 16 emphasises the state's role as enforcer of the peace, an over-reliance on state actors to quell violence and promote peace can, in some contexts, have negative environmental and social outcomes, including the assertion of state control through the conversion of forests into commercial agriculture and/or repression of political dissent. The literature reveals varying effects of armed conflict on forest cover, resulting in either a net gain or loss of cover. This variation is observed both within and between countries, as witnessed in Case Study 16.1 on Colombia and Peru. Post-conflict periods are often associated with forest loss, through state-sponsored settlement schemes and other forms of securitisation of frontier areas, as well as the expansion of agricultural land and resource extraction fuelled by economic recovery and growth.

Regarding the second SDG theme – the rule of law, accountability, transparency, and access to justice – the impacts on forests and people of realising these goals will depend on a wide range of contextual factors, including pre-existing formal and informal institutions, and the scales of the actors and markets involved. National laws governing forests and land use in many countries favour political elite and/or large-scale industry actors, as do government policies encouraging commercial expansion into forest frontiers. Law enforcement without legal and policy reforms to dis-incentivise land conversion and to improve local resource rights and benefit capture could drive forest loss and the displacement of local communities. Furthermore, the focus of international initiatives on international transparency may legitimise large producers and industries based on their ability to control and track their supply chains, while rendering local actors, local markets and local livelihoods illegitimate or invisible (Scott 1998). This could further disempower rural communities and undermine their access to natural resources, while contributing to increased urban and international commodity consumption. Nevertheless, in some cases local actors may leverage international attention to successfully defend their interests against the state, particularly if their cause is aligned with international norms and agendas.

It is not inevitable that the implementation of SDG 16 will reinforce existing inequities. As illustrated by Case Study 16.2 of the FLEGT VPA and REDD+ in Ghana, the way in which governance is understood and approached in any given intervention can either further entrench existing power dynamics or lead to more transformative change. Approaches to legal reform which recognise the informal rules and norms underpinning sustainable resource management are more likely to support the wider SDG objectives than approaches that reinforce state laws designed to benefit large private entities and states.

Our analysis of the third theme – inclusiveness and participation – highlights the prevalence of the concept in international discourse and the diversity of ways in which it has been operationalised in the forest sector. These include the legal recognition of some form of community control over increasing areas of the world's forests. As highlighted in the Indonesian case, participation is time-consuming, costly and contentious. It can become more burdensome than empowering if it is accompanied by increasing accountability demands from states and other actors, and if these demands are not balanced with adequate benefits. Ambitious targets dictated at national or international levels to hand over forests to communities can overwhelm capacities for meaningful implementation.

The Swedish case unpacks the stereotypes of the Global North as the referential role model of democracy and participation for the Global South. Assumptions about Sweden as a developed country that has solved participation issues through formal legal means preclude attention to questions of participation and gender in resource decision-making on the ground. By contrasting Swedish policies at home and abroad, Arora-Jonsson (2009, 2018) deconstructs the development divide – the modern state in the Global North versus the not-quite modern state in the Global South – and how this might be used to further particular interests in either place. All the SDGs aim to motivate countries to self-assess and improve their efforts to reach common goals; this may open up the space for debating these important transnational relationships.

SDG 16 reflects international agreement on the importance of participatory processes and local decision-making. In the context of forests, this means recognising that forest use, management and governance can fail some people, and that long-term solutions to social and environmental problems, while global in scope, may result in local injustices. Furthermore, current research indicates that the heavily state-centric focus of the SDG 16 targets and indicators as a means to achieve good governance risks reinforcing these injustices. This suggests that SDG 16 may best be served through greater acknowledgement of non-state actors and institutions at multiple scales – from traditional governance systems to global-scale initiatives – and adjusting the appropriate mix of governance approaches to individual national, sub-national and local contexts. This also argues against a heavy focus on the standardised, quantitative metrics associated with SDG 16's reporting requirements. Instead, what is needed are placed-based assessments of the power dynamics of particular governance systems, and the identification of actions to address inequities in ways that are locally and contextually appropriate.

While the achievement of a truly inclusive and sustainable development, with justice for all and at all levels, is a laudable goal, it must not obscure the inevitable trade-offs inherent in governance, and, indeed, in the very concept of equity. It is critical to keep a power-informed perspective in the implementation of SDG 16 that acknowledges these trade-offs and leaves room for healthy conflicts and struggles among state and non-state actors alike. Creating adequate space for conflict and dispute is integral to the pursuit of individual and social welfare, peace and fulfilment.

References

Agarwal, B. 2009. Rule making in community forestry institutions: The difference women make. *Ecological Economics* 68:2296–308.

Agarwal, B. 2010. *Gender and green governance: The political economy of women's presence within and beyond community forestry*. Oxford: Oxford University Press.

Álvarez, M. 2003. Forests in the time of violence: Conservation implications of the Colombian War. *Journal of Sustainable Forestry* 16(3–4):47–68.

Arora-Jonsson, S. 2009. Discordant connections: Discourses on gender and grassroots activism in two forest communities in India and Sweden. *Signs: Journal of Women in Culture and Society* 35(1):1–28.

Arora-Jonsson, S. 2014. Forty years of gender research and environmental policy: Where do we stand? *Women's Studies International Forum* 47:295–308.

Arora-Jonsson, S. 2017. Blind spots in environmental policy-making: How beliefs about science and development may jeopardize environmental solutions. *World Development Perspectives* 5:27–29.

Arora-Jonsson, S. 2018. Across the development divide: A North–South perspective on environmental democracy. In Marsden, T. (ed.) *Sage handbook of Nature*. Sage Publications, pp. 1–24.

Arora-Jonsson, S. Westholm, L., Petitt, A. and Temu B. J. 2016. Carbon and cash in climate assemblages: The making of a new global citizenship. *Antipode* 48(1):74–96.

Arts, B. 2014. Assessing forest governance from a 'Triple G' perspective: Government, governance, governmentality. *Forest Policy and Economics* 49:17–22.

Arumingtyas, L. 2018. *Pengakuan Hutan Adat Minim: Perlu Terobosan Baru di 2018*. Mongabay (8 January). Available at: www.mongabay.co.id/ (Accessed 15 January 2018).

Asare, R. A. 2014. *Understanding and defining climate-smart cocoa: Extension, inputs, yields, and farming practices*. Accra, Ghana: Climate-Smart Cocoa Working Group, Nature Conservation Research Centre and Forest Trends.

Baynes, J., Herbohn, J., Smith, C., Fisher, R. and Bray, D. 2015. Key factors which influence the success of community forestry in developing countries. *Global Environmental Change* 35:226–38.

Beeko, C. and Arts, B. 2010. The EU-Ghana VPA: A comprehensive policy analysis of its design. *International Forestry Review* 12(3):221–30.

Benhin, J. K. and Barbier, E. B. 2004. Structural adjustment programme, deforestation and biodiversity loss in Ghana. *Environmental and Resource Economics* 27(3):337–66.

Blom, A. and Yamindou, J. 2001. *A brief history of armed conflict and its impact on biodiversity in the Central African Republic*. Washington, DC: Biodiversity Support Program, USAID/WWF.

Bodansky, D. 1999. The legitimacy of international governance: A coming challenge for international environmental law? *American Journal of International Law* 93(3):596–624.

Bowler, D. E., Buyung-Ali, L. M., Healey, J. R., et al. 2012. Does community forest management provide global environmental benefits and improve local welfare? *Frontiers in Ecology and the Environment* 10(1):29–36.

Burgess, R., Miguel, E. and Stanton, C. 2015. War and deforestation in Sierra Leone. *Environmental Research Letters*. doi:10:095014.

Büscher, B. and Ramutsindela, M. 2015. Green Violence: Rhino poaching and the war to save Southern Africa's Peace Parks. *African Affairs* 115:1–22.

Campese, J. 2016. *Natural resource governance framework assessment guide: Learning for improved natural resource governance*. IUCN/CEESP NRGF Working Paper, Gland, Switzerland: IUCN and CEESP.

Castells, M. 2008. The new public sphere: Global civil society, communication networks, and global governance. *The Annals of the American Academy of Political and Social Science* 616(1):78–93. doi:10.1177/0002716207311877.

Castro-Nuñez, A., Mertz, O., Buritica, A., Sosa, C. C. and Lee, S. T. 2017a. Land related grievances shape tropical forest-cover in areas affected by armed conflict. *Applied Geography* 85:39–50.

Castro-Nuñez, A., Mertz, O. and Sosa, C. C. 2017b. Geographic overlaps between priority areas for forest carbon-storage efforts and those for delivering peacebuilding programs: implications for policy design. *Environmental Research Letters* 12(5):054014.

Chamarbagwala, R. and Morán, H. E. 2011. The human capital consequences of civil war: Evidence from Guatemala. *Journal of Development Economics* 94(1):41–61.

Coleman, E. A. and Mwangi, E. 2013. Women's participation in forest management: A cross-country analysis. *Global Environmental Change* 23:193–205.

Collier, P. and Hoeffler A. 2002. *Greed and grievance in Civil War*. The World Bank Policy Research Working Paper 2355. Washington, DC: World Bank.

Corntassel, J. and Holder, C. 2008. Who's sorry now? Government apologies, truth commissions and Indigenous self-determination in Canada, Australia, Guatemala and Peru. *Human Rights Review* 9(4):1.

COWI 2018. *Feasibility study on options to step up EU action against deforestation*. EU, ECOFYS, Milieu, COWI.

Cutler, A. C. 2001. Critical reflections on the Westphalian assumptions of international law and organization: a crisis of legitimacy. *Review of International Studies* 27:133–150.

CVR (Comissión de la Verdad y Reconciliación) 2003. *Peru Truth and Reconciliation Commission Final Report*. Lima, Peru. Available at: www.cverdad.org.pe/ingles/ifinal/index.php (Accessed 20 February 2019).

Dávalos, L. M. 2001. The San Lucas mountain range in Colombia: How much conservation is owed to the violence? *Biodiversity & Conservation* 10(1):69–78.

de Jong, W., Donovan, D. and Ken Ichi, A. (eds.) 2007. *Tropical forests and extreme conflicts*. Dordrecht: Springer.

de Koning, R. 2007. Greed or grievance in West Africa's forest wars? In de Jong, W., Donovan, D. and Abe, K. (eds.) *Extreme conflicts and tropical forests*. Dordrecht: Springer, pp. 37–56.

de Koning, R., Capistrano, D, Yasmi, Y. and Cerutti, P. 2008. *Forest-related conflict: Impact, links, and measures to mitigate*. Washington, DC: Rights and Resources Initiative.

Donovan, D., de Jong, W. and Abe, K. 2007. Tropical forest and extreme conflict. In de Jong, W., Donovan, D. and Abe, K. (eds.) *Extreme conflicts and tropical forests*. Dordrecht: Springer, pp. 1–15.

Doyle, M. W. and Sambanis, N. 2000. International peacebuilding: A theoretical and quantitative analysis. *American Political Science Review* 94(4):779–801.

EC 2003. *Forest Law Enforcement, Governance and Trade (FLEGT) Proposal for an EU Action Plan*. European Commission.

EC 2005. *Council Regulation (EC) No 2173/2005 on the establishment of a FLEGT licensing scheme for imports of timber into the European Community*. European Commission.

EC 2010. *Regulation (EU) No 995/2010 of the European Parliament and of the Council of 20 October 2010 laying down the obligations of operators who place timber and timber products on the marketText with EEA relevance*. European Commission.

Ece, M., Murombedzi, J. and Ribot, J. 2017. Disempowering democracy: Local representation in community and carbon forestry in Africa. *Conservation and Society* 15(4):357–70.

Eghenterr, C., Suhardjito, D., Purwanto, E. et al. 2017. Mewujudkan Hak Rakyat: Reformasi Penguasaan Tanah dan Pengelolaan Hutan. *Proceedings of Academic Papers for Tenure Conference 2017*, Jakarta.

Fairhead, J., Leach, M. and Scoones, I. 2012. Green grabbing: A new appropriation of nature? *The Journal of Peasant Studies* 39(2):237–61.

Fearon, J. D. 2011. *Governance and civil war onset*. Washington, DC: World Bank. Available at: https://openknowledge.worldbank.org/handle/10986/9123 (Accessed 29 July 2019). License: CC BY 3.0 IGO.

Ferguson, J. 1994. The anti-politics machine: Development, depoliticization and bureaucratic power in Lesotho. *The Ecologist* 24:176–81.

Gaynor, K. M., Fiorella, K. J., Gregory, G. H. et al. 2016. War and wildlife: Linking armed conflict to conservation. *Frontiers of Ecological Environments* 14(10):533–42.

Gilmour, D. 2016. *Forty years of community-based forestry. A review of its extent and effectiveness*. FAO Forestry Paper 176. Rome: FAO.

Global Witness 2018. *At what cost? Irresponsible business and the murder of land and environmental defenders in 2017*. London.

Gnych, S., Lawry, S., Monterroso, I. and Adhikary, A. 2018. *Common benefits: Is community tenure facilitating investment in the commons for inclusive and sustainable development?* Paper presented at the World Bank Conference on Land and Property, Washington, DC, 4 March.

Gockowski, J. and Sonwa, D. 2011. Cocoa intensification scenarios and their predicted impact on CO_2 emissions, biodiversity conservation, and rural livelihoods in the Guinea rain forest of West Africa. *Environmental Management* 48(2):307–21.

Government of Ghana 2005. *Growth and Poverty Reduction Strategy (2006–2009)*. Republic of Ghana, Accra.

Gunningham, N. 2009. Environment law, regulation and governance: Shifting architectures. *Journal of Environmental Law* 21(2):179–212.

Hansen, C. P., Damnyag, L., Obiri, B. D. and Carlsen K. 2012. Revisiting illegal logging and the size of the domestic timber market: The case of Ghana. *International Forestry Review* 14(1):39–49.

Hansen, C. P., Lund. J. F. and Treue, T. 2009. Neither fast, nor easy: The prospect of Reduced Emissions from Deforestation and Degradation (REDD) in Ghana. *International Forestry Review* 11 (4):439–55.

Hansen, C. P., Rutt, R. and Acheampong, E. 2018. 'Experimental' or business as usual? Implementing the European Union Forest Law Enforcement, Governance and Trade (FLEGT) Voluntary Partnership Agreement in Ghana. *Forest Policy and Economics* 96:75–82.

Hansen, C. P. and Treue, T. 2008. Assessing illegal logging in Ghana. *International Forestry Review* 10 (4):573–90.

Harwell, E. 2010. *Forests in fragile and conflict-affected states*. Washington, DC: Program on Forests (PROFOR).

Hirons, M., McDermott, C., Asare, R. et al. 2018. Illegality and inequity in Ghana's cocoa-forest landscape: How formalization can undermine farmers' control and benefits from trees on their farms. *Land Use Policy* 76:405–413. doi:1.1016/j.landusepol.2018.02.014.

Hojen, L. 2015. *Colombia's 'invisible crisis': Internally displaced persons. Council on Hemispheric Affairs*. Available at: www.coha.org/colombias-invisible-crisis-internally-displaced-persons/ (Accessed 20 February 2019).

HUMA (Association for Law Reform Based on Community and Ecology) 2015. *Penetapan Hutan Adat Menuju Pengakuan Hak Masyarakat Adat*. Jakarta: HUMA.

Hyden, G., Court, J. and Mease, K. 2004. *Making sense of governance: Empirical evidence from sixteen developing countries*. London: Lynne Reiner.

Kaag, M. and Zoomers, A. 2014. *The global land grab: Beyond the hype*. Chicago: Zed Books.

Khan, M. H. 1998. The role of civil society and patron–client networks in the analysis of corruption. In *Corruption and integrity improvement initiatives in developing countries*. New York: UNDP, Management Development and Governance Division, pp. 111–28.

Khatun, K., Gross-Camp, N., Corbera, E., Martin, A., Ball S. and Massao, G. 2015. When participatory forest management makes money: Insights from Tanzania on governance, benefit sharing, and implications for REDD+. *Environmental and Planning A* 47:2097–112. doi:10.1177/0308518X15595899.

Kull, C. A., Arnauld de Sartre, X. and Castro-Larrañaga, M. 2015. The political ecology of ecosystem services. *Geoforum* 61:122–34.

Lund, J. F. and Saito-Jensen, M. 2013. Revisiting the issue of elite capture of participatory initiatives. *World Development* 46:104–12.

Lunstrum, E. 2014. Green militarization: Anti-poaching efforts and the spatial contours of Kruger National Park. *Annals of the Association of American Geographers* 104(4):816–32.

Maryudi, A., Devkota, R. R., Schusser, C. et al. 2012. Back to basics: Considerations in evaluating the outcomes of community forestry. *Forest Policy and Economics* 14(1):1–5.

McDermott, C. L. 2014. REDDuced: From sustainability to legality to units of carbon – the search for common interests in international forest governance. *Environmental Science and Policy* 35:12–19, doi:10.1016/j.envsci.2012.08.012.

McDermott, M., Mahanty, S. and Schreckenberg, K. 2013. Examining equity: A multidimensional framework for assessing equity in payments for ecosystem services. *Environmental Science and Policy* 33:416–27. doi:10.1016/j.envsci.2012.10.006.

McNeely, J. A. 2003. Biodiversity, war, and tropical forests. *Journal of Sustainable Forestry* 16(3–4):1–20.

Melander, E., Pettersson, T. and Themnér, L. 2016. Organized violence, 1989–2015. *Journal of Peace Research* 53(5):727–42. doi:10.1177/0022343316663032.

Messer, K. D. 2010. Protecting endangered species: When are shoot-on-sight policies the only viable option to stop poaching? *Ecological Economics* 69:2334–2340.

MoEF (Ministry of Environment and Forestry) 2017. *Capaian percepatan perhutanan sosial 2017*. MoEF presentation at the tenure conference in Jakarta, Indonesia, October 2017.

Moeliono, M., Thuy, P. T. and Wong, G. Y. 2017. A comparison of policies in Vietnam and Indonesia. *Forest Policy and Society* 1(2):1–20.

Mutersbaugh, T. 2005. Fighting standards with standards: Harmonization, rents, and social accountability in certified agrofood networks. *Environment and Planning A* 37(11):2033–51. doi:10.1068/A37369.

Myers, R., Intarini, D., Sirait, M. T. and Maryudi, A. 2017. Claiming the forest: Inclusions and exclusions under Indonesia's 'new' forest policy on customary forest. *Land Use Policy* 66:205–17.

Nunan, F., Menton, M., McDermott, C. and Schreckenberg, K. 2018. Governing for ecosystem health and human wellbeing. In Schreckenberg, K., Mace, G. and Poudyal, M. (eds.) *Ecosystem services for poverty alleviation: Trade-offs and governance*. London: Routledge, pp. 159–73.

Obiri, B. D., Bright, G. A., McDonald, M. A., Anglaaere, L. and Cobbina, J. 2007. Financial analysis of shaded cocoa in Ghana. *Agroforestry Systems* 71(2):139–49.

OECD 2015. *OECD Public Governance Review. Indonesia Highlights*. Available at: www.oecd.org/gov/open-gov-review-indonesia.pdf (Accessed 12 January 2018).

Ordway, E. M. 2015. Political shifts and changing forests: Effects of armed conflict on forest conservation in Rwanda. *Global Ecology and Conservation* 3:448–60.

Paret, M. 2015. Violence and democracy in South Africa's community protests. *Review of African Political Economy* 42(143):107–23. doi:10.1080/03056244.2014.995163.

Peluso, D. 2018. Traversing the margins of corruption amidst informal economies in Amazonia. *Culture, Theory and Critique* 59:400–18.

Persha, L. and Andersson, K. 2014. Elite capture risk and mitigation in decentralized forest governance regimes. *Global Environmental Change* 24:265–76.

Pettersson, T. and Wallensteen, P. 2015. Armed conflicts, 1946–2014. *Journal of Peace Research* 52(4):536–50.

Price, S. V. 2003. War and tropical forests: Conservation in areas of armed conflict. *Journal of Sustainable Forestry Special Issue* 16:3–4.

Price, S., Donovan, D. and de Jong, W. 2007. Confronting conflict timber. In de Jong, W., Donovan, D., and Abe, K. (eds.) *Extreme conflicts and tropical forests*. Dordrecht, Netherlands: Springer, pp. 117–32.

Ravikumar, A., Sears, R. R., Cronkleton, P., Menton, M. and Pérez-Ojeda del Arco, M. 2017. Is small-scale agriculture really the main driver of deforestation in the Peruvian Amazon? Moving beyond the prevailing narrative. *Conservation Letters* 10(2):170–77.

Ribot, J. C., Lund, J. F. and Treue, T. 2010. Democratic decentralization in sub-Saharan Africa: Its contribution to forest management, livelihoods, and enfranchisement. *Environmental Conservation* 37(1):35–44.

Ribot, J. C. and Peluso, N. L. 2003. A theory of access. *Rural Sociology* 68:153–81.

Rosenau, J. N. 1995. Governance in the twenty-first century. *Global Governance* 1(1):13–43.

RRI (Rights and Resources Institute) 2013. *Tenure data and tools*. Available at: https://rightsandresources.org/ (Accessed 22 January 2018).

Rudel, T., DeFries, R., Asner, G. P. and Laurance, W. 2009. Changing drivers of deforestation and new opportunities for conservation. *Conservation Biology* 23(6):1396–1405.

Ruf, F. O. 2011. The myth of complex cocoa agroforests: The case of Ghana. *Human Ecology* 39(3):373–88.

Rustad, S., Rød, J. K., Larsen, W. and Gleditsch, N. P. 2008. Foliage and fighting: Forest resources and the onset, duration, and location of civil war. *Political Geography* (27):761–82.

Samii, C., Lisiecki, M., Kulkarni, P., Paler, L. and Chavis, L. 2015. Decentralised forest management for reducing deforestation and poverty in low- and middle- income countries: A systematic review. *Systematic Review 16*. London: International Initiative for Impact Evaluation (3ie).

Sánchez-Cuervo, A. M. and Aide, T. M. 2013. Consequences of the armed conflict, forced human displacement, and land abandonment on forest cover change in Colombia: A multi-scaled analysis. *Ecosystems* 160:1052–70.

Scheidel, A. and Work, C. 2018. Forest plantations and climate change discourses: New powers of 'green' grabbing in Cambodia. *Land Use Policy* 77:9–18.

Scott, J. C. 1998. *Seeing like a state: How certain schemes to improve the human condition have failed.* Yale Agrarian Studies. New Haven: Yale University Press.

Setyowati, A. 2012. Ensuring that women benefit from REDD+. *Unasylva* 239(63):57–62.

Shanee, N. and Shanee, S. 2016. Land trafficking, migration, and conservation in the 'no-man's land' of northeastern Peru. *Tropical Conservation Science* 9(4):1–16. doi:1940082916682957.

Sikor, T. 2013. REDD+: Justice effects of technical design. In *The justices and injustices of ecosystem services*. London: Routledge, pp. 47–68.

Slotin, J. and Elgin-Cossart, M. 2013. *Why would peace be controversial at the United Nations? Negotiations toward a post-2015 development framework.* New York University. New York: Center on International Cooperation.

Social Forestry Working Group 2017. *Strategi Percepatan Implementasi Perhutanan Sosiial, Terobosan dalam Mengelola Tantangan dan Hambatannya*. Paper presentation at Tenure Conference, October 2017.

Staver, C., de Jong, W., Kaimowitz, D. 2007. Nicaragua's frontier: The Bosawas Biosphere Reserve. In de Jong, W., Donovan, D. and Abe, K. (eds.) *Extreme conflicts and tropical forests.* Dordrecht: Springer., pp. 51–74.

Stewart, F. 2009. *Horizontal inequalities as a cause of conflict*. Bradford Development Lecture: November 2009. Available at: www.academia.edu/28567760/Horizontal_inequalities_as_a_cause_of_conflict (Accessed 27 July 2019).

Transparency International. 2016. *Corruption perception index 2016*. Available at: www.transparency.org (Accessed 15 January 2018).

UN 2015. *The challenge of sustaining peace: Report of the advisory group of experts for the 2015 review of the United Nations peacebuilding architecture*. New York: United Nations.

UN Climate Summit 2014. *New York Declaration on Forests Declaration and Action Agenda*. New York: United Nations.

Wade, A. S. I., Asase, A., Hadley, P. et al. 2010. Management strategies for maximizing carbon storage and tree species diversity in cocoa-growing landscapes. *Agriculture, Ecosystems & Environment* 138(3–4):324–34.

Wood, D. M. 2013. What is global surveillance? Towards a relational political economy of the global surveillant assemblage. *Geoforum* 49:317–26.

Woodman, J. 2014. India's rejection of Vedanta's bauxite mine is a victory for tribal rights. *The Guardian*. 14 January. Available at: www.theguardian.com/global-development/poverty-matters/2014/jan/14/india-rejection-vedanta-mine-victory-tribal-rights (Accessed 20 February 2019).

Chapter 17 SDG 17: Partnerships for the Goals – Focus on Forest Finance and Partnerships

David Humphreys*, Benjamin Singer, Kathleen McGinley, Roy Smith, Jessica Budds, Mónica Gabay, Shonil Bhagwat, Wil de Jong, Helen Newing, Charlotte Cross and Poshendra Satyal

Key Points

- Funding for forests from official development assistance and other sources has trended upward since 2000, providing reason for cautious optimism. However, finance for REDD+ is in decline.
- Private-sector investment remains important. Impact investment, which aims to solve pressing environmental and social problems, could make a significant contribution to the sustainability agenda.
- Not all sustainable development finance promotes forest conservation. SDG 2 (Zero Hunger) aims to increase funding for agricultural production, which can incentivise the conversion of forests to farmland.
- The policy of zero net deforestation is leading to some important partnerships, including with the financial sector, that aim to ensure deforestation-free commodity supply chains of key agricultural commodities.
- Partnerships for sustainable development exist within a neoliberal global economic order, in which net financial flows from the Global South to the Global North negate financial flows for sustainable development.

17.1 Introduction

Successful realisation of SDG 17 is vital for attaining the other SDGs, all of which depend on securing means of implementation and forging durable partnerships for sustainable development. It is one of the most comprehensive goals as the means of implementation encompass finance, information and communication technology, capacity-building, international trade and data monitoring. SDG 17 contains a broad range of targets and indicators (see Table 17.1), some of which are analysed here. To examine the complex relationships between SDG 17 and forests, an extensive literature review and

* Lead author.

synthesis was undertaken to identify policy papers and analyses on forest-related means of implementation and partnerships for sustainable development. Websites of actors working in these areas were trawled and links followed to identify additional source material and 'grey literature'.

This chapter explores ways to strengthen the means of implementation. Section 17.2 focuses in depth on financial assistance and partnerships. We do not focus on international trade, which is examined in Chapter 10. Section 17.3 examines the distinction between sustainable and unsustainable forest

Table 17.1 SDG 17 targets

Target 17.1:	Strengthen domestic resource mobilisation, including through international support to developing countries, to improve domestic capacity for tax and other revenue collection
Target 17.2:	Developed countries to implement fully their official development assistance commitments, including the commitment by many developed countries to achieve the target of 0.7 per cent of gross national income for official development assistance (ODA/GNI) to developing countries and 0.15 to 0.20 per cent of ODA/GNI to least-developed countries
Target 17.3:	Mobilise additional financial resources for developing countries from multiple sources
Target 17.4:	Assist developing countries in attaining long-term debt sustainability through coordinated policies aimed at fostering debt financing, debt relief and debt restructuring, as appropriate, and address the external debt of highly indebted poor countries to reduce debt distress
Target 17.5:	Adopt and implement investment promotion regimes for least-developed countries
Target 17.6:	Enhance North–South, South–South and triangular regional and international cooperation on and access to science, technology and innovation, and enhance knowledge-sharing on mutually agreed terms
Target 17.7:	Promote the development, transfer, dissemination and diffusion of environmentally sound technologies to developing countries on favourable terms
Target 17.8:	Fully operationalise the technology bank and science, technology and innovation capacity-building mechanism for least-developed countries by 2017 and enhance the use of enabling technology, in particular information and communications technology

Table 17.1 (cont.)

Target 17.9: Enhance international support for implementing effective and targeted capacity-building in developing countries to support national plans to implement all the Sustainable Development Goals
Target 17.10: Promote a universal, rules-based, open, non-discriminatory and equitable multilateral trading system under the World Trade Organisation
Target 17.11: Significantly increase the exports of developing countries, in particular with a view to doubling the least-developed countries' share of global exports by 2020
Target 17.12: Realise timely implementation of duty-free and quota-free market access on a lasting basis for all least-developed countries, consistent with World Trade Organisation decisions
Target 17.13: Enhance global macroeconomic stability, including through policy coordination and policy coherence
Target 17.14: Enhance policy coherence for sustainable development
Target 17.15: Respect each country's policy space and leadership to establish and implement policies for poverty eradication and sustainable development
Target 17.16: Enhance the Global Partnership for Sustainable Development, complemented by multi-stakeholder partnerships that mobilise and share knowledge, expertise, technology and financial resources
Target 17.17: Encourage and promote effective public, public–private and civil society partnerships, building on the experience and resourcing strategies of partnerships
Target 17.18: By 2020, enhance capacity-building support to developing countries to increase significantly the availability of high-quality, timely and reliable disaggregated data
Target 17.19: By 2030, build on existing initiatives to develop measurements of progress on sustainable development that complement gross domestic product, and support statistical capacity-building in developing countries
Source: Adapted from https://sustainabledevelopment.un.org/SDG17

financing and their different impacts on forests. In particular, subsidies that incentivise the expansion of agricultural land can have a deleterious impact on forests. Section 17.4 addresses zero net deforestation (ZND), examining whether tensions exist among different SDGs. We argue that careful attention should be paid to the agro-forestry interface, with sustainable agriculture a prerequisite for achieving ZND. Building on this, Section 17.5 looks at some of the more innovative partnership arrangements that promote sustainable forest-related development. Section 17.6 briefly examines the broader structure of global economic governance, and how this negates efforts to increase the means of implementation in developing countries. Section 17.7 presents the conclusions.

17.2 Strengthening the Means of Implementation through Increased Financing

Target 17.3 aims to 'Mobilise additional financial resources for developing countries from multiple sources'. The United Nations Conference on Trade and Development (UNCTAD) estimates that achieving the SDGs will require investments in developed and developing countries of USD 5–7 trillion per year (UNCTAD 2014). For developing countries, the estimate is USD 3.3–4.5 trillion. At today's level of investment – public and private – an annual shortfall of USD 2.5 trillion is estimated for developing countries. Hence, strengthening the means of implementation, including implementing the Addis Ababa Action Agenda on financing for development (endorsed by the UN General Assembly in 2015), is essential for achieving the SDGs. In terms of forests, SDG 17 promotes the need to increase financing levels for sustainable forest management (SFM) and to enhance cooperation among public, private and non-governmental stakeholders. It has been estimated that halving deforestation rates in developing countries will cost USD 20 billion per year (Boucher 2008, Forest Trends 2017).

TWO COMPLEMENTARY TYPOLOGIES

In the absence of a universally recognised definition of SFM, quantifying financing levels is a daunting challenge, further complicated by the lack of financial statistics available for sustainable investments in general (Holopainen and Wit 2008). Singer (2016) suggests two typologies of SFM financing. The first, based on *sources*, is inspired by the fivefold categorisation of the United Nations Sustainable Development Agenda: international public financing, domestic public financing, international private financing, domestic private financing and – as a residual category – blended and innovative financing (Figure 17.1).

The second typology is based on the cross-sectoral nature of forests, in particular the distinction between forest financing and SFM financing. *Forest financing* refers to all financial sources that benefit the forest sector. Many of these sources, however, do not support sustainable forms of

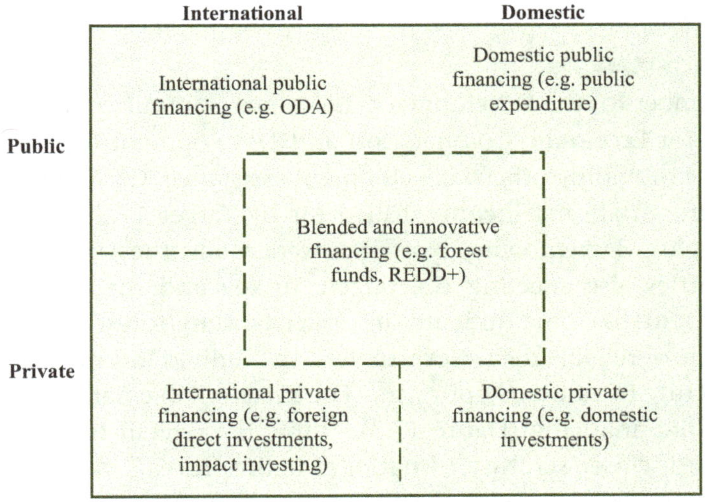

Figure 17.1 SFM financing by type of flow. Source: UN 2014; see also Singer 2016.

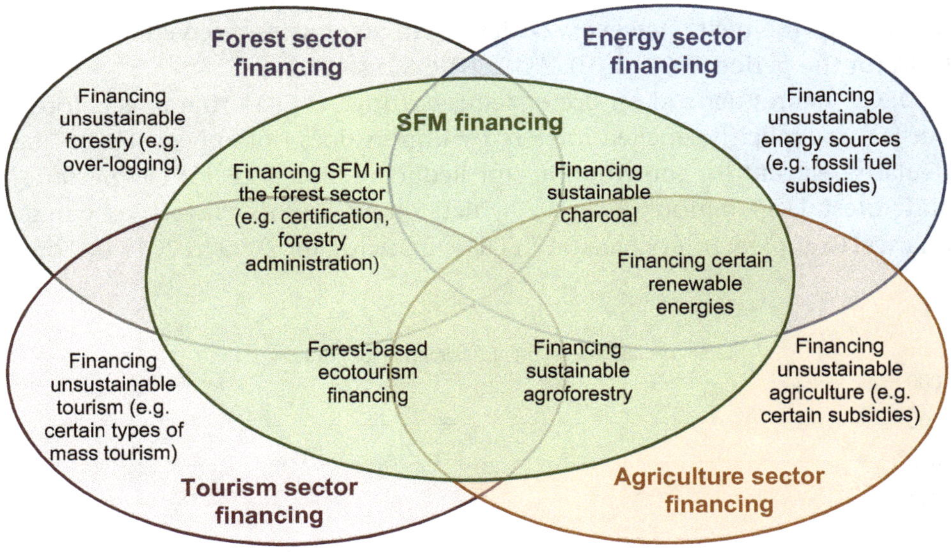

Figure 17.2 SFM financing as a cross-sectoral category. For simplicity, additional sectors which may impact upon SFM financing are not depicted; neither is financing in other sectors that do not impact upon SFM either positively or negatively. Source: Singer 2016.

forest management: some forest sector investments – whether public, private, domestic or international – incentivise unsustainable management, such as overharvesting. *SFM financing* is a cross-sectoral category that overlaps partially with forest financing. While SFM financing comes largely from the forest sector, there are also sources outside this sector (Figure 17.2).

17.2.1 Trends in Increased Finance

INTERNATIONAL PUBLIC FINANCING

In 2017 only five members of the Development Assistance Committee (DAC) of the Organisation for Economic Co-operation and Development (OECD) met the UN's target of providing official development assistance (ODA) equal to 0.7 per cent of gross national income, called for by Target 17.2. These countries were Denmark, Luxembourg, Norway, Sweden and the UK, with two non-DAC countries also meeting the target: Turkey and the United Arab Emirates (OECD 2018a). One difficulty in understanding forest-related financing is a shortage of reliable data. Nevertheless, the limited information available paints a picture of cautious optimism. The most reliable data focus on international public financing (Figure 17.3). While the overall trend in recent years is upward, closer scrutiny of national figures reveals volatility over space and time (Singer 2016). While only a handful of countries receives most forestry ODA each year, which countries these are changes over time and thus forestry ODA per country may increase or decrease several-fold from one year to the next. However, 27 developing countries received no forestry ODA for the period 2002–2010 (AGF 2012, Singer 2016).

OECD figures depend on donor self-reporting, so ODA that affects forests but is not explicitly labelled forestry by donors does not appear. This is particularly relevant for contributions for Reducing Emissions for Deforestation and forest Degradation (REDD+), which many donors classify as climate finance. Complementary data on REDD+ financing for 2009 to 2014 by Forest

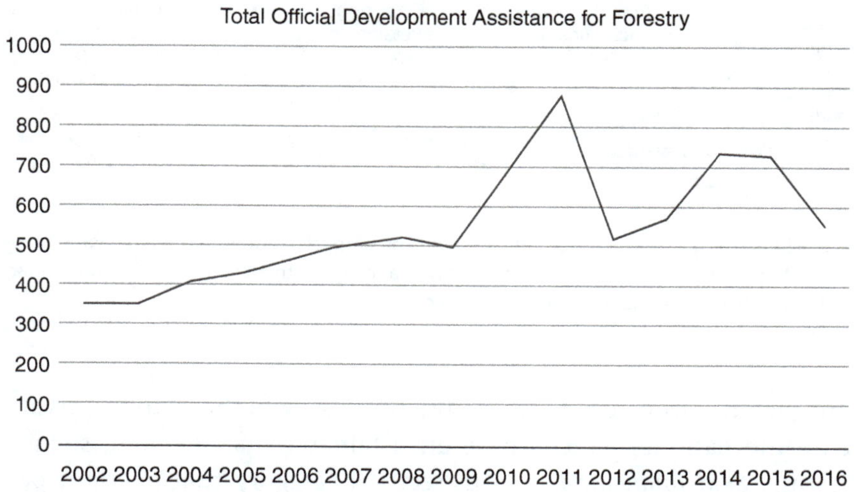

Figure 17.3 Global ODA for forestry (gross disbursements) 2002–2016 in USD millions of constant 2016 USD. Source: OECD 2018b.

Trends show significant variability: USD 1.6 billion in 2009, but just USD 0.3 billion in 2014 (Silva-Chávez et al. 2015).

ODA for the global forest sector, including assistance for forestry development, education and research, generally increased, with annual fluctuations between 2000 and 2015, with a low of USD 400 million in the early 2000s and a high of USD 1.15 billion in 2011 (OECD 2017a, 2017b). An uptick in commitments occurred in the early 2010s, associated with the fast-start climate finance promised by developed countries in the climate negotiations and increased funding commitments during the International Year of Forests in 2011 (OECD 2017a). At the Paris climate summit in 2015, Germany, Norway and the UK collectively committed to provide more than USD 5 billion from 2015 to 2020 to forest countries demonstrating verified emission reductions, with the summit agreeing on a collective goal of USD 100 billion by 2025 (Nakhooda et al. 2016).

Commitments, however, do not always translate into dollars invested on the ground. Since 2011, REDD+ financing has slowly dwindled. The carbon bubble that led to speculation among private investors was short-lived. The Forest Carbon Partnership Facility (FCPF), which helps developing countries implement REDD+ through financial and technical assistance, has 17 financial contributors with total commitments of more than USD 1.1 billion (FCPF 2017). However, the FCPF, UN-REDD and other REDD+ funders have been criticised for promoting a narrow focus on just one forest public good, namely carbon sequestration, while neglecting others – what has been termed the 'climatisation' of the global forest regime (Singer and Giessen 2017). This reflects a tension between SDG 13 (Climate Action) and SDG 15 (Life on Land). By the end of 2016, 73 per cent of the 2008–2016 commitments had been deposited, of which just 36 per cent were approved for project disbursement (CFU 2017). This gap between pledged and disbursed funds reflects the challenges in moving beyond capacity-building to implementation and the impact of the global financial crisis on public-sector finances (Watson et al. 2016, Norman and Nakhooda 2014).

Verchot (2015) compared the popularity of REDD+ with Gartner's Hype Cycle; it has gone from a 'peak of inflated expectations' into a 'trough of disillusionment'. While the Paris Agreement (UN 2015, Article 5.2) is the first international agreement to recognise REDD+, subsequent funding pledges have been slow to materialise. In 2017 the Green Climate Fund launched a USD 500 million request for proposals on REDD+ implementation. The pledge by developed countries at COP15 in Copenhagen to mobilise USD 100 billion in climate finance by 2020, however, seems unlikely to be fulfilled (Roberts and Weikmans 2016). The likelihood of leveraging sufficient funding to meet REDD+ implementation requirements is uncertain.

INTERNATIONAL PRIVATE FINANCING

A consistent REDD+ figure over the years has been the low proportion of private REDD+ finance – approximately 10 per cent (Silva-Chávez et al. 2015, UN Environment 2016) – confirming that the central role once envisaged for the private sector has failed to materialise, leaving public donors at the forefront of REDD+ financing. Data on SFM financing from private sources are scarce. The World Bank (2008) estimated annual private investment in the forestry sector in developing countries at nearly USD 15 billion – i.e. 40 times the forestry ODA disbursed that same year. Castrén et al. (2014) calculated private investment in forest plantations to be USD 1.8 billion per annum; no systematic data were found on private investments in tropical natural forests. Because these figures include investments along the broad spectrum of forest management, from unsustainable to sustainable, it is near impossible to place a figure on the proportion of these investments that would support SDG 17.

From 2009 to 2014 the private sector gave USD 35 million to support national REDD+ initiatives, with a further USD 381 million for carbon offset projects through the voluntary carbon market. During this period the private sector contributed approximately 10 per cent of REDD+ finance (Environmental Defense Fund and Forest Trends 2018). Government spending on REDD+ can help to leverage additional private-sector finance through public–private partnerships. An area of increasing importance, and one that can help meet Target 17.5, is *impact investment* – namely, an investment made with the specific intention to help solve the world's most pressing environmental and social problems while also generating financial returns for investors. According to the Business and Sustainable Development Commission (2017), achieving the SDGs could provide USD 12 trillion of investment opportunities and create 380 million new jobs by 2030. Examples of impact investment funds include the Mirova Land Degradation Neutrality Fund, launched in 2017, which aims to provide USD 300 million for SDG 15, including sustainable agriculture and forestry. The fund involves contributions from the private sector and government donors (Global Impact Investor Survey 2018). Institutional investors – now the main market participants in developing countries, with more than a thousand pension funds, foundations, insurance companies and others (DANA 2011, Glauner et al. 2012) – show increasing interest in investing in SFM.

Forest Trends (2017) note a dramatic recent increase in conservation investments, intended to generate both a financial return and a measurable environmental result. From an average of USD 0.2 billion for the period 2004–2008, annual private investments increased tenfold by 2015 to USD 2 billion. Of this, 80 per cent went to sustainable food and fibre, 18.5 per cent to habitat conservation and 2.5 per cent to water quality and quantity. These figures indicate the growing interest of some private investors in sustainability, mainly in the USA and to a lesser extent in Europe, although investments in emerging countries are increasing.

Despite this, most private finance in developing countries continues to be directed at developing forest plantations; Brazil is an example. Although they represent only 1.3 per cent of the country's forests, plantations produce 78 per cent of Brazil's sawlog and veneer (Tomaselli et al. 2012). Plantations do not necessarily provide high returns, although risk-adjusted returns are higher than for natural forests. The main reason for investing in plantations is that they are much more productive systems and contribute more quickly to closing the fibre gap. Tropical forests, particularly natural forests, continue to suffer from low levels of sustainable private investment due to macroeconomic instability, weak governance systems and a lack of enabling conditions, including:

- *Natural forest policies and legislation:* Contradictory pressures from timber industries, public opinion and international organisations lead to incoherent policies and laws.
- *Land tenure:* Lack of clear tenure adds to the risks posed by political and economic volatility, discouraging domestic and foreign investors.
- *Low risk-adjusted returns:* While timber exploitation in tropical natural forests is a lucrative business, profits fall once SFM is applied because of the low productivity and relatively high management costs of tropical natural forests.
- *Reputation and information:* The technical complexity of the timber sector combined with the continued (and incorrect) portrayal of the sector as the main cause of deforestation may discourage new investments.

The idea of using private capital to achieve forest-related SDGs is not without controversy. Writing about the 'corporate capture of biodiversity', Lovera (2017) objects to the role of corporations. She argues that they offer financial support to the SDGs to conceal their attempts to undermine them, since safeguards and standards would limit profitability. However, as private-sector corporations are publicly pressured to adopt, or altruistically seek, sustainability certification, they are likely to become an increasingly important partner in achieving forest-related SDGs. Furthermore, corporate–community partnerships can facilitate market access for commodities in ways that support community-driven forest development (Katila et al. 2017).

DOMESTIC FINANCING

Target 17.1 – strengthen domestic resource mobilisation – can be met through improved tax collection, tax incentives, subsidies and payments for environmental services (PES). Whether domestic financing supports SFM depends in

Box 17.1 Domestic Forest Financing in Cameroon

In Cameroon, debates over forest revenue focus on domestic financing. Cameroon has a thriving timber industry, yet until the 1990s the state received minimal revenue as company profits were generally underdeclared and repatriated abroad (Eba'a Atyi et al. 2013). In 1994, the Forest Law introduced major changes, including an auction system for allocating timber concessions and a tax increase on timber production. Revenue increased fivefold before settling to an annual USD 52–63 million (Karsenty et al. 2006). Revenue distribution, however, has been more problematic. As stipulated in the 1994 law, half of the annual area fee goes to local municipalities and communities, yet poverty alleviation has been minimal due to financial mismanagement. The case highlights the vast potential of tax reforms to increase domestic financing, and the need for effective allocation of tax revenues to receive equal attention.

large part on fair and effective implementation as well as other potentially countervailing policies. Domestic finance, whether public or private, is difficult to track because it varies widely among countries, with data compilation depending on the capacity and reliability of national statistics agencies. This may explain why, despite being identified as a critical source of financing for development (UN 2015), it continues to receive limited attention from analysts and decision makers. However, domestic private-sector financing is the most important source of forest-related investment in many Latin American countries, especially Brazil. In Africa, an important example is Cameroon (Box 17.1).

Many countries in the tropics, and elsewhere, have systems for allocating public timber resources, with harvest and/or area-based tax schemes intended to generate forest revenue for the state. However, these exhibit varying degrees of success in terms of rent capture and the equitable and effective distribution and use of funds, particularly in terms of activities that might be associated with the SDGs. Corruption and bribery often thwart potentially positive outcomes.

17.2.2 The Bigger Picture: Coherence and Coordination

This brief overview leads to an obvious conclusion: increasing levels of financing – whether public or private, domestic or international – is only half the battle. Effective coherence, as called for by Target 17.14, is key to ensuring that financing is allocated to optimise SDG implementation.

Public financing alone will not realise the SDGs. Private financing can help close the gap, but it is generally attracted to activities with high returns, declining once these returns fall below a certain threshold due to low productivity or high risk. Public financing could leverage additional private finance in two ways: (1) focus on forest-related areas of SFM with low returns, such as conservation and community forestry that can have a positive effect on forests and people; and (2) guarantee a minimum return for private investments to compensate for low returns or high risks. Public financing is also vital for: (1) creating the enabling conditions for sustainability (e.g. related to land and governance reform, jurisdictional planning processes and capacity-building); (2) developing and piloting new approaches that, once established, may attract private investments; and (3) facilitating new partnerships.

One means of coordinating different sources of financing is national forest financing. The United Nations Forum on Forests (UNFF) has developed a four step strategy for SFM:

1. *Mapping priorities and needs:* Identify priorities in terms of goals, objectives and financing needs.

2. *Mapping existing and potential sources of financing:* Identify all existing sources and potential new financing sources, such as new taxes or payments for ecosystem services.

3. *Matching priorities and needs with sources:* Match objectives and activities with different financing sources according to criteria such as donor preferences, profitability and risk. Activities can be funded by more than one source.

4. *Creating a roadmap for mobilising finance:* Match each activity with one or more stakeholder(s) responsible for implementation. Budget for the financing needs quantified in Step 1 (Singer 2017).

Depending on the level of country ownership and donor support, national forest-finance strategies could form an effective tool in mobilising finance and implementing the SDGs. Where financing shortfalls are identified, there needs to be a mechanism for prioritising resource allocation.

17.3 Sustainable versus Unsustainable Financing

One cannot assess SFM financing without comparing it with financing for unsustainable forms of land management. The impacts of other land uses on forests are well documented, such as Myers' (1981) hamburger connection. In recent years, researchers have started quantifying these cross-sectoral linkages. This is particularly important since it relates to the trade-offs among SDGs, explored in this chapter and elsewhere in this book. Lawson et al. (2014) calculate that commercial agriculture caused more than two-thirds of illegally cleared forests between 2010 and 2012, with Brazil and Indonesia accounting for 71 per cent of the global tropical forest area illegally converted to commercial agriculture. Most tropical deforestation is driven by four forest-risk commodities, namely palm oil, soy, cattle and timber products (including paper). Persson et al. (2014) estimate that between 2000 and 2009 these four commodities accounted for a third of tropical deforestation across eight countries (Argentina, Bolivia, Brazil, Paraguay, Democratic Republic of Congo, Indonesia, Malaysia and Papua New Guinea).

If national governments and international organisations are to reduce forest loss, they clearly must counter the deforesting effects of producing these four forest-risk commodities. An Overseas Development Institute study reveals the difficulty of doing so (McFarland et al. 2015). With a focus on Brazil and Indonesia, which together lost 78 million ha of forest between 1990 and 2010 (FAO 2010), the authors calculate that public subsidies to beef and soy in Brazil and to palm oil and timber in Indonesia totalled USD 47.242 billion per year between 2009 and 2012. By comparison, in a period when REDD+ funding was at an all-time high, both countries received a combined USD 323 million a year for REDD+ and a mere USD 26 million a year in forestry ODA (McFarland et al. 2015). The study compares national-level agricultural subsidies against international finance. It does not include national public-sector forest finance or consider the extent to which forest subsidies were included within agricultural subsidies.

Even so, and assuming that REDD+ financing and forestry ODA are additional, the main drivers of deforestation in both countries received as much as a staggering 136 times more domestic public funding than international public finance for forests over this period. Adding private investments to these highly lucrative agricultural commodities would further increase this figure. This is not to say that all cattle, soy, palm oil and timber production, and certainly not all related subsidies, generate deforestation (McFarland et al. 2015). Some agricultural subsidies also tackle social and environmental issues, including forest conservation and support for sustainability (e.g. through crop intensification and land-use change restrictions). Nevertheless, if even a fraction of the financial weight of subsidies for these commodities were to generate deforestation pressures, it would dwarf public international financial support to SFM in both countries (Figure 17.4, Table 17.2 and Box 17.2).

This is especially relevant for two SDGs with both potential synergies and potential conflicts with forestry. First, SDG 2 (Zero Hunger) includes Target 2.A, to 'increase investment ... to enhance agricultural productive capacity in developing countries, in particular in least developed countries'. The abovementioned subsidies could lead to forest loss, although this is less a problem

Table 17.2 Comparing REDD+ finance received with domestic expenditure on biofuel and agriculture subsidies (average annual USD million)

	REDD+ Finance (2006–2014 Annual Average)	Agricultural Subsidies (2010–2012 Annual Average)	Biofuel Subsidies (2009)
Brazil	158	11 082	2 700
Chile	0	709	n/a
China	9	160 023	500
Indonesia	165	27 072	79
Mexico	12	7 880	n/a
Total	**344**	**206 766**	**3 279**

Sources: REDD+ finance (Norman and Nahkooda 2014), Agricultural subsidies (OECD 2014a), Biofuel subsidies (Gerasimchuk et al. 2012).

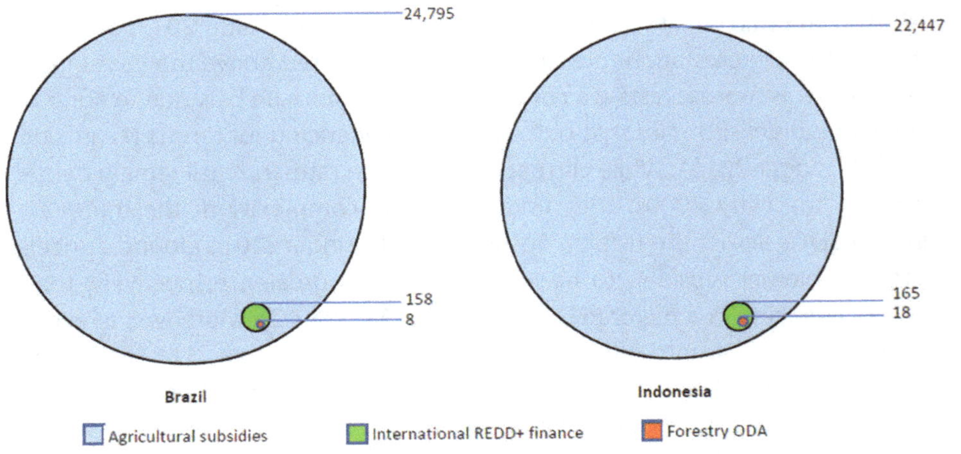

Figure 17.4 Annual subsidies to agricultural commodities (beef and soy in Brazil; palm oil and timber in Indonesia) compared to annual international REDD+ finance and forestry ODA in Brazil and Indonesia, 2009–2012 (USD million). Sources: McFarland et al. 2015, OECD 2017b.

of unsustainable forest management and more one of unsustainable land use, illustrating that progress to achieve policy coherence for sustainable development as called for by Target 17.14 has been limited.

Second, SDG 7 aims to 'ensure access to affordable, reliable, sustainable and modern energy for all' (without defining *sustainable*). While it does not

Box 17.2 Forest Financing in Latin America

Many Latin America and Caribbean (LAC) countries are active in REDD+. Twenty-one LAC countries accounted for 56 per cent (USD 819 million) of the total funds approved for REDD+ project implementation globally between 2008 and 2016 (USD 1.45 billion) (Watson et al. 2016, CFU 2017). Brazil alone accounted for 45 per cent of REDD+ funding approved for implementation during this period (69 per cent of the LAC total), followed by Mexico, Colombia, Peru and Chile (CFU 2017). Brazil's Amazon Fund has received and disbursed the largest share of REDD+ financing (Amazon Fund 2017). By the end of 2016, it had received USD 1.747 billion in pledged funding, including more than USD 1 billion from Norway through a performance-based agreement to slow forest loss and reduce greenhouse gas emissions from deforestation. By the end of 2016, USD 1.037 billion of total pledged funds had been deposited to the fund, of which USD 576 million had been disbursed to projects (CFU 2017). Amazon Fund policies to reduce forest loss and enhance forest sustainability are credited with measurable improvements in the forest sector, with deforestation declining from 27 772 km^2 in 2004 to 4403 km^2 in 2012, but are associated with significant declines in agricultural commodity prices in the mid-to-late 2000s (Arima et al. 2014, Fearnside 2017, Nepstad et al. 2014). Deforestation rates in Brazil have increased above the lows of the early 2010s as more forests are converted to agriculture, in part due to agriculture and biofuel subsidies that outpace climate financing for forests (Fearnside 2017, Kissinger 2015). While current deforestation rates in Brazil remain below those of the early 2000s, they underscore the complexity of the trade-offs among SDGs across the tropics. For example, Kissinger (2015) found agriculture and biofuel subsidies to be 600 and 9 times greater, respectively, than REDD+ financing in 5 major REDD+ countries (Table 17.2). Moreover, a review of more than 40 countries found that REDD+ readiness projects rarely include specific actions to address intersectoral conflicts relating to two or more SDGs, or to eliminate subsidies incentivising forest loss (Salvini et al. 2014).

explicitly mention them, biofuels are often considered sustainable since the carbon emitted by their combustion is theoretically sequestered when crops consumed are replaced by new crop growth for future consumption. In this respect, biofuels can help achieve SDG 7 on affordable and green energy since they theoretically provide additional energy to meet growing demand while aiding climate-change mitigation. However, their large-scale adoption – e.g. in Brazil (sugarcane) and Indonesia (palm oil) – could have a significant impact on forest cover and on forest peoples in tropical countries (Acheampong et al. 2017).

This section has assembled a body of evidence from several sources that leads to some important conclusions for achieving SDG 17. In particular, the financial incentives for the conversion of forests to alternative land uses such as agriculture are a fraction of those available for forest conservation and sustainable management through public and private international and domestic financing. The relationship between SFM and agricultural production is relevant to policies for ZND, to which attention now turns.

17.4 Zero Net Deforestation Commitments

While agriculture may contribute to rural economic development, food security and other SDGs, forests conversion into agricultural land remains the leading cause of deforestation in many countries. To counter this, a growing number of companies and governments have committed to eliminate deforestation from production processes and supply chains through ZND over the last decade. ZND seeks to secure production of agricultural commodities without deforesting primary forests, although deforestation that is compensated by afforestation planting elsewhere may be acceptable. Through examining ZND, this section considers the role that the agricultural and financial sectors can play in promoting SDG 17.

ZND emerged in 2008 during the Bonn Conference of Parties to the Convention on Biological Diversity when the World Wildlife Fund (WWF) led a campaign supported by 67 countries calling for ZND by 2020 (WWF 2009). WWF's commitment to ZND is significant, as the organisation has a history of forging innovative partnerships to promote ambitious targets and international rules that other actors later adopt. The WWF was one of the policy leaders behind the creation of the Forest Stewardship Council (FSC), with the World Bank later adopting operational policies that drew directly from FSC principles (Humphreys 2006). ZND commitments require companies to identify the sources of their commodities and make supply chains traceable and transparent.

Global Forest Watch Commodities supports efforts to monitor forest activity in commodity supply chains (WRI 2017b). The Consumer Goods Forum, with some 400 member companies, has also backed ZND. The Tropical Forest Alliance 2020, launched in 2012 at Rio+20 as a global public–private partnership to reduce tropical deforestation, reduce greenhouse emissions, improve smallholder livelihoods and conserve natural habitats, has also pledged support for ZND. The New York Declaration on Forests of 2015 includes commitments from several governments and companies to remove deforestation from commodity supply chains. The idea of ZND commodity chains has grown in popularity among donors. In 2015 the Global Environment Facility (GEF) announced a USD 500 million programme to remove deforestation from commodity supply chains (GEF 2015). In 2017, Norway created a USD 400 million fund to support this initiative, aiming to raise more than USD 1.6 billion in deforestation-free agricultural investments (GEF 2017).

While ZND commitments from the private sector have transformative potential, many companies publicly committed to ZND are failing to demand that their suppliers adopt a ZND policy. Many businesses with deforestation-related commitments lack time-bound, actionable plans, and the majority do not publicly report on compliance with their own policies, making independent verification of progress difficult (Climate Focus 2016). Some business targets are aspirational only. Donofrio et al. (2017) analyse 760 commitments by 447 companies to reduce deforestation in palm oil, soy, cattle, timber and pulp supply chains. Difficulties in measuring and meeting stated goals, including lack of corporate transparency (e.g. withholding information), led to about a quarter of commitments being either dormant or delayed. Furthermore, the voluntary self-regulatory nature of many commitments means that implementation gaps may emerge (Jopke and Schoneveld 2018).

On a global scale about 27 per cent of deforestation is caused by permanent land-use change for commodity production (Curtis et al. 2018). In Latin America some two-thirds of deforestation is driven by commercial agriculture (Kissinger et al. 2012). In particular, production of the four forest-risk commodities (Section 17.3) has caused extensive tropical deforestation and is the source of widespread conflict between agriculture companies and local people (Abram et al. 2017).

Achieving ZND requires an agriculture sector based on deforestation-free commodity chains, particularly for the forest-risk commodities. Approximately 40 per cent of global demand for the four risk commodities is accounted for by emerging producer-consumers (Brazil and Indonesia) and emerging major importers (China and India) (TFA 2018b). Effective SFM thus requires the active support of the governments of these four countries and their leading agricultural corporations. Without robust and verifiable, sustainable sourcing of these risk commodities, future expansion in their international trade will generate further deforestation pressures. This is particularly pressing given the emphasis in Target 17.11 to significantly increase the developing countries' exports, which would enable developing countries to increase hard currency earnings. Expanding the production of agricultural commodities could contribute towards SDG 2 (Zero Hunger) but would conflict with SDG 15 (Life on Land). It seems clear that ZND targets cannot be achieved unless integrated action is taken at the agriculture-forestry interface.

Global Canopy selected 250 companies, 150 financial institutions and other actors (the 'Forest 500') that are at risk of being linked to tropical deforestation through potential exposure to forest-risk commodity chains and that have the greatest influence within the political economy of tropical deforestation. Their report on the Forest 500 shows that progress towards ZND has been limited. For example, although cattle production is the most important forest-risk commodity, only 17 per cent of cattle companies surveyed have a policy for forest protection, while just 8 of the 150 financial institutions surveyed have a policy for all four forest-risk commodities (Rogerson 2017). A UN Environment Programme (UNEP) report found that none of the companies it surveyed have a process to quantify the risks associated with investment portfolios in forest-related agricultural commodities (UNEP 2015). The importance of the forestry–agriculture interface suggests that the transition to ZND requires a dramatic shift in investments from the drivers of deforestation towards sustainable agriculture and forestry (Climate Focus 2017).

While banks and other financial institutions that lend to or invest in companies engaged in harvesting and trading in forest-risk commodities are themselves exposed to the financial and reputational risks of deforestation, only a limited number have made progress in integrating these risks into their management structures. The important role of investment suggests a crucial role for banks and investment companies. Uptake of certification – forest certification such as the FSC and agricultural products such as the Roundtable on Sustainable Palm Oil (RSPO) – is low in many tropical areas. However, financial institutions may foster further uptake by insisting that client companies be members of certification schemes or that the schemes be used to set minimum standards for loans (TFA 2018a). Financial institutions need to look beyond reputational risks and better understand how funding forest commodities can expose them to financial risks, especially given the growing interest of many institutional investors in impact investment (TFA 2018a). Options include introducing new financial products linked to ZND, such as green bonds and sustainable landscape bonds.

Banking-sector engagement in forest issues includes the Banking Environment Initiative (BEI), a University of Cambridge initiative of 12 leading banks that seeks to direct investment capital towards ZND business models. The BEI has partnered with the Consumer Goods Forum on the Soft Commodities Compact, which promotes partnerships between agricultural businesses and the financial sector to transform commodity supply chains of the forest-risk commodities to achieve ZND (Cambridge Institute for Sustainability Leadership 2018). Another important initiative is the Principles for Responsible Investment, which includes an Investor Initiative for Responsible Forests focused mainly on cattle supply chains (PRI 2018).

These examples suggest the need to broaden sustainable development partnerships to involve new actors, including banks and investment companies, regional and national governments, and national and international non-governmental organisations (NGOs). The active promotion of deforestation-free commodity chains by the financial sector would mean that companies continuing to trade in products produced by deforestation would find it difficult to raise capital. Governmental involvement may be necessary to offset financial incentives discouraging the sustainable sourcing of agricultural products. In China, for example, soybean producers wishing to adopt RSPO standards may face a cost increase of USD 3–4 per metric tonne, a significant cost in a country where profit margins are thin (TFA 2018b). Government underwriting of sustainability standards (e.g. subsidies) may help overcome such market barriers.

The financial sector can thus play an important role in incentivising deforestation-free commodity chains. Noting that the most important indirect causes of deforestation are found in global financial and commodity markets, the World Resources Institute (WRI) proposes that more effective use be made of financial data and corporate governance to hold corporations accountable for how well they implement their supply chain commitments, including ZND and elimination of illegal deforestation. This requires greater corporate transparency, including providing access to relevant data (Graham et al. 2018). Financial markets fail to distinguish between commodities produced according to ZND principles and those generating a deforestation footprint. There are no 'deforestation free' commodities listed on the world's financial markets, limiting both the incentives for companies to produce such commodities and gain a price premium from them and the opportunities for responsible investors to reward companies committed to ZND (Graham et al 2018). Many companies will find it disadvantageous to market deforestation-free products when doing so increases their costs and erodes their competitive advantage relative to more unscrupulous businesses. The two Amsterdam Declarations of 2015 – on deforestation and sustainable palm oil – aim to address this problem by generating demand for sustainable commodities and supporting the implementation of private-sector commitments to deforestation-free commodity supply chains (Partnership for Forests 2017, 2018).

One mechanism that could enable agricultural businesses to internalise the financial risks of producing deforestation-free products is a new global data platform on corporate data and forest risks. This could be structured around the Accountability Framework, which provides a set of definitions and core principles for establishing, implementing and monitoring ethical supply chain commitments (Accountability Framework 2018). Such a database would document the financial risks of investing in commodities produced through deforestation. It could also document the procedures that key financial institutions expect from client businesses involved in trading forest-risk commodities and could collate company data on the performance of investors in financing deforestation-free commodity chains. This would be consistent with Target 17.8 to enhance the use of enabling technologies, including information technology for sustainable development.

The High Carbon Stock Approach (HCSA) is a multi-stakeholder initiative designed to standardise the implementation of commitments to ZND in palm oil, pulp and paper. Its members include eight of the world's largest palm oil, pulp and paper companies, together with consumer goods manufacturers, environmental and human rights organisations, and technical organisations, including the Union of Concerned Scientists. The HCSA offers a standard approach for fulfilling ZND commitments, including a field methodology for identifying forests with a high carbon stock (HCS forests) that must be conserved. HCSA also has protocols related to the rights and livelihoods of Indigenous peoples and local communities, including the need for free, prior and informed consent (FPIC). The commitments enshrined in the HCSA are impressive but raise issues concerning the interplay between environmental protection and community rights. For example: What should happen when communities want plantation development but the HCSA requirements make it unacceptable? How can conservation of HCSA forests on community lands be reconciled with the right to FPIC? Should there be restrictions on local people's access to and use of HCS forests? If so, what incentives and benefits are there for communities to collaborate in conservation (Colchester et al. 2016)? It is not yet clear whether a system of this kind can provide a significant level of accountability, but in the absence of legislation, the HCSA has the potential to advance norms on acceptable social and environmental practice. The case of ZND makes it clear that innovative partnerships for sustainable development can both generate innovative sources of finance and promote integrated sustainability strategies between the forest sector and other sectors.

17.5 Partnerships for Sustainable Development

Targets 17.16 and 17.17 stress the importance of partnerships and the contributions they can make to sustainable development. Advantages of sustainable development partnerships include managing complexity (Visseren-Hamakers 2013); filling governance gaps where governments are unable or unwilling to act (Visseren-Hamakers et al. 2011, Visseren-Hamakers and Glasbergen 2006, Von Moltke 2002); addressing deficits in regulation, participation and implementation (Biermann et al. 2007); and regularising interactions, including placing previously informal interactions on a more formal, perhaps legal, footing (Visseren-Hamakers et al. 2012).

There is nothing inherently sustainable about partnerships. Partnerships are discursive battlefields that reflect power imbalances among actors grappling with different values and principles (Arévalo and Ros-Tonen 2009). Some partnerships may promote sustainable practices, others may not. For Andonova and Levy (2003), the popularity of partnerships as a form of governance originates from the disengagement from sustainable development of public authorities who have 'franchised' environmental governance to other actors. The UNFF initiative on financing (Paramaribo Initiative) argues that because stakeholders have different levels of power, governments must establish the rules governing partnerships to ensure that the interests of weaker stakeholders, such as Indigenous communities and small enterprises, are equitably represented (Paramaribo Initiative 2008). Partnerships that include local communities are essential for achieving the SDGs (SDIA 2013, 2015, CCAFS 2017).

As discussed, a number of forest-related partnerships help achieve the SDG targets on strengthening the means of implementation, such as the FCPF, UN-REDD (Section 17.2) and the Partnerships for Forests, which supports a range of national-level forest partnerships with investment models, for example, on sustainable palm oil development as part of ZND commitments (Section 17.4). This section considers some of the further roles partnerships can play to promote the SDGs, examining three global partnerships, three regional partnerships and one public-private partnership.

17.5.1 Global Partnerships

The *International Labour Organisation (ILO) Sectoral Policies Department (SECTOR)* promotes the ILO's Decent Work Agenda to advance Target 8.8 on protecting labour rights and promoting safe and secure working environments.[1] The ILO's promotion of decent work in forestry includes interventions to support the transition from the informal economy (e.g. illegal logging) to the formal economy, promoting employment creation, enhancing training and skills development and improving working conditions (ILO 2017). Together with FAO and the United Nations Economic Commission for Europe (UNECE), the ILO formed the Joint Experts Network on Green Jobs in Forestry, which fosters international cooperation on the technical, economic and organisational aspects of forest management, working techniques and training forest workers. The network contributes to the integrated

[1] Information in this paragraph from Sabine de Bruijn, ILO Secretariat, email, 7 June 2017.

work programme of the Committee on Forests and the Forest Industry and the European Forestry Commission, in particular for green jobs in the forest sector and the social and cultural aspects of SFM (ILO 2018). Other organisations engaged in partnerships on forest workers' rights and, more broadly, the rights of Indigenous peoples and forest communities include the International Union for Conservation of Nature (IUCN), especially through its Commission on Environmental, Economic and Social Policy; the Forest Peoples Programme; the Centre for People and Forests; and the International Model Forest Network.

The *Collaborative Partnership on Forests* (CPF) is an interagency partnership among 14 international organisations (including IUFRO, CIFOR, ICRAF, GEF, FAO and IUCN). Among the SDG targets that the CPF contributes to are Target 17.6, on enhancing access to science, technology, innovation and knowledge sharing, and Target 17.8, on operationalising capacity-building mechanisms in science, technology and innovation. The CPF aims to streamline and align the work of member organisations and find ways to improve forest management (including conservation, production and trade of forest products). One of the most important initiatives of CPF member organisations, especially IUFRO, is the Global Forest Expert Panels, which serve as an international boundary mechanism that mediates the transfer of state-of-the-art knowledge across the science–policy interface. The knowledge these panels generate is disseminated at international forest policy bodies such as the World Forestry Congress and the UNFF and is widely accepted by the forest policy community as authoritative (Humphreys 2009).

The *World Business Council for Sustainable Development* has a Forest Solutions Group (FSG) that aims to provide a global platform for collaboration across value chains for forest products. The FSG's emphasis on expanding markets for responsible forest products and sustainability performance (WBCSD 2018) contributes to SDG 12 on responsible production and consumption. Businesses signing up with the FSG agree to adhere to a set of membership responsibilities on sustainable development that are measured by key performance indicators, including resource efficiency and climate and water stewardship.

17.5.2 Regional Partnerships

Initiative 20x20 is the first regional commitment to at-scale forest and landscape restoration in Latin America. Participants include the WRI, International Centre for Tropical Agriculture (CIAT), Tropical Agricultural Research and Higher Education Center (CATIE), IUCN and Natural Capital Project, organisations in national and regional governments and the private sector (WRI 2017a). Its work promotes Target 15.2 on ending deforestation and restoring degraded forests. The initiative has secured commitments from 11 countries, 3 states and 4 NGOs to restore 27.7 million ha of land by 2020, and has secured private investments of USD 1.15 billion (WRI 2017a). These commitments directly contribute to the SDGs and can generate co-benefits for people, economy and ecosystems. However, challenges exist in turning commitments into measurable restoration, particularly where there are limitations in institutional capacities, financial architectures and local participation. Efforts that rely on planting trees to achieve restoration goals are unlikely to result in vast expanses of reforested areas given the costs, time required, low survival rates of planted trees and forestry departments with limited resources (Reij and Winterbottom 2017).

The *Asia Pacific Economic Cooperation (APEC)* is a partnership established in 1989 between 21 Pacific Rim economies. Among the forest-related SDGs to which APEC contributes is SDG 15.2 on increasing forest cover. In 2007 APEC adopted the 2020 Forest Cover Goal to restore 20 million ha of forests. The Sydney APEC Leaders' Declaration on Climate Change, Energy Security and Clean Development includes the commitment 'to achieve an APEC-wide aspirational goal of increasing forest cover in the region by at least 20 million ha of all types of forests by 2020' (APFNet 2015). If achieved, this would store approximately 1.4 billion tonnes of carbon – about 11 per cent of annual global emissions. However, the goal, endorsed by all 21 APEC members, is voluntary, with no enforcement mechanisms to assure compliance. Since its adoption, planted forests have increased by slightly more than 20 million ha across APEC countries. However, the net increase of forest cover was only 15.4 million ha due to a 7.9 million ha decrease in forest cover in Indonesia, Peru and Australia (APFNet 2015).

The *SAMOA Pathway* is an initiative of small island developing states (SIDS). SAMOA stands for SIDS Accelerated Modalities of Action, a pathway approach to sustainable development with several priority areas pertaining to forest-related SDGs (SIDS 2014). The main sustainable development concern of the SAMOA Pathway is to build resilience to counter sea-level rise related to climate change (consistent with Target 13.1), a key priority for all low-lying atoll states. For the larger SIDS, such as Papua New Guinea and the Solomon Islands, forests provide a revenue stream as well as non-economic benefits such as tourism. Regional organisations such as the Caribbean Community and the Pacific Islands Forum play an important role in information sharing and coordination. In December 2015, the General Assembly established the SIDS Partnership Framework, in accordance with the SAMOA Pathway, to monitor and ensure the implementation of pledges through partnerships for SIDS.

17.5.3 Public–Private Partnerships

The *Global Partnership on Forest and Landscape Restoration* (GPFLR) is an international network that brings together governments, research institutes, communities and individuals. Launched in 2003 by the IUCN, the WWF and the UK's Forestry Commission, the partnership of 25 governments and NGOs aims to restore 350 million ha of deforested and degraded land by 2030 (GPFLR 2017). It aims to respond to the Bonn Challenge (consistent with Target 15.2) on restoring deforested land and degraded forests by promoting forest and landscape restoration (FLR), defined as 'a process that aims to regain ecological functionality and enhance human well-being in deforested or degraded landscapes' (Besseau et al. 2018). The focus of FLR is landscapes (rather than individual forest sites), with FLR taking place both within and across landscapes in order to create interacting land uses and management systems.

This brief survey makes clear some of the roles partnerships can play. In addition to raising and disbursing finance, roles include generating and disseminating scientific knowledge, pooling expertise, promoting innovative solutions (such as FLR), protecting workers' rights and promoting sustainability practices among forest-related businesses in support of the SDGs. Local umbrella organisations play an essential role in forging partnerships for sustainable development. Local concerns can be channelled into global processes by organisations representing local groups and communities. Examples from Latin America include the Coordinator of Indigenous Organizations of the Amazon River Basin and the Mesoamerican Alliance of People and Forests (AMPB). These organisations help ensure that delivery of the SDGs and related commitments is respectful of Indigenous peoples and local communities. For instance, AMPB organised local consultations through members to identify advocacy priorities in climate-change negotiations to inform their 'If Not Us, Then Who?' campaign, which called for the recognition of resource rights and FPIC. Community–company partnerships for timber and non-timber forest products, including partnerships for PES, can provide communities with income and other benefits. Although PES is often presented as a win–win scenario that raises additional finance for sustainable development, contributes to conservation goals and enables land-based poor groups to benefit from additional income (Duncan 2006, Pagiola et al. 2005, Wunder 2008), care must be taken to ensure respect for the rights and ancestral lifestyles of Indigenous peoples.

Different partnerships can generate different outcomes. Foreign direct investment from a multinational logging company will bring short-term gain to an indebted economy that relies on export earnings. Yet the communities most impacted by logging often see little of the market value of timber production while bearing the socio-economic and environmental costs. This underlines an inherent tension in the dynamics of partnerships: relationships tend to be asymmetrical, with a clear hierarchy of power and influence. In the context of the tropical timber trade such relationships operate between local landowners and national governments, which, in turn, are connected to the power dynamics between the more developed 'core' economies and those that trade with them. This dynamic is overlaid, and often reinforced, by the private sector, with many multinational corporations aiming to maximise profit extraction while social and environmental costs are often borne by local communities. We now turn to the question of asymmetrical power relationships in the global economy.

17.6 The Broader Structure of Economic Governance

If it is to be comprehensive, the discussion in this chapter on strengthening the means of implementation and revitalising partnerships for sustainable development cannot focus solely on forests and forest-related sectors. With Target 17.13 stressing the need to enhance macroeconomic stability, including policy coordination and coherence, and Target 17.14 emphasising the importance of policy coherence for sustainable development, the broader political and economic context within which efforts to promote sustainable development occur must be considered.

Here an international political economy framework is helpful. For international political economists, understanding global power structures requires comprehension of both politics and economics. Those who wield economic power, such as business executives and financial elites, must take into account political factors, such as government policy, while the exercise of political power is shaped to a large degree by the economic context. Hence, there is a complex and iterative relationship between political and economic power.

Contrary to what SDG 17 aims to promote – a global partnership for sustainable development – a political economy view argues that in as much as a global partnership may be said to exist, it is one founded on neoliberal principles, such as the expansion of international trade and economic growth, with relatively limited attention to environmental conservation. In this view, the triumvirate of international economic and financial institutions – the World Bank, the International Monetary Fund and the World Trade Organisation – provide a neoliberal normative framework that favours the interests of transnational corporations and powerful states, primarily from the Global North (Humphreys 2006). This framework promotes the liberalisation of goods and services and the restructuring of economies in the Global South to enforce the repayment of debts. This can be seen as a neoliberal business-based constitutionalism in which corporate rights and capitalist expansion dominate at the expense of the environment and human welfare (Gill 2002, Derber 2002).

According to this view, the pursuit of sustainable development cannot succeed because the normal and routine functioning of the global economy generate unsustainability, negating any gains that may be realised through the promotion of the SDGs. For example, in 2012 (the most recent year with reliable data) developing countries received about USD 2 trillion in ODA, foreign investment and trade. However, for this same year, some USD 5 trillion flowed from developing countries to the Global North in the form of debt repayments, capital flight, repatriation of profits, payment of intellectual property rights and illicit outflows (Hickel 2017). In other words, the poorer countries of the global economy made a net transfer of approximately USD 3 trillion to the richer countries. According to another estimate, in 2012 the governments of developing countries repaid USD 182 billion to their creditors but received only USD 133 billion in ODA. Remittances from emigrants grossed an estimated USD 350 million, while multinational corporations made about USD 678 billion in profits, most of which was repatriated to their headquarters in developed countries (Gottiniaux et al. 2015). According to Global Financial Integrity (GFI), the cumulative total of net South-to-North financial transfers since 1980 is USD 26.5 trillion (GFI 2015, Hickel 2017). These figures illustrate the exacerbation of a problem that SDG 10 seeks to address: reducing income inequalities between countries.

Debt servicing ratios as a percentage of exports of goods and services are once again trending upward, indicating extra pressure on forests and other natural-resource sectors to increase exports to earn hard currency to service external debts. From 2000 to 2011, debt service fell from 12.9 per cent to 3.6 per cent in lower-middle-income countries before increasing to 6.1 per cent in 2015 (UN 2017, UN 2016), a trend that runs counter to Target 17.4 to assist developing countries attain long-term debt sustainability. Although African countries annually receive USD 161.6 billion inflow through loans, remittances and aid, they incur net losses of about USD 203 billion through trade misinvoicing, debt payments and resource extraction (Curtis and Jones 2017). Donoso Game (2018) estimates that between 1990 and 2004 Latin America paid USD 1.9 trillion in debt services (i.e. about USD 126.9 billion per year).

Llistar Bosch (2009) coins the term *transnational interference* to denote interventions from outside a country that directly or indirectly affect the internal dynamics of a social group, community or country. These interferences take the form of normative transmissions through transnational mechanisms, such as loan agreements and sovereign debt-repayment schemes that impose conditions on developing countries. Not all transnational interferences are negative. International cooperation that promotes SFM or that addresses illegal logging are examples of positive transnational interference. For Llistar Bosch (2009), however, most transnational interference is negative: economic support to developing countries is influenced by geopolitical realities that correspond more to donor interests than those of the beneficiaries.

As well as focusing on legal flows of finance and natural resources, it is also necessary to examine those that take place illegally. It is estimated that, since 1980, illicit outflows account for 82 per cent of South-to-North net resource transfers (GFI 2015). Hickel contrasts the GFI estimates with more cautious OECD ones: 'there is general consensus that illicit financial flows likely exceed aid flows and investment in volume' (OECD 2014b: 15, cited in Hickel 2017: 334). These illicit financial flows include criminal activities in the logging sector, resulting in revenue losses that could be reinvested in forests. The World Bank has estimated that as much as 10 per cent of the value of the global timber trade is from illegal sources (World Bank 2006), with the figure for some countries as high as 90 per cent (Pereira Goncalves et al. 2012). The World Bank has also estimated that the global loss of revenue from illegal logging is at least USD 10 billion annually, about eight times greater than the total ODA flows to forests (World Bank 2013) and possibly as high as USD 15 billion (Pereira Goncalves et al. 2012). According to GFI, USD 1.09 trillion flowed illegally from developing countries in 2013, compared to USD 465.3 billion in 2004 (Kar and Spanjers 2015). Research from UNCTAD reveals that the widespread illicit practice of trade misinvoicing – the practice of deliberately misreporting the value of imports or exports on invoices to enable capital flight, usually to an offshore account – is weakening the capacity of developing countries to implement sustainable development. This problem is widespread in primary commodity sectors in many developing countries (UNCTAD 2016). Between 1980 and 2012, developing countries lost USD 13.4 trillion through leakages in the balance of payments and trade misinvoicing (Centre for Applied Research 2015). Trade misinvoicing is an important route for capital flight from timber industries in many countries, including Cameroon (Mpenya et al. 2016). Countering illegal logging and other forest-related crimes relates to Target 16.4 to 'significantly reduce illicit financial and arms flows, strengthen the recovery and return of stolen assets and combat all forms of organised crime'. However, progress towards achieving this target is mixed.

17.7 Conclusions

The evidence assembled in this chapter provides reason for cautious optimism on SDG 17. Significant progress has been made in generating additional funding for implementing forest-related sustainable development, with funding for forests from ODA and other sources trending upward. However, while a focus on increasing SDG implementation may lead to some tangible gains for forests, it may simultaneously reinforce some potential contradictions among SDGs. For example, much forest finance, in particular REDD+, has been targeted at the carbon sink function of forests. While this helps to attain SDG 13 (Climate Action), without strong safeguards for other forest goods and services it may run counter to realising SDG 15 (Life on Land). A further example includes the relationship between agriculture and forests. In several countries, international forest-related ODA is dwarfed by domestic subsidies for agricultural production. This provides a structural incentive for the conversion of forests to agricultural land, in particular for the four forest-risk commodities. Both these examples suggest that realising forest-related sustainable development depends on policy coherence between the forest sector and forest-related SDGs, as well as addressing the complex conflicts and synergies between them. While strengthening the means of implementation is to be welcomed, it should not be seen as a panacea. It can only lead to SFM when the intersectoral causes and consequences are fully understood, so that the broad range of public and private goods forests provide are conserved.

The case of ZND illustrates the importance of the duality of SDG 17, which promotes strengthening the means of implementation and revitalising partnerships for sustainable development. New partnerships can help generate additional finance, yet if finance is to be spent in ways that enhance SFM, then cross-sectoral partnerships extending beyond forests and promoting integrated strategies are necessary. Only with such partnerships can the synergies and trade-offs among SDGs and the implementation of SFM be effectively addressed. However, such efforts are a work in progress, and there is a need to pay more attention to the promotion of finance and investments for sustainable land use, particularly sustainable agriculture and livestock. Here, impact investment may make a positive contribution. The underlying logic of impact investment is that there is a positive-sum game among carefully targeted investments that generate added value in terms of both sustainability and profits for investors. Where there is a need from a sustainability standpoint but no prospect of returns for investors, then the need must be met from national or international public finance.

Evidence has been presented which suggests that a focus solely on the SDGs is insufficient when contemporary global economic governance, including forest-related crimes, negates the gains from the sustainability agenda. It is argued that efforts to strengthen the means of implementation and forge innovative partnerships are taking place within a global economic system governed by neoliberal principles rather than the principles of conservation or sustainability. Major constraints to sustainable development lie in deep-rooted structures and practices that continue to generate and reinforce unsustainable practices in forestry and other extractive industries, with widening inequalities and power disparities severely curtailing the ability of governments and other actors to pursue sustainability.

The dominant global political agenda remains focused on economic growth and the liberalisation of trade and investment rather than conservation and sustainable development. We therefore finish on a cautionary note: a focus on just forest-related financial flows and forest-related partnerships misses the bigger picture. Financing for sustainable development is negated by net South-to-North financial flows and vast inequalities of power that undermine the capacity of many countries to conserve the ecological life-support functions on which present and future generations depend. Achieving genuinely durable and long-term sustainability requires turning our attention to the environmentally degrading effects of the broader structures of economic governance.

References

Abram, N. K., Meijaard, E., Wilson, K. A. et al. 2017. Oil Palm – community conflict mapping in Indonesia: A case for better community liaison in planning for development initiatives. *Applied Geography* 78:33–44.

Accountability Framework 2018. *Overview of the Accountability Framework*. Available at: https://accountability-framework.org/framework/ (Accessed 8 December 2018).

Acheampong, M., Ertem, F. C., Kappler, B. and Neubauer, P. 2017. In pursuit of Sustainable Development Goal (SDG) 7: Will biofuels be reliable? *Renewable and Sustainable Energy Reviews* 75:927–37.

AGF 2012. *2012 Study on Forest Financing*. Unpublished report. Available at: www.un.org/esa/forests/wp-content/uploads/2014/12/AGF_Study_July_2012.pdf (Accessed 23 July 2019).

Amazon Fund 2017. *Amazon Fund/Purposes and management*. Available at: www.amazonfund.gov.br/en/home/ (Accessed 23 July 2019).

Andonova, L. B. and Levy, M. A. 2003. Franchising global governance: Making sense of the Johannesburg Type II Partnerships. In Stokke, O. S. and Thommessen, O. B. (eds.) *Yearbook of International Co-operation on Environment and Development*. London: Earthscan, pp. 19–32.

APFNet 2015. *Assessment of progress towards the APEC 2020 forest cover goal: Synthesis of economy reports and additional research*. Beijing: China Forestry Publishing House.

Arévalo, E. B. and Ros-Tonen, M. A. F. 2009. Discourses, power negotiations and Indigenous political organization in forest partnerships: The case of Selva de Matavén, Colombia. *Human Ecology* 37:733–47.

Arima, E., Barreto, P., Araujo, E. and Soares-Filho, B. 2014. Public policies can reduce tropical deforestation: Lessons and challenges from Brazil. *Land Use Policy* 41:465–73.

Besseau, P., Graham, S. and Christophersen, T. (eds.) 2018. Restoring forests and landscapes: The key to a sustainable future. *Global Partnership on Forest and Landscape Restoration*. Available at: www.forestlandscaperestoration.org/sites/forestlandscaperestoration.org/files/resources/GPFLR_FINAL%2027Aug.pdf (Accessed 4 December 2018).

Biermann, F., Chan, M-S., Mert, A. and Pattberg, P. 2007. *Multi-stakeholder partnerships for sustainable development: Does the promise hold?* Paper presented at 2007 Amsterdam Conference on the Human Dimensions of Global Environmental Change, Vrije Universiteit, Amsterdam. 24–26 May.

Boucher, D. 2008. *Estimating the cost and potential of reducing emissions from deforestation, Briefing Paper 1*, Tropical Forests and Climate. Washington, DC: Union of Concerned Scientists. Available at: www.ucsusa.org/sites/default/files/legacy/assets/documents/clean_energy/Briefing-1-REDD-costs.pdf (Accessed 23 July 2019).

Business and Sustainable Development Commission 2017. *Better business better world*, January. Available at: http://report.businesscommission.org/uploads/BetterBiz-BetterWorld_170215_012417.pdf (Accessed 27 November 2018).

Cambridge Institute for Sustainability Leadership 2018. *The Banking Environment Initiative (BEI) and Consumer Goods Forum (CGF)'s 'Soft Commodities' Compact*. Available at: www.cisl.cam.ac.uk/business-action/sustainable-finance/banking-environment-initiative/programme/sustainable-agri-supply-chains/soft-commodities (Accessed 26 November 2018).

Castrén, T., Katila, M., Lindroos, K. and Salmi, J. 2014. *Private financing for sustainable forest management and forest products in developing countries: Trends and drivers*. Washington, DC: Program on Forests (PROFOR).

CCAFS 2017. *Power of partnerships: Annual Report 2016*. Wageningen, Netherlands: CGIAR Research Programme on Climate Change, Agriculture and Food Security (CCAFS). Available at: https://ccafs.cgiar.org/publications/annual-report-2016-power-partnerships#.Ww6urViWzmQ (Accessed 8 April 2018).

Centre for Applied Research – Norwegian School of Economics, Global Financial Integrity, Jawaharlal Nehru University, Instituto de Estudos Socioeconômicos, Nigerian Institute of Social and Economic Research 2015. *Financial flows and tax havens: Combining to limit the lives of billions of people*. Available at: www.gfintegrity.org/wp-content/uploads/2016/12/Financial_Flows-final.pdf (Accessed 28 November 2018).

CFU (Climate Funds Update) 2017. *Data dashboard*. Available at: www.climatefundsupdate.org/data (Accessed 31 January 2018).

Climate Focus 2016. *Progress on the New York Declaration on Forests – Achieving collective forest goals. Updates on Goals 1–10*. Prepared by Climate Focus in cooperation with the NYDF Assessment Coalition with support from the Climate and Land Use Alliance and the Tropical

Forest Alliance 2020. Available at: www.climatefocus.com/sites/default/files/2016-Updates-on-Goals-1–10-Report.pdf (Accessed 27 May 2018).

Climate Focus 2017. *Progress on the New York Declaration on Forests. Finance for forests. Goals 8 and 9 Assessment Report*. Prepared by Climate Focus in cooperation with the New York Declaration on Forest Assessment Partners with support from the Climate and Land Use Alliance, October. Available at: https://climatefocus.com/sites/default/files/NYDF%20report%202017%20FINAL.pdf (Accessed 24 November 2018).

Colchester, M., Anderson, P., Nelson, J. et al. 2016. *How can 'zero deforestation' policies accommodate the rights and livelihoods of local communities and Indigenous peoples? Lessons from the field*. Paper presented at the 2016 World Bank Conference on Land and Poverty, 14–18 March. Available at: www.forestpeoples.org/sites/fpp/files/news/2016/02/LandConference2016Colchester_356.pdf (Accessed 27 May 2018).

Curtis, M. and Jones, T. 2017. *Honest Accounts 2017: How the world profits from Africa's wealth*. Research funded by Global Justice Now. Building on previous work by Health Poverty Action and partners. 2nd. ed. July. Available at: www.globaljustice.org.uk/sites/default/files/files/resources/honest_accounts_2017_web_final_updated.pdf (Accessed 9 September 2018).

Curtis, P. G., Slay, C. M., Harris, N. L., Tyukavina, A. and Hansen, M. C. 2018. Classifying drivers of global forest loss. *Science* 361:1108–11.

DANA 2011. *International timberlands ownership and investment review*. Rotorua, New Zealand: DANA.

Derber, C. 2002. *People before profit: The new globalisation in an age of terror, big money and economic crisis*. London: Souvenir Press.

Donofrio, S., Rothrock, P. and Leonard, J. 2017. *Supply change: Tracking corporate commitments to deforestation-free supply chains*. Washington, DC: Forest Trends.

Donoso Game, A. 2018. De deudores a acreedores. In Delen, B. (ed.) *Territorios y recursos naturales: el saqueo versus el buen vivir*. Quito: Alai, pp. 32–8.

Duncan, E. 2006. *Payments for environmental services: An equitable approach for reducing poverty and conserving nature*. Zeist, Netherlands: WWF.

Eba'a Atyi, R., Lescuyer, G., Ngouhouo Poufoun, J. and Moulendè Fouda, T. (eds.) 2013. *Étude de l'importance économique et sociale du secteur forestier et faunique au Cameroun: Rapport final*. Bogor, Indonesia: CIFOR.

Environmental Defense Fund and Forest Trends 2018. *Mapping forest finance: A landscape of available sources of finance for REDD+ and Climate Action in Forests*. February. Available at www.edf.org/sites/default/files/documents/EDF101-REDD%2BFinance.pdf (Accessed 17 November 2018).

FAO 2010. *Global forest resources assessment 2010: Main report*. FAO Forestry Paper 163. Rome: FAO.

FCPF 2017. *About FCPF*. Available at: www.forestcarbonpartnership.org/about-fcpf-0 (Accessed 8 December 2017).

Fearnside, P. 2017. Business as usual: A resurgence of deforestation in the Brazilian Amazon. *Yale Environment* 360. Available at: https://e360.yale.edu/features/business-as-usual-a-resurgence-of-deforestation-in-the-brazilian-amazon (Accessed 6 March 2018).

Forest Trends 2017. *State of private investment in conservation 2016: A landscape assessment of an emerging marke.* Washington, DC: Forest Trends. Available at: www.forest-trends.org/documents/files/doc_5474.pdf# (Accessed 17 December 2017).

GEF 2015. *Taking deforestation out of commodity supply chains. A GEF integrated approach pilot.* Washington DC: GEF. Available at: www.thegef.org/sites/default/files/publications/GEF_SDG_Deforestation_r6_1.pdf (Accessed 22 November 2017).

GEF 2017. $400 million fund launched in Davos to stop tropical deforestation and boost farming. *GEF News.* Available at: www.thegef.org/news/400-million-fund-launched-davos-stop-tropical-deforestation-and-boost-farming (Accessed 22 November 2017).

Gerasimchuk, I., Bridle, R., Beaton, C. and Charles, C. 2012. *State of play on biofuel subsidies: Are policies ready to shift? Global Subsidies Initiative Research Report.* Geneva: International Institute for Sustainable Development.

GFI 2015. *Financial flows and tax havens: Combining to limit the lives of billions of people.* December. Available at: www.gfintegrity.org/wp-content/uploads/2016/12/Financial_Flows-final.pdf (Accessed 31 January 2018).

Gill, S. 2002. *Power and resistance in the new world order.* London: Palgrave Macmillan.

Glauner, R., Rinehart, J. A. and D'Anieri, P. 2012. Can timberland investments in emerging markets secure forest sustainability? *European Tropical Forest Research Network (ETFRN) News* 54:3–9.

Global Investor Impact Survey 2018. *Annual investor impact survey 2018.* USAID and UKAid, June, 8th ed. Available at: https://thegiin.org/assets/2018_GIIN_Annual_Impact_Investor_Survey_webfile.pdf (Accessed 17 November 2018).

Global Partnership on Forest and Landscape Restoration 2017. *About the partnership.* Available at: www.forestlandscaperestoration.org/about-partnership (Accessed 6 March 2018).

Gottiniaux, P., Munevar, D., Sanabria, A. and Toussaint, E. 2015. *World debt figures 2015.* Committee for the Abolition of Illegitimate Debt (CADTM). Available at: www.cadtm.org/World-Debt-Figures-2015 (Accessed 9 September 2018).

Graham, P., Thoumi, G., Drazen, E. and Seymour, F. 2018. Mining global financial data to increase transparency and reduce drivers of deforestation, *World Resources Institute Working Paper*, Washington, DC. Available at: www.wri.org/publication/ending-tropical-deforestation-mining-global-financial-data-increase-transparency (Accessed 18 November 2018).

Hickel, J. 2017. *The divide: A brief guide to global inequality and its solutions.* London: William Heinemann.

Holopainen, J. and Wit, M. (eds.) 2008. *Financing sustainable forest management.* Wageningen, Netherlands: Tropenbos International.

Humphreys, D. 2006. *Logjam: Deforestation and the crisis of global governance.* London: Earthscan.

Humphreys, D. 2009. Working across boundaries: Science-policy interfaces and international forest politics. *Journal of Integrative Environmental Sciences* 6(3):163–74.

ILO 2017. *Decent work in forestry: Decent work in the rural economy – policy guidance notes*. ILO, Geneva. Available at: www.worldcat.org/title/decent-work-in-forestry-decent-work-in-the-rural-economy-policy-guidance-notes/oclc/935517206&referer=library_profile_recentitems&returnRegistryId=&libraryname=International%20Labour%20Office (Accessed 23 July 2019).

ILO 2018. *Decent work and the 2030 Agenda for Sustainable Development*. Available at: www.ilo.org/global/topics/sdg-2030/lang–en/index.htm (Accessed 23 July 2019).

Jopke, P. and Schoneveld, G. C. 2018. Corporate commitments to zero deforestation: An evaluation of externality problems and implementation gaps, *Centre for International Forestry Research (CIFOR) Occasional Paper* 181. Available at: www.cifor.org/publications/pdf_files/OccPapers/OP-181.pdf (Accessed 17 November 2018).

Kar, D. and Spanjers, J. 2015. *Illicit financial flows from developing countries: 2004–2013*. Washington, DC: Global Financial Integrity. Available at: www.gfintegrity.org/wp-content/uploads/2015/12/IFF-Update_2015-Final-1.pdf (6 January 2018).

Karsenty, A., Roda, J. M., Milol, A. and Fochivé, E. 2006. *Audit économique et financier du secteur forestier au Cameroun. Rapport final*. Economic audit carried out for the Cameroonian Ministry of Economy and Finance.

Katila, P., de Jong, W., Galloway, G. Pokorny, B. and Pacheco, P. 2017. *Building on synergies: Harnessing community and smallholder forestry for Sustainable Development Goals (Policy Brief)*. Available at: www.iufro.org/news/article/2017/05/08/harnessing-the-potential-of-community-and-smallholder-forestry-for-sustainable-development-goals/ (Accessed 8 December 2018).

Kissinger, G. 2015. *Fiscal incentives for agricultural commodity production: Options to forge compatibility with REDD+*. UN-REDD Programme Policy Brief 7.

Kissinger, G., Herold, M. and de Sy. V. 2012. *Drivers of deforestation and forest degradation: A synthesis report for REDD+ policymakers*. Vancouver: Lexeme Consulting.

Lawson, S., Blundell, A., Cabarle, B. et al. 2014. *Consumer goods and deforestation: An analysis of the extent and nature of illegality in forest conversion for agriculture and timber plantations*. Forest Trends and UK Aid. Available at: www.forest-trends.org/publications/consumer-goods-and-deforestation/ (Accessed 23 July 2019).

Llistar Bosch, D. 2009. *Anticooperación. Interferencias Norte-Sur. Los problemas del Sur Global no se resuelven con más ayuda internacional. Antrazyt 309*. Barcelona: Icaria Editorial.

Lovera, S. 2017. SDG15: Trends in the privatization and corporate capture of biodiversity. In Civil Society Reflection Group on the 2030 Agenda for Sustainable Development (ed.) *Spotlight on Sustainable Development 2017: Reclaiming Policies for the Public. Privatization, Partnerships, Corporate Capture and their Impact on Sustainability and Inequality*. Available at: www.2030spotlight.org/sites/default/files/download/spotlight_170626_final_web.pdf (Accessed 27 November 2017).

McFarland, W., Whitley, S. and Kissinger, G. 2015. Subsidies to key commodities driving forest loss: implications for private climate finance. *ODI Working Paper*. London: ODI. Available at:

www.odi.org/sites/odi.org.uk/files/odi-assets/publications-opinion-files/9577.pdf (Accessed 27 November 2017).

Mpenya, H. T. A., Metseyem, C. and Epo, B. N. 2016. Natural resources and capital flight in Cameroon. *African Development Review* 28:88–99.

Myers, N. 1981. The hamburger connection: How Central America's forests became North America's hamburgers. *Ambio* 10:3–8.

Nakhooda, S., Watson, C. and Schalatek, L. 2016. The global climate finance architecture. Climate Finance Fundamentals 2. *Climate Funds Update*. Washington DC: Overseas Development Institute [ODI] and Heinrich Boll Siftung North America (HBS). Available at: www.odi.org/sites/odi.org.uk/files/resource-documents/11021.pdf (Accessed 27 May 2018).

Nepstad, D., McGrath, D., Stickler, C. et al. 2014. Slowing Amazon deforestation through public policy and interventions in beef and soy supply chains. *Science* 344(6188):1118–23.

Norman, M. and Nakhooda, S. 2014. *The state of REDD finance*. Working Paper 378. Washington DC: Centre for Global Development and ODI. Available at: www.cgdev.org/sites/default/files/CGD-Norman-Nakhooda-Climate-Forests-5-REDD-Finance.pdf (Accessed 23 July 2019).

OECD 2014a. *Agricultural policies and support:* Producer *and consumer support estimates database* [online]. Available at: www.oecd.org/tad/agricultural-policies/producerandconsumersupportestimatesdatabase.htm (Accessed 19 August 2014).

OECD 2014b. *Illicit financial flows from developing countries: Measuring OECD responses*. Paris: OECD.

OECD 2017a. QWIDS Query Wizard for International Development Statistics Database. Agriculture and Rural Development Official Development Assistance. Available at: https://bit.ly/2K3nbrw (Accessed 17 October 2017).

OECD 2017b. Creditor Reporting System Database. Available at: http://stats.oecd.org/ (Accessed 12 December 2017).

OECD 2018a. *Development aid stable in 2017 with more sent to poorest countries*. Available at: www.oecd.org/development/development-aid-stable-in-2017-with-more-sent-to-poorest-countries.htm (Accessed 8 December 2018).

OECD 2018b. Creditor Reporting System Database. Available at: http://stats.oecd.org/ (Accessed 6 November 2018).

Pagiola, S., Arcenas, A. and Platais, G. 2005. Can payment for environmental services help reduce poverty? *World Development* 33(2):237–53.

Paramaribo Initiative 2008. *Country-led initiative on financing for sustainable forest management in support of the UN Forum on Forests*, 8–12 September 2008, Paramaribo, Suriname. Paper circulated at the United Nations Forum on Forests, 2008.

Partnership for Forests 2017. *From commitment to implementation: The African palm oil initiative enters a new phase*. Available at: https://partnershipsforforests.com/2017/03/09/commitment-implementation-africa-palm-oil-initiative-enters-new-phase/ (Accessed 14 September 2018).

Partnership for Forests 2018. *The Amsterdam Declarations*. Available at: https://partnershipsforforests.com/partnerships-projects/the-amsterdam-declarations/ (Accessed 26 November 2018).

Pereira Goncalves, M., Panjer, M., Greenberg, T. S. and Magrath, W. B. 2012. *Justice for forests: Improving criminal justice efforts to combat illegal logging*. Washington, DC: World Bank. Available at: https://siteresources.worldbank.org/EXTFINANCIALSECTOR/Resources/Illegal_Logging.pdf (Accessed 6 December 2018).

Persson, M., Henders, S. and Kastner, T. 2014. *Trading forests: Quantifying the contribution of global commodity markets to emissions from tropical deforestation*. CGD Working Paper 384. Washington, DC: Center for Global Development.

PRI 2018. *PRI and Ceres expand Sustainable Forest Initiative to include dialogue with companies in soy value chain*. 9 July. Available at: www.unpri.org/news-and-press/pri-and-ceres-expand-sustainable-forest-initiative-to-include-dialogue-with-companies-in-soy-value-chain/3386.article (Accessed 25 November 2018).

Reij, C. and Winterbottom, R. 2017. Can we restore 350 million hectares by 2030? *World Resources Institute Blog*. Available at: www.wri.org/blog/2017/02/can-we-restore-350-million-hectares-2030 (Accessed 27 May 2018).

Roberts, T. and Weikmans, R. 2016. *Roadmap to where? Is the '$100 billion by 2020' pledge from Copenhagen still realistic?* Washington, DC: Brookings. Available at: www.brookings.edu/blog/planetpolicy/2016/10/20/roadmap-to-where-is-the-100-billion-by-2020-pledge-from-copenhagen-still-realistic/ (Accessed 17 December 2017).

Rogerson, S. 2017. Forest 500 – Achieving 2020: How can the private sector meet global goals of eliminating commodity-driven deforestation? *Global Canopy*. Available at: www.globalcanopy.org/publications/achieving-2020-how-can-private-sector-meet-global-goals-eliminating-commodity-driven (Accessed 18 November 2018).

Salvini, G., Herold, M., De Sy, V., Kissinger, G., Brockhaus, M. and Skutsch, M. 2014. How countries link REDD+ interventions to drivers in their readiness plans: Implications for monitoring systems [online]. *Environmental Research letters* 9. Available at: www.cifor.org/publications/pdf_files/articles/ABrockhaus1401.pdf (Accessed 31 May 2018).

SDIA (Sustainable Development in Action) 2013. Voluntary commitments and partnerships for sustainable development. *Special Report of the SD in Action Newsletter*, Issue 1, July.

SDIA 2015. *Special report on voluntary multi-stakeholder partnerships and commitments for sustainable development*. Sustainable Development in Action. Available at: https://sustainabledevelopment.un.org/content/documents/1855SD%20in%20Action%20Report%202015.pdf (Accessed 8 January 2018).

SIDS 2014. *SIDS Partnership Framework: Ensuring the full implementation of pledges and commitments through partnerships for small island developing states*. Available at: http://sids2014.org/partnerships/ (Accessed 23 July 2019).

Silva-Chávez, G., Schaap, B. and Breitfeller, J. 2015. *REDD+ finance flows 2009–2014: Trends and lessons learned in REDDX countries*. Washington DC: Forest Trends. Available at: www.forest-trends.org/documents/files/doc_5029.pdf (Accessed 17 December 2017).

Singer, B. 2016. Financing sustainable forest management in developing countries: The case for a holistic approach. *International Forestry Review* 18(1):96–109.

Singer, B. 2017. *National forest financing strategies*. PowerPoint presentation given at the national capacity-building workshop on forest financing in Madagascar, 2–6 October, Antananarivo, Madagascar (8 slides).

Singer, B. and Giessen, L. 2017. Towards a donut regime? Domestic actors, climatization and the hollowing-out of the international forests regime in the Anthropocene. *Forest Policy and Economics* 79:69–79.

TFA (Tropical Forest Alliance) 2018a. *The roadmap to financing deforestation-free commodities*. Available at: www.weforum.org/whitepapers/the-roadmap-to-financing-deforestation-free-commodities (Accessed 18 November 2018).

TFA 2018b. *Emerging market consumers and deforestation: Risks and opportunities of growing demand for soft commodities in China and beyond*. Available at: www.tfa2020.org/wp-content/uploads/2018/09/47530_Emerging-markets_consumers_and_deforestation_report_2018.pdf (Accessed 23 July 2019).

Tomaselli, I., Hirakuri, S. R. and Penno Saraiva, G. B. 2012. Increasing the competitiveness of the Brazilian forest sector. *European Tropical Forest Research Network (ETFRN) News* 54:42–50.

UN 2014. *Report of the Intergovernmental Committee of Experts on Sustainable Development Financing*. Document A/69/315. Available at: www.un.org/ga/search/view_doc.asp?symbol=A/69/315&Lang=E (Accessed 27 November 2017).

UN 2015. *Paris Agreement*. Bonn, Germany: United Nations Framework Convention on Climate Change, Available at: http://unfccc.int/files/essential_background/convention/application/pdf/english_paris_agreement.pdf (Accessed 27 November 2017).

UN 2016. *The Sustainable Development Goals report 2016*. Available at: www.un.org.lb/Library/Assets/The-Sustainable-Development-Goals-Report-2016-Global.pdf (Accessed 17 January 2018).

UN 2017. *The Sustainable Development Goals report 2017*. Available at: https://unstats.un.org/sdgs/files/report/2017/TheSustainableDevelopmentGoalsReport2017.pdf (Accessed 6 January 2018).

UNCTAD 2014. *World Investment Report – Investing in the SDGs: an action plan* [online]. Geneva: United Nations. Available at: https://unctad.org/en/PublicationsLibrary/wir2014_en.pdf (Accessed 8 December 2018).

UNCTAD 2016. *Trade misinvoicing in primary commodities in developing countries: The cases of Chile Côte d'Ivoire, Nigeria, South Africa and Zambia*. Geneva: United Nations.

UN Environment 2016. *How to attract investment to sustainable management of forests and REDD+: Experts from 12 countries met in Panama*. UN-REDD Programme. Available at: www.un-redd.org/single-post/2017/07/11/How-to-attract-investment-to-sustainable-management-of-forests-and-REDD-experts-from-12-countries-meet-in-Panama (Accessed 17 October 2017).

UNEP 2015. *Bank and investor risk policies on soft commodities: A framework to evaluate deforestation and forest degradation in the agricultural value chain*, July. Available at: www.unccleam.org/sites/default/files/inventory/unep07092015.pdf (Accessed 18 November 2018).

Verchot, L. 2015. *CIFOR's REDD+ research*. Presentation at the Paris Global Landscapes Forum, 5–6 December, Paris, France. Available at: www.landscapes.org/cifors-redd-research/ (Accessed 27 November 2017).

Visseren-Hamakers, I. J. 2013. Partnerships and sustainable development: The lessons learned from international biodiversity governance. *Environmental Policy and Governance* 23:145–60.

Visseren-Hamakers, I. J., Arts, B. and Glasbergen, P. 2011. Interaction management by partnerships: The case of biodiversity and climate change. *Global Environmental Politics* 11(4):89–107.

Visseren-Hamakers, I. J. and Glasbergen, P. 2006. Partnerships in forest governance. *Global Environmental Change* 17(3–4):408–19.

Visseren-Hamakers, I. J., Leroy, P. and Glasbergen, P. 2012. Conservation partnerships and biodiversity governance: Fulfilling governance through interaction. *Sustainable Development* 20:264–75.

Von Moltke, K. 2002. Governments and international civil society in sustainable development: A framework. *International Environmental Agreements: Politics, Law and Economics* 2:341–59.

Watson, C., Patel, S. and Schalatek, L. 2016. Climate finance thematic briefing: REDD+ finance. Climate Funds Update. *Climate Finance Fundamentals* 5. Available at: www.odi.org/sites/odi.org.uk/files/resource-documents/11030.pdf (Accessed 27 May 2018).

WBCSD 2018. *Forest Solutions Group: Shared commitment, concerted action and transparent progress*. World Business Council for Sustainable Development. Available at: www.wbcsd.org/Sector-Projects/Forest-Solutions-Group (Accessed 5 December 2018).

World Bank 2006. *Strengthening forest law enforcement and governance: Addressing a systemic constraint to sustainable development*. Report No. 36638-GLB. Washington, DC: World Bank. Available at: http://siteresources.worldbank.org/INTFORESTS/Resources/WB_Rpt_36638_Forest_Law.pdf?resourceurlname=WB_Rpt_36638_Forest_Law.pdf (Accessed 17 December 2017).

World Bank 2008. *Forest Source Book: Practical Guidance for Sustaining Forests in Development Cooperation*. Washington DC: World Bank.

World Bank 2013. *Forest law enforcement and governance*. Available at: http://siteresources.worldbank.org/INTFORESTS/Resources/WB_Rpt_36638_Forest_Law.pdf?resourceurlname=WB_Rpt_36638_Forest_Law.pdf (Accessed 6 December 2018).

WRI 2017a. *Initiative 20x20*. World Resources Institute. Available at: www.wri.org/our-work/project/initiative-20x20/restoration-commitments#project-tabs (Accessed 8 December 2017).

WRI 2017b. *Global Forest Watch Commodities*. www.wri.org/resources/websites/global-forest-watch-commodities (Accessed 8 December 2017).

Wunder, S. 2008. Payment for environmental services and the poor: Concepts and preliminary evidence. *Environment and Development Economics* 13:279–97.

WWF 2009. *Zero net deforestation by 2020*. Available at: wwf.panda.org/?181181/Zero-Net-Deforestation-for-2020 (Accessed 8 December 2017).

Chapter 18 Synergies, Trade-Offs and Contextual Conditions Shaping Impacts of the Sustainable Development Goals on Forests and People

Wil de Jong, Glenn Galloway, Carol J. Pierce Colfer, Pia Katila, Georg Winkel and Pablo Pacheco

18.1 Introduction

The 2030 Agenda for Sustainable Development and its 17 Sustainable Development Goals (SDGs) have been adopted as a follow-up to the 8 Millennium Development Goals (MDGs) that were pursued from 2000 until 2015. There are similarities between the SDGs and the MDGs, but also important differences. Agenda 2030 and the SDGs address a wider array of development challenges, linking them to the broad sustainability agenda, including equitable economic development, social inclusion and environmental protection. Pursuing 17 goals and 169 targets simultaneously implies tremendous challenges in terms of commitment, planning and coordination (Allen et al. 2018). This was recognised when the goals were being formulated, and multiple actors who support the implementation of Agenda 2030 have reiterated the need for integration in implementation. The possibility of positive and negative interactions, or synergies and trade-offs, between both goals and targets were pointed out (ICS 2017).

A body of literature is developing ideas on how to foster successful implementation of Agenda 2030. Authors try to evaluate possible interactions among SDGs when they are being implemented concurrently. Some authors frame this in terms of SDG interactions (ICS 2017; Le Blanc 2015, Nilson et al. 2016), synergies and trade-offs (Katila et al. 2017), while others highlight the need for integration of SDG implementation (Nunes et al. 2018, Stafford-Smith et al. 2017, Waage et al. 2015). Allen et al. (2018) identify the need for methods, models and tools that can capture and address the intrinsic complexities of the SDGs, including interactions, synergies and trade-offs, or options for their integration.

The present volume is a contribution to facilitate successful implementation of Agenda 2030 and the SDGs, focusing specifically on forests and the people who live in or near them and who depend on them for their material, social, cultural and emotional well-being (from here on referred to as *forests and people*). The focus on SDGs and forests is not new. Forests constitute several of the major global biomes; they are major contributors of ecosystem services, second only to oceans (Constanza et al. 2014). Forests are important not only for the 1.6 billion people who depend directly on them for their economic and non-material well-being (Chao 2012), but also for the entire human race because of the multiple services they provide, such as sequestering and storing large amounts of carbon and regulating weather patterns. In line with these assertions, most discussions on SDGs and forests address how forests or enhanced forest management can contribute to achieving the SDGs (FAO 2018, Seymour and Busch 2017). In this volume the analysis is reversed in order to explore potential impacts of the SDGs on forests and people.

A few prior studies already warn of likely adverse impacts of SDG implementation on forests (Swamy et al. 2018): for instance, the expansion of road building into forest areas under SDG 9 (Industry, Infrastructure and Innovation) or the expansion of agriculture into forests to achieve SDG 2 (Zero Hunger). These undesirable outcomes are to be expected since some SDGs are linked to factors commonly recognised as direct or indirect drivers of deforestation or forest degradation (Geist and Lambin 2002).

In each of the 17 SDG-focused chapters of this volume, the authors address the questions posed in Chapter 1: How does the pursuit of Agenda 2030, through the implementation of the 17 SDGs and their 169 targets, affect forests and the people who live at forest fringes and who depend on them for their material, social, cultural and emotional well-being? What are the positive and negative interactions among the SDGs and Targets? What are the probable impacts on forests and people? What are the projected repercussions of the possible synergies and likely trade-offs? Finally, what contextual conditions shape all these interactions, and how do they interact with the SDGs?

This chapter is organised as follows. Section 18.2 develops a conceptual model that illustrates the different dimensions of SDG implementation: positive and negative interactions, relevant contextual factors, and their impacts on forests and people. Section 18.3 summarises the evidence provided by the chapters on the impact of individual SDG implementation on forests. Section 18.4 does the same, but shifts the attention to impacts on people. Section 18.5 addresses potential synergies and trade-offs implicit in the implementation of the SDGs, and how contextual conditions influence interactions among SDGs and their subsequent impact on forests and people.

18.2 Multi-Layered Impact Pathways of the SDGs on Forests and People

The purpose of this section is to understand the dynamics or mechanisms by which SDGs influence forests and people. Human agency is a significant factor in the SDGs and how they link with forests and people. We chose to represent these linkages as processes of SDG implementation; when SDGs are implemented, their impacts on forests and people can vary. By 'SDG implementation' we mean the efforts pursued by multiple actors with the specific purpose of achieving one or multiple SDGs or targets. We structure the discussion around different implementation pathways. By 'SDG implementation pathway' we refer to efforts that begin at higher public administrative levels, either international or national, leading to efforts at lower public administrative levels all the way to the lowest level where efforts are expected to have direct impacts. We do recognise that there may be initiatives contributing to Agenda 2030 that do not follow such a linear pathway; however, we argue that these may be in response to what is referred to below as Pathway of Influence 2, shaped by discourses and ethical incentives (Bernstein and Cashore 2012).

One other conceptual positioning that we make here is that we chose to view Agenda 2030 as a global policy and the 17 SDGs as policy instruments to achieve this policy. We recognise that alternative lenses could have been utilised in this analysis – for instance, a systems dynamic perspective or a network conceptual framework (ISC 2017). Agenda 2030 and the SDGs represent global efforts initiated and implemented under the umbrella of the United Nations (UN), endorsed by its member countries and implemented nationally, as well as by multiple international agencies and other groups. As such, Agenda 2030 and the SDGs constitute a global policy regime (c.f. Krasner 1983). International regime theorists are particularly concerned with measuring impacts of international regimes (Humphreys 2016) and have proposed ways to measure such impacts (Sprinz and Helm 1999). In recent years, global governance has gradually taken over from regime theory as the dominant theoretical framework to reflect multiple efforts to govern issues of public interest at global scales. Global governance academic inquiry is more concerned with the complex interactions of actors and international instruments, such as UN conventions and the organisations with mandates to implement. The actions of these organisations are influenced by multiple constituencies, especially groups that are not part of statutory government.

A framework that has been proposed to analyse the impact of international policy instruments is the Pathways of Influence Framework (Bernstein and Cashore 2012, Cashore et al. 2016). This framework distinguishes four types of pathways: (1) direct compliance with international policy instruments; (2) following norms that emerge concurrently with the enactment of instruments and that are transmitted via discourses; (3) market-based incentives that encourage compliance with instruments; and (4) direct efforts to support implementation from international actors, such as international non-governmental organisations (NGOs) or overseas development assistance (ODA) organisations (Cashore et al. 2016).

The four Pathways of Influence – as well as international regime theory and global governance theory – recognise that governance efforts of international origin are implemented at national and subnational levels, and that impact analysis needs to reflect this. Formal policy impact monitoring is a well-developed field of modern public administration. Such monitoring is quite sector specific: for instance, monitoring of agricultural policies is quite different and is carried out independently from health-sector policy monitoring. This suggests that monitoring progress of SDG implementation will often be carried out on a piecemeal basis, differentiated by single SDGs. For the purpose of this chapter, however, our interest is not in learning the general impacts of SDG implementation but, specifically, their projected impacts on forests and people.

Agenda 2030 places major responsibility for SDG implementation on national governments. Countries are invited to prepare annual voluntary reports on progress with Agenda 2030 and SDG implementation (Kindornay 2018). To date, these are the best officially available accounts of how SDG implementation is taking place. The voluntary reports submitted so far demonstrate a few relevant developments. There is a widespread verbal embrace of Agenda 2030, but countries go about implementing the agenda and the goals via different approaches. Countries prioritise selected SDGs: the majority of countries that submitted voluntary progress reports in 2017 reported on progress on a subset of goals. While the key role of national governments is widely acknowledged, a myriad of other initiatives are underway to contribute to the SDG agenda. Consistent with the prevailing growing interest for the sustainable development idea, the private sector is encouraged to take leadership in this global effort, as is the academic sector through research and education. Many countries still have to adopt the underlying principles of Agenda 2030. For instance, many have not adjusted or integrated national development agendas with these principles. They have yet to identify linkages between SDGs and ongoing national policies, or to adopt integrated policies covering multiple goals (Kindornay 2018).

This state of affairs has important implications for understanding the impacts of Agenda 2030 and the SDGs on forests and people. To date, there is a paucity of empirical evidence to trace implementation, let alone to identify the degree to which integrated implementation of multiple SDGs leads to synergies or trade-offs related to forests and people. However, as is done in many of the SDG chapters of this volume, it is possible to draw on prior experiences with similar processes.

We propose a conceptual model that represents how the pursuit of Agenda 2030 and the SDGs will likely impact forests and people. The model is structured as follows. Agenda 2030 and the SDGs follow multiple pathways, and we locate these within the Pathways of Influence Framework. Agenda 2030 and the goals themselves serve as the starting points for the impact pathways. There are multiple agents along the influence pathways, with quite a few operating at the international level with coordinating or supporting functions. This coordination and support targets national governments as they are the mandated bodies to implement Agenda 2030 and the SDGs within each country. There is also a plethora of NGOs operating at multiple public administrative levels that support SDG implementation. We believe it is relevant to consider influences as far down as single forest landscapes to foster understanding of how SDG implementation actually affects forests and people – our primary interest in this volume.

As observed in the voluntary SDG implementation reports (Kindornay 2018), the actors who influence SDG implementation are primarily government actors and civil society organisations who have been given or have assumed co-responsibilities to implement the SDGs. Other actors – for instance, the private sector and forest-dependent community organisations – are likewise adopting the SDG discourse and playing an increasingly important role as they respond to incentives and constraining regulations that emerge from SDG implementation (Pathway 1) and to new discourses emerging from Agenda 2030 (Pathway 2). Private sector actors respond to market signals, related to economic opportunities, but also costs (Pathway 3). In addition, different actors along implementation pathways are experiencing direct influences (Pathway 4); for instance, when external agents undertake projects or training related to SDG implementation.

Key elements of this conceptual model (Figure 18.1) are contextual conditions influencing choices and decisions made along SDG impact pathways that lead to actions and eventual outcomes. According to this model, at the national level governments or their agencies prioritise SDGs, develop coherent and integrated implementation strategies, and allocate resources, responsibility and authority taking, into consideration priorities relating to economic development, nature conservation and social inclusion. The comparative weight placed on these interrelated dimensions of sustainability influences on SDG priority setting would ideally be manifested in: the development of integrated implementation strategies; the actual implementation of separate goals; the successful integration and coherence of efforts that

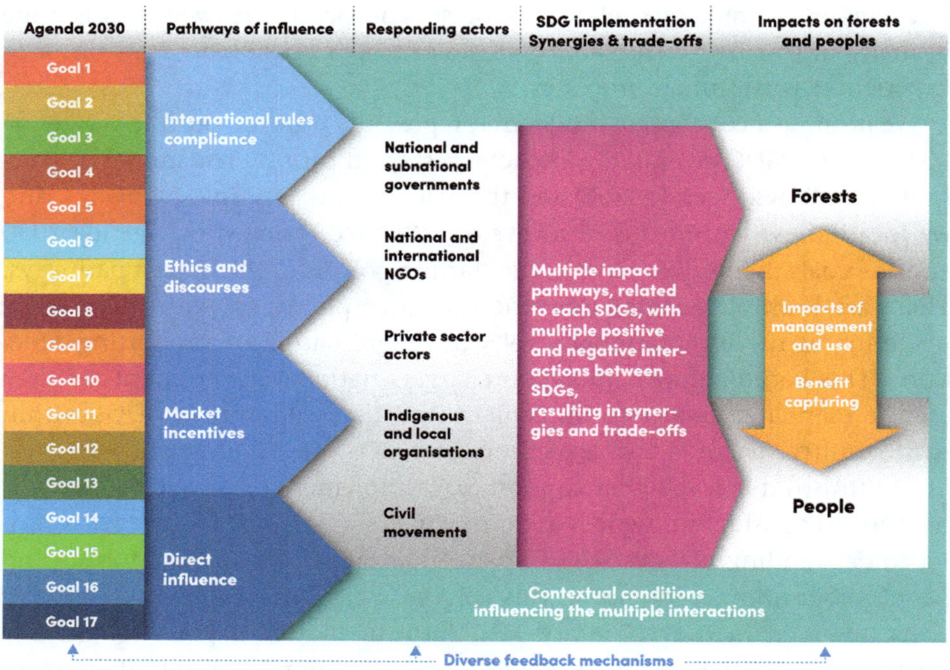

Figure 18.1 Conceptual model representing how the pursuit of Agenda 2030 and the SDGs will likely impact forests and people.

collectively seek to achieve the SDGs; or, even more broadly, the integration and coherence among multiple sectoral policies such that they align within the overarching Agenda 2030.

Contextual conditions play a key role along the SDG implementation pathway. These include a country's national development status and trajectory, or the overall condition of forests. Since contextual conditions vary widely, efforts to understand them are key to developing effective implementation strategies. Because of their recognised importance, many of these conditions are actually targeted by the SDG agenda. Like the SDGs, these conditions do not operate in isolation, but, rather, interact in complex ways. This results in a web of interactions of SDGs and contextual conditions, leading to observed and projected impacts on forests and people. Examples of these interactions are provided in Section 18.5. First, however, we provide a synthesis of how single SDGs and SDG groups are expected to impact forests (Section 18.3) and people (Section 18.4).

Finally, we do suggest feedback mechanisms within the model described above and presented graphically in Figure 18.1. These mechanisms have been built since Agenda 2030 was developed. We do not, however, address these further in this chapter.

18.3 SDG Implementation Impacts on Forests

Agenda 2030 and the 17 SDGs are gaining much momentum. The plethora of communications and initiatives at international and national levels indicate that the agenda and goals are being taken up along a normative pathway of influence (Pathway 2). Agenda 2030 is also gaining traction along Pathway 1, the international rules pathway. All of the countries that endorsed Agenda 2030 are taking action to implement at least part of the agenda, corresponding to a variable number of SDGs and associated targets. Increasing momentum is manifested in voluntary reports on SDG implementation and in numerous other reports on how countries envision SDG implementation (e.g. Schandl et al. 2016 for Malaysia) or on the administrative structure that has been set up to implement the agenda and goals (e.g. Dag Hammerskjold Foundation 2018). The latter example illustrates how Agenda 2030 is achieving influence through Pathway 4, the direct access pathway – for example, a Norwegian NGO is providing support to set up an administrative structure for SDG implementation in Colombia (Dag Hammerskjold Foundation 2018).

18.3.1 SDGs Contributing to Deforestation and Forest Degradation

We now turn to the following question: what is the influence of SDG implementation on important drivers of deforestation and forest degradation? The SDGs that can be linked to deforestation or forest degradation are primarily SDGs 1 (No Poverty), 2 (Zero Hunger), 7 (Affordable and Clean Energy) and 9 (Industry, Infrastructure and Innovation). Three less obvious SDGs that may contribute to deforestation and forest degradation are SDGs 5 (Gender Equality), 10 (Reduced Inequalities) and 14 (Life below Water).

The SDG that will arguably have the strongest effect on deforestation and forest degradation is SDG 9. We recognise that large-scale agricultural projects to produce oil palm or cattle are a major contributor to deforestation, while logging and fuelwood collection remain major culprits of forest degradation (FAO 2016). Expanding oil palm and cattle production, however, are not the result of agriculture production that would result from addressing SDGs 1 and 2. They might be a result of efforts carried out under SDG 9, if they are related to any SDG at all. We do recognise that small-scale agriculture can also contribute to deforestation, for instance in sub-Saharan Africa, and this may have implications for how SDGs 1 and 2 are addressed in those regions (Chapters 1 and 2, FAO 2016). The largest threat to forests linked to SDG 9 implementation is undoubtedly infrastructure expansion, which often encourages mega agro-industrial projects such as those referred to earlier. Plans for infrastructure expansions are far reaching, especially in Asia, South and Central

America and Africa. Likely, many countries will report ongoing infrastructure expansion plans as part of SDG 9 implementation. Infrastructure implementation will directly impact swathes of forests. Moreover, it will increase access to forests currently intact due to their remoteness. For instance, China is implementing the Belt and Road Initiative in much of Asia, but also in Africa and Europe (Rolland 2017). In sub-Saharan Africa, 33 development corridors are planned (Dulac 2013). The Initiative for the Integration of Regional Infrastructure of South America is continuing to expand roads, water ways and hydropower basins across the Amazon (Kis Madrid et al. 2011). These projects, among many others, are having major impacts on forests.

The ultimate impact of such programmes on forests (and on people) will depend greatly on how capacities and interests evolve to preserve or sustainably manage forests made accessible and vulnerable through infrastructure expansion, and the degree to which social and cultural impacts of the programmes are considered and addressed (Allen et al. 2018). If that happens, negative impacts on forests can be significantly reduced.

It is expected that SDG 9 will increase energy consumption from forests, especially when the number and activities of small and medium-sized enterprises are pursued vigorously in low- to middle-income countries (Chapter 9). This will contribute significantly to forest degradation. However, SDG 9 implementation may potentially result in the transformation of value chains that rely on wood, increase economic returns for the use of wood and thereby lead to a reduction of wood harvesting (Chapter 9).

The impact of SDGs 1 and 2 on forests is projected through an expansion of agricultural production in many places, which leads to deforestation and forest degradation. Land-use change driven by agricultural expansion is mostly linked to SDG 2. Achieving SDG 2 implies improving food provision to about a billion undernourished people, a number expected to increase; much of this will require facilitating agricultural production, especially in regions where food insecurity is problematic. To offset this possible environmental cost of SDG 2, Target 2.4 calls for an expansion of food production in such a way that natural habitats are not negatively affected. A major question remains about the extent to which progress can be made to boost the capacities or access to inputs among people who suffer from undernourishment so they are able to boost agricultural outputs while adopting methods that are in line with other SDGs, such as 13 (Climate Action) and 15 (Life on Land). Another challenge is to motivate public administrations charged with facilitating these tasks (Chapter 2).

Global food demand is expected to increase for years to come. According to some estimates, it is expected to double by 2050, driven by increased wealth among large numbers of people especially in Asia. Doubling food production in the next 30 years can be expected to result in the conversion of substantial areas of forests into agricultural lands, unless alternative options can be found (FAO 2016). Technological innovation will likely alter these estimates (Henry et al. 2018); however, this may not suffice to meet production needs in locations where undernourishment is most problematic. A growing population in forest areas may be another factor that complicates the relationship between food production and forest protection (Chapter 3), although in tropical forest regions such as the Amazon, for instance, it is the urban population that increases, while the rural population remains largely stable (FAO 2016).

While SDG 1 is not directly focused on food production, it is likely to lead to similar impacts on forests because a high proportion of the global poor live in rural areas, and supporting their emergence from poverty can most readily be achieved by boosting income from agriculture or other natural resource-based activities. Chapter 1 highlights two important points with regards to efforts to reduce poverty and how these may impact forests. The first focuses on how agriculture-centred strategies exert an influence on forests. This largely depends on where along the forest transition curve the agriculture-based poverty reduction is taking place, and the degree to which forest conservation and the mitigation of negative impacts on forests are specifically considered in national poverty-reduction strategies (Chapter 1).

The second point made in Chapter 1 relates to proposed direct money transfers to people below the poverty line, a widely supported measure under SDG 1. A significant portion of these payments are expected, at least in the short term, to contribute to the expansion of agriculture, resulting in further forest conversion. Remittances (another type of cash transfer) that people receive from relatives who have migrated overseas or to cities can be expected to have the same impact. SDG 10 includes among its targets improving facilities to streamline international remittances. If accomplished, increased investments in agricultural production are to be expected, including forested areas (Chapter 10).

SDG 7 (Affordable and Clean Energy) may result in significant impacts on forests due to mechanisms similar to those described for SDG 9. SDG 7 implementation may have negative and positive impacts on forest cover and forest condition (Chapter 7). Its successful implementation will reduce consumption of woodfuels, offset by increases in the use of hydrocarbon-based fuels or other cleaner energy sources. A reduction in traditional woodfuel use would reduce the negative impact of forest exploitation for energy, which is still significant in many parts of the world (Hosonuma et al. 2012). Future energy trends include turning to improved woodfuels, such as wood pellets, and the use of liquid biomass fuels, such as palm oil-based biodiesel, whose production may happen at the expense of forests. These options will replace hydrocarbon consumption to reduce greenhouse gas emissions, thus contributing to SDG 13 (Climate Action, Chapter 13). Switching to these energy sources and energy pathways can in principle be undertaken through sustainable production of biofuels. However, as Henry et al. (2018) suggest, achieving sustainable food production and sustainable bioenergy production is likely not to occur within the planetary boundary limit of 15 per cent of ice-free land designated for bio-production.

18.3.2 SDGs Contributing to Forest Restoration

The undesirable impacts described as resulting from trade-offs implicit in the pursuit of different SDGs is counteracted by two of the 17 SDGs: SDG 13 (Climate Action) and SDG 15 (Life on Land). The implementation of these two Goals is primarily expected to have positive impacts on forests, while the impact on forest peoples is less clear. SDG 13 adopts the Paris Agreement of the Parties to the United Nations Framework Convention on Climate Change as its implementation vehicle. The Paris Agreement considers forests as central to the pursuit of reducing carbon emissions into the atmosphere, as well as in the effort to support adaptation efforts where impacts of climate change are felt strongly. The Paris Agreement aims to reduce forest carbon emissions through the Reducing Emissions from Deforestation and forest Degradation and the role of conservation, sustainable management of forests and enhancement of forest carbon stocks in developing countries (REDD+) programme, achieving zero deforestation supply chains, but also through multiple efforts to restore forests and forest landscapes. While not strictly part of the Paris Agreement, multiple global commitments and initiatives are aiming for forest restoration, including the Bonn Challenges, the New York Declaration on Forests, the UN's Forum on Forests Program, the Asia-Pacific Economic Cooperation 2020 Forest Goals and many others (see Chapters 13 and 17).

The extent to which the implementation of SDGs 13 and 15 actually curbs negative impacts on forests brought on by the expansion of agricultural production lands and the expansion of infrastructure remains to be seen. These parallel agendas can be expected to play out and interact in complex ways that will vary greatly by context. Sayer et al. (Chapter 15) are sceptical that efforts planned under SDG 15, primarily intended to boost natural habitat protection and conservation of natural forests, will lead to the transformational change required in a dynamic world. To achieve this, SDG 15 should encourage innovative strategies and governance arrangements that transcend traditional approaches, which have often fallen short of expectations. The authors point out that SDG 15 and the associated targets fail to consider options that result in multiple benefits being captured by those (including forest peoples) who are in a position to assure forest integrity (Chapter 15).

Impacts of the implementation of other SDGs on forests can be mixed, depending on the specific SDG and contextual conditions. The pursuit of SDG 6 (Clean Water and Sanitation) will largely have a positive impact on forests, as forests are conserved in upper watersheds for the sake of assuring stable downstream water supply. However, forests may become consumers of huge quantities of water, when fast-growing, dense plantations are grown in watersheds. Forest restoration may not be pursued in some locations to avoid water consumption by forest plantations. Hence, this can be a negative outcome of the implementation of SDG 6 (Chapter 6). Efforts towards achieving sustainable cities (SDG 11) can have negative impacts on forests when there is an increased demand for timber or recreation in sensitive areas is boosted, (Chapter 11), but could be positive if green areas, including water catchments, are established or conserved. Success with the implementation of SDG 3 (Good Health and Well-being) will change demographics in locations where livelihoods rely on agriculture, which may increase the pressure on forests (Chapter 3). The achievement of better health for local people may result in greater fertility and thus increased population pressure; on the other hand, the emphasis on access to women's reproductive rights (Chapters 3, 5) may result in a gradual reduction of family size, as has been observed in many parts of the world. Lopez (2008) documents the adverse effects of HIV/AIDS on forest management in southern Africa. The implementation of SDG 4 (Quality Education) may improve environmental awareness regarding the impacts of modern societies on natural habitats, including forests (Chapter 4). This in turn may have a positive impact on forests, since knowledge and awareness precedes desired changes in behaviour. Similarly, the implementation of SDG 5 (Gender Equality) may boost concerns for the environment and forests where choices must be made that imply trade-offs between material benefits and environmental health (Chapter 5). At the local level, gender-balanced groups tend to do a better job of forest management than either all-male or all-female groups (Sun et al. 2011); the adverse effects on forest management of ignoring women's knowledge, goals and experience have been repeatedly shown. Similar reasoning applies to reducing inequalities within and across countries (SDG 10). Forest management can be improved when the concerns of various stakeholders are addressed and traditional rights are respected (Chapter 10).

Implementation of SDG 16 (Peace, Justice and Strong Institutions) can be expected to have largely positive outcomes on forests, as conflict, injustice, illegality and institutional vacuums are observed to be major indirect drivers of deforestation and forest degradation in many places (Geist and Lambin

2002). However, SDG 16 implementation primarily supports governments and upholding the rule of law. The extent to which this actually enhances broad, inclusive forest governance is questioned (Chapter 16). In addition, progress to be achieved under SDG 16 can in some cases lead to detrimental impacts on forests, for instance when armed conflicts are resolved and land use that was constrained during conflict becomes possible or even encouraged as part of peace agreements, as is occurring now in some parts of Colombia (Chapter 16).

A final comment is reserved for SDG 17 (Partnerships for the Goals), which aims to mobilise the means to implement the SDGs by promoting partnerships and cooperation. SDG 17 largely strives to facilitate the implementation of other SDGs, and thus its impact on forests depends very much on which SDG resources will be mobilised. In this light, partnerships and support to further sustainable forest management, forest restoration and forest emission reduction could be tracked. Insofar as such reviews are undertaken and conclusions relevant to forest conservation are integrated into practice, the impact of SDG 17 on forests should largely be positive (Chapter 17).

18.4 SDG Implementation Impacts on Forest People

As with the impacts on forests, the pursuit of Agenda 2030 and the 17 SDGs will impact forest-dependent peoples in both negative and positive ways. Changes in forest extent and condition will influence goods and services that people obtain from forests linked to livelihoods and well-being. The impacts that SDG implementation will have on forest people requires a separate discussion because the impact pathways from SDGs to forest-dependent people are substantially different from the impact pathways from SDGs to forests.

We focus primarily on the impact that SDG implementation has on people living in or near forests and who draw importantly from forest ecosystem services for their material or non-material well-being. We consider various pathways through which SDG implementation affects the benefits that people derive from forests. The first refers to SDG implementation with significant (positive or negative) impacts on forest conditions that, in turn, affect goods and services that forests provide to meet livelihood or well-being needs. An example of positive impacts on forests, but negative impacts on people, would be when priority is given to carbon stocks or biodiversity conservation and this results in reduced legal access to forest goods and services by local people. If SDG implementation has negative impacts on forests, that is likely to negatively affect goods and services provision.

The 17 SDG chapters of the book provide examples of this first pathway. Related to people's health, Chapter 3 points out that deforestation and forest degradation may, in some instances, increase exposure to infectious diseases. Deforestation and forest degradation influence the abundance and distribution of infectious disease vectors. Ebola, malaria and AIDS are examples of this dynamic (Gonzalez et al. 2008, Pattanayak et al. 2008). Expected changes in vector population dynamics resulting from climate change may exacerbate these trends.

The chapters also identify positive linkages between SDG implementation, the impact on forests and people's well-being. Improved water management (SDG 6) can result in better health for local people, reduced energy use for women who must carry water and improved life expectancy for babies. During the implementation of SDG 11 (Sustainable Cities and Communities), if surrounding forest landscapes are included in urban and peri-urban environmental management, this may have beneficial outcomes for both forests and forest residents. As pointed out earlier, poorly implemented city expansion may result in just the opposite effect and have negative outcomes for both forests and people.

A similar argument can be made for the implementation of SDG 13 (Climate Action) as well as SDG 15 (Life on Land). The implementation of both these SDGs, as already discussed, is likely to result in positive outcomes for forest cover (SDG 13) and forest conditions (SDG 15). However, both outcomes may very well be accomplished at the expense of forest people's well-being. Preserving forests or expanding forests for the sake of safeguarding or enhancing terrestrial carbon stocks, proposed under SDG 13, does have significant potential to generate benefits to forest residents. However, there is justified reason for concern that programmes such as REDD+ or large-scale forest restoration efforts may have negative impacts on forest residents whose forest uses may be unfairly constrained (Brown 2011, Scoones et al. 2013). The same holds true in the case of SDG 15 implementation. Command and control conservation has resulted in declining access to forest benefits among forest residents, or even eviction from locations where people have had long-term ties. Until SDG 15 implementation results in the transformational changes needed, these concerns are certainly warranted (Chapter 15).

We differentiate a second pathway of SDG impact on forest people. These are impacts from the implementation of SDGs that focus on people's conditions and capacities, specifically, SDGs 3, 4, 5, 8, 10 and 16 (Good Health and Well-being; Quality Education; Gender Equality; Decent Work and Economic Growth; Reduced Inequalities; Peace, Justice and Strong Institutions). These goals all aim to address conditions that directly affect people and their ability to beneficially engage in the economic, social, political and cultural life of the societies in which they live. The impact of the pursuit of these SDGs is primarily on people and less on habitats (e.g. forests). As with impacts on forests, impacts of SDG implementation on people depend very much on context. Contexts for forest people often include life in remote regions of countries where public administration and public services, performance of institutions and the rule of law differ from less-remote regions. Public institutional presence is often weak, and public services are commonly poor or even lacking altogether. Finally, special interests may apply to these marginalised regions because of conservation interests or wider interests in exploitable forest resources, prospective land for non-forest land use or other natural resources.

A clear challenge for Agenda 2030 and the SDGs relates to the difficulty of service provision to forest people. Notably, this is the case for healthcare and education. The provision of these services is more costly in remote regions and beneficiaries have less capacity to pay for services provided. Working conditions for teachers and healthcare providers are often poor and may even include security risks. At the same time, it can be argued that forest settings offer unique opportunities for both health and education. As for health, the availability of forest-based pharmacopoeia can, and often does, contribute to traditional healthcare provision in remote forest settings. This has already been recognised by many researchers on the topic, but also by healthcare providers. However, the actual adoption of local traditional knowledge into mainstream healthcare has been slow (Cunningham et al. 2008). Equally, traditional knowledge related to the environment could become an important contribution to curriculum development and improved, targeted education among forest people.

Another challenge that emerges from forest people's unique location is related to SDG 5 (Gender Equality). The degree to which gender inequality is a problem varies greatly from one forest context to another, with some Indigenous groups remarkably gender equitable. On the other hand, forest societies each have their own particular social relationships and power relations, including related to gender. Efforts undertaken under SDG 5 to address gender inequality may actually have negative outcomes if the local context is not adequately recognised (Chapter 5). In addition, a major obstacle related to moving forward with gender equality among forest people is that the forest sector, timber industry as well as forest administration, have traditionally been heavily male dominated. This can result in resistance to programmes seeking to enhance gender equality.

The remoteness of forest people jeopardises progress in SDG 8 (Decent Work and Economic Growth). There are fewer economic opportunities in locations that are less connected to mainstream economies. This is especially a challenge in forest settings that are in transition, from largely endogenous economies to economies that progressively rely on economic interactions with external markets, including labour markets. In that sense, forest settings pose major challenges for the implementation of SDG 10 (Reduced Inequality). Although remoteness can interfere with efforts to stimulate greater equity, increased interaction between remote populations and dominant national groups can in fact often lead to increased inequity. Dominant groups, such as higher castes in India or Malays in Malaysia, can import hierarchical social views, with local forest dwellers seen as the lowest rung on the social ladder (Bose 2011, Lin 2008).

As for governance issues, the focus of SDG 16, these are more difficult to address in more remote locations, for reasons similar to those given above. The governance issues pertaining to people living in tropical forests are legion (from formal governmental corruption to power plays by 'land grabbing' companies to inequities related to conservation projects). In addition, many of the violent conflicts that continue today play out in forested regions and particularly affect remote forest residents.

18.5 Impacts across SDGs: Drivers of Synergies and Trade-Offs

The SDGs and associated targets can have positive and negative impacts on forests and people. Agenda 2030 and the SDG framework have been proposed and designed to achieve positive outcomes for all, including forests and people. As we illustrate, there will be tensions among some goals and targets. While this has to do with how SDG implementation is planned and executed, there is a fair degree of inevitability of negative interactions among some SDGs. If an increase in agriculture production is an unavoidable choice to achieve SDGs 1 (No Poverty) and 2 (Zero Hunger), then this likely will have negative implications for SDG 15 (Life on Land).

Nevertheless, it is also widely recognised that there is tremendous potential for positive or synergistic interactions among SDGs and targets. The chapters in this volume provide ample evidence of this. Decent work (SDG 8) and progress in education (SDG 4) will contribute to improving people's health (SDG 3), for instance. Multiple similar positive interactions among the 17 SDGs and their targets can be recognised.

These positive interactions, however, do not necessarily imply positive outcomes for forests and people, which is the topic of interest here. We are very much interested in cases when the implementation of multiple SDGs will lead to probable synergistic outcomes for forests and people, and when there will likely be trade-offs. Understanding these interactions has great relevance to enhance capacity to take SDG implementation decisions that result in win–win outcomes. Furthermore, once the unavoidability of trade-offs can be fully recognised, more attractive alternatives can be chosen to mitigate negative impacts (Cashore et al. 2016: 31).

The authors of Chapters 1–17 were asked to explore synergies and trade-offs related to the particular SDG the chapter addressed, as related to the impact on forests and people. While each chapter carried out this exercise, some did so more than others. Synergies and trade-offs were also extensively discussed during our second author workshop in September 2018 in Alghero, Sardinia. The major overarching conclusion from this rich discussion is that synergies and trade-offs will depend very specifically on the particular impact pathway and the context in which the implementation takes place.

This section summarises what Chapters 1–17 and the participants at the Alghero workshop say about synergies and trade-offs related to SDG implementation and the impacts on forests and people, but also how the synergies and trade-offs and their impacts are influenced by contextual conditions. The section first reflects briefly on general positive and negative interactions among SDGs linking them to forests and people; it then summarises how synergies and trade-offs among SDGs are influenced by context conditions.

18.5.1 Positive and Negative Interactions among SDGs Related to Impacts on Forests and People

There is the possibility of negative interactions if the SDGs are looked at pair-wise, as has been done in Figure 18.2. This figure suggests how the interactions between two SDGs are expected to have an impact on forests or people. The figure was composed with information provided by the chapter authors, who assessed these interactions from the perspective of the SDG focused on in their chapter. The interactions may be synergistic (green cells), imply trade-offs (red cells), or possibly result in both synergies and trade-offs (yellow cells). The nature of the interactions is strongly influenced by the contextual conditions discussed in Section 18.5.2. The chapter authors' interpretations are reflected in the rows. The figure suggests, for instance, that the authors of Chapter 1 expect synergistic interactions between the implementation of SDG 1 (No Poverty) and six other SDGs (6: Clean water and Sanitation; 7: Affordable and Clean Energy; 10: Reduced Inequalities; 12: Responsible Consumption and Production; 13: Climate Action; 15: Life on Land). The likelihood of trade-offs is strong with SDGs 2 (Zero Hunger) and SDG 9 (Industry, Innovation and Infrastructure). The interactions with other SDGs may result in synergies, but also require trade-offs. Rows and columns in Figure 18.2 are not identical, for instance because the authors of Chapter 2 (SDG 2 Zero Hunger) expected both synergies and trade-offs resulting from interactions between SDG 2 and SDG 1.

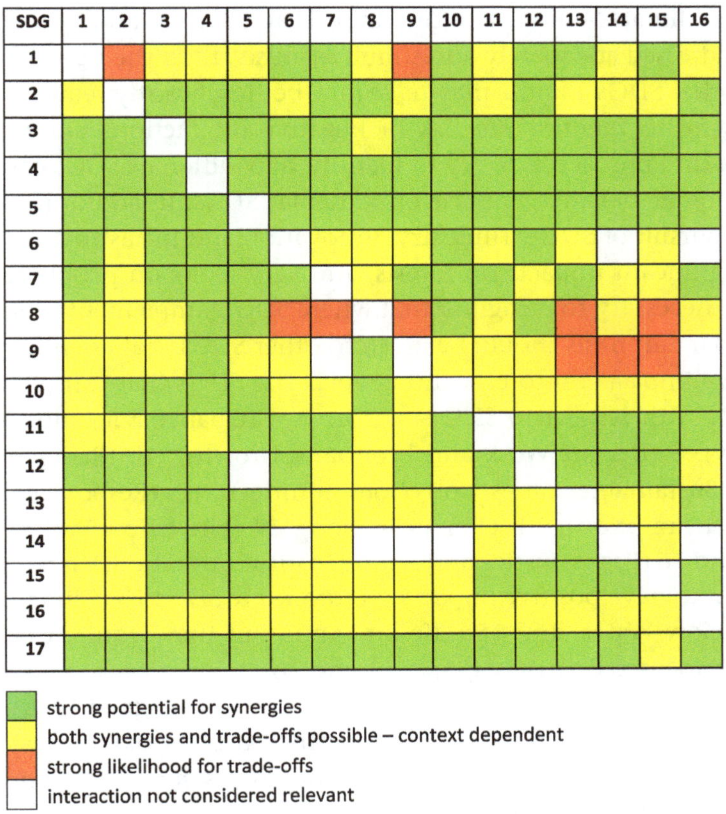

Figure 18.2 Potential synergies and trade-offs among the SDGs in relation to forests and people.

One positive conclusion that can be drawn from Figure 18.2 is that, at least according to the chapter authors, the majority of the interactions between SDG pairs is likely to be positive for forests or people, as is suggested by the 139 of the 289 squares in the table that are green, and the 97 squares that are yellow, which suggests synergistic interactions but also possibly trade-offs. These conclusions, however, are tentative, as the colours in Figure 18.2 reflect neither the magnitude of these interactions nor any implications for other SDGs. A red square may very well imply consequences for a swath of green squares (Katila et al. 2014).

More complex multiple linkages can be recognised among SDGs. For instance, the pursuit of SDG 2 (Zero Hunger) is linked to many other SDGs. Progress in this SDG will contribute to progress in many of the other SDGs simply because well-fed people are more capable of undertaking actions that contribute to other SDGs. The pursuit of SDG 1, but also SDG 2, can both have important negative impacts on forests. But, if several related SDGs, such as SDGs 4, 11, 13, 15 and 16 (Quality Education; Sustainable Cities and Communities; Climate Action; Life on Land; Peace, Justice and Strong Institutions) are addressed adequately and simultaneously, then the negative impacts on forests that SDGs 1 and 2 may have may be significantly reduced.

Rather than trying to identify the risk of negative interactions among SDGs, a more relevant approach is to try to identify how different SDGs can be mobilised to mitigate negative impacts of particular SDGs. In addition to SDGs 1 and 2, the pursuit of SDG 9 (Industry, Innovation and Infrastructure) is likely to have significant impacts on forests, and very likely on people as well. It is therefore necessary to recognise that where SDG 9 implementation is planned, particular emphasis should be given to other SDGs that are likely to mitigate negative impacts on forests and people. The SDGs that can play this mitigation role with respect to SDG 9 are SDGs 4 (Quality Education), 5 (Gender Equality), 8 (Decent Work and Economic Growth), 10 (Reduced Inequalities), 11 (Sustainable Cities and Communities), 12 (Responsible Production and Consumption), 13 (Climate Action), 14 (Life below Water), 15 (Life on Land) and 16 (Peace, Justice and Strong Institution). The SDGs that have great potential to have positive impacts on forests are SDGs 13 (Climate Action), 14 (Life below Water) and 15 (Life on Land), but they require that special attention be focused on SDGs 4, 5, 8, 11 and 16.

Summarising the evidence so far, and considering the full range of SDGs, it is possible to differentiate one cluster of SDGs with clear potential for positive synergistic interactions resulting in positive outcomes for forests and people. These are SDGs that primarily address institutions and governance conditions, as well as social conditions. They include SDGs 1 (No Poverty, when focusing on tenure and social protection), 3 (Good Health and Well-being), 4 (Quality Education), 5 (Gender Equality), 10 (Reduced Inequalities), 12 (Responsible Production and Consumption), 13 (Climate Action related policies) and 16 (Peace, Justice and Strong Institutions). All these SDGs have high numbers of green cells in Figure 18.2. On the other hand, there is one cluster of SDGs that affect land use directly: SDGs 2 (Zero Hunger), 6 (Clean Water and Sanitation), 7 (Affordable and Clean Energy), 8 (Decent Work and Economic Growth), 9 (Industry, Innovation and Infrastructure), 14 (Life below Water) and 15 (Life on Land). Among SDGs that have important implications for land use, there is a strong likelihood of trade-offs, but also of synergies – as between SDG 6 and SDG 15, or between SDG 7 and SDG 15. This suggests that the outcomes of Agenda 2030 on forests and people will importantly be defined by how the SDGs belonging to the SDG cluster that relates to institution, governance and social conditions are implemented and how much this can steer the implementation of and the interactions among the other SDGs to sustainable outcomes. This will to an important degree imply how potential trade-offs within the second cluster of SDGs can be managed and what the results will be for forests and people.

18.5.2 Contextual Conditions, SDGs Impacts, Synergies and Trade-Offs

The impacts of the SDGs on forests and people, as well as the positive and negative interactions among SDGs and how those affect forests and people, are influenced by what this chapter refers to as contextual conditions. These conditions differ between administrative levels where SDG-related actions take place: national, prefectural or provincial, district or municipal. The latter level in many cases represents the forest landscape scale.

Individual countries are developing their own national SDG implementation strategies while parallel, less-structured initiatives contribute to Agenda 2030. Which strategy a government decides upon will greatly depend on where they are along the national development curve. Countries with low per capita gross domestic product (GDP) or low human development index (HDI) will invest in pursuing SDG 1 (No Poverty), and likely they will try to achieve this by focusing on productive sectors, in which case much effort will also be made to move forward with SDG 9 (Industry, Innovation and Infrastructure) and SDG 8 (Decent Work and Economic Growth). The magnitude of impacts of such Agenda 2030 strategies will depend on how much institutional and governance issues are addressed – i.e. how much attention is given to SDGs 1, 3, 4, 5, 10, 12, 13, 16. We hypothesise that several key national indicators, such as per capita GDP, HDI, foreign direct investment and good governance indicators, eventually might explain the SDG implementation strategies that countries have chosen and their outcomes. Future empirical evidence will be required to prove or disprove this hypothesis.

Causal links can be hypothesised between the impact of SDG implementation and multiple indicators of socio-economic development. The same countries that score high on indicators related to education (SDG 4), good governance (SDG 16), and national equity (SDGs 5, 10) may see fewer negative impacts of those SDGs that are more directly linked to material wellbeing. Often, however, low scores on education, good governance and equity indicators are closely correlated with low per capita GDP, Gini coefficient or low HDI scores (Oliver and Oliver 2018). One might expect that a mitigating effect of education, good governance and equity may become evident when such countries do increase investment in these and related SDGs.

A second set of conditions that we expect will eventually turn out to shape SDG implementation and their impacts on forests and people is the relative influence of the four different pathways of influence in the Bernstein and Cashore (2012) framework. The four different pathways – International rule compliance, Discourses and ethics, Markets incentives and Direct influence (see Section 18.2) – can already be seen to have much relevance for SDG agendas and strategies in many countries. They influence the processes of SDG strategy development, and they also strongly influence what happens next. Countries will respond to pressures to implement Agenda 2030 because they will need to report back on progress; national governments or designated agencies will pass this on to lower levels of the administration (Pathway 1). There is a tremendous discursive influence of Agenda 2030 and the SDG framework, at the national level and at subnational levels; this is already mobilising enthusiasm and, ultimately, actions of multiple actors and constituencies in society (Pathway 2). While these forces appear to be unleashing significant bottom-up processes (e.g. in the USA, where they have contributed to the New Green Deal movement), it remains to be seen whether the same will play out in regions where Agenda 2030 is likely to have the greatest impacts on forests and people.

As is the case with the two previous pathways, much of what happens to national SDG strategy implementation will depend on the degree to which market forces influence processes on the ground towards progress on the SDGs (Pathway 3). Finally, countries that are offered and accept external support, not only for national SDG strategy development but also during implementation (Pathway 4), are likely to pay more attention to the institution and governance of SDGs as well as to the other SDGs that improve material well-being.

Implementing the SDGs will be a pluralistic response to different contexts. In any country, there is great variation in what forest landscapes look like, and within these landscapes, great variation exists in socio-cultural, biophysical and economic-institutional realities. These factors, as well as the characteristics of the forest landscape itself, influence how multiple actions related to the SDGs will be planned and implemented and eventually will impact forests and people. The type of forest, its extent and the biophysical condition of the landscape carrying the forest are of major influence. Colfer (2005) concludes that the following conditions seriously affect how external interventions affect forests and people: ownership of forests and land, population pressure, social and cultural diversity, social capital and related conflicts over land, forests or other resources, the mix of management goals that occur within a single forest landscape and forest type. There are many nuances that can be added to these overarching landscape and human conditions and how they possibly influence how pursuing the 17 SDGs impacts forests and people.

Besides the search for overarching conditions that influence the implementation of SDGs, we must also consider the notion that some governments may recognise the value of progressively involving wider constituencies such as local organisations, communities and their government structures as partners in the implementation of the SDGs. Insofar as that happens, the diverse contexts of forest peoples and forests will certainly be more meaningfully taken into account, the SDG implementation outcomes will be more finely tuned and appropriate for local conditions, and hence, likely to be more efficacious in terms of impacts on forests and people.

18.6 Conclusions

Agenda 2030 and the SDGs are a major global collective action to achieve equitable economic development and social inclusion for all, while ensuring environmental protection. The SDGs are meant to achieve these overarching interrelated goals in a concerted fashion, but collective progress will be challenged by inevitable interactions among goals. These interactions will, at times, result in mixed outcomes – for example, if progress on one goal gives rise to negative outcomes for forests and people. This volume seeks to shed light on the types of contextual conditions and circumstances that lead to undesirable trade-offs and those that lead to positive, synergistic outcomes. It does so in hopes of avoiding or mitigating the former and enhancing and reinforcing the latter.

Agenda 2030 and the SDGs can be understood as part of global governance or as an international regime. The Pathway of Influence Framework (Bernstein and Cashore 2012, Cashore et al. 2016) was utilised to illustrate that Agenda 2030 and the SDGs achieve influence along multiple pathways, involving governments striving to implement the SDGs and other actors motivated by the growing emphasis on sustainable development ethics, emerging market incentives and the presence of international development agencies and civil society organisations trying to support efforts domestically and across borders. The framework makes it possible to do justice to multiple efforts to pursue Agenda 2030, in addition to the formal government efforts.

Key central questions serve to guide the development of this volume. How is Agenda 2030 likely to impact forests and people? What are possible synergies and trade-offs among goals and targets? What are the contextual conditions that shape these various interactions? These questions are addressed from the viewpoint of single SDGs in Chapters 1–17 of this volume and the answers are summarised in this chapter (Sections 18.3 and 18.4). The analysis also explores potential positive and negative interactions among SDGs and contextual conditions that influence these interactions (Section 18.5).

Two broad groups of SDGs were differentiated. The first group primarily focuses on institutional, governance and social conditions (1: No Poverty; 3: Good Health and Well-being; 4: Quality Education; 5: Gender Equality; 10: Reduced Inequalities; 12: Responsible Production and Consumption; 13: Climate Action; 16: Peace, Justice and Strong Institutions). Many of these SDGs contribute to what has been termed an enabling environment for inclusive forest management and conservation with associated livelihood benefits (Katila et al. 2014). The second group of SDGs affect land use directly and thus are expected to directly impact forests (2: Zero Hunger; 6: Clean Water and Sanitation; 7: Affordable and Clean Energy; 8: Decent Work and Economic Growth; 9: Industry, Innovation and Infrastructure; 14: Life below Water; 15:

Life on Earth). Collective progress among the first group of SDGs is expected to result in positive (synergistic) interactions, favouring positive outcomes for forests and peoples. However, particularly poor performance in one or more of these SDGs can undermine progress on the other SDGs, highlighting the non-linear nature of the interactions among these SDGs.

With respect to the second group of SDGs, the potential for trade-offs is high, with important repercussions for forests and people. Understanding the potential for these trade-offs is essential in order to avoid implementation pathways that favour a small subset of these SDGs at the expense of the others. From the perspective of this volume, pursuit of the SDGs should foster the creation of an enabling environment that favours sustainable landscapes on which forests and peoples thrive.

References

Allen, C., Metternicht, G. and Wiedmann, T. 2018. Initial progress in implementing the Sustainable Development Goals (SDGs): A review of evidence from countries. *Sustainability Science*, published online May 2018. https://doi.org/10.1007/s11625-018-0572-3.

Bernstein, S. and Cashore, B. 2012. Complex global governance and domestic policies: four pathways of influence. *International Affairs* 88(3):585–604.

Bose, P. 2011. Forest Tenure Reform: Exclusion of Tribal Women's Rights in Semi-arid Rajasthan, India. *International Forestry Review* 13(2):220–32.

Brown, M. I. 2011. *Redeeming REDD: Policies, Incentives and Social feasibility for Avoided Deforestation*. London and New York: Routledge and Earthscan.

Cashore, B., Visseren-Hamakers, I., Caro Torres, P. et al. 2016. *Can legality verification enhance local rights to forest resources? Piloting the policy learning protocol in the Peruvian forest context.* International Union of Forest Research Organizations (IUFRO) and Yale University's Governance, Environment and Markets (GEM) Initiative.

Chao, S. 2012. *Forests peoples: Numbers across the world*. Moreton-in-Marsh: Forest Peoples Programme.

Colfer, C. J. 2005. *The complex forest: communities, uncertainty, and adaptive collaborative management*. Washington. Washington, DC: Resources for the Future and Center for International Forestry Research (CIFOR).

Constanza, R., de Groot, R. and Sutton, P. et al. 2014. Changes in the global value of ecosystem services. *Global Environmental Change* 26:152–8.

Cunningham, A. B., Shanley, P. and Laird, S. 2008. Health, Habitats and Medicinal Plant Use. In Colfer, C. J. (ed.) *Human health and forests: A global overview of issues, practice and policy*. London: Earthscan, pp. 35–62

Dag Hammerskjold Foundation 2018. Localising the 2030 Agenda in Colombia. *Development dialogue paper 25*. Uppsala: Dag Hammerskjold Foundation.

Dulac, J. 2013. *Global land transport infrastructure requirements: Estimating road and railway infrastructure capacity and costs to 2050*. Paris: International Energy Agency.

FAO 2016. *The State of the World's Forests. Forests and agriculture: land use challenges and opportunities*. Rome: FAO.

FAO 2018. *The State of the World's Forests. Forest pathways to sustainable development*. Rome: FAO.

Geist, H. J. and Lambin, E. F. 2002. Proximate causes and underlying driving forces of tropical deforestation. *Bioscience* 52(2):143–50.

Gonzalez, J. P., Ar Gouilh, M. Reynes, J. M. and Leroy, E. 2008. Bat-Borne Viral Diseases. In Colfer, C. J. (ed.) *Human health and forests: A global overview of issues, practice and policy*. London: Earthscan, pp. 161–96.

Henry, R. C. Engstroem, K., Olin, S. et al. 2018. Food supply and bioenergy production within the global cropland planetary boundary. *PLoS ONE* 13(3): e0194695. https://doi.org/10.1371/journal.pone.0194695

Hosonuma, N., Herold, M., De Sy, V. et al. 2012. An assessment of deforestation and forest degradation drivers in developing countries. *Environmental Resource Letters* 7:044009.

Humphreys, D. 2016. Integers, integrants and normative vectors: The problem of environmental policy integration under neoliberalism. *Environment and Planning C: Government and Policy* 34 (3):433–447.

International Council for Science 2017. *A guide to SDG Interactions: From science to implementation*. Paris: ICS.

Katila, P., Galloway, G., de Jong, W., Mery, G., Pacheco, P. (eds.) 2014. Forests under pressure: Local responses to global issues. *IUFRO World Series* Volume 32. Vienna. IUFRO Special. Available at: www.iufro.org/science/special/wfse/forests-pressure-local-responses/ (Accessed 3 March 2019).

Katila, P., de Jong, W., Galloway, G., Pokorny, B. and Pacheco, P. 2017. *Policy brief. Building on synergies: Harnessing community and smallholder forestry for Sustainable Development Goals*. Helsinki: IUFRO-WFSE.

Kindornay, S. 2018. *Progressing national SDGs implementation: An independent assessment of the voluntary national review reports submitted to the United Nations High-level Political Forum on Sustainable Development in 2017*. Ottawa: Canadian Council for International Co-operation.

Kis Madrid, C., Hickey, G. M. and Bouchard, M. A. 2011. Strategic environmental assessment effectiveness and the initiative for the integration of regional infrastructure in South America (IIRSA): A multiple case review. *Journal of Environmental Assessment Policy and Management* 13(4):515–40. https://doi.org/10.1142/S1464333211003997.

Krasner, S. D. (ed.) 1983. *International Regimes*. Ithaca: Cornell University Press.

Le Blanc, D. 2015. Towards integration at last? The Sustainable Development Goals as a network of targets. *DESA Working Paper* No. 141. ST/ESA/2015/DWP/141

Lin, C. Y. O. 2008. Autonomy Reconstituted: Social and Gendered Implications of Resettlement on the Orang Asli of Peninsular Malaysia. In Resurreccion, B. P. and Elmhirst, R. (eds.) *Gender and Natural Resource Management: Livelihoods, Mobility and Interventions*. London: Earthscan, pp. 109–26.

Lopez, P. 2008. The Subversive Links between HIV/AIDS and the Forest Sector. In Colfer, C. J. (ed.) *Human health and forests: A global overview of issues, practice and policy*. London: Earthscan, pp. 221–37.

Nilsson, M., Griggs, D. and Visbeck, M. 2016. Map the interactions between Sustainable Development Goals. *Nature* 435:320–2.

Nunes, A. R., Lee, K. and O'Riordan T. 2018. The importance of an integrating framework for achieving the Sustainable Development Goals: the example of health and well-being. *BMJ Global Health* 2016;1:e000068. doi:10.1136/bmjgh-2016-000068.

OECD 2018. *Supporting an integrated implementation of the SDGs: Tools for addressing SDG connections and enhancing policy and institutional coherence*. Available at: www.oecd.org/governance/pcsd/Learning%20session%20HLPF%202018_Tools.pdf (Accessed 3 March 2019).

Oliver, C. and Oliver, F. A. 2018. *Global Resources and the Environment*. Cambridge: Cambridge University Press.

Pattanayak, S. K. and Yasuoka, J. 2008. Deforestation and Malaria: Revisiting the Human Ecology Perspective. In Colfer, C. J. (ed.) *Human Health and Forests: A Global Overview of Issues, Practice and Policy*. London: Earthscan, pp. 197–217.

Rolland, N. 2017. *China's Eurasian Century? Political and Strategic Implications of the Belt and Road Initiative*. National Bureau of Asian Research. Available at: https://muse.jhu.edu/ (Accessed 23 February 019).

Schandl, H., West, J., Baynes, T. and Hosking, K. 2016. *Natural Resources and the SDGs: A country snapshot for Malaysia*. Canberra: CSIRO.

Scoones, I., Fairhead, J. and Leach, M. 2013. Green Grabbing: A New Appropriation of Nature. *The Journal of Peasant Studies* 39(2):237–261.

Seymour, F. and Busch, J. 2017. *Why forests? Why now? The science, economics and politics of tropical forests and climate change*. Washington, DC: Center for Global Development.

Sprinz, D. F. and Helm, C. 1999. The Effect of Global Environmental Regimes: A Measurement Concept. *International Political Science Review* 20(4): 359–69. https://doi.org/10.1177/0192512199204003

Stafford-Smith, M., Griggs, D. Gaffney, O. et al. 2017. Integration: The key to implementing the Sustainable Development Goals. *Sustainability Science* 12:911–19. DOI 10.1007/s11625-016–0383-3.

Sun, Y., Mwangi, E. and Meinzen-Dick, R. 2011. Is Gender an Important Factor Influencing User Groups' Property Rights and Forestry Governance? Empirical Analysis from East Africa and Latin America. *International Forestry Review* 13(2):205–19.

Swamy, L., Drazen, E., Johnson, W. R. and Bukoski, J. J. 2018. The future of tropical forests under the United Nations Sustainable Development Goals. *Journal of Sustainable Forestry* 37(2):221–56. doi:10.1080/10549811.2017.1416477.

Waage, J., Yap, C., Bell, S. et al. 2015. Governing Sustainable Development Goals: interactions, infrastructures, and institutions. In Waage, J. and Yap, C. (eds.) *Thinking Beyond Sectors for Sustainable Development*. London: Ubiquity Press, pp. 79–88. http://dx.doi.org/10.5334/bao.i

Chapter 19 The Impacts of the Sustainable Development Goals on Forests and People – Conclusions and the Way Forward

Georg Winkel, Glenn Galloway, Carol J. Pierce Colfer, Wil de Jong, Pia Katila and Pablo Pacheco

19.1 Introduction

Since time immemorial, forests have served as a foundation for human development. Across the globe, the evolution of societies, and their welfare, has been closely connected to various uses of forests. Through this close association, humans have shaped and transformed forest landscapes in many parts of the world. The forest transition literature is illustrative of this connection, focusing on the interrelation of humans and forests along a development trajectory, and seeking to explain the regularity of a particular pattern across societies: that forest cover declines until a turning point – transition – after which gains in forest cover due to natural regeneration and plantations overtake deforestation losses (Angelsen and Rudel 2013, Mather 1992; see also Chapter 1).

Drawing on an analogy to ecosystem research, the forest–people interface can be perceived as an ongoing (socio-ecological) co-evolution process. This co-evolution of people and forests is characterised by constant change. Relationships range from symbiosis, to co-existence, to sometimes predatory exploitation that can even lead to extinction (i.e. permanent forest destruction). As with co-evolution in ecosystems, abundance and mutation are important driving factors of the process. For the forest–people interface, abundance relates to how many people are present in relation to a forest area, possibly transforming the demand for ecosystem services from a sustainable pattern into an unsustainable one. Mutation acts as an analogy for both the evolving human needs and desires towards forests (e.g. related to economic growth and accumulation) and the new technical capacities people develop to more efficiently manage, but also exploit or destroy, forests. Both factors determine the human side of this socio-ecological system.

On the side of forest ecosystems, such human interventions have led to adaptations, major changes of the forest landscape with shifting species compositions, altered biological dynamics and processes, and, taken to the extreme, have led to their transformation into agricultural or urban ecosystems.

The SDGs may impact this co-evolution process from multiple perspectives. They merge several previously separated policy concepts into a comprehensive – though not necessarily coherent – development agenda, a 'snapshot' of different elements of development thinking at a certain time. This agenda is so broad that it allows many societal groups with partially contradicting demands and interests to identify with it. At the same time, it provides space for interpretation and adaptation at those levels where trade-offs need to be resolved and priorities need to be agreed upon.

In this chapter, recalling the main purpose of this book – to explore the impacts of the SDGs on forests and people – the attempt is made to offer broad conclusions on key lessons learnt from the comprehensive assessment carried out in this book. This is done with a view to providing guidance for the future co-evolution of people and forests in a changing world, in a way that acknowledges dependency and stewardship and works towards symbiosis over parasitism.

19.2 Lessons Learnt

1) While forests play a critical role in sustainable development, the SDGs will impact forests and the people dependent on them in a multitude of ways

This first conclusion is at the heart of this book and is supported by the comprehensive evidence presented throughout. Forests are critically important for human development, not only from a historical perspective but in today's world as well. They are one of the three major human life-supporting ecosystems of the planet, along with agricultural lands and oceans. They currently cover one-third of the global land area, contain 80 per cent of terrestrial biomass and provide habitat for more than half of the world's known terrestrial plant and animal species (Aerts and Honnay 2011; see also Chapter 15). Approximately 1.6 billion people directly depend on forests and the ecosystem services they provide for their livelihoods, and a staggering 2.8 billion people, mostly in the developing world, burn woodfuel for their basic energy supply (see Chapter 7). Furthermore, everyone on this planet depends on the world's forests for their essential role in global carbon, water and nutrient cycles, which impact climate, food and urban systems (see Chapters 13, 6, 2 and 11, respectively). Clearly, human well-being and development hinges on the existence of forests and their ecosystem services. This needs to be acknowledged and considered by policymakers who pursue progress towards sustainable development around the world.

While research has shown the universal scale of dependency of humanity on forests, this book has shed light on the huge variety of contextual settings of how the connections between forests and people play out on the ground. Relations range from emotional and spiritual connectedness, e.g. in relation to recreation that is perceived being a priority in several urbanised agglomerations (Chapter 11), up to total dependence for livelihoods in remote rural areas. Significantly, the comprehensive assessment carried out in this book shows that the needs and interests of people do not always align with the vision to maintain permanent forests. While forests provide critical ecosystem services for societies, there are also cases where potential 'disservices' exist, e.g. relating to specific forest-borne human diseases (Chapter 3). Furthermore, increasing populations as well as economic and social development have been and often continue to be accompanied by the exploitation, degradation and loss of forests, particularly in the early stages of the forest transition process. Historic and ongoing global deforestation processes may serve as an indication for a higher economic and sometimes also sociocultural appreciation of other land use systems in the context of growing populations and economies. In turn, the resurgence of forest (area) in North America, Europa and East Asia indicates a new appreciation of forests after 'development' has advanced, albeit that various reasons drive this observable reforestation process (Forest Europe 2015).

Hence, societal attitudes and needs regarding forests and decisions impacting them are highly context dependent and continually evolving. Overarching trends with respect to forests in specific contexts will often reflect predominant societal (and political) conceptualisations of forests in these contexts, through the lens of socio-economic needs and demands, including the degree to which ecosystem services are recognised and valued. An important caveat relates to the fact that different people in the same country or even region possess quite distinct perceptions regarding forests, their value and people's needs with regard to their ecosystem services (Racevskis and Lupi 2006, Tyrväinen et al. 2003). When forest exploitation and loss intensify, for example, traditional forest-dependent groups will be under pressure to adapt, i.e. might be forced to change their way of life (Hobley 2005, Meyfroidt et al. 2013). Moreover, they often lack the power and political clout to influence decisions impacting their livelihoods. Meanwhile, other sectors of society – for example farmers, or those in urban centres or consumers in countries located far away – through global trade and markets, might at least temporarily benefit from the additional resources mobilised and profits generated (Angelsen and Kaimowitz 1998, Angelsen and Wunder 2003). The frequent reality of 'winners' and 'losers' creates challenging questions relating to who has the right to determine the contribution of forests to human development, particularly in cases of conflicting forest management and land use options. It also calls for the necessity to agree or compromise across scales, if winners and losers are to be found in different spaces. The complexity and contested nature of decisions regarding forest management and land-use change will be further increased when a biocentric perspective is considered, emphasising the necessity to preserve forests for their natural beauty and ecosystem value per se. The same holds true for considering aspects of intergenerational justice that lies at the heart of the sustainability concept (UN 1987).

From the perspective of the SDGs, it can first be concluded that human needs shape the value people place on forests; these needs are again affected by different contextual factors and are interrelated to societal and political settings. This might influence the implementation of the SDGs, and with that the impact of the SDGs on forests and people in a specific context. There is no perfect forest for all, and there is no perfect forest–people interface. While humans are dependent on forests, the relationship between humans and forests is characterised by an ongoing co-evolution process that can also be dominated by antagonism. Second, people and their interests are very diverse. The implementation of one or more SDGs will, in many cases, result in winners and losers, depending on the impacts on forests. In these situations, societal groups with more resources and power to influence investments, markets and policy decisions may prevail over other groups, including possibly those whose livelihoods are most dependent on forests. Since the benefits of the SDGs are meant to be inclusive and equitable, efforts must be made to create awareness of potential trade-offs and transparent mechanisms to address them. Third, the assumption of an a priori positive correlation between forest conservation and societal development is misleading. While one critique this book offers is that the SDGs rarely mention forests explicitly and seem to underestimate the importance of forests for human sustainable development – or take it for granted – this criticism should not be taken to imply that forests should be prioritised in every case. Expanding forest area is not always the best answer to complex development needs. Some of the SDGs might result in forest loss but drive social and economic development, e.g. through agricultural expansion or more space for housing and infrastructure. What is crucially important is that potential trade-offs implicit in the SDGs with respect to forests and other land uses are understood and are made transparent to societies, and that the forest–people interconnectedness is fully accounted for in societal and policy decisions. This must include thinking across different scales and generations. It must also include giving voice to forest-dependent people, who are at risk of being disregarded by efforts meant to advance the SDG agenda. Focusing attention on the potential trade-offs associated with the implementation of the SDGs and how they will impact the forest–people interface is a primary contribution this book makes to the global sustainable development debate.

2) The SDGs include distinct and partially conflicting visions for forests and people, involving the necessity to consider trade-offs and set priorities

The main chapters of this book analyse the complex interrelations among the SDGs and their potential impacts on forest and people. These interrelations are sometimes synergistic, and other times conflictive in nature, reflecting the breadth and complexity of the Agenda for Sustainable Development. In fact, the SDGs integrate various and partly competing policy discourses, paradigms or worldviews within one framework as an outcome of a participatory negotiation process and political lobbying. These include, for instance, convictions regarding: (a) the importance of equity versus freedom/competition; (b) the importance of domestic versus global consumers; (c) the necessity to have economic growth; (d) the diverse and sometimes contradicting consumer and producer motivations and behaviour; (e) the importance of markets and rules (e.g. relating to free trade versus regulations/protectionism), and (f) the emphasis on individual versus common welfare. The SDGs and their targets can be clustered around the classical three dimensions of sustainable development and related worldviews emphasising them, distinguishing social (e.g. SGD 1: No Poverty, SDG 3: Good Health and Well-Being, SDG 4: Quality Education, SDG 5: Gender Equality, SDG 10: Reduced Inequalities, SDG 16: Peace, Justice and Strong Institutions), economic (e.g. SDG 7: Affordable and Clean Energy, SDG 8: Decent Work and Economic Growth, SDG 9: Industry, Innovation and Infrastructure, SDG 12: Responsible Consumption and Production) and environmentally focused SDGs (e.g. SDG 13: Climate Action, SDG 14 Life below Water: SDG 15 Life on Land). Major lines of potential trade-offs develop between these SDG clusters. It could be argued that the most critical underlying fault line in the SDGs, when it comes to forests and people, concerns the unresolved question of the degree to which economic growth is the solution to achieve sustainable development, or, rather, the problem that will prevent it. The following statistics are emblematic of this paradigmatic dilemma: between 1970 and 2010, the global economy tripled in size (UNEP 2016, see Chapter 9); hundreds of millions of people have escaped poverty, and economic power is distributed in a more multipolar manner across the planet than in the past. At the same time, 1 per cent of the world population is estimated to own more than half of the planet's wealth (see Chapter 9 referencing Oxfam 2016). Between 1990 and 2015, global forest cover decreased by 3.1 per cent, to 30.6 per cent of the global land area, with most forest loss occurring in the tropics (FAO 2015, see Chapter 15); 4 of 10 planetary boundaries have been crossed and others are seriously at risk (Steffen et al. 2015). The quandary is the increasing tension between economic development (growth) and an ambivalent and inequitable social development, as well as largely negative environmental impacts: how can all this be addressed under the umbrella of a holistic, sustainable development paradigm? While this question extends well beyond concerns relating to forests and people, processes that play out in the forest–people interface exemplify the challenge.

The SDGs remain ambiguous on how to proceed here. Some SDG goals are more in line with a general economic growth paradigm (e.g. SDGs 8: Decent Work and Economic Growth, and 9: Industry, Innovation and Infrastructure). They promote an ecological modernisation approach in line with assumptions that economic growth, trade and technological progress will eventually resolve the trade-offs between increasing wealth and environmental exploitation through greater efficiency, accompanying better regulation and increasing shares of renewable energy and materials (see Chapter 9). Other SDG goals and their targets can be read as more 'transformational'. They address shortcomings of the current economic system and world trade, emphasising unjust distribution and environmental degradation (e.g. SDGs 1: No Poverty, 10: Reduced Inequalities, and 16: Peace, Justice and Strong Institutions). The potentially tremendous friction and contradictions between both worldviews remain unresolved.

These worldviews correspond with global forest-related discourses, which again relate to broader environmental and development discourses (Arts et al. 2010). These forest-related discourses set different priorities regarding the preservation or sustainable use of forests: finding the most feasible approach to conserve them by integrating them into (increasingly responsible) markets or shielding them from markets (assumed to remain irresponsible). Worldviews inherent to the SDGs also resonate with the 'cultural biases' in forest policy suggested by Sotirov and Winkel (2016) in reference to the Cultural Theory's global typology of society–nature interrelations, ranging from laissez faire/free-market individualism to rule-based sustainable forest management or participatory environmental protection (Thompson et al. 1990).

It is certainly beyond the scope of this book to resolve such basic ideological controversies. Yet there is a need to acknowledge their importance, to investigate related trade-offs and conflicts implicit in the SDGs and envision measures to deal with these. Different worldviews foresee different pathways towards the sustainability of the forest–people interface. This book often conveys a rather sceptical assessment of the potential effectiveness of ecological modernisation thinking, which would assume in a simplified manner the possibility of green (economic) growth, to achieve the goals of the 2030 Agenda. Calls for a more transformative approach, possibly moving away from economic growth thinking to implement the SDGs, or even redesigning the SDGs themselves, figure prominently in some chapters. This criticism may, at least, be seen as an evidence-based statement that a business-as-usual development pathway will not suffice to achieve sustainable development, particularly when considering the environmental dimension and the evidence of growing inequality. It may also show a certain scepticism about whether the SDGs (and ecological modernisation thinking) have enough transformational power and ambition to overcome the inertia of the current pathway. Yet, how far ecological modernisation can bring the co-evolution of forests and people onto a more sustainable track in different settings remains to be seen. An alternative mix of fewer social and environmental science authors and more engineers and neoclassical economists might have assessed the status quo and prospects for forests and people in a more positive or optimistic fashion, e.g. regarding the potential for forests to contribute to a low-carbon economy and society of the future through efficient use (e.g. Hetemäki et al. 2017).

Ultimately, it will be at least partially a matter of opinion how far ecological modernisation thinking can go and whether (more) transformational approaches are needed to balance the sustainable development of forests and people. Combining this with the initial finding herein – that there is no perfect forest for all demands, and that trade-offs are frequent among different forest ecosystem services, and more so, among different land use types – calls for the need to engage in the SDG implementation across ideological boundaries. This will include making the inherent frictions between major governance, related land use and forest management paradigms visible, but also investigating how different paradigms play out in a specific implementation context. Yet, this also needs a better understanding of what implementing the SDGs might actually mean – a point for reflection under the next lessons learnt point.

3) The SDGs impacts on forests and people are context dependent and may be locked into path-dependent socio-economic development trajectories, but global interdependencies remain

In an ideal world, wise men and women in government would sit together and develop a coherent implementation plan for the SDGs that engages all sectors of society in order to achieve the SDG targets in a well-coordinated implementation process. However, taking into consideration early, often-cited lessons learnt from policy implementation research (Pressman and Wildavsky 1984), a more realistic view recognises that SDG policy on paper will not necessarily translate into corresponding implementation practices on the ground.

While theoretical assumptions about pathways of implementation/influence can be made (see the model in Chapter 18), and national reports indicate that several activities are taken up at different levels to implement the SDGs, a key finding of this book is that the real effects of SDG implementation – and the interplay among them – are often not (yet) known. The assessment of each SDG for potential impacts on forests and people had to assume it would be implemented, and/or to estimate how implementation could play out, and then anticipate impacts of such assumed implementation in different contexts. In addition, authors assessed evidence on trends that are currently occurring, but without being able to connect them (at this point in time) causally to the SDGs.

This limitation of the book is unavoidable, as there is no better evidence yet available. Clearly, there is a need for research to better understand alternative impact pathways of how the SDGs unfold in different contexts. This could, for instance, take the format of a comparative SDG study across several countries. Nevertheless, this book has provided a substantive basis for estimating real-world impacts of the SDGs. In this regard, one key finding from this book is that the same SDGs and the same targets will often lead to different effects in different contexts. Countries will set their own priorities, and people at the forest–people interface will harbour different attitudes regarding these priorities. Furthermore, the prioritisation will probably merge with past socio-economic and political development trajectories that determine forest governance and management. Frankly, it would be naive to assume that the SDGs could be implemented without fitting them into pre-existing sectoral development pathways. This integration into the existing policy and socio-economic context will largely determine their potential impacts. It will also result in different spatial-temporal responses, and related impact patterns, in relation to the SDGs.

A highly interesting question, then, is if it is possible to detect similar context patterns resulting in comparable SDG implementation routes and impact pathways. In other words, will countries with similar socio-economic and ecological situations set the same priorities regarding the SDGs, resulting in similar outcomes? Moreover, can we then assume a certain temporal succession in the SDGs' impact over time, e.g. along the forest transition curve? This could, for instance, play out through a shifting focus from social SDGs towards environmental ones regarding forests – from food or fighting poverty to biodiversity – with economic ones likely remaining important throughout the transition process. If such a sequential order exists, it poses the uncomfortable question of whether (some) forest destruction needs to be accepted when it enables development and decreases poverty, and this way again later might enable a less exploitative forest–people interface, following the forest transition thinking.

Yet, development pathways and directions and the impact of contextual factors do not seem to follow one consistent pattern. Taking a closer look at the forest–people interface in several countries reveals that many different development states relate to different forest-management and land-use situations. Context dependency results in distinct forest–people interface co-evolution pathways that can be identified across countries: e.g. a 'boreal timber economy trajectory', an Anglo-Saxon 'splitting the land in conserved and heavily used forests trajectory', a Central European 'integrated forest management trajectory', or a 'tropical deforestation and then forest restoration through plantations and natural regeneration trajectory'. These trajectories relate specific ecological forest settings to specific socio-economic situations and cultural paradigms and convictions about the forest–people interface. Depending on a country's trajectory, the SDGs may not only be interpreted and implemented differently, but may also result in quite distinct impacts on both forests and people. Major processes along the forest transition timeline may then add to these spatially and contextually distinct development pathways.

Context can play out even more dramatically. In several forest-rich world regions, poverty, rapid population growth or even violent conflicts are prevalent. In such contexts, the SDGs – if recognised at all – will be interpreted differently, with emphasis on those basic values that ensure survival and livelihoods. Forest (over-)exploitation may seem more justified in these contexts if it is meant to address local basic needs, yet this does not lessen the environmental footprint. With that comes an important message: while context will be key to the SDGs impacts on forests and people, the spillovers and interconnections across scales by means of ecological process and footprints, trade, political collaboration and partnership call for continued, if not intensified, engagement across contexts and scales. Specifically, it calls for international exchange, collaboration and cooperation. Moreover, recalling the picture of the forest transition process, the crucial question remains how far the environmentally destructive patterns of the process can be mitigated, or even reversed, without curbing development. This may, however, also include the necessity to reflect on what socio-economic development means in relation to its environmental impact. The SDGs do not provide a consistent response to this major strategic question. Responses will likely differ depending on the specific forest–people interface co-evolution pathway at hand.

4) There are fundamental values and principles that may guide sustainable development related to forests and people

In the first three lessons learnt, we have focused on the multifaceted nature of the SDGs and their relation to forests and people. We have shown that the SDGs set a normative framework, but one full of potential contradictions. We have emphasised that there is no perfect forest for all people and demands. We have hence indicated the necessity to regulate trade-offs relating to values, interests and convictions about how to best govern sustainable development at the forest–people interface.

In this section, we partially reverse this relativism and seek to (1) elicit some fundamental consensual values of the SDGs that might guide their implementation in relation to forests and people globally, and (2) consider if there are related fundamental procedural principles that can guide decision-making in more detail during the implementation.

First, it is critical to recall fundamental principles of global importance such as human survival, human rights, peace, justice, welfare and democracy, or referring to the Agenda 2030: peace, planet, prosperity, people and partnership. Most on this planet will agree that these principles need to be respected (which does not mean that they always are). Implementing them in the context of the forest–people interface should be beyond dispute.

Second, forest-specific considerations may be added to the list. These may refer to the necessity to consider the long-term cycles of forest ecosystems, connected to the challenge of irreplaceability. This means, for instance, that it can be nearly impossible to restore a primary forest once it has been destroyed, and it is impossible to regain a species that has been lost. This also relates to aspects of intergenerational justice and equity. In a more classical forester's understanding, this includes the basic axiom that a massive overuse of forest biomass will lead to an undersupply in the future. To this list, aspects of space and related justice need to be added – e.g. the regional benefit of cutting down a forest may jeopardise climate change mitigation, a globally important ecosystem service. Moreover, the multiple values of forests for society need to be considered. These considerations may lead to a fundamental forest-related SDG implementation principle: to maintain resilient forest landscapes that, now and in the future, provide ecosystem services that respond to various societal demands originating from people across multiple locations, and with evolving demands over time.

Third, connecting this forest-specific principle to the fundamental human-related ones leads to acknowledging the special rights of those half-a-billion people that directly depend on forest ecosystem services, including the right to maintain forest-based livelihoods.

Fourth, recalling the contradicting targets and worldviews that are conceptually embedded in the SDGs, another fundamental principle could address the process of implementing the SDGs. Here, postulating a basic principle of good governance in resolving the trade-offs inherent to the SDGs is recommendable. This includes the necessity to accept that a diversity of viewpoints, societal needs, cultural values and interests exists, and that these should be considered in forest policy and management decision-making. Different groups should have access to the policymaking and implementation process at all levels; transparent decision-making processes perceived by everyone as fair and effective could be the procedural key recommendation facing non-resolved value and interest conflicts.

5) Sustainable development at the forest–people interface will not happen without recognising and tackling major driving forces of (un)sustainable development in general, and integrating forest governance and management into a broader landscape perspective

One important conclusion from this book is that the fate of the world's forests, and of forest-dependent people, will mostly be decided outside the forests, and mostly not by forest-dependent people. The continuing expansion of agriculture, for food and feed production, is by far the largest direct driver of deforestation, causing approximately 80 per cent of forest loss (see Chapter 2). This means, first, that the agricultural sector and the larger context of factors driving that sector are crucial for the sustainable co-evolution of forests and people. Secondly, recalling Figure 18.1 in Chapter 18, the entire socio-ecological system determining the planet's land use needs to be assessed to understand the impacts of the SDGs on the forest–people interface. Essentially, this means looking into major driving forces such as population and economic growth, consumption and its environmental footprint, climate change, corruption and failing governance, technology (access and funding), the fundamental role of culture and religion in driving behavioural patterns, and the prevalence of wars and violent conflicts in many world regions. The assessments done in this book raise questions regarding the degree to which the SDGs are suited to address these overarching driving forces of (un-)sustainable development. For instance, Chapter 3 emphasises the critical importance of global population growth on the depletion of the planet's natural resources while pointing out that this issue is not really addressed in the SDGs for being politically inappropriate. Regarding economic growth, the ambivalent positioning of the SDGs on this driver was discussed earlier. At the same time, the planetary boundaries theory (Steffen et al. 2015) stresses that the impact of increasing demands for resources by a growing population is already exceeding the ecological capacities of the planet, with forests playing a prominent role in providing these (diminishing) capacities. It may be the 'big three' driving forces – (1) ongoing (albeit flattening) population growth, (2) economic growth together with a fixation on short-term profit, and (3) the steadily increasing global environmental footprint and environmental pressures resulting from it, all three interwoven with a complex array of socio-economic path dependencies, cultural and religious barriers to change and learning – that will determine the ultimate outcomes of (un-)sustainable development on this planet. These elephants in the room are also at the core when it comes to the forest–people interface: as long as forests are largely considered (and needed?) as a resource to be converted into often intense food and commodity production systems, and as long as global pressure drives this land-use change, deforestation and loss of forest ecosystem services will continue.

One more elephant in the room may also be of critical importance here: sustainable investment and financing. As long as vastly more money is invested in activities driving large-scale forest conversion, mostly with the aim of expanding industrial agriculture, than is invested in forest conservation, including management and/or small-scale agroforestry (Chapter 17), it is obvious that attempts to achieve sustainable management of forest landscapes will be difficult to accomplish. Ultimately, it is also the lack of recognition of the full economic potential of forests, including their wide spectrum of often non-valorised ecosystem services, that leads to the disregard and elimination of forests for the purpose of economic development and profits.

The challenges are thus (a) how to manage and possibly curb the major growth driving forces (greater numbers of people with a desire to live well and prosper), and (b) how to decouple demographic and economic growth from environmental impacts, including carbon emissions. Forests may play an important role in such a decoupling, e.g. through innovation and valuation of the ecosystem services they provide as pillars for a green and sustainable economy (Winkel 2017). At an operational scale, innovative, cross-sectoral governance arrangements and management will be needed to develop and promote forest landscapes that sustainably integrate a mosaic of different uses. This calls for an integrated landscape approach that goes beyond narrow sectoral perspectives and reaches out to include the various value chains associated with the use of forest ecosystem services. With their many links to different sectors and people, forests play a critical role in facilitating such an approach, if their multifaceted importance is correctly accounted for. Research that investigates the full societal and economic importance of forests through accounting for their full ecosystem services spectrum can support this transformation. It will be important to involve adjacent powerful societal sectors into this approach, increasing the possibility to diversify forest-related development, and increasing the likelihood for forests to remain a permanent part of the landscape in the future.

The implementation of such an integrated landscape approach will also mean taming one last elephant that inhabits the forest habitat and is difficult to approach: the multifaceted informal/illegal activities taking place in the world's forest. Estimates indicate that the greatest proportion of the forest-related economy operates in the informal sphere (World Bank 2016). The spectrum ranges from informal local economies and livelihoods based on customary rights connected to traditional knowledge and forest uses to corruption and criminal forest exploitation through large-scale illegal logging. This reality characterising many forested regions in the world diverges from the idea of a systematic implementation of the SDGs and associated targets through government in line with the ideal of Plato's just state. Yet, research has repeatedly shown that attempts to formalise the informal sector have often had ambivalent outcomes. In some cases, enforcement not only targets criminal behaviour, but also traditional use forms and customary community rights harmfully impacting the very people the SDGs are supposed to benefit (McDermott et al. 2019). Addressing the informal dimensions of the forest–people interface in any approach aiming to advance sustainable development is essential. It must be done, however, with caution, recognising the importance of this sector for local livelihoods, traditional forest uses and related forest knowledge.

6) The planet is changing rapidly – key elements of the success of the SDGs will require learning that results in adaptation of targets and their implementation

While the SDGs are a comprehensive development agenda, they are unavoidably bound to the specific perspective resulting from their negotiation. Moreover, their potential impacts during implementation will evolve in relation to changing socio-economic and ecological conditions. In the case of forests, for instance, climate change is resulting in changing disturbance patterns and changing species distributions. Those will alter the capacity of forests to provide ecosystem services for society, and of people to manage these forests. Also, societal needs and demands towards forests will change, inter alia, through processes of urbanisation and globalisation. These developments cannot be projected with precision for periods of a decade or more.

Together with the frequently stressed context-dependence of the SDGs, they can be read as a call for an adaptive implementation process of the SDGs. Such a process is characterised by monitoring and evaluation to leverage progress through continual learning, and to put into place feedback loops to adjust political targets and measures to attain them. A focus on meaningful, partially qualitative indicators will be needed, complementing quantitative ones. Moreover, much emphasis should be given to the processes of how indicators are developed, how monitoring is conducted and how feedback is processed. Participatory bottom-up approaches that are mindful of specific forest contexts and that empower local people to assume a key role in monitoring could favour the advancement of locally suitable sustainable development trajectories. Recalling the challenging question of the importance of economic growth as a driver of or challenge for sustainable development, measurements of economic growth and well-being could be broadened to take into account environmental and social impacts. This may include going beyond GDP growth by accounting for natural capital stock in combination with broadened socio-economic metrics (see Chapter 9). Knowledge and education, specifically the willingness to share knowledge and learn, will be critical to enhance informed decision-making in the SDG implementation process. New means in information technology could hold substantial promise for increasing transparency regarding the co-evolution process of the forest–people interface locally, but also at the global level. They can, inter alia, bring consumers and producers closer to each other, and thus be fundamental for fostering greater awareness and shared responsibility for achieving shared goals. In such a setting, the SDG implementation ideally becomes a continuous learning process of how to adapt and improve sustainable development at the forest–people interface based on traceable outcomes on the ground. This book seeks to encourage such a learning process by providing empirical evidence on past efforts, and evoking interest to explore unknown territories that will inevitably await practitioners that seek to further the attainment of the SDGs, while advancing sustainable development at the forest–people interface.

19.3 Outlook: The Way Forward

What can a reader who wants to help advance sustainable development at the forest–people interface take away from this book? In short, this chapter concludes that:

- Forests are a key base for sustainable development.
- The SDGs will impact forests and the people dependent on them in many ways.

- The SDGs include partially conflicting visions for forests and people, making it necessary to consider trade-offs and to set priorities.
- The SDGs' impacts on forests and people are highly context dependent.
- There are fundamental values and principles that may guide sustainable development related to forests and people.
- There are major driving forces of (un)sustainable development that need to be tackled to advance forest governance and management.
- There is a necessity to continuously learn, and adapt, in the process of working towards sustainable forest management at the forest–people interface.

These lessons learnt are critically important for advancing sustainable development at the forest–people interface. They are, however, in no way meant to constrain creativity, inspiration and action at this interface. The issues at stake are often terribly complex, and ambiguity is at every corner. However, though the magnitude of some of the challenges is new, several patterns of the challenges ahead are not new to humanity. It is not the first time that societies have faced limits of growth; that deforestation, overexploitation of forest resources and dwindling forest ecosystem services are raising concerns; or that there is striking inequality. All of these patterns have existed repeatedly – at the regional scale – in the history of humanity. Scarcities and paucities have frequently been the source of what is perhaps the best humanity has to offer: creativity, the ability to innovate and to develop further. Importantly, never before have the technological, scientific and information-related capacities been more advanced to support such innovation. This is not meant to naively promote green growth and ecological modernisation. The point we make here is that it is critical to trust in the ability of mankind to overcome a development challenge, one with a magnitude and impact never before experienced, but with often familiar basic patterns. Human creativity, confidence and readiness to innovate is needed for policymakers to work across sectoral silos (e.g. connecting agriculture and forestry), for scientists to critically ask the right questions (e.g. thoroughly monitoring the interdependencies of people and forests) and to communicate evidence frankly, and for forest practitioners and local people to innovate on the ground and try new approaches for integrating manifold societal demands. Obviously, there are frequently policy silos, power games, economic interests and reform-resistant or sometimes even corrupt structures dominating the land use sectors. This requires visionary policymakers to overcome these hurdles, and path-breaking business entrepreneurs to think beyond conventional economic pathways. Policy, business, civil society, science and forest-practice innovators are needed to advance the co-evolution of forests and people on a sustainable track, and their thinking needs to go beyond well-established channels to tackle the major issues at hand.

Hence, it is at the human part of the co-evolution process at the forest–people interface where changes need to happen and innovation needs to advance. Rich forests have existed on this planet since long before humans arrived; it seems unlikely that humans would survive a future without these rich forests given their tremendous importance for the planet's ecological and socio-economic systems. For a reader pessimistic about the human capacity to overcome the danger of our own impact, this book perhaps offers little more than a detailed exploration of where those impacts lie regarding forests. For the confident reader, however, this book can hopefully be a huge source of inspiration and guidance for possible changes towards sustainable development for both forests and people. In this sense, such readers may feel very much encouraged to engage in a holistically understood, but locally shaped, sustainable development of societies and their natural resource base. Such sustainable development, as this book has shown, will be impossible to envision without one of the most astonishing and admirable part of nature on our planet – the world's unique and beautiful forests.

References

Aerts, R. and Honnay, O. 2011. Forest restoration, biodiversity and ecosystem functioning. *BMC Ecology* 11(1):29.

Angelsen, A. and Kaimowitz, D. 1998. *Economic models of tropical deforestation. A review*. Bogor: Center for International Forestry Research.

Angelsen, A. and Rudel, T. K. 2013. Designing and implementing effective REDD+ policies: A forest transition approach. *Review of Environmental Economics and Policy* 7(1):91–113.

Angelsen, A. and Wunder, S. 2003. *Exploring the forest – poverty link: key concepts, issues and research implications*. CIFOR Occasional Paper 40. Bogor: Center for International Forestry Research.

Arts, B. J. M., Appelstrand, M., Kleinschmit, D. et al. 2010. Discourses, actors and instruments in international forest governance. In Rayner, J., Buck, A., and Katila, P. (eds.) *Embracing complexity: Meeting the challenges of international forest governance. A global assessment report*. IUFRO World Series no. 28. Vienna: International Union of Forest Research Organizations (IUFRO), pp. 57–74.

FAO 2015. *Global Forest Resources Assessment 2015. How are the world's forest changing?* Rome: FAO. Available at: www.fao.org/3/a-i4793e.pdf (Accessed 28 March 2019).

Forest Europe 2015. *State of Europe's Forests 2015*. Madrid: Ministerial Conference on the Protection of Forests in Europe, Forest Europe Liaison Unit.

Hetemäki, L., Hanewinkel, M., Muys, B. et al. 2017. *Leading the way to a European circular bioeconomy strategy*. From Science to Policy 5. Joensuu: European Forest Institute.

Hobley, M. 2005. The impacts of degradation and forest loss on human well-being and its social and political relevance for restoration. In Vallauri, D., Dudley, N., and Mansourian, S. (eds.) *Forest restoration in landscapes*. New York: Springer, pp. 22–30.

Mather, A. S. 1992. The forest transition. *Area* 24 (4): 367–79.

McDermott, C. L., Hirons, M. and Setyowati, A. 2019. The Interplay of Global Governance with Domestic and Local Access: Insights from the FLEGT VPAs in Ghana and Indonesia. *Society and Natural Resources*: 1–19.

Meyfroidt, P., Vu, T .P. and Hoang, V. A. 2013. Trajectories of deforestation, coffee expansion and displacement of shifting cultivation in the Central Highlands of Vietnam. *Global Environmental Change* 23(5):1187–98.

OXFAM 2016. *An Economy for the 1% (Summary)*. OXFAM Briefing Papers 210.

Pressman, J. L. and Wildavsky, A. 1984. *Implementation: How great expectations in Washington are dashed in Oakland; Or, why it's amazing that federal programs work at all, this being a saga of the Economic Development Administration as told by two sympathetic observers who seek to build morals on a foundation*. Berkeley: University of California Press.

Racevskis, L. A. and Lupi, F. 2006. Comparing urban and rural perceptions of and familiarity with the management of forest ecosystems. *Society and Natural resources* 19(6):479–95.

Sotirov, M. and Winkel, G. 2016. Toward a cognitive theory of shifting coalitions and policy change: linking the advocacy coalition framework and cultural theory. *Policy Sciences* 49(2):125–54.

Steffen, W., Richardson, K., Rockström, J. et al. 2015. Planetary boundaries: Guiding human development on a changing planet. *Science* 347(6223):1259855.

Thompson, M., Ellis, R. and Wildavsky, A. 1990. *Cultural theory*. Boulder: Westview Press,

Tyrväinen, L., Silvennoinen, H. and Kolehmainen, O. 2003. Ecological and aesthetic values in urban forest management. *Urban Forestry & Urban Greening* 1(3):135–49.

UN 1987. *Report of the World Commission on Environment and Development: Our common future*. New York: United Nations.

UNEP 2016. *Global material flows and resource productivity*. Assessment Report for the UNEP International Resource Panel.

Winkel, G. (ed.) 2017. *Towards a sustainable European forest-based bioeconomy – assessment and the way forward. What Science Can Tell Us 8*. Joensuu: European Forest Institute.

World Bank 2016. *Forests create jobs and wealth*. Available at: www.worldbank.org/en/news/infographic/2016/03/16/forests-create-jobs-infographic (Accessed 28 March 2019).

For EU product safety concerns, contact us at Calle de José Abascal, 56–1°, 28003 Madrid, Spain or eugpsr@cambridge.org.

www.ingramcontent.com/pod-product-compliance
Lightning Source LLC
LaVergne TN
LVHW062159200926
848484LV00006B/1288
9781108486996